Mineral Resources, Economics and the Environment

STEPHEN E. KESLER

UNIVERSITY OF MICHIGAN

MACMILLAN COLLEGE PUBLISHING COMPANY, INC.
NEW YORK

MAXWELL MACMILLAN CANADA
TORONTO

MAXWELL MACMILLAN INTERNATIONAL
NEW YORK   OXFORD   SINGAPORE   SYDNEY

To Judy, Sarah, and David
for coming along

Available to Adopters:

Printed Test Bank
Slide Set

Requests should be directed to
Robert A. McConnin, Senior Editor,
Macmillan College Publishing Company,
866 Third Ave.
New York, N.Y. 10022

Editor: Robert A. McConnin
Production Supervisor: bookworks
Production Manager: Aliza Greenblatt
Text Designer: Robert Freese
Cover Designer: Hothouse Designs
Cover Art: F. Carmichael: *A Northern Silver Mine*
Illustrations: Dale Austin

This book was set in Palatino by The Clarinda Company, and
printed and bound by Semline.
The cover was printed by Lehigh Press.

Macmillan College Publishing Company
866 Third Avenue, New York, New York 10022

Macmillan College Publishing Company is part of
the Maxwell Communication Group of Companies.

Maxwell Macmillan Canada, Inc.
1200 Eglinton Avenue East
Suite 200
Don Mills, Ontario M3C 3N1

Library of Congress Cataloging-in-Publication Data

Kesler, Stephen E.
    Mineral resources, economics and the environment / Stephen E.
Kesler.
        p. cm.
    Includes bibliographical references and index.
    ISBN 0-02-362842-1
    1. Mines and mineral resources. 2. Mineral industries.
3. Mineral industries--Environmental aspects. I. Title.
TN145.K44 1994
333.8'5—dc20   93-25488
                                                        CIP

Printing: 1 2 3 4 5 6 7   Year: 4 5 6 7 8 9 0

Preface

At the end of the twentieth century, we are faced with two closely related threats. First, there is the increasing rate at which we are consuming mineral resources, the basic materials on which civilization depends. Although we have not yet experienced global mineral shortages, they are on the horizon. Second, there is the growing pollution caused by the extraction and consumption of mineral resources, which threatens to make Earth's surface uninhabitable. We may well ponder which of these will first limit the continued improvement of our standard of living.

These threats have generated a wide range of opinions about mineral resources and the environment. At one end of the spectrum are those who advocate a dramatic reduction in mineral production with recycling and conservation providing for the future. At the other end are those who feel that vigorous exploration will always find new minerals and that the old ways of mineral production and consumption are good enough for the future. Both camps are on perilous ground. Many mineral commodities, such as oil and fertilizers, cannot be recycled and the enormous populations of the less developed countries are poised to consume any minerals that are conserved by more developed countries. Similarly, we cannot ignore the environmental catastrophes that have been caused by past mineral production or the impending problems related to increasing global mineral consumption. Unless we are willing to make a dramatic reduction in our standard of living, however, we must find a way to produce and consume the enormous volumes of minerals that we need without significant degradation of the environment. In other words, we must find a middle ground in these arguments.

This book provides an introduction to the factors that will control our efforts to solve Earth's mineral and related pollution problems. It deals with mineral resources, the engineering and economic factors that govern their availability, and the environmental effects of their production and use. It is intended for use as both a college text and a primer for anyone with an interest in the technical, economic, or policy aspects of mineral resources. Mineral professionals who seek a broader view of their field will also find it useful. Because this audience has such a wide range of backgrounds, an effort has been made to make the book a self-contained document, in which all terms and concepts are explained. Introductory material on geology, chemistry, engineering, economics, and accounting have been included, all of which is linked through a glossary of terms. Appendices containing information on elements, minerals, rocks, mineral commodities, units of weight and measure, and mineral reserves and resources, have also been included, as have references to recent literature on most subjects. (Although some readers prefer citations to original work on a subject, more recent studies are cited here to help identify earlier studies on each topic.) In keeping with their wide use throughout the world, metric (SI) units are used in most parts of this book, including the term *tonne* for metric ton. Other units, such as flasks and troy ounces, are also used where dictated by convention.

iii

This book deals with controversial subjects about which opinions must be expressed and solutions suggested. I have tried to do this on a case-by-case basis, without following any specific agenda or point of view. It is encouraging in that respect that the book has been cited as too "industry oriented" by some and too "environmentally oriented" by others. Hopefully, each camp will find much that is familiar and friendly, but also much that challenges assumptions and encourages factual debate intended to solve problems and produce a consensus. We will all find many areas in which more data are needed before final decisions can be reached.

Although this book has one author, it is the product of many minds. I am very grateful to the numerous geologists, mining engineers, metallurgists, mineral economists, and other professionals who have allowed me access to their projects or operations and shared their thoughts with me. Long-term association with Stan Bearden, Pierce Carson, Jack Frost, Pete Gray, Fred Humphrey, Bill Kelly, Norm Russell, Mike Seaward and my father, Tom Kesler, have been particularly helpful. I am equally grateful to the many students, particularly those in GS280, who have been a constant source of new information and challenging questions. Several people in the Departments of Geological Sciences, Nuclear Engineering, and Public Health at Michigan, including John Chesley, Eric Essene, Joe Graney, Hank Jones, John Lee, K.C. Lohmann, Phil Meyers, Jim O'Neil, Henry Pollack, Ed Van Hees, Torsten Vennemann, Lynn Walter, and Bruce Wilkinson read various versions of the text and made many helpful comments and corrections. Similarly useful input came from a host of others, particularly many personnel of the U.S. Bureau of Mines. All or parts of the book were read by the following reviewers whose comments were very helpful:

Greg B. Arehart, Argonne National Laboratory
G. Arthur Barber, Littleton, Colorado
Philip E. Brown, University of Wisconsin
William F. Cannon, U.S. Geological Survey
Christine E. Clynes, Rio Algom Ltd.
John H. DeYoung, U.S. Geological Survey

James E. Elliott, U.S. Geological Survey
Jacqueline R. Evanger, Mineral Information Institute
Richard Franks, Rio Algom Ltd.
John Greene, Louisiana Land and Exploration
James B. Hedrick, U.S. Bureau of Mines
Gerald W. Houck, U.S. Bureau of Mines
Alfred G. Hoyle, Mineral Information Institute
John F. Joity, Exxon Exploration Company
Janice L.W. Jolly, U.S. Bureau of Mines
Thomas S. Jones, U.S. Bureau of Mines
Jean D. Juilland, Bureau of Land Management
Thomas L. Kesler, Bellevue, Washington
J. Roger Lobenstein, U.S. Bureau of Mines
John M. Lucas, U.S. Bureau of Mines
Jeffrey Mauk, University of Auckland
Karr McCurdy, Mellon Bank
Bill McKechnie, De Beers Group
Michael A. McKibben, University of California, Riverside
G. David Menzie, U.S. Geological Survey
Fred W. Metzger, Arco Oil Company
P.A. Meyer, Mineral Information Institute
David W. Mogk, Montana State University
L.J.P. Muffler, U.S. Geological Survey
John F. Papp, U.S. Bureau of Mines
Colin J. Patterson, South Dakota School of Mines
Erich U. Petersen, University of Utah
Patricia A. Plunkert, U.S. Bureau of Mines
Juan Proaño, Interamerican Development Bank
Robert G. Reese, U.S. Bureau of Mines
Norman Russell, American Pacific Mining Co.
James P. Searles, U.S. Geological Survey
Morris Viljoen, University of the Witwatersrand
Richard Viljoen, Gold Fields of South Africa

Many others who helped obtain photos or information are mentioned in the captions. Paula Kunde helped with typing, formatting and printing much of the text, Dale Austin drew all of the illustrations, which is rarely done for a book these days, and Sarah Kesler helped with research on Chapters 5 and 6. I am grateful to all of these people for their assistance but absolve them of responsibility for any errors or opinions expressed here.

Contents

1

Mineral Resources

OUR MINERAL RESOURCE CRISIS

We are facing a global mineral resource crisis. Earth's finite supply of minerals is being used by a population that is growing faster than at any time in history (Figure 1-1). To make matters worse, mineral consumption is growing even faster than the population. Most minerals are used in the *more developed countries (MDCs)* with relatively small consumption in the *less developed countries (LDCs)* (Table 1-1). Although the MDCs account for only 16% of world population, they consume 70% of world aluminum, copper, and nickel, 58% of world oil, 48% of the natural gas, and 37% of world coal. As the standard of living increases in the LDCs, they will begin to consume their share of minerals, causing world per capita mineral consumption to rise even faster.

This creates a dilemma. Although we need more minerals to supply civilization, we are becoming increasingly aware that their production and use are polluting the planet. Effects that were once local in scale have become truly global, with mineral consumption implicated strongly in global warming, acid rain, and the destruction of ozone in the upper atmosphere. Just when we need to expand mineral production, there is concern that Earth is reaching its limit of mineral-related pollution.

We cannot ignore this crisis. Our civilization is based on mineral resources. Most of the machines, appliances, and furniture that make life comfortable are made of metals and powered by energy from fossil fu-

els. Large-scale production of food for urban populations depends on mineral fertilizers. The buildings in which we live and work are made almost entirely of mineral material that was extracted from Earth. Even the gems and gold that we use for adornment and to support global trade come from minerals. Although a return to Walden Pond might free some of us from mineral dependency, most of Earth's 5.3 billion inhabitants are actively seeking the comforts that mineral consumption can provide. If global population increases as rapidly as many estimates suggest, the pressure to find and produce minerals will be enormous, as will the potential pollution related to their extraction and use.

Although the magnitude of our impending mineral shortfall is easy to see, we have become dangerously complacent about it. This would have been unimaginable to the authors of *Limits to Growth* (Meadows et al., 1972), who alerted the world in 1972 to its finite mineral supplies and soaring consumption. The collision between these forces had been developing for almost a century, as world living standards improved. Between 1900 and 1973, world oil consumption grew by more than 7% annually, with each succeeding decade using about as much oil as had been consumed throughout all previous history. World oil supplies were on their way to exhaustion by the turn of the century (Petersen and Maxwell, 1979; Bartlett, 1980A). With steel, aluminum, coal, and other commodities following similar trends, it appeared that we were about to witness the end

1

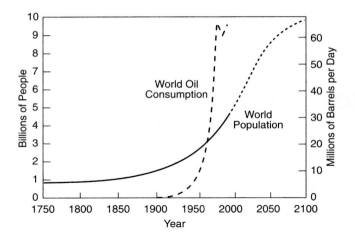

FIGURE 1-1

Increase in world population and oil consumption, showing that oil consumption has grown more rapidly than population. This relation holds for almost all mineral commodities and is both a comforting sign of improving global standard of living and an alarming warning of impending shortage.

of a brief mineral-using era in the history of civilization.

But, this did not happen. Instead, world economic patterns were jolted by the huge increase in the price of oil, which began as *Limits to Growth* was published. As oil prices rose, they pulled along the prices of coal, natural gas, and uranium (Figure 7-1), and this made everything else more expensive to produce, including other mineral commodities (Figures 8-1, 9-1, 10-1, 11-1, 12-1). Erroneously interpreting rising prices as confirmation of mineral shortages, exploration fanned out to the ends of the globe, dramatically increasing reserves for most mineral

commodities (Figure 1-2). This led to increased production capacity. As all this was taking place, the strain of increased commodity prices slowed the world economy and demand for minerals declined. Thus, just when we were supposed to feel the cold breath of mineral shortages and rising prices, the world saw excess supplies and plummeting prices. All of this occurred in response to the initial rise in the price of oil, providing strong testimony to the tremendous power that mineral supplies have over the world economy.

Unfortunately, the mineral supply shortage has only been postponed, and in the process it has become more complex. Only a short time ago, our mineral supplies were determined largely by geologic, engineering, and economic factors. Their relation to Earth's mineral endowment was usually depicted as shown in Figure 1-3. Here you can see that the most important part of the mineral endowment consists of *reserves*, material that has been identified geologically and that can be extracted at a profit at the present time. The *reserve base* includes reserves as well as already identified material of lower geologic quality that might be extractable in the future, depending on economic factors. *Resources* include the reserve base plus any undiscovered deposits, regardless of economic or engineering factors. Our impending mineral supply crisis can be averted only by converting resources to reserves. Appendix I, which tabulates reserve data for most mineral commodities, will form the basis for much of what we have to say in later chapters.

But this must now be done in the context of environmental constraints, the newest factor that controls the conversion of resources to reserves (Plate 1-1). Environmental costs impact the economic axis of Figure 1-3, thereby controlling the overall profitability of ex-

TABLE 1-1

High-income countries, termed *more developed countries (MDCs)* in this book, listed in order of decreasing per capita GNP. All other countries are referred to here as *less developed countries (LDCs)*. This list does not include Nauru, Luxembourg, Iceland, Bahrain, Brunei, and Qatar, which have high GNP's but do not consume significant minerals (from World Bank, *World Bank Tables, 1993,* Baltimore, The Johns Hopkins University Press).

Switzerland	$32,310	Austria	$20,240	Saudia Arabia	$14,870
Canada	$26,522	United Arab Emirates	$19,840	New Zealand	$12,770
Japan	$25,840	France	$19,620	Singapore	$12,630
Finland	$24,580	Netherlands	$17,850	Israel	$11,160
Denmark	$22,440	Belgium	$17,580	Spain	$11,010
Sweden	$23,780	Italy	$16,940	Ireland	$10,390
Norway	$22,830	Australia	$16,333	Taiwan	$10,110
Germany	$22,630	Kuwait	$26,210		
United States	$21,910	United Kingdom	$16,020		

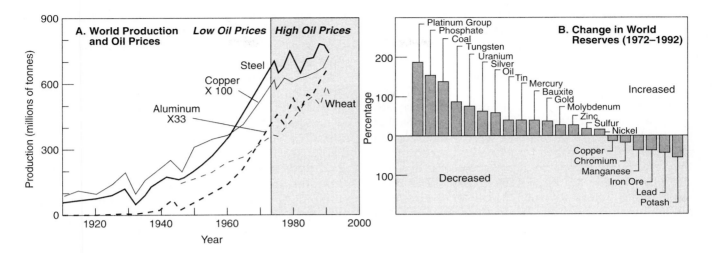

FIGURE 1-2

A. Decrease in world production rates for steel, copper, and aluminum coinciding with steep rise in oil prices in the mid-1970s. Note that wheat production continued to rise at the same rate through this period. **B.** Change in world reserves of important mineral commodities after publication of *Limits to Growth* (from data of U.S. Bureau of Mines and Department of Energy).

traction. Just as important, however, and more difficult to show in the diagram are government regulation and public opinion. It is no exaggeration to say that mineral deposits cannot be extracted in most MDCs until the proposed operation has been approved by environmental regulators *and* accepted by the public.

Thus, the nature and extent of our global mineral endowment is no longer controlled strictly by market forces and administered by mineral professionals who make decisions on the basis of geologic, engineering, and economic factors. Instead, it is in the hands of a broader constituency with a more complex agenda focused largely on the environment, but with additional concerns about the distribution of wealth. More and more of us are expressing opinions about our mineral supplies and in doing so, we incur an obligation to understand the factors that control their distribution, extraction, and use. That is what this book is about. As an introduction, we will start with a brief review of the four major factors that control mineral availability.

FIGURE 1-3

Mineral resource classification system of the U.S. Bureau of Mines and U.S. Geological Survey (McKelvey, 1973; USGS, 1980).

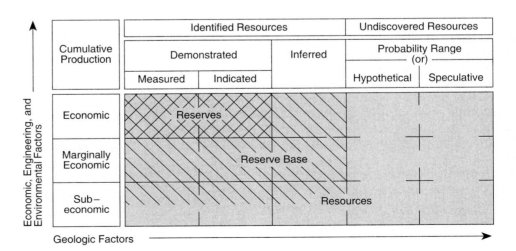

FACTORS CONTROLLING MINERAL AVAILABILITY

Geologic Factors

Our mineral supplies come from *mineral deposits,* which are concentrations of elements or minerals that formed by geologic processes. Where something can be recovered at a profit from these concentrations, they are referred to as *ore deposits.* Mineral deposits can be divided into four main groups. The most basic group comprises soil and water, here called *essential resources.* These lack the excitement of gold and oil but have been essential to civilization from its beginning. *Energy resources* can be divided into the fossil fuels, including crude oil, natural gas, coal, oil shale, and tar sand, the nuclear fuels, including uranium and thorium, and geothermal power. As interesting as they may be for the future, wind, tidal, and solar power are not important factors in our present energy supplies and, along with hydroelectric power, have been omitted from this discussion in order that we can concentrate on minerals, as the title suggests. *Metal resources* range from structural metals such as iron, aluminum, and titanium to ornamental and industrial metals such as gold, platinum, and gallium. *Industrial mineral resources,* the least publicized of the four groups, include more than 30 commodities such as salt, asbestos, and sand that are critical to our modern agricultural, chemical, and construction industries.

The essential resources, soil and water, require special consideration in our discussion of mineral resources. Our interest in most of the other mineral resources discussed here deals with the balance between the benefits that we derive from them and the environmental damage that they cause. In contrast, soil and water have become the main dumping grounds for most of the wastes that are produced by modern society, including those related to mineral resources. Thus, the essential resources become the context in which we assess environmental cost-benefit ratios of other mineral resources. Rather than being the focus of a single chapter, their role in world mineral extraction and use must be discussed throughout the text.

Mineral deposits have two geologic characteristics that make them a real challenge to modern civilization. First, almost all of them are *nonrenewable resources;* they form by geologic processes that are much slower than the rate at which we exploit them. Whereas balanced harvesting of fishery and forest resources might allow them to last essentially forever, there is little likelihood that we will be able to grow mineral deposits at a rate equal to our consumption of them. Second, mineral deposits have a *place value.* We cannot decide where to extract them; nature made that decision for us when the deposits were formed. The only decision that we can make is whether to extract the resource or leave it in the ground.

The erratic distribution of minerals around the globe adds another dimension of complexity. Wars usually involve mineral-rich ground, either directly or indirectly. More recently, battles have become legal and have focused on whether exploration or extraction should even take place on potentially mineral-rich ground. These altercations are in fact admissions that mineral deposits are a fundamental function of Earth's complex geologic history. As we will see throughout this book, there is a close relation between the type of mineral resource found in an area and its geologic setting. Just as common sense tells us not to look for oil in the crater of a volcano, our study of Earth has taught us to look for minerals in favorable geologic environments. As population pressures place more demand on land, geologic controls on the distribution of mineral deposits will become increasingly important in land-use decisions.

Engineering Factors

Engineering factors affect mineral availability in both technical and economic ways. Technical constraints are imposed when we simply cannot do something regardless of desire or funding. An example is the extraction of iron from Earth's core, which is too deep and hot to be reached by any known mining methods. Economic factors constrain mineral availability only when we judge the cost of a project to be too great. We could build the necessary equipment to mine the moon, for instance, but the cost of the equipment and the mining expedition would far exceed any benefit that the minerals might afford us.

Engineering considerations place important limits on our ability to extract minerals from Earth. Mining does not extend below about 2.3 kilometers in most areas and the gold mines of South Africa, the deepest in the world, reach depths of only about 3.7 kilometers. Wells extend to deeper levels; some oil and gas production comes from depths of about 8 kilometers. Experimental wells extend to 11 kilometers, but there is little likelihood that significant production will come from these depths in the near future simply because few rocks at these levels have holes from which fluids can be pumped. Additional engineering constraints are imposed by the need to process most raw minerals to produce forms that can be used in industry and by the need to handle wastes efficiently and effectively.

Environmental Factors

Environmental concerns about mineral resources focus on two main problems. First to be recognized was pollution associated with mineral extraction and processing. Extraction of raw materials, including minerals, timber, crops, and animals, produces over 2 billion tonnes of wastes annually in the United States, by far the largest amount of waste generated in the economic cycle (Figure 1-4). Study of older mineral extraction sites has shown that elements and compounds were dispersed into the environment around them for distances of many kilometers. In an effort to prevent future calamities of this type, laws and regulations have been developed to control the generation and disposal of waste products from mineral exploration and production. The cost of compliance with these regulations has increased enormously and has become a growing factor in determining whether a mineral deposit can be extracted profitably.

We have been slower to recognize the importance of wastes associated with mineral consumption. These wastes are more widely dispersed and it has required longer periods of observation and better analytical techniques to demonstrate that the soil,

water, and air around us are changing as a consequence of our activities (Figure 1-5). This recognition has produced legislation to remove lead from gasoline, to decrease the amount of SO_2 emitted from smelters, and to limit the release of salt and fertilizers from storage areas, all of which add to the cost of using minerals.

The globalization of environmental concerns presents complex ethical problems that we have just begun to face. Just what right does any country have to pollute the atmosphere and ocean, when that pollution affects other countries? MDCs are at least trying to limit damaging emissions, but LDCs such as China are major polluters (Figure 3-11). Conversely, MDCs can "export" pollution by importing raw or processed minerals from LDCs. How far do the interests of major mineral-consuming countries go in defining the role of mineral suppliers? In a world with finite resources and growing demand, the decision not to exploit one deposit requires that another be exploited to supply world demand. What might have happened, for instance, if Kuwait had responded to the environmental damage of Iraqi sabotage of its oil wells during the 1991 war by limiting production to just enough for its own energy needs? Would the MDCs have accepted that and increased their domestic exploration and production, or do they expect environmental sacrifices from supplier countries which they are not willing to make themselves?

Economic Factors

Economic factors that control mineral production include those on the *supply side,* which are largely engineering and environmental costs related to extraction and processing, and those on the *demand side,* which include commodity prices, taxation, land tenure, and other legal policies of the host government. Although the balance between these forces can be considered from many political and economic perspectives, it is impossible to avoid the fact that the cost of producing a mineral must be borne by the deposit from which it comes or from some other segment of the host economy.

In a free market, costs and prices are usually part of a global system, which places similar constraints on all countries. However, legal, tax, and environmental regulations differ from country to country. The overall importance of these factors to mineral availability is shown by the positive correlation between number of minerals produced and land area for countries with high-income, stable economies (Figure 1-6A). This correlation supports the notion that large areas of Earth are more likely to have lots of important mineral re-

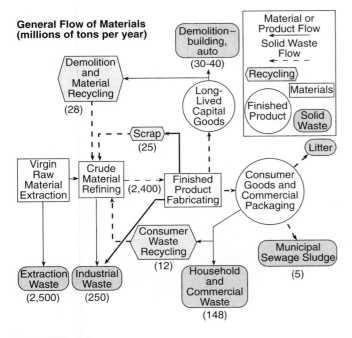

FIGURE 1-4

Flow of materials in the United States showing large volume of wastes generated by primary production of mineral, forest, agricultural, and animal products (Council on Environmental Quality, 1989).

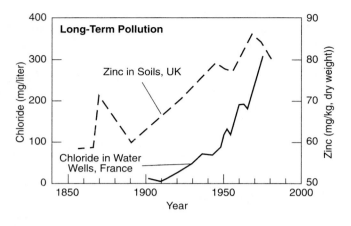

FIGURE 1-5

Increase in zinc in archived soils from Broadbalk, United Kingdom, and chloride in water wells from Colmar, France (Jones et al., 1987; GEMS, 1989).

gross domestic product (GDP). As can be seen in Figure 1-6B, raw mineral production makes up less than 2% of GDP in the United States and Japan but is greater than 25% in Kuwait and Saudi Arabia. It is a mistake to conclude that countries are unimportant mineral producers just because raw mineral production makes up a small percentage of GDP, however. The United States, for instance, is the leading world producer of 19 mineral commodities.

MINERALS AND GLOBAL ECONOMIC PATTERNS

The impact of minerals on the global economy is enormous. World primary (not including recycled material) fuel and metal production are worth about $700 and $500 billion, respectively, and primary industrial mineral production is worth about $150 billion (Figure 1-7). The value of world raw and processed mineral exports ranges from $400 to $600 billion annually, as much as a quarter of all exports.

Classical theory holds that economic activity depends on domestic mineral resource availability (Hewett, 1929). According to this scheme raw mineral exports occur early in a nation's development, as mineral deposits are discovered (Figure 1-8A). Profits from these exports are used to build an industrial in-

sources than are small areas. A similar correlation is not seen for low- and middle-income countries. In view of the lack of environmental regulations in most LDCs, the lack of a correlation in these countries must reflect legal and tax considerations, which make investments less attractive (Govett and Govett, 1977).

A good indication of the role of mineral production in economic activity in any country can be obtained by comparing the value of mineral production and

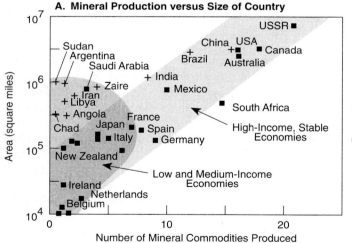

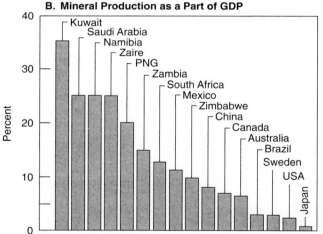

FIGURE 1-6

A. Relation between land area and number of mineral commodities produced for more developed countries (MDCs; good correlation) and less developed countries (LDCs; no correlation). **B.** Mineral production as a percentage of GDP for some important mineral-producing countries.

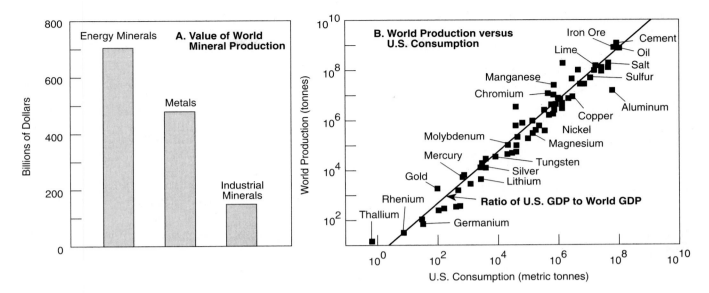

FIGURE 1-7

A. Value of world primary mineral production (not including recycled material). World production values for industrial minerals not tabulated by the U.S. Bureau of Mines were estimated from U.S. consumption data by relations shown in part B. Exclusion of steel would drop the value of metal production (including iron ore) to about $200 billion. **B.** Constant relation between world mineral production and U.S. mineral consumption. Note that U.S. and world GDP have the same relation (data from U.S. Bureau of Mines Mineral Commodity Summaries, 1992). For comparison, value of world production of other important raw materials are (in billions) cattle ($570), rice ($150), plastics and resins ($100), hogs ($85), wheat ($80), corn ($45), cotton ($25), tobacco ($25), sugar ($25), and soybeans ($22).

frastructure, which supports growing exports of goods manufactured from domestic raw materials. As mineral reserves dwindle, imports rise to support continued manufacturing. Many LDCs, such as Angola and Zaire, have bogged down at the start of this evolution and their national budgets and overall welfare are highly dependent on raw mineral prices. Because raw mineral prices vary unpredictably, these countries cannot control their revenues, a factor that limits stable development. This situation is a universal sore spot, with almost all countries wishing to sell more finished goods and less raw minerals. Even Canada, which occupies an enviable position in a global context, agonizes about its role as "hewer of wood and drawer of water" for the world.

The key question in world mineral trade is the effect of increasing *import reliance*, the fraction of mineral needs supplied by imports. Classical mineral economic theory predicts disaster, brought on by a lack of manufactured exports and increasing trade deficits. The rise in energy prices has brought this question into prominence for many oil-importing countries. In the United States, for instance, net crude oil imports between 1970 and 1990 were valued at $751 billion, a large fraction of the trade deficit. Other minerals such as chromium, potash, and platinum, for which the United States has a high import reliance, have also contributed to the deficit.

The miracle exception to classical theory is Japan, which has a strong positive balance of trade in spite of an annual deficit of about $15 billion in mineral imports and strongly declining raw mineral production (Figure 1-8B). Japan is the world's largest importer of aluminum, coal, copper and nickel concentrates, iron ore, liquified natural gas, phosphate, and industrial salt and the second largest importer of crude oil. It imports 90% to 100% of its aluminum, antimony, bauxite, chromium, cobalt, columbium, copper, oil, fluorine, ilmenite, iron ore, lithium, manganese, molybdenum, natural gas, nickel, platinum-group elements, rare-earth elements, rutile, strontium, tantalum, tin, and zirconium. Lower wages and higher domestic

Evolution of Mineral Production and Consumption

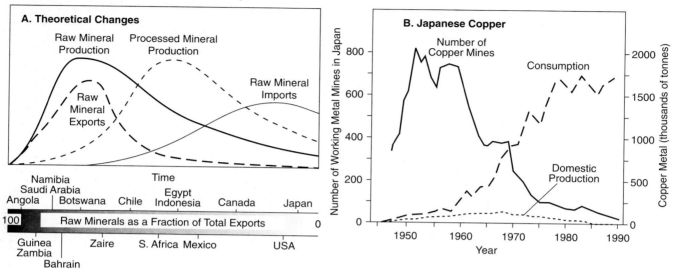

FIGURE 1-8

A. Classical relation between economic development and mineral supplies showing position of several important mineral-producing countries, as indicated by proportion of minerals in their total exports. **B.** Change in copper mining and production in Japan since the 1940s (Ishihara, 1992).

productivity are commonly cited reasons for success in the face of such adversity. Just as important, and less widely recognized, have been Japanese raw mate rial trade policies. During the last two decades Japan has invested in mineral extraction projects throughout the world. Most of these investments have involved agreements to buy some or all of the production, thus assuring an orderly supply of minerals.

Japan's emergence as a dominant manufacturing power without a domestic mineral base has mirrored two important shifts in the world mineral picture. First, whereas earlier generations have attempted to secure adequate domestic mineral supplies, economically successful countries of the late twentieth century have bought what they needed on world mineral markets (Cammarota, 1992). Some feel that the success of this approach has weakened the concept of *strategic minerals,* which holds that the security of a country depends on its mineral supplies, particularly those that are necessary for defense needs. In fact, the 1991 Gulf war showed clearly that MDCs will fight for access to mineral supplies, probably the strongest endorsement possible of the strategic minerals concept. Full appreciation for this fact is important as we enter the new era in mineral trade brought about by collapse of the U.S.S.R. Many of the CIS countries (as we will refer

here to the states that emerged from this change) are attempting to engage in world trade largely through offerings of raw materials. As noted at several points later in this book, mineral trade between MDCs and the CIS countries is increasing and it is almost certain that the next decade will see an increased dependence of MDCs on CIS minerals.

Global mineral trade has eroded the power of mineral cartels by promoting *market transparency,* in which production and consumption data are shared by producers throughout the world. The second and related development is the introduction of *modern materials* such as plastics, polymers, ceramics, and composites. These materials offer competition to conventional minerals in the form of substitutes, but they also offer new markets. They are particularly attractive to countries with manufacturing economies but limited raw materials because they offer alternative materials in the event of raw material supply problems.

THE NEW ERA OF WORLD MINERALS

Mineral resource availability is entering a new era, one in which traditional geologic, engineering, and economic constraints are joined by limitations imposed by environmental considerations. Dealing with these many factors and the uncertainties that they involve, while moving ahead to supply the next generation with minerals, will require compromises based on a full understanding of the issues. As a first step in this direction, this book explores the ramifications and interrelations of geologic, engineering, economic, and environmental constraints on global mineral resources.

2

Origin of Mineral Deposits

IT IS EASY TO FORGET ABOUT THE NEED FOR MINERAL DEPOSITS when you hear that a cubic kilometer of average rock contains billions of dollars worth of mineral commodities (Table 2-1). Although these amounts might make you think that we could supply our mineral needs from average rock, and get rich at the same time, this notion is easily dispelled by a look at the costs involved. A tonne of average rock contains only a few cents worth of gold versus the several dollars that you would have to pay to get it out. Unfortunately, you cannot get around this constraint by extracting other metals from the same tonne of average rock because most mineral commodities require different processes, each with its own high cost. To help keep things in perspective, remember that it takes several dollars just to buy a tonne of gravel, which takes essentially no processing. Thus, as long as we operate under market economic constraints we will continue to depend on mineral deposits for our needs. We define those deposits from which we can extract minerals at a profit as ore deposits.

The distinction between ore deposits and mineral deposits is not fixed; it is a dynamic function of economic, engineering, political, and environmental factors, which are the focus of this book. For example, the need to use catalytic converters to clean automobile exhaust made platinum deposits around Rustenburg, South Africa, into profitable ore deposits. Development of new "heap-leach" extraction methods allowed the large Gold Quarry deposit in Nevada to become the largest gold producer in the

United States (Plate 8-3). The increase in oil prices during the 1970s provided new incentive to mine the extensive tar sands of Alberta. On the other side of the coin, SO_2 recovery costs closed many old smelters and litigation over asbestos has bankrupted most of its original producers.

With all of these factors to consider, you might lose sight of the fundamental fact that we cannot get minerals unless Earth forms mineral deposits in the first place. Without mineral deposits, all other considerations are moot. Thus, we begin our discussion of mineral resources, economics, and the environment with a review of the geologic factors that control the setting and formation of mineral deposits.

GEOLOGIC SETTING OF MINERAL DEPOSITS

Earth consists of four global-scale divisions: the atmosphere, hydrosphere, biosphere, and lithosphere (Figure 2-1). Mineral deposits are part of the lithosphere, which is made up largely of rocks and minerals. *Minerals*, which control the distribution of elements in Earth, are naturally occurring solids with a characteristic crystal structure and definite chemical composition. They are divided into groups on the basis of their chemical compositions (Appendix II, Table A2-1). Although about 2,000 minerals are known, only about 30 minerals make up almost all common rocks, another 50 or so account for most metal ores, and about

TABLE 2-1
Value of selected mineral commodities in a tonne and a cubic kilometer of average crustal rock based on 1992 prices. The total value of just these commodities in a cubic kilometer of average rock is almost $400 billion!

Element	Value of 1 Km^3 of Crust ($ million)	Value of 1 Tonne of Crust ($)	Average Crustal Abundance (ppm)*
Aluminum	$260,000	$97.00	82,300
Titanium	$115,425	$42.75	5,700
Iron	$9,677	$3.58	56,000
Nickel	$1,300	$0.48	75
Copper	$317	$0.12	55
Platinum	$288	$0.11	0.005
Gold	$227	$0.08	0.0038
Lead	$24	$0.009	12.5
Molybdenum	$18	$0.007	1.5

*1 ppm = 0.0001%

100 comprise the industrial minerals (Appendix III). Chemical compositions are a fundamentally important part of geology and mineral resources, and you should refer to Appendices II and III if you are unsure of the composition of a mineral or rock discussed here.

Rocks consist of grains of minerals that hold together well enough to be thrown across a room. They are divided into groups on the basis of their origin and composition (Appendix II; Table A2-2). *Igneous rocks* form by crystallization or solidification of molten rock called *magma,* which is generally less dense than the lithosphere and rises through it. Where magma reaches the surface it is called *lava* and forms *extrusive* rocks, including volcanoes. Where magma solidifies below the surface, it forms *intrusive* rocks, including large masses known as *stocks* and *batholiths* and tabular bodies known as *sills* and *dikes.* Igneous rocks are divided into *felsic, intermediate, mafic,* and *ultramafic* groups on the basis of their mineral and chemical compositions. *Sedimentary rocks* consist of material that was deposited by water, wind, or glaciers. They can contain clasts or fragmental material, compounds precipitated from water, or products of organic activity. *Metamorphic rocks* form when other rocks are buried in the crust where high heat and pressure recrystallize original minerals, forming new minerals and textures.

The solid Earth has been divided into the core, mantle, and crust. The *core,* which consists largely of iron and related elements, is one big mineral deposit, although engineering constraints prevent us from extracting it, as noted earlier. The *mantle,* which forms a thick shell around the core, consists of ultramafic rocks

that contain locally important concentrations of chromium, cobalt, and nickel. Even the mantle is too deep to be within reach of mining, although we do extract ore from it where mantle rocks have been moved to the surface by plate tectonics, as discussed later.

Most of our mineral deposits are in the *crust,* which covers the mantle and contains the greatest variety of Earth's rocks. It is divided into two parts: the *ocean crust,* which is 5 to 10 kilometers thick and consists of mafic igneous rocks, largely basalt, and the *continental crust,* which is 20 to 70 kilometers thick and consists of felsic igneous and metamorphic rocks overlain by sedimentary rocks. Even the ocean crust is difficult to reach from an engineering standpoint, and we extract mineral resources from it only where it has been moved onto the continents by faulting. It is the continental crust that hosts the vast majority of our mineral resources.

Mineral deposits are a direct result of *plate tectonics,* which involves the movement of *lithospheric plates* about Earth's surface (Sawkins, 1990B). Geologists use the term *lithosphere* for the rigid outer 100 kilometers of Earth, including the continental and ocean crust and the uppermost mantle. Lithospheric plates move about on the underlying, more plastic mantle, which is known as *asthenosphere,* and they have three types of margins. *Divergent margins* form where plates are moved apart by convection in the mantle (Figure 2-2). They are located at *mid-ocean ridges* where the mantle rises and melts partially. This process produces basalt magma, which flows onto the ocean floor creating new ocean crust that migrates away from the ridges, gradually becoming covered by a thin veneer of sediment. Similar processes beneath the continental crust create

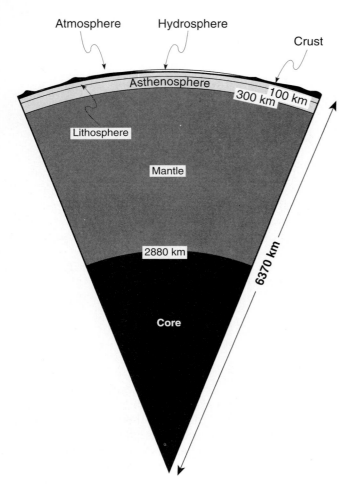

FIGURE 2-1

Distribution of the lithosphere, hydrosphere, and atmosphere (biosphere is too small to be shown at this scale) and major divisions of the solid Earth (core, mantle, asthenosphere, and lithosphere). Note that the term *lithosphere* as used by environmental geochemists refers to the entire solid part of the planet rather than the outer 100-km-thick plate that geologists call the lithosphere.

rifts such as the Dead Sea, which enlarge to form new oceans such as the Red Sea. *Convergent margins* form where plates move together. Most convergent margins include *subduction zones,* where ocean crust sinks back into the mantle causing melting and the production of magmas of intermediate to felsic composition that rise to the surface creating chains of volcanoes known as *island arcs.* Where continents collide, neither plate sinks and one rides onto the other along *obduction zones. Transform margins* form where two plates move horizontally past each other along faults, such as the San Andreas in California.

Plate tectonics drives a wide range of processes that form mineral deposits. As discussed in more detail later, mineral deposits are concentrated by subsurface chemical processes related to magmas and hot water, as well as by near-surface chemical and physical processes such as erosion and evaporation. These processes are much more common on the continental crust and their products are better preserved there because the continents are buoyant features that float on the mantle. In contrast, ocean crust is not bouyant and sinks back into the mantle at subduction zones. In fact, the oldest known ocean crust is only about 200 million years (Ma) old, whereas the oldest rocks on the continents are about 4 billion years old. Even these old continental rocks are significantly younger than the 4.65 billion year (Ga) age of Earth because the oldest rocks have been destroyed by erosion or covered by younger ones. Thus, even on the continents, rocks of the four major divisions of Earth history, the *Hadean, Archean, Proterozoic,* and *Phanerozoic* eons (Appendix II, Table A2-3), become progressively less abundant with increasing age. In most cases, rocks of Hadean and Archean age form stable cores for the continents

Plate Tectonics and Mineral Deposit Environments

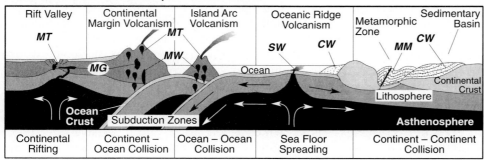

FIGURE 2-2

Plate tectonic environments, showing their relation to ore-forming magmatic and hydrothermal processes discussed in this chapter.

(Figure 2-3), which are known as *cratons* or *shields*, and Proterozoic and Phanerozoic rocks are found largely in deformed belts surrounding them.

The perspective of plate tectonics provides useful insight into the ultimate limit on civilization's mineral resources. Plate tectonics moves the lithospheric plates at the agonizingly slow rate of a few centimeters per year. This rate, in turn, controls magma formation, erosion, and sedimentation, all of which form mineral deposits. The resulting rate at which individual mineral deposits form is considerably slower than the rate at which we extract them. We are able to satisfy our need for resources only because so many deposits are forming at one time. As mineral consumption rates rise, we approach the limit set by the rate at which Earth forms

mineral deposits. Thus, it behooves us to understand how these deposits form.

GEOLOGIC CHARACTERISTICS OF MINERAL DEPOSITS

Ore deposits are important to civilization because they represent work that nature does for us. For instance, Earth's crust contains an average of about 55 ppm (parts per million[1]) of copper, whereas copper ore deposits must contain about 0.5% (5,000 ppm) copper be

[1]Parts per million (ppm) is a notation used to express very small concentrations. 1,000,000 ppm = 100%.

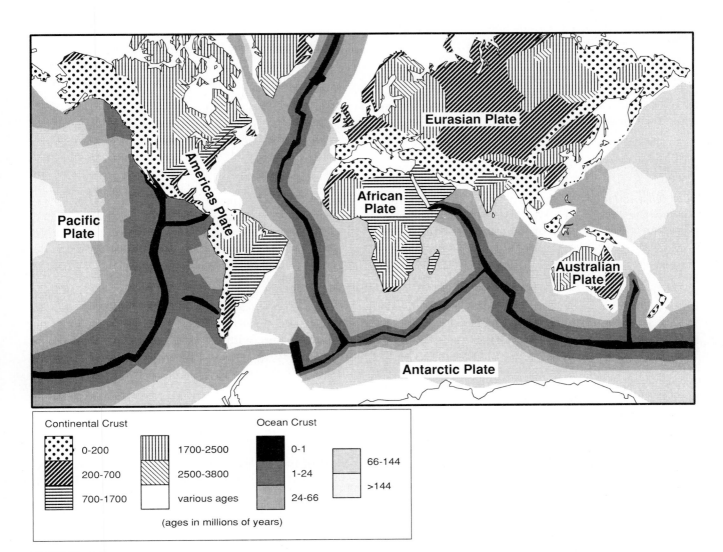

Continental Crust		Ocean Crust	
0-200	1700-2500	0-1	66-144
200-700	2500-3800	1-24	>144
700-1700	various ages	24-66	

(ages in millions of years)

FIGURE 2-3
Distribution of lithospheric plates. Note that continental crust is much older than ocean crust and ages of ocean crust increase away from spreading ridges (modified from Broecker, 1985).

FIGURE 2-4
The dark cubic crystal of galena grew on top of smaller crystals of white dolomite. Whereas the dolomite contains almost no lead, galena contains as much as 86% by weight and is our principal source of lead.

fore we can mine them. Thus, geologic processes need to concentrate the average copper content of the crust by at least 100 times to make a copper ore deposit that we can use (Table 2-1). We then use industrial processes to convert copper ore into pure copper metal, an increase of about 200 times. By a fortunate natural coincidence, elements and materials that we use in large amounts need less natural concentration than those that we use in small amounts. Thus, we are likely to have larger deposits of mineral materials that we use in large amounts. As long as energy costs remain high, the relation between work that we can afford to do and work that we expect nature to do will control the lower limit of natural concentrations that we can exploit, which puts very real limits on our global mineral reserves.

Most ore deposits contain an *ore mineral* or *compound* in which the element or substance of interest has been concentrated. For instance, the common lead ore mineral, *galena* (PbS), contains about 86% (860,000 ppm) lead. In rocks that contain galena, the job of recovering lead from Earth is greatly simplified because grains of galena can be separated from the ore to produce a smaller volume of material for further treatment to obtain lead metal (Figure 2-4). Lead does not form galena everywhere, however. Instead, it *substitutes* for other, more abundant atoms, such as potassium in silicate minerals that make up most of the crust. Ore deposits form only where geologic pro-

cesses concentrate elements into an ore mineral or compound, and then concentrate enough of the ore mineral or compound into a single location. Even crude oil and natural gas ore deposits form when organic material dispersed through sedimentary rock undergoes reactions that release liquid or gaseous hydrocarbons that accumulate in traps.

The importance of ore minerals or compounds is well illustrated by molybdenum and germanium which have approximately the same abundances in the crust, about 1.5 ppm. In spite of this, world molybdenum consumption of 110,000 tonnes is over 1,300 times greater than that for germanium. This difference reflects the fact that molybdenum forms the common ore mineral, *molybdenite* (MoS_2), whereas germanium forms no common minerals.

Ore minerals such as galena and molybdenite are rarely found in massive clumps that can be extracted directly. Instead, they are found in mixtures with *gangue minerals,* which are minerals that have no commercial value. The concentration of the element or compound of interest in an ore is referred to as the *grade* of a deposit, and the minimum concentration needed to extract the ore at a profit is known as the *cut-off grade* (Lane, 1988). Many lead deposits have an average grade of about 5% lead, almost all of which is in galena. Grades for natural gas, crude oil, and other fluids are not usually quoted as percentages. Instead, they are given as the amount of fluid that can be recovered from a given volume of rock.

In addition to grade, a mineral deposit must attain a minimum *size*. No matter how high its grade, a deposit must contain enough ore to pay for the equipment, labor, and costs of extraction. The size of solid mineral deposits is almost always given as the *weight* of ore, which we will specify here in metric tons *(tonnes)* and the size of fluid deposits is given as the *volume* of fluid, with the understanding that it can be recovered from a given volume of rock. Few deposits except groundwater and gravel contain ore worth less than a million dollars.

ORE-FORMING PROCESSES

Most of the geologic processes that form mineral deposits involve chemical changes in rocks and minerals (Table 2-2). In these changes, elements or compounds that were dispersed through large volumes of rock are collected and concentrated to form ore minerals or compounds. The most effective agents for chemical changes of this type are water and magmas, both of which dissolve elements and crystallize new minerals. Because hot water is a better solvent than cold water, waters that form many mineral deposits

TABLE 2-2
Geologic processes that form mineral deposits, with examples of deposits formed by each process and elements concentrated in them.

Type of Process	*Types of Deposits Formed/Minerals Concentrated*
Surface Processes	
Weathering	Laterite deposits—nickel, bauxite, gold, clay
	Soil
Physical Sedimentation	
Flowing water (stream or beach)	Placer deposits—gold, platinum, diamond, ilmenite, rutile, zircon, sand, gravel
Wind	Dune deposits—sand
Chemical Sedimentation	
Precipitation from or in water	Evaporite deposits—halite, sylvite, borax, trona
	Chemical deposits—iron, volcanogenic massive sulfide deposits, sedex deposits
Organic Sedimentation	
Organic activity or accumulation	Hydrocarbon deposits—oil, natural gas, coal
	Other deposits—sulfur, phosphate
Subsurface Processes	
Involving Water	Groundwater and related deposits—uranium, sulfur
	Basinal brines—Mississippi Valley type, sedex
	Seawater—volcanogenic massive sulfide, sedex
	Magmatic water—porphyry copper-molybdenum, skarn
	Metamorphic water—gold, copper
Involving Magmas	Crystal fractionation—chromium, vanadium
	Immiscible magma separation—nickel, copper, cobalt, platinum-group elements

are hot and are known by the special name of *hydrothermal solutions*. The formation of ore minerals and compounds by magmas and water can take place at Earth's surface or at depth in the crust or mantle. In many cases, these chemical processes make a mineral deposit that is rich enough to be an ore deposit. In other cases, further concentration is needed by an additional chemical process or a physical process. In the following discussion, we divide ore-forming processes into those that take place at or near Earth's surface and those that take place at depth.

Surface and Near-Surface Ore-Forming Processes

Weathering and Soils

Weathering, the process by which rocks and minerals are decomposed at Earth's surface, includes chemical and physical actions of water, plants, and animals. It produces the *regolith,* a layer of partly decomposed rock and soil, that covers most of the land surface to depths of several hundred meters. The term *soil* has many different definitions. Some civil engineers refer to it as unconsolidated material at Earth's surface, a definition that includes not only the regolith but also young, unlithified sediments below it. To the farmer, soil is the upper part of the regolith that will support marketable plant life. Most geologists and environmentalists extend this definition to take in the upper part of the regolith throughout the world, including materials somewhat deeper than the soil of agriculture. This layer contains most of the minerals that are produced by weathering and it bears the brunt of environmental pollution (Foth, 1984; Brady, 1990; Buol et al., 1989).

In warm, humid climates, soil forms largely by *chemical weathering* involving the dissolution of common rock-forming minerals. This puts elements such as sodium, potassium, calcium, and magnesium into solution and on their way via streams to the sea. Although some minerals, such as *halite* (NaCl), are readily soluble in water, most silicate minerals are not very soluble. Thus, chemical weathering depends heavily on acid water, which is generated by dissolution of atmospheric gases such as CO_2 and SO_2, as discussed in the next chapter (see Table 3-4). This acidity is further enhanced by bacterial activity in soil, which boosts CO_2 concentrations to levels of several percent versus only 0.03% in the atmosphere, by decomposition of pyrite and other sulfide minerals to form sulfuric acid, and by acids formed during decay of organic matter. Some elements, such as iron and carbon, are oxidized during weathering, further contributing to the decomposition of minerals containing them, and plant roots create special chemical environments that enhance mineral dissolution. The ease with which rock-forming minerals decompose during weathering is roughly proportional to the temperature at which they formed originally, with high-temperature igneous minerals decomposing most readily.

Physical weathering involves processes that actually break the rock into pieces. Root growth has this effect, as does ice, which has a volume 9% greater than an equivalent amount of water. Growth of salt and calcite crystals when water evaporates in cracks can also fracture a rock. Diurnal (daily) and seasonal temperature changes cause expansion and contraction that gradually promote fracturing, as does the removal of overlying sediment by erosion. These effects dominate only in arid and arctic environments, where chemical processes are limited by the absence of moisture or by low temperatures.

Soils produced by chemical and physical weathering are usually classified on the basis of their vertical zonation. The generalized *soil profile* shown in Figure 2-5, which is typical of humid environments, consists of an upper O horizon rich in organic material from vegetation, which grades downward into an A horizon dominated by silicate and oxide minerals formed by rock weathering. Acid rain, augmented by acid from decomposition of organic material, leaches material from the lower part of the A horizon, which is

sometimes given the separate name E horizon. These dissolved constituents move downward and are deposited in the B horizon, which becomes enriched in soluble material. Underlying all of these is the C horizon, which is partly decomposed bedrock.

Weathering and soil formation create mineral deposits in two ways. *Removal of soluble constituents* leaves behind relatively insoluble elements and minerals of value in *residual deposits*. For instance, iron sulfide minerals containing small inclusions of gold weather to produce iron oxides and small grains of gold, which can be extracted more easily. Similar processes form concentrations of barite, mica, and other insoluble minerals when their enclosing rocks dissolve and disintegrate. *Redistribution of soluble constituents in the soil* also creates mineral deposits, most of which are known as *laterites*. To most soil scientists, laterites are soils that form when an iron-rich B horizon is exposed by erosion and made harder by desiccation. To the geologist, a laterite is any strongly leached soil rich in iron, aluminum, nickel, and related elements.

We should not lose sight of the fact that soil is one of our most important mineral resources in its own right. It is the basis for agriculture as well as the principal host of many microorganisms that carry out chemical reactions essential to life. By its very nature, soil is not coherent or lithified like rock and is therefore highly subject to erosion. Soil is held in place largely by plant roots and where they are removed, soil that has taken thousands of years to form can be removed in a few hours. About 6 billion tonnes of soil are eroded by wind and water each year in the United States, far more than the amount of waste from raw

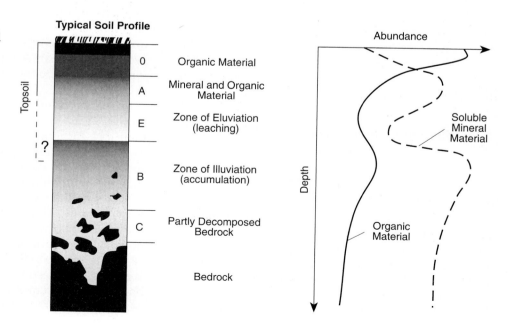

FIGURE 2-5

Typical profile of a well-drained soil in a humid temperate region showing zones in which soluble mineral and organic matter are leached and accumulated.

material extraction (Figure 1-3A). Although some erosion is natural and necessary, a disproportionate share affects agricultural land, with as much as 40% of U.S. crop land suffering unacceptable rates of erosion. Estimates by the U.S. Department of Agriculture put annual soil loss from the intensely farmed corn belt in Iowa and adjacent states as high as 600 million tonnes. Soil is also subject to *desertification,* which occurs naturally on the margins of deserts, but is also caused by overgrazing, improper cultivation, and mining without proper reclamation. *Salinization* of soils is a problem in areas of excessive irrigation, where soluble salts concentrate near the surface as the water moves upward by capillary action and is evaporated. Major changes in soil chemistry can be also caused by removal of vegetation, which produces *laterization,* particularly the formation of iron-rich laterite soils where vegetation is removed from tropical areas. Finally, soil is the locus of most of the pollution that affects the lithosphere, as discussed in the next chapter.

Sedimentary Processes

The simplest type of mineral deposits that form by *clastic sedimentary processes* are stream gravels and beach or dune sands. Clastic deposits consisting of heavy grains that were deposited where flowing water was slowed by obstructions are known as *placer deposits* (Figure 2-6). The most familiar placer deposits contain *gold,* which has a specific gravity about six

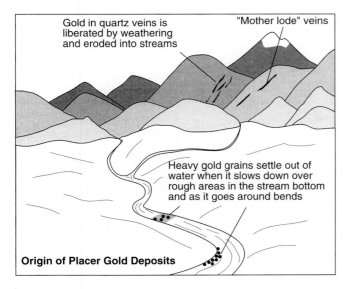

Gold in quartz veins is liberated by weathering and eroded into streams

"Mother lode" veins

Heavy gold grains settle out of water when it slows down over rough areas in the stream bottom and as it goes around bends

Origin of Placer Gold Deposits

FIGURE 2-6
Origin of placer deposits, which are concentrations of heavy clastic grains in stream, lake, or ocean sediment. As shown here, minerals liberated from quartz veins by weathering and erosion are carried downstream and deposited where the water slows.

times greater than average rock. Placer deposits that have been buried and preserved in older sedimentary rocks are known as *paleoplacers.* Paleoplacers older than about 2.4 Ga contain the heavy minerals *uraninite* and *pyrite,* which dissolve in modern oxygen-rich, weathering environments. Their preservation in paleoplacers suggests that older weathering environments contained less oxygen.

Chemical sedimentation is an important ore-forming process only where the influx of clastic sediment is very limited, such as in isolated arms of the sea, inland lakes, or the open ocean far from land. The simplest and most widespread ores that form by chemical sedimentation are *marine evaporites,* which are the result of evaporation of seawater. Marine evaporites contain elements that are abundant in seawater and they are our main source of salt, gypsum, potassium, and bromine. *Nonmarine evaporites* are usually found in arid regions of the continents where water flows into enclosed valleys with intermittent saline lakes and evaporates to form salt pans known as *playas.* The composition of lake-hosted, or *lacustrine,* evaporites usually reflects the composition of the local drainage basin, and includes a complex array of sodium, magnesium, boron, nitrogen, and iodine minerals. Pore spaces in buried evaporite deposits and their host clastic sediments contain aqueous solutions more saline than seawater, which are known as *brines,* and which contain high concentrations of similar elements.

Other metal-rich chemical sediments, which are enriched in various combinations of iron, manganese, copper, zinc, lead, and barium, did not form by direct precipitation from seawater. Most of them formed where hot metal-rich hydrothermal solutions flowed into the ocean from submarine hot spring vents, precipitating minerals as they cooled. Others formed where metal-rich seawater accumulated in the deep ocean (possibly enriched by submarine hot spring waters) and then flowed up into shallower, oxygen-rich water to deposit metals.

Organic sedimentation also forms mineral deposits. Coral reefs, which consist largely of the calcium carbonate minerals *calcite* and *aragonite,* are mined for *limestone.* Skeletal material in other organisms consists of the calcium phosphate mineral, *apatite,* which is also mined. Some mineral deposits form by modification of organic material in sediments. For instance, chemical changes associated with burial beneath younger sediment transform organic matter rich in carbon and hydrogen into *coal, oil,* and *natural gas.* Bacteria facilitate these changes, usually because they derive energy from them. For instance, some bacteria convert dissolved sulfate ions (SO_4^{-2}) into H_2S gas, which is later reduced to the elemental form, known as *native sulfur*. Many such changes take place below Earth's surface and merge with the subsurface processes discussed below.

Subsurface Ore-Forming Processes

Subsurface processes that form mineral deposits involve aqueous, hydrocarbon, and magmatic fluids. *Aqueous fluids* consist of water and dissolved solids and gases, including fresh water, steam, and brine. *Hydrocarbon fluids* consist of crude oil and natural gas, and *magmatic fluids* include magmas of all compositions. These fluids form mineral deposits in two ways. Most commonly, they precipitate ore minerals from aqueous solution or crystallize them from magma. In other cases, they form fluid concentrations that constitute ore in their own right, such as groundwater, oil and gas accumulations, and some special metal-rich magmas.

Discussion of these subsurface fluids and their ore-forming processes requires a brief review of terminology. *Groundwater*, for instance, really means all water in the ground but is widely applied to near-surface water that is consumed for municipal, agricultural, and other uses. We will use this more restrictive definition here and discuss groundwater resources separately. *Hydrothermal solutions*, which are hot groundwaters, are usually distinguished as a separate type of natural fluid, a practice that we will follow here. Hydrocarbon fluids (oil and gas) are associated locally with hydrothermal solutions and will be discussed together with them. The distinction between groundwater and hydrothermal solutions is reflected in the terms *supergene*, which refers to water that moves downward through the crust, and *hypogene*, which refers to water that moves upward. In general, hypogene water is of hydrothermal origin, whereas supergene water is precipitation on its way down into the crust. Finally, magmas and their role in formation of mineral deposits will be discussed separately. Although this division is imperfect and imposes categories where nature has not, it allows us to discuss crustal fluids in a way that reflects their relation to mineral resources.

Aqueous and Hydrocarbon Ore-Forming Systems

Movement of Aqueous and Hydrocarbon Fluids in the Crust The movement of aqueous and hydrocarbon fluids in the crust is controlled by *porosity,* the proportion of the rock that consists of pores or open spaces, and *permeability,* a term that describes the degree to which these pores connect together to form a pathway for migrating fluids (Figure 2-7). Rocks with high porosity and permeability that contain water are *aquifers* and those that contain oil and gas are *reservoirs* (Domenico and Schwartz, 1990). Permeabilities of rocks for water differ by many orders of magnitude (Figure 2-7B), and those for oil and gas are usually much lower and much higher, respectively. In fact, crude oil is very difficult to pump from the ground.

The flow of underground fluids is described by *Darcy's Law* (Table 2-3), which says that the amount of fluid that will move through a porous rock per unit of time is directly proportional to *head gradient* and *fluid conductivity.* The head gradient reflects the eleva-

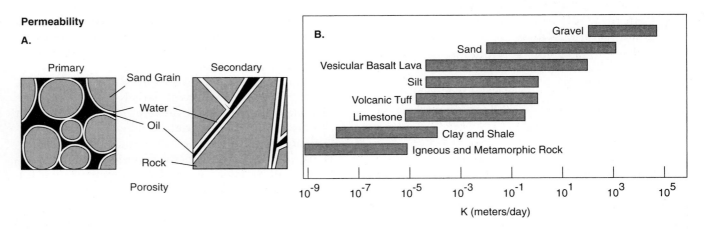

FIGURE 2-7

A. Typical primary and secondary porosity, showing the relation between water and oil in pores. **B.** Hydraulic conductivity (permeability to water) of various common rock types under near-surface conditions (modified from Driscoll, 1986).

TABLE 2-3
Brief summary of Darcy's law, the relation that explains the movement of fluids through rocks. It says, in general, that the amount of fluid that will move through a rock is directly proportional to fluid conductivity and the head gradient forcing the fluid to flow.

Darcy's law

$q = -K * dh/dl$

Definition of terms used in Darcy's law

q = volume of fluid flow per unit of area per unit time
K = fluid conductivity
dh/dl = head gradient
$K = k * F * R$
k = constant
F = fluid term = density of fluid/viscosity of fluid
R = rock term = permeability (proportional to square of average diameter of grains)

tion of fluid along the flow path and any additional pressure forcing it along. Fluid conductivity describes both the rock and the fluid, and is known as *hydraulic conductivity* if the fluid is water. The rock is described by its permeability and the fluid by its density and viscosity.

Porosity includes all types of openings in a rock, whether they are coeval with the rock (*primary porosity*) or formed later (*secondary porosity*). Primary porosity is most common in sedimentary rocks and is largely absent from igneous and metamorphic rocks. Clastic sedimentary rocks such as sandstone and conglomerate consist of well-rounded, spherical grains with as much as 30% effective porosity. *Reefs*, which consist of an intergrowth of marine animal skeletons, also have a high effective porosity, as does the debris shed from these reefs, which is known as *reef talus*. Clay minerals, the dominant component of most shales, also pack together in a way that produces a high porosity, but their plate-like form blocks the connections between pores, giving them a low effective porosity and therefore a low permeability. They also lose porosity as they become compacted by burial beneath younger sediments. Rocks of this type which will not transmit fluid are referred to as *impermeable*, a characteristic of most igneous and metamorphic rocks. *Vesicles*, which are small holes in the shape of gas bubbles that exsolved from lava, provide some primary porosity in volcanic rocks.

Secondary porosity is created in all types of rocks by faulting, fracturing, and chemical reactions. In many igneous and metamorphic rocks, fault zones are the only areas with high porosity and permeability. Chemically corrosive water can invade otherwise im-

permeable rocks by dissolving them, a process that is most common in limestones where descending rainwaters or supergene solutions form caves.

In the upper crust, rocks have high porosities and permeabilities that allow widespread fluid flow. Consequently, mineral deposits are more common there. The downward decrease in porosity and permeability in the upper crust is caused by compaction of rocks in response to increasing pressure and by deposition of new minerals in pores, a process known as *cementation*. Fractures that fill with quartz and other minerals to form *veins* are a special type of cementation. Below depths of about 10 to 20 km, deformation and metamorphic mineral growth obliterate most porosity and permeability and hydrothermal mineral deposits are rare (Nesbitt et al., 1989).

Groundwater Systems   Most of the upper part of the continental crust is filled with fresh *groundwater* (Freeze and Cherry, 1979; Driscoll, 1986). Groundwater comes from *recharge*, the fraction of precipitation that flows into the ground (Figure 2-8). This water flows downward through an unsaturated zone, known as the *vadose zone*, to the *water table*, beneath which rock is saturated and can constitute an *aquifer*. Groundwater aquifers range from regolith and sediments just below the surface to sedimentary layers or fracture zones that extend to depths of many kilometers. Recharge can also come from seawater or waters made saline by evaporation in desert areas. These saline and fresh waters do not mix easily and less dense fresh water usually floats on more dense saline water (Figure 2-8).

Where groundwater aquifers are undisturbed, they reach an equilibrium between the amount of recharge

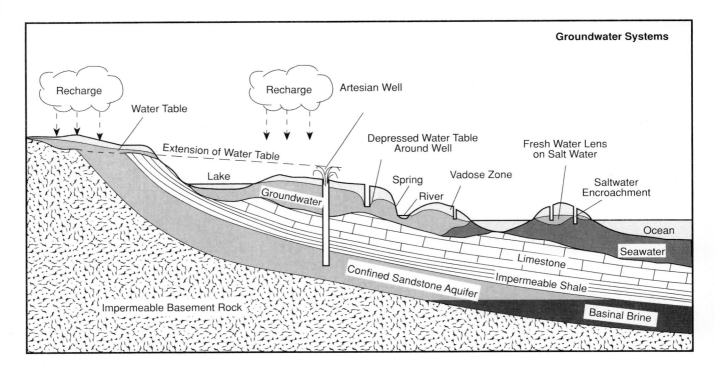

FIGURE 2-8

Schematic illustration of groundwater systems showing both unconfined and confined aquifers. Unconfined aquifer shown here consists of a layer of glacial deposits that covers the entire surface. Note that the water table mimics surface topography and forms lakes, springs, and rivers where it meets the surface. Fresh water floats on salt water where the aquifer extends beneath sea level. A confined aquifer, which is recharged in the hills to the left, supports an artesian (flowing) well. As long as the water table in the confined aquifer is above the top of the well, it will continue to flow.

and the amount of water leaving the system at springs, lakes, and streams. Use of groundwater has disrupted this equilibrium, leading to problems in some areas. Most commonly, *excessive withdrawal* from a well or well field lowers the local water table, sometimes drying up shallow wells in the vicinity. Excessive withdrawal from wells throughout a large area can cause regional water table declines. In a 1985 study, the U.S. Geological Survey showed that the water table in regions from New Jersey to California had declined 1 to 3 meters annually over the last few decades (Figure 2-9). In coastal regions and on ocean islands, excessive withdrawal causes *saltwater encroachment,* a process by which seawater invades and contaminates the fresh-water aquifer. This can even occur relatively far inland, as in the eastern part of San Francisco Bay, where excessive withdrawals have been made for agriculture and pipeline export to the southern parts of the state. In areas like southern Florida, saltwater encroachment is aggravated by excavation of ocean-access channels that permit the influx of seawater from above. Irrigation also exacerbates saltwater en-

croachment and salinization when evaporation increases the salt content of irrigation waters that recharge local aquifers. Excessive withdrawal, which causes *subsidence* of the land surface (Figure 2-10), has been observed in California, Nevada, Arizona, Texas, Louisiana, and South Carolina (Figure 2-9). Subsidence of only a few inches can cause big problems by greatly enlarging areas subject to flooding and by reversing the slope of sewer and water pipes. In areas underlain by shallow cave systems, such as central Florida, excessive withdrawal of groundwater can cause formation of sinkholes.

Excessive withdrawal constitutes *groundwater mining.* The largest area of groundwater mining in the United States is the High Plains aquifer, which consists largely of the Miocene-age Ogallala Sandstone, a layer of sand and gravel that eroded from the Rocky Mountains (Figure 2-9). The High Plains aquifer contained at least 4,000 trillion liters of water before large-scale farming began in the area in the late 1800s (Gutentag et al., 1984). Since that time, about 5% of the water in the aquifer has been removed, largely to

FIGURE 2-9

Distribution of groundwater problems and water-transfer systems in central North America (compiled from U.S. Geological Survey Water Supply Paper 2250; U.S. Geological Survey Professional Paper 1985-B, and other sources).

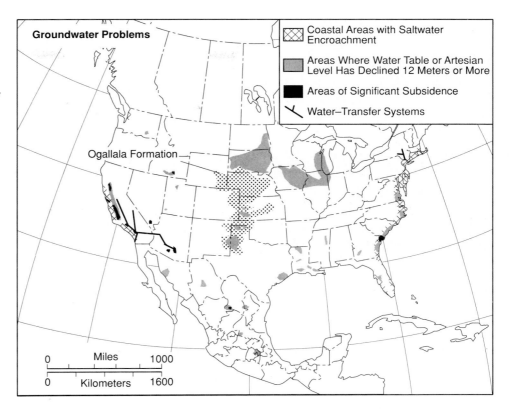

irrigate agricultural land. Annual withdrawal has ranged as high as 100 times the natural recharge in parts of the aquifer, leading to prodigious declines in the water table, reaching a maximum of about 60 meters in the Texas panhandle (Figure 2-9). Declines of more than 16 meters have been measured in an area covering at least 31,000 square kilometers. Continued fall of the water table has put some farmers out of business because of the increased cost of pumping water from these great depths and has caused the Internal Revenue Service to treat land in the area as a depreciable asset, an approach similar to the depletion allowance discussed in the section on mineral economics. Largely in response to groundwater mining of this type, many states and smaller jurisdictions have systems to control withdrawals from large aquifers.

Water shortages are widespread around the world but nowhere are they more acute than in the Middle East. The Tigris and Euphrates Rivers, which flow from Turkey through Syria and Iraq on their way to the Persian Gulf, are more heavily used even than the Colorado River in the southwestern United States. Where surface water is not available at all, especially in Kuwait and Saudi Arabia, water supplies are obtained from a combination of desalinization and groundwater withdrawal. Desalinization, although very effective, costs as much as ten times more than current water rates in the United States and will only

become more expensive as energy costs rise. Groundwater withdrawal in Saudi Arabia, in particular, is proceeding at an extremely rapid pace, with total exhaustion predicted within less than a century. Although a similarly gloomy scenario can be envisioned for some parts of western North America, particularly Arizona and Sonora, international cooperation will likely solve the problem by water transfer from the north. The same is not true for the Middle East, where present politics are already affected by water needs. Many observers believe the next war in the area may well be over water rather than oil.

An obvious solution to such problems would be to import water. *Long-distance water transfers,* whether surface or groundwater, have become more and more common. The pioneer in all of this is southern California, which imports water from the northern part of the state, as well as the Colorado River (Figure 2-9). Most residents of wetter parts of the world are not enthusiastic about such long-distance water transfers. The logic of this stance is elusive in view of our willingness to transfer oil, gas, and other mineral fluids around the world. No one says that all of us have to live near the oil wells. The same logic probably applies to water, as long as we can afford to move it to market.

Hydrothermal Systems Hydrothermal solutions are widespread in the upper crust, where they form iso-

FIGURE 2-10
Subsidence around a water well such as seen here can be caused by withdrawal of
water from the underlying aquifer or by compaction of surrounding sediments caused
by water spilled from the well (courtesy of U.S. Department of Commerce, National
Oceanic and Atmospheric Administration).

lated systems that come and go as geologic features change (White, 1974). Most of these systems underlie and mix upward into groundwater, as defined earlier. In fact, the water in some hydrothermal systems is groundwater that percolated deeper into the crust and became heated. Other hydrothermal solutions were released from magmas, sedimentary basins, or rocks undergoing metamorphism. Beneath the oceans, seawater also flows down into the crust to form hydrothermal systems. Hydrocarbon fluids (oil and gas), on the other hand, come largely from organic material buried in sedimentary basins.

The importance of a hydrothermal system to the formation of mineral deposits lies in the geometry of water flow, in which water from widely separated areas coalesces into a small, high-volume outflow zone. Most hydrothermal solutions react with the enclosing rock to form new minerals throughout their flow path in a process known as *hydrothermal alteration* (Plate 1-1). During hydrothermal alteration, hydrothermal solutions often dissolve metals and other elements of interest from the original minerals and carry them into the restricted outflow zones where they are precipitated to form mineral deposits. Deposition of ore minerals from hydrothermal solutions can take place by open-space filling or replacement. *Open-space filling* refers to the precipitation of minerals in existing porosity. If the fluid is flowing through small pores in a rock,

this process is analogous to cementation. If it is flowing through a fracture, the new minerals can form a vein. Where rock is highly fractured and consists of many intersecting veins and veinlets, it is known as a *stockwork*. *Replacement* refers to the process by which new minerals actually take the place of preexisting minerals almost atom for atom. This process involves many materials in addition to ores. In petrified wood, carbon and hydrogen have been replaced in this way by opal, chalcedony, or some other form of silica.

Only groundwater, crude oil, natural gas, and geothermal resources come from active fluid systems. All other hydrothermal mineral deposits come from systems where the fluid has departed (Guilbert and Park, 1986). In fact, recognition of *extinct hydrothermal systems* is one of the main challenges that we face in exploring for mineral resources and in classifying land for future use. The main features that aid in recognition of ancient hydrothermal systems are hydrothermal alteration, which can extend for many kilometers outward from a mineral deposit, and *fluid inclusions,* which are small imperfections in crystals that have trapped samples of the fluid from which their host crystal grew (Plate 1-2). Measurements made on fluid inclusions provide information on the temperature, composition, and character of ancient hydrothermal systems.

FIGURE 2-11

Schematic illustration of
geologic environments for major
types of hydrothermal systems.
Relation between these systems
and larger plate tectonic
environments is shown in
Figure 2-2.

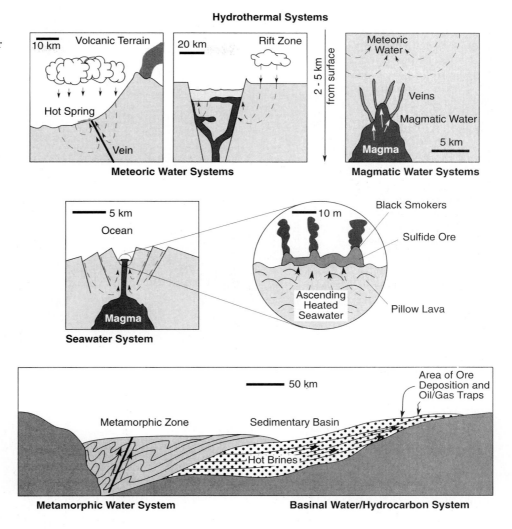

The widespread occurrence of hydrothermally altered rocks and fluid inclusions proves that large
amounts of water have moved through the crust.
These waters appear to have had several origins,
which are divided into meteoric, seawater, basinal,
magmatic, and metamorphic. In reading these descriptions, it should be kept in mind that few waters
are of a single origin, and that waters probably mix
extensively throughout the crust.

Meteoric water is simply water that has passed
through the atmosphere as rain or snow. It is, in fact,
the recharge for groundwater systems, and *meteoric hydrothermal systems* are the lower, hotter parts of
groundwater systems. These systems form where meteoric water percolates downward, is heated, and becomes buoyant enough to rise along more focused
zones (Figure 2-11). Meteoric water systems are best
developed in volcanic terranes (Plate 1-3), where shallow intrusions underlying the volcanoes provide heat

to drive convection of the water (Norton, 1978). Where
the intrusion is near the surface, meteoric systems often form *hot springs* (Figure 2-12A) or episodic foun-
tains of hot water known as *geysers*. Temperatures in
such systems reach a maximum of about 350°C at
depths of a kilometer or so, but many are only 100° to
200°C. In areas with no volcanic activity, even cooler
meteoric systems form along large faults where water
is made buoyant by the natural increase in heat downward in the crust, which is known as the *geothermal
gradient* (see Figure 4-16). The most common type of
mineral deposits formed by meteoric hydrothermal
systems are epithermal vein deposits, which are an important source of gold and silver.

Seawater hydrothermal systems are the submarine
equivalent of meteoric hydrothermal systems in the
subaerial environment (Figure 2-11). They usually
form along mid-ocean ridges at divergent margins
where rifts allow seawater to flow downward into hot

FIGURE 2-12

Hydrothermal systems. **A.** Hot springs, which form the upper part of the meteoric system at Soufriere, St. Lucia, consist of pools of hot water that appear to be boiling because of large amounts of CO_2 gas that rise through the water from beneath. **B.** Black smoker at the top of the seawater hydrothermal system on the East Pacific Rise near 12°50′N at 2,620-meter depth, as taken from submersible CYANA during the Cyantherm cruise (courtesy of Roger Hékinian, IFREMER, Centre de Brest, France).

A.

B.

rocks surrounding basalt magma chambers (Cathles, 1981). Seawater that circulates through these systems returns to the surface to form spectacular hot-spring vents that have been observed by research submarines (Figure 2-12B). Water leaves the vents at temperatures as high as 350°C and is prevented from boiling only by the high pressures that prevail at the 5-kilometer water depths where they are found. These hot-spring vents emit water with dissolved metals and H_2S, which react to precipitate minute particles of metal sulfide as the vent water mixes with cold seawater, a process that has earned them the name *black smokers*. Cooler seawater hydrothermal systems have been found away from active submarine volcanic areas and areas without any sources of heat are probably saturated with normal seawater, just as the upper part of the continents is saturated with groundwater. Although seawater hydrothermal systems can form veins, they usually form *exhalative deposits* consisting of metal sulfides deposited at the vents. The form of these deposits depends on the density of the hydrothermal solution that is emitted into seawater. Where it is more dense than seawater, it ponds in depressions in the sea floor, forming saucer-shaped deposits. Where it is less dense, it rises to form smokers that coalesce to form mounds (Rona, 1988).

Basinal fluid systems include all fluids in or derived from sedimentary basins such as the thick prism of sediment accumulating off the Gulf coast of Louisiana and Texas. Much of the water in sedimentary basins consists of seawater that was trapped in sediments when they were deposited, a type of water known as *connate water*. Basins also contain meteoric water that flows in from above and some water is released when clays and other hydrous minerals undergo metamorphism at the higher temperatures caused by burial of the sediment. Changes in organic material during burial release hydrocarbon fluids, largely oil and gas. During burial, water moves through sediments, becoming *formation water*, a nonspecific term for water in sedimentary rocks. Formation water that is very saline is known as *basinal brine* and probably formed by evaporation of seawater at the surface or dissolution of buried evaporite layers consisting of gypsum and halite (Hanor, 1987). Basinal waters frequently circulate to deep enough levels to reach temperatures of more than 100°C and the brines, in particular, are potent solvents because they contain abundant chloride, which combines with metals to form soluble complex ions.

Movement of basinal fluids takes place largely in response to gravity and the geological evolution of the basin. Brines usually accumulate at the bottom of the basin because they are most dense. Oil and gas, which are least dense, rise toward the surface. Some reaches the surface to form oil springs and tar pits, but most is trapped in reservoir rocks, forming oil and gas fields (Plate 1-4). Compaction of sediment during burial closes pores, driving fluids outward and upward from the basin. If the basin is tilted, fluids flow out the low side, often pushed by recharge from the high side. Finally, deformation of a basin, as might happen at collisional plate margins, causes fluids to be squeezed outward. During this outward fluid migration, oil and gas form fields where they intersect appropriate traps. Other mineral deposits, particularly Mississippi Valley-type lead-zinc deposits, form where basinal brines precipitate dissolved elements in their flow path (Plate 1-5). Finally, some basinal brines are expelled as hot springs to form sedimentary exhalative (sedex) deposits, a special class of the exhalative deposits mentioned earlier.

Magmatic hydrothermal systems consist of fluids that separate from a magma as it cools and crystallizes at depths of several kilometers (Figure 2-11). All magmas contain dissolved water, as well as gases such as CO_2, SO_2, H_2S, and HCl. When the magma crystallizes, some of these fluids become incorporated into magmatic minerals such as amphiboles, but most remain in magma, gradually exerting greater pressure as their concentration increases. When this pressure exceeds that of the surrounding rocks, the water and gases exsolve from the magma to become magmatic hydrothermal solutions with temperatures of at least 600°C. These solutions are enriched in chlorine and metals, which also concentrate in residual magmas (Holland, 1972). As would be expected, magmatic hydrothermal systems are best developed in and around intrusions. They are usually associated with felsic magmas, which dissolve more water and gases than mafic magmas (Figure 2-13). Mineral deposits formed by these solutions take two main forms. Small felsic intrusions that reach high levels in the crust exsolve magmatic fluids in sudden, explosive events that shatter large volumes of rock, producing stockwork zones that host porphyry copper and molybdenum deposits. Where felsic magmas intrude limestone, reactions at and near the contact form *skarn,* a rock consisting of calcium silicates with metal oxide and sulfide minerals.

Metamorphic hydrothermal systems consist of fluids that were expelled from rocks during metamorphism (Figure 2-11). As temperature and pressure increase during metamorphism, minerals such as clay, mica, and calcite, which contain water, CO_2, and other gases, react to form minerals such as feldspar that do not contain water or gases (Fyfe, 1978). At the same time, organic matter remaining after the formation of crude oil continues to break down to form methane. The water and gas released by these reactions are driven out of the rock, with their outflow usually along fault zones. The reactions that drive out water and gases usually occur at temperatures of 200° to 600°C, and if the fluid release occurs in a single large pulse, it can produce a significant fluid event in the rocks. Fluids of this type appear to have formed quartz veins with gold in faults that cut metamorphosed sedimentary and volcanic rocks above some Mesozoic and Cenozoic subduction zones and in regional faults that cross Archean cratons. Elsewhere, rift valleys on the continents are filled with basalt, which was metamorphosed to release copper-rich water that flowed outward and formed deposits at the edges of the basins.

Geochemistry of Hydrothermal Systems It might have occurred to you in reading about the types of hydrothermal systems that water is water. How can one water molecule be distinguished from another? By far the best way to do this is through analysis of the *isotopic composition* of the water, which involves measurement of the relative abundances of isotopes of hydrogen and oxygen that make up water. Waters in different geologic environments have different hydrogen and oxygen isotopic compositions and it is possible to use these differences to determine the source of water in many hydrothermal systems (Figure 2-14). Information about water that formed extinct systems can be obtained by analysis of fluid inclusions or minerals that contain hydrogen and oxygen. These measurements have provided extremely important insights into the nature of hydrothermal systems, but they are not foolproof. Complications result because isotopic compositions of waters of different types overlap and, worse still, because the isotopic composition of many minerals is changed by other waters that pass by after the deposits form. Thus, efforts are still underway to sort out the distribution and history of fluid movement in Earth's crust.

You might also have wondered why a hydrothermal solution that had dissolved metals or other elements and transported them for a great distance would deposit them. That is, how do hydrothermal

FIGURE 2-13
Copper, lead, zinc, and silver ores at Concepción del Oro, Zacatecas, Mexico are concentrated along the contact between a felsic intrusion (on left) and limestone (on right), which runs up the valley. Piles of overburden from open pit mines are at the bottom of the valley and a smaller pile removed from an underground mine is at the head of the valley.

systems actually form hydrothermal mineral deposits? These processes, which are known as *depositional mechanisms,* reflect changes that take place in hydrothermal solutions as they flow outward from their source. Most minerals become less soluble with decreasing temperature; thus, they should precipitate as the solutions rise from deep, hot levels of the crust toward the cooler surface environment. However, cooling occurs gradually over a long flow path and usually produces widespread, weak mineral concentrations rather than rich ore deposits. More localized chemical changes that can produce rich ore occur when the hydrothermal solution boils, mixes with another type of water, or reacts with rocks such as limestone (Figure 2-15).

Magmatic Ore-Forming Systems

Although some igneous rocks such as crushed stone and dimension stone are mineral deposits in their own right, most magmas require some internal process to form mineral concentrations that are of economic interest. These concentrations form by two main processes, crystal fractionation and magmatic immiscibility (Figure 2-16).

Magmas crystallize over a range of temperatures, with different minerals crystallizing in different parts of this range. This permits the magma to undergo *crystal fractionation,* a process by which early-crystallized minerals sink to the bottom of the magma chamber to form a type of "magmatic sediment" known as *cumu-*

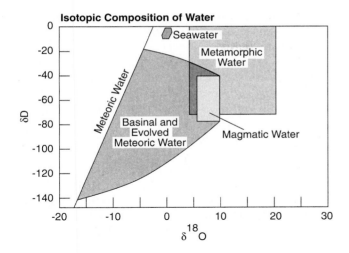

FIGURE 2-14

Oxygen and hydrogen isotopic compositions of some important types of hydrothermal systems. $\delta^{18}O$ indicates the ratio of ^{18}O to ^{16}O in the sample versus that in a standard and δD indicates the same ratio of ^2H to ^1H (deuterium). These isotopes are stable (that is, they are not produced by radioactive decay), but their ratios are affected by physical and chemical reactions involving oxygen and hydrogen, particularly in water.

late layers. Layers of this type are particularly important in large mafic intrusions known as *layered igneous complexes.* The largest of these is the Bushveld Complex in South Africa, a major source of chromium, platinum, and vanadium.

Although the general process of crystal fractionation is clear enough, most magmas crystallize more than one mineral at a time, making it difficult to form thick cumulate layers consisting of only one mineral.

The most likely explanation for these special cumulate layers appears to be magma mixing, in which a batch of new magma is injected into a magma chamber that still contains some old magma. If these two magmas have different densities and compositions, one will probably float on the other, producing a contact along which chemical interaction allows crystallization of a single type of mineral that settles to the bottom of the magma chamber.

Some magmas split into two separate, *immiscible magmas* of different composition at some point in their crystallization history. In all cases, the main magma is a relatively typical silicate magma, such as forms most igneous rocks. The other magma, however, is usually rich in metal sulfides or metal oxides. This process appears to occur in two main types of magma. Some felsic magmas separate an immiscible iron oxide magma and some ultramafic magmas separate an immiscible copper-nickel-iron sulfide magma. Immiscible magmas of both types usually sink to the bottom of their magma chamber because they are heavier than the silicate magma. However, some iron oxide magmas are very rich in gases which exsolve and make them buoyant enough to extrude onto the surface like lava. Iron deposits of this type are often associated with copper, uranium, and fluorite that were probably deposited from hydrothermal solutions formed from these gases.

A few magmas contain enough valuable minerals to be extracted in bulk. *Kimberlites,* some of which are the main source of the world's diamonds, are potassium-rich, ultramafic fragmental rocks that appear to have drilled their way up through the crust from the mantle. The rapid ascent of these magmas is usually ascribed to a high gas content, which probably expanded into overlying rock, breaking it and making a

FIGURE 2-15

Processes that deposit ore minerals from hydrothermal solutions include *reaction* between solution and chemically reactive wallrock such as limestone, *boiling* that releases gases from the hydrothermal solution, and *mixing* with descending, cool meteoric waters.

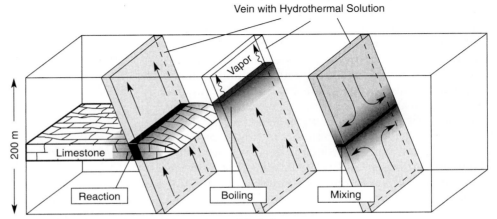

Depositional Mechanisms in Hydrothermal Solutions

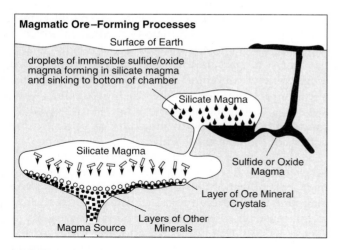

FIGURE 2-16

Schematic diagram showing magmatic processes that form mineral deposits by crystal fractionation and immiscibility. (All processes probably do not take place in the same magma.)

path for the magma. *Carbonatites*, which are sources of vanadium, columbium, and tantalum, are igneous rocks that consist almost entirely of calcite and other carbonate minerals. They are also thought to originate in the mantle. Both of these unusual igneous rocks are usually found in old cratons.

DISTRIBUTION OF MINERAL DEPOSITS IN SPACE AND TIME

Ore-forming processes are closely related to plate tectonic activity. In the ocean crust, seawater hydrothermal systems are formed along the mid-ocean ridge by the heat of rising basalt magmas. Magmatic processes beneath the ridge also generate cumulate deposits, although such deposits are seen only in a few obduction zones (Figure 2-2).

Most other ore-forming processes take place in continents. Meteoric hydrothermal systems are found at rift zones, where volcanism is abundant and magmas are near the surface. Magmatic hydrothermal systems are common above subduction zones, where felsic and intermediate magmatism is widespread, but are not as strongly developed along mid-ocean ridges, probably because basalt magmas contain less dissolved water. Sedimentary basins and related waters, oil, and gas form along passive margins of continents, around volcanic arcs, and in rift zones. Some basins even form in the central parts of continents, possibly in response to the collapse of old rift zones in the underlying continent. Metamorphic waters are released wherever

sedimentary or volcanic rocks undergo metamorphism, usually along collisional margins.

Even ore-forming processes that take place at Earth's surface are affected by plate tectonics. Placer deposits form where mountains have been uplifted, largely along collisional margins. Evaporite and laterite deposits are related to plate tectonics because weather patterns are a function of the size of a continent and its position with respect to the poles. Continents in polar regions are more likely to accumulate an ice cap, which favors the formation of glaciers to scrape away valuable near-surface deposits, as happened only a few thousand years ago throughout most of northern North America, Europe, and Asia. Fortunately, glaciation often exposes deposits in the underlying bedrock, making them easier to find.

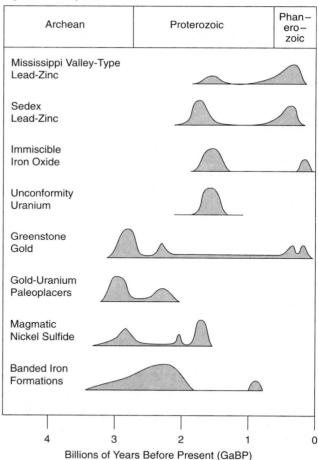

FIGURE 2-17

Formation of important types of mineral deposits through geologic time (modified from Meyer, 1981 and Barley and Groves, 1992).

Because of their close link to global geologic processes, mineral deposits also reflect important changes in Earth history (Figure 2-17). For instance, during Archean time, the upper mantle was hotter and partial melting took place at higher temperatures, extracting unusual ultramafic magmas that formed rocks known as *komatiite*. Komatiite magmas were rich in sulfur and nickel and formed immiscible magmatic sulfide magmas more readily than the basalt magmas that come from the mantle today. Metamorphism and intrusion in volcanic arcs in Archean time apparently produced larger amounts of hydrothermal fluids as well.

By late Archean and early Proterozoic time, sizeable continental masses had developed. Erosion of them produced a new geologic environment consisting of large sedimentary basins with shallow shelf areas. Where these basins are preserved, they contain large chemical and clastic sedimentary deposits, including much of Earth's iron, gold, and uranium ores. These deposits also reflect the changing composition of the atmosphere because they apparently did not form after middle Proterozoic time when the oxygen content of the atmosphere became high enough to change the geochemical behavior of iron and uranium at the surface.

By Phanerozoic time, the continents had grown large enough to host even bigger sedimentary basins, which contained limestone and evaporite minerals. The appearance of marine organic life enhanced the probability of forming crude oil and natural gas and the appearance of vascular land plants enabled the formation of coal deposits. Hot brines expelled from these basins formed Mississippi Valley-type lead-zinc deposits, which are rare in pre-Phanerozoic rocks. The increased concentration of oxygen in the atmosphere permitted more effective dissolution of uranium during weathering of the rocks and allowed the creation of uranium deposits related to groundwater movement. Finally, numerous collisional margin volcanic arcs developed on the newly formed continents, leading to the formation of mineral deposits associated with meteoric and magmatic hydrothermal systems.

CONCLUSIONS

It should be apparent from this discussion that mineral deposits are not randomly distributed in Earth. They are a direct product of their geologic environment and of the time in Earth history at which they formed. It follows that the mineral endowment of any country is a function of its geologic setting. The good news from this is that we can use geologic relations to predict the type of undiscovered mineral deposits that might be present in an area. The bad news is that no amount of government action or popular effort will change these geologic constraints. Our mineral endowment is fixed by geologic factors. The ray of hope in all of this is that we do not yet know all about the geology of the world or about all of the types of mineral deposits that it contains. Thus, if we continue to study Earth's geology and explore for mineral concentrations, we can hope that there will be some pleasant surprises that will help us with the growing problem of dwindling resources.

3

Environmental Geochemistry and Mineral Resources

THE FIXED LOCATION OF MINERAL DEPOSITS LEADS TO ENVIRON-
mental controversy. On the one hand, we must extract
the minerals where they are. On the other hand, we
fear environmental degradation caused by their extraction and use. Earlier in history when populations
were more dispersed, mineral deposits more abundant, and the demand smaller, environmental concerns were minor. Now they dominate the resource
scene and require informed debate to reach the best
solution.

Most environmental debates require an understanding of the fundamental principles of geochemistry that
govern the movement of pollutants. These principles
control environmental decisions in the same way that
geological principles control the distribution of mineral deposits. Failure to appreciate them creates the
risk of advocating impossible solutions, such as disposing of radioactive waste by burning it. Thus, before beginning our discussion of mineral resources, we
must review the chemical controls that determine their
environmental effects.

PRINCIPLES OF ENVIRONMENTAL GEOCHEMISTRY

The fundamental building blocks of Earth are *atoms,*
which consist of a positively charged *nucleus* sur-

rounded by a cloud of negatively charged *electrons.*
The nucleus is made up of *neutrons* that have no
charge and *protons* with a positive charge equal to
that of an electron. Different atoms, each with a
specific number of protons, make up the *elements,*
which are arranged in the *periodic table* (Figure 3-1).
Elements lose or gain electrons to become *ions.* Those
with fewer electrons than protons have a net positive
charge and are known as *cations;* those with more
electrons than protons have a net negative charge
and are known as *anions.* Ions that lose electrons are
said to be *oxidized,* whereas those that gain electrons
are *reduced.* The *valence* or *oxidation state* of the ion
refers to the number of electrons that it has lost or
gained. Elements with different numbers of neutrons
but the same number of protons are known as
isotopes. They are commonly referred to by their
number of protons plus neutrons, or *mass number,*
and are written in the form ^{235}U.

Atoms form *chemical bonds* with other atoms to produce *compounds,* substances with a definite chemical
composition. A group of atoms is known as a *molecule*
if it has no net charge, such as CO_2, or a *complex ion* if
it has a net charge, such as NH_4^+ or CO_3^{-2}. Compounds are classified as *organic* if they contain carbon
in association with hydrogen, oxygen, nitrogen, sulfur, phosphorous, chlorine, or fluorine, or *inorganic* if
they do not contain carbon. Compounds that form
during processes occurring in Earth are considered

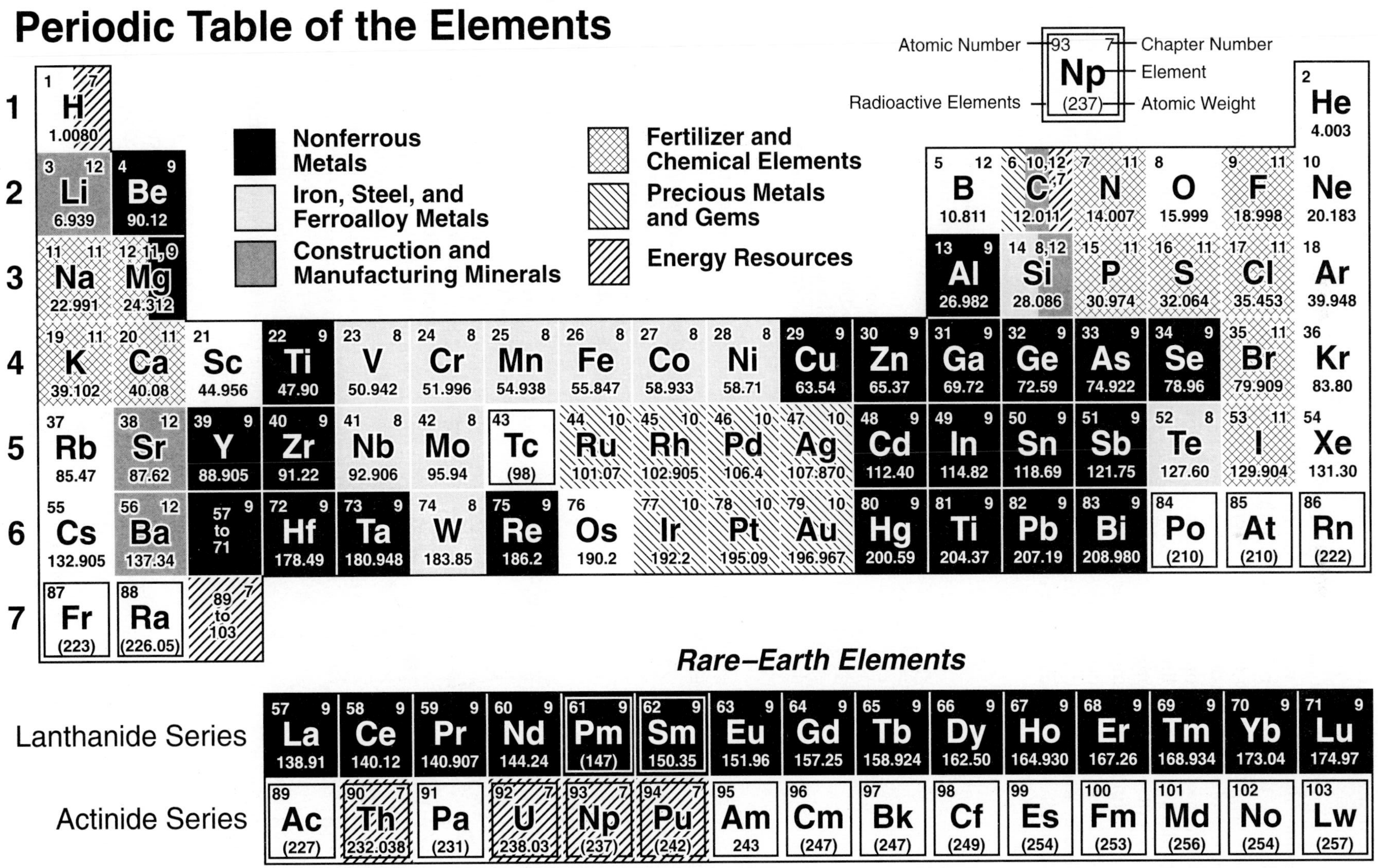

Periodic Table of the Elements

Atomic Number — 93
Element — Np
Atomic Weight — (237)
Chapter Number — 7

Legend:
- Nonferrous Metals
- Iron, Steel, and Ferroalloy Metals
- Construction and Manufacturing Minerals
- Fertilizer and Chemical Elements
- Precious Metals and Gems
- Energy Resources
- Radioactive Elements

Period																		
1	1 H 1.0080																	2 He 4.003
2	3 Li 6.939	4 Be 90.12										5 B 10.811	6 C 12.011	7 N 14.007	8 O 15.999	9 F 18.998	10 Ne 20.183
3	11 Na 22.991	12 Mg 24.312										13 Al 26.982	14 Si 28.086	15 P 30.974	16 S 32.064	17 Cl 35.453	18 Ar 39.948
4	19 K 39.102	20 Ca 40.08	21 Sc 44.956	22 Ti 47.90	23 V 50.942	24 Cr 51.996	25 Mn 54.938	26 Fe 55.847	27 Co 58.933	28 Ni 58.71	29 Cu 63.54	30 Zn 65.37	31 Ga 69.72	32 Ge 72.59	33 As 74.922	34 Se 78.96	35 Br 79.909	36 Kr 83.80
5	37 Rb 85.47	38 Sr 87.62	39 Y 88.905	40 Zr 91.22	41 Nb 92.906	42 Mo 95.94	43 Tc (98)	44 Ru 101.07	45 Rh 102.905	46 Pd 106.4	47 Ag 107.870	48 Cd 112.40	49 In 114.82	50 Sn 118.69	51 Sb 121.75	52 Te 127.60	53 I 129.904	54 Xe 131.30
6	55 Cs 132.905	56 Ba 137.34	57 to 71	72 Hf 178.49	73 Ta 180.948	74 W 183.85	75 Re 186.2	76 Os 190.2	77 Ir 192.2	78 Pt 195.09	79 Au 196.967	80 Hg 200.59	81 Tl 204.37	82 Pb 207.19	83 Bi 208.980	84 Po (210)	85 At (210)	86 Rn (222)
7	87 Fr (223)	88 Ra (226.05)	89 to 103															

Rare–Earth Elements

Lanthanide Series

57 La 138.91	58 Ce 140.12	59 Pr 140.907	60 Nd 144.24	61 Pm (147)	62 Sm 150.35	63 Eu 151.96	64 Gd 157.25	65 Tb 158.924	66 Dy 162.50	67 Ho 164.930	68 Er 167.26	69 Tm 168.934	70 Yb 173.04	71 Lu 174.97

Actinide Series

89 Ac (227)	90 Th 232.038	91 Pa (231)	92 U 238.03	93 Np (237)	94 Pu (242)	95 Am 243	96 Cm (247)	97 Bk (247)	98 Cf (249)	99 Es (254)	100 Fm (253)	101 Md (256)	102 No (254)	103 Lw (257)

FIGURE 3-1
Periodic table of the elements showing elements divided into groups discussed in the text.

natural and those that are made in the laboratory are *synthetic.*

The degree of combination of atoms, molecules, and ions determines the *state* of matter. *Gases* contain atoms or molecules that are far apart, with little mutual attraction. *Liquids* consist of loosely bonded clusters of atoms, molecules, or compounds with no long-range frameworks. *Solids* consist of the same basic building blocks that are linked by stronger bonds that form ordered structures, and *crystalline solids* have a well-defined, long-range *atomic structure* (Figure 3-2). Solidified liquids such as *glass* lack a well-defined atomic structure and are referred to as *amorphous.* The margins of all solids consist of broken chemical bonds that can combine with ions or molecules in surrounding liquids and gases, a process known as *adsorption.*

Most natural materials are mixtures. Air is a *homogeneous mixture* of several gases, including oxygen and nitrogen, and rocks are *heterogeneous mixtures* of minerals and other solids. A *colloid* is a special type of mixture in which very small particles consisting of a few atoms or molecules are suspended in a solid, liquid, or gas. Homogeneous mixtures become scarcer as the long-range structure of the material increases, and are therefore least common in solids where one element can *substitute* for another only if it has very similar properties.

Matter undergoes physical, chemical, and nuclear changes, all of which can be described by an equation with *reactants* on one side and *products* on the other. In *physical changes,* a compound retains the same composition and simply changes from one state to another, as from liquid to vapor when water is boiled. In *chemical changes,* atoms in the reactants are reorganized to form new product substances, as when methane oxidizes (burns) in oxygen to form carbon dioxide and water, or when salt dissolves in water to produce sodium and chloride ions. Physical and chemical changes are subject to the *law of conservation of matter,* which states that these reactions do not create or destroy matter.

Some substances are *stable* in their physical and chemical environment and others are *unstable.* In natural systems, it is not possible to determine whether a substance is stable without considering all aspects of its physical and chemical environment. For instance, liquid water is stable at atmospheric pressure and surface temperature but converts to vapor above a temperature of 100°C. The iron sulfide mineral, pyrite, is stable in an argon atmosphere because it will not react with argon. But, in contact with water and oxygen, pyrite reacts to form acid water containing dissolved iron. Definitions of stability are complicated by the rate or *kinetics* of these changes. Although some reactions proceed at a very rapid rate, many do not, and some never produce stable compounds. Compounds that persist even though they are not stable are described as *metastable.* Most organic compounds, including those in your body, are metastable in the atmosphere.

Nuclear changes, which involve the nuclei of atoms, convert one isotope to another isotope. Reactions of this type include natural radioactivity, nuclear fission, and nuclear fusion. All nuclear changes obey the *law of conservation of matter and energy,* which states that the sum of matter and energy in the reaction is constant. Of the 2,000 or so known isotopes, only 266 are stable. The remainder are naturally *radioactive* and

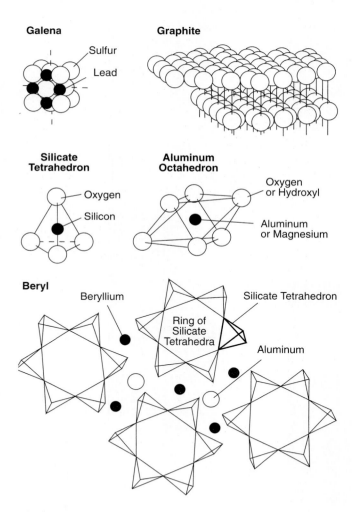

FIGURE 3-2

Crystal structure of galena (PbS) and graphite (C). Also shown are the basic crystal units of many silicate minerals, the silicate tetrahedron and aluminum octahedron. Beryl ($Be_3Al_2Si_6O_{18}$), a silicate mineral, consists of rings of six silicate tetrahedra that are linked by beryllium and aluminum.

have a finite, measurable probability of forming a different isotope by *radioactive decay.* The fixed period of time required for half of the amount of the radioactive isotope to decay is known as the *half-life.* Half-lives vary from milliseconds to billions of years for the different isotopes.

Products of radioactive decay include *alpha particles,* which are the nuclei of helium atoms, *beta particles,* which are electrons, and *gamma rays,* which are similar to X-rays, but with a shorter wavelength. These decay products are called *ionizing radiation* because they dislodge electrons from atoms in surrounding compounds, damaging their structures and producing reactive ions. *Nuclear fission,* the basis for nuclear power, takes place when isotopes of elements with large mass numbers, such as ^{235}U, are struck by neutrons and split into isotopes of lower mass number, more neutrons, and energy. Most of the isotopes created by nuclear fission are radioactive and undergo further decay, creating our problems with nuclear waste disposal. *Nuclear fusion* involves the combination of light isotopes such as ^2H (deuterium) to form a heavier isotope (e.g., one with a higher mass number) plus energy. While this reaction takes place in stars, controlled fusion remains an elusive goal for civilization.

GEOCHEMICAL RESERVOIRS—THEIR NATURE AND COMPOSITION

Earth can be divided into regions of relatively similar average composition, which are known as *reservoirs* (Horne, 1978). The largest of these are the atmosphere, hydrosphere, biosphere, and lithosphere, which have very different compositions (Table 3-1), and which can be further divided into smaller reservoirs such as mineral deposits and human bodies. Environmental geochemistry deals with the chemical interaction between the biosphere and any of the other reservoirs, and the environmental importance of a substance is determined by its effect on life. Access of a substance to the biosphere is controlled largely by its ability to be concentrated in the lithosphere, particularly in the soil, to dissolve in the hydrosphere or to evaporate into the atmosphere. The capacity of substances to move from one reservoir to another is termed *mobility,* and the degree to which substances concentrate in living organisms is known as *bioaccumulation.*

Lithosphere

The lithosphere is where it all started. The original Earth probably consisted only of the lithosphere. The atmosphere and hydrosphere formed largely from water and other gases released by magmas that rose from the mantle, and the biosphere probably began with the formation of organic compounds in the early hydrosphere. The composition of all of these reservoirs has continued to evolve with time. The lithosphere, for instance, separated into the core, mantle, and crust, all of which continue to undergo changes related to plate tectonics.

The most important part of the lithosphere from the standpoint of environmental geochemistry is *soil,* which is in intimate contact with the other reservoirs. Soil is a heterogeneous mixture of water, minerals, and living and dead organic matter. Soil minerals include

TABLE 3-1
Ten most abundant elements in Earth's major reservoirs. Composition given here for biosphere includes water in living organisms; composition for lithosphere is that of the crust. In addition to the single-element gases tabulated here, the atmosphere also contains 1.86 ppm CH_4, 0.5 ppm NO, 0.04 to 0.2% CO, 0.002% SO_2, 0.0015% NO_2, and 0.00006% H_2S.

Atmosphere		*Hydrosphere*		*Biosphere*		*Lithosphere*	
				(all abundances in parts per million)			
Nitrogen	755,100	Oxygen	857,000	Oxygen	523,400	Oxygen	464,000
Oxygen	231,900	Hydrogen	108,000	Carbon	302,800	Silicon	282,000
Argon	12,800	Chlorine	19,000	Hydrogen	67,500	Aluminum	82,300
Carbon	100	Sodium	10,500	Nitrogen	5,000	Iron	56,000
Neon	12.5	Magnesium	1,350	Calcium	3,700	Calcium	41,000
Krypton	2.9	Sulfur	885	Potassium	2,300	Sodium	24,000
Helium	0.72	Calcium	400	Silicon	1,200	Magnesium	23,000
Hydrogen	0.45	Potassium	380	Magnesium	980	Potassium	21,000
Xenon	0.36	Bromine	65	Sulfur	710	Titanium	5,700
Sulfur	0.001	Carbon	28	Aluminum	555	Hydrogen	1,400

quartz, calcite, iron-manganese oxides, micas, and clay minerals. Of these, clay minerals are most important in terms of both abundance and reactivity. *Clay mineral* structures are based on the *silicate tetrahedron,* a pyramid-shaped unit consisting of a silicon ion bonded to four oxygen ions, and a similar octahedral unit with a central aluminum or magnesium ion and six oxygen or hydroxyl (OH^-) ions. These units combine to form a wide variety of crystal structures on which the many silicate minerals are based, including the sheets that make up the clay minerals (Figure 3-3). Clay minerals are divided into two groups, the *kaolinite group,* which contains alternating sheets of tetrahedra and octahedra, and the *smectite (or montmorillonite) group,* which contains an octahedral sheet sandwiched between two tetrahedral sheets (Figure 3-3). The composition of kaolinite is essentially constant, but the composition of smectite can vary tremendously, with different ions substituting for silicon and aluminum.

Many clay minerals in soils have poorly developed crystal structures and some iron oxides and manganese oxides are almost amorphous. Thus, they have a large surface area consisting of many broken bonds, which give them an enormous capacity to adsorb elements and compounds from their surroundings, a process that is the key to pollution of the lithosphere. In smectite clays, for instance, substitution of Al^{+3} for Si^{+4} in a silicate tetrahedron creates an additional negative charge that can be satisfied by substitution

of another positively charged ion such as sodium (Na^+) in the structure. Sometimes the new ion is simply adsorbed on the surface of the mineral rather than being included in the structure. Substitution or adsorption sites can exchange one ion for another as concentrations of the ions change in the surrounding water, a process known as *ion exchange.* Even water undergoes exchange and adsorption reactions because it is an asymmetric ion with a residual charge, as discussed later. Clays of the smectite group are known as *swelling clays* because they adsorb water between their layers, with a resultant increase in volume (Buol et al., 1989).

Exchange and adsorption involve competition for available sites on a mineral. The winners are determined by size and bonding characteristics, and by their abundance. In soil water, for instance, an ion that is present in large amounts has a better chance of finding a place on a soil mineral than one that is dissolved in only small amounts. Hydrogen ion (H^+) competes very effectively because it is small, and it takes up most of the cation-exchangeable sites on minerals in acid solutions (Brady, 1990). In neutral and alkaline soils, the competition is open to more ions and in an interesting twist, small, highly charged metal ions have a better chance than large ions such as sodium that have a low charge. Thus, metal ions concentrate on soils even when they are present in relatively low concentrations in natural water, a factor that can cre-

FIGURE 3-3

Crystal structure of the two major clay mineral groups: kaolinite and montmorillonite. Exchangeable cations and water are found largely between the sheets, with montmorillonite containing much larger amounts.

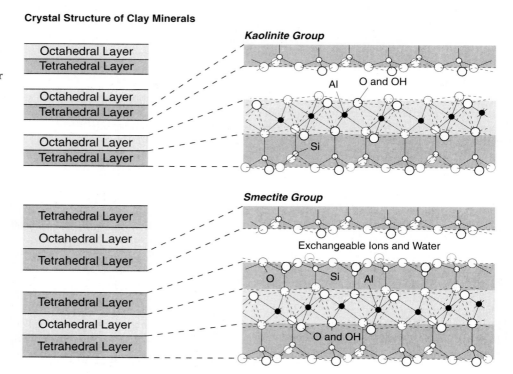

ate pollution problems or be used to clean them up (Plate 2-1).

Hydrosphere

The hydrosphere consists of water that is not chemically bound in minerals, along with everything that the water can dissolve (Westall and Stumm, 1980; Broecker, 1983). Water is one of the few compounds that forms a stable solid, liquid, and vapor under surface conditions. Changes between these states form the basis for the *water cycle,* the repetitive pattern of evaporation and condensation that allows us to pour our filth into water and hope that nature will clean it up for us. About 97.2% of the hydrosphere consists of *seawater,* which contains about 3.5% (by weight) dissolved material (Table 3-1). As noted earlier, evaporite minerals formed from seawater are major mineral resources. Efforts to recover scarcer elements directly from seawater have not been very successful, although the use of selective adsorption methods offers promise for the future. Most of the rest of the hydrosphere is *fresh water,* which is found in glaciers and ice caps (2.1%), groundwater (0.6%), rivers and lakes (0.01%), and the atmosphere (0.001%). These supply almost all of our municipal, industrial, and agricultural needs. Saline waters derived from the evaporation of fresh water make up only about 0.01% of the hydrosphere.

Water is an extremely effective solvent because it consists of two hydrogen atoms that extend outward from one side of an oxygen atom. This configuration creates a molecule with a positive charge on one side and a negative charge on the other that attracts other charged compounds into solution. Water also dissociates to produce ionic H^+ and OH^-, which determine its *pH* (the negative logarithm of the H^+ concentration). At normal surface temperatures, water has equal concentrations of H^+ and OH^- at a pH of 7. Natural rainwater is more acid than this because it dissolves carbon, sulfur, and nitrogen gases from the atmosphere, as discussed later. In general, acid solutions are more effective solvents than neutral solutions and most natural materials become more soluble as the temperature increases.

The *mobility* of a substance in the hydrosphere is controlled by its *solubility,* which is the amount that can be dissolved. Water that cannot dissolve any more of a substance is said to be *saturated.* At room temperature, natural saturated water can dissolve up to 264,000 ppm halite (NaCl) compared to only about 20 ppm quartz (SiO_2). Many ionic substances *dissociate* into more than one ion when they dissolve, making their solubility dependent on the concentrations of these ions in solution. For instance, halite dissolves largely by dissociating into Na^+ and Cl^- ions, and its solubility will therefore be greater in pure water than in water with dissolved sylvite (KCl), which already contains Cl^-. Water also contains a wide range of organic and inorganic *suspended material,* much of which is in the form of colloids. Colloids remain suspended because they have highly charged surfaces and electrostatic repulsion prevents them from aggregating into larger grains that would sink. Colloids are more abundant in fresh water because dissolved ions such as Na^+ and Cl^- in saline water nullify their surface charges, causing them to aggregate into larger grains, a process known as *flocculation.*

In general, the hydrosphere is oxidizing except where special conditions have consumed all available oxygen. This takes place most commonly where water bodies become *stratified.* In temperate areas, lakes more than about 10 meters deep are stratified throughout most of the year but undergo two periods of mixing or *overturn* in spring and fall (Figure 3-4). While they are stratified, the upper part of the lake, known as the *epilimnion,* mixes with the overlying atmosphere to maintain a high concentration of oxygen. The underlying *hypolimnion* is isolated from the atmosphere, however, and becomes more reducing with time. During overturn, reduced material from the hypolimnion is oxidized by contact with the atmosphere (Horne, 1978). The overall evolution of the hydrosphere also reflects a gradual change from reducing to oxidizing conditions. Elements with more than one natural oxidation state, such as sulfur and iron, were reduced in the early ocean, whereas they are oxidized in the modern ocean (Holland and Schidlowski, 1982).

Atmosphere

The atmosphere consists of gases (Table 3-1) and particulate matter known as the atmospheric aerosol (Ingersall, 1983; Eiden, 1990). The concentration of *gases* in the atmosphere is determined by their vapor pressures and abundances. As can be seen in Figure 3-5A the main atmospheric gases have high vapor pressures. Water has a low vapor pressure and is an important part of the atmosphere only because it is an abundant liquid at Earth's surface. Organic compounds with high vapor pressures, such as ethane (C_2H_6), propane (C_3H_8), acetylene (C_2H_2), and ethylene (C_2H_4), are not common at Earth's surface and are correspondingly rare in the atmosphere, although we have begun to change that by the extraction and use of crude oil and natural gas.

The *atmospheric aerosol* includes liquid particles called *mist* or *fog,* solid particles from combustion known as *smoke,* and windblown rocks and minerals

Chapter 3

Thermal Stratification of Dimictic Lakes

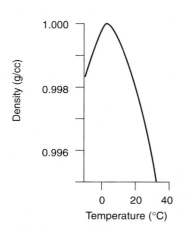
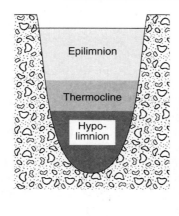
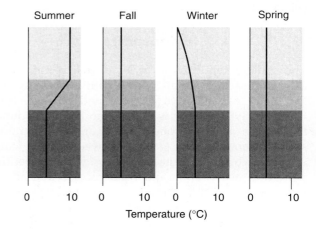

FIGURE 3-4

Thermal stratification in dimictic lakes, showing change in the density of water with temperature and its relation to seasonal variation of thermal stratification. Note that stratification occurs because water is most dense at 4°C.

known as *dust* (Figure 3-5B). *Smog* consists of atmospheric gases that were condensed into particles by photochemical processes based on light energy. SO_2, for instance, forms calcium sulfate particles by reacting with calcium in airborne dust. Thus, gases that enter the atmosphere can be deposited as dissolved gas or aerosol particles. These two forms of deposition are sometimes referred to as *dry deposition* (aerosol particles) and *wet deposition* (acid from dissolved gas). Pollen, spores, bacteria, and other organic material and condensates can be important locally, and volcanic eruptions are very important periodically (Rampino et al., 1988). The aerosol is dominated by large dust particles in unpopulated continental areas, and by smaller combustion and gas condensate particles in populated areas (Figure 3-5B). Over the ocean, the aerosol is largely ocean spray, which diminishes landward from the coasts. The aerosol is removed by precipitation and sedimentation, with particles larger than about 2 micrometers in diameter forming most nuclei for raindrops, and smaller particles having a longer residence time in the atmosphere.

The atmosphere has changed composition through Earth's history. The early atmosphere was more reducing than it is now and lacked abundant oxygen. The concentration of oxygen was probably increased by the loss of hydrogen from water vapor in the upper atmosphere and by respiration of land plants, which began in early Paleozoic time. Oxygen is necessary for most life on Earth and the appearance of abundant life in the fossil record at the beginning of Paleozoic time (0.57 Ga) is thought to mark the point at which atmospheric oxygen neared its present concentration (Holland, 1984).

Biosphere

The biosphere consists of both living and dead organic matter that is rich in carbon, hydrogen, and other light elements (Table 3-1). *Living organic material* can be divided into skeletal material, body fluids, and soft tissue or biomaterial. *Skeletal material* in animals consists of *calcite* or *aragonite,* which are different crystal forms of calcium carbonate, various forms of *silica,* and the calcium phosphate mineral, *apatite.* Most apatite found in skeletal material, including that which makes up human teeth and bones, contains fluorine and is known as *fluorapatite.* Although many plants do not have true skeletal material, vascular land plants are supported by *lignin,* a complex hydrocarbon consisting of ring structures with attached OH^- ions. The *body fluid* in plants and animals is largely water with dissolved inorganic and organic substances, including hemoglobin and chlorophyll as discussed later, and the *biomaterial* includes a wide range of organic molecules, such as *cellulose,* as well as organs (Buol et al., 1989).

Living plants and animals have a tremendous effect on the chemical and physical nature of soil. Earthworms pass as much as a full body-weight of soil through themselves each day, enhancing soil porosity and permeability. The cylinder of soil surrounding plant roots, which is known as the *rhizosphere,* is much more acidic than adjacent soil and has a higher con-

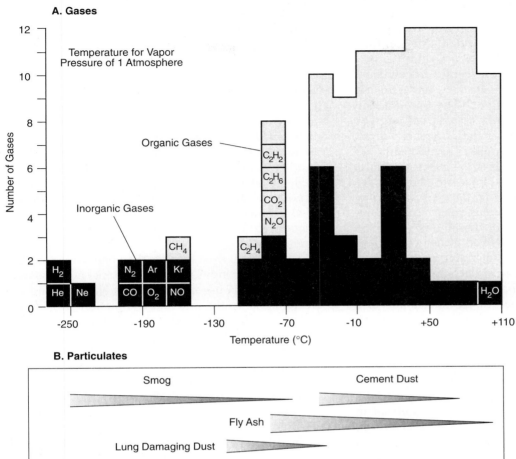

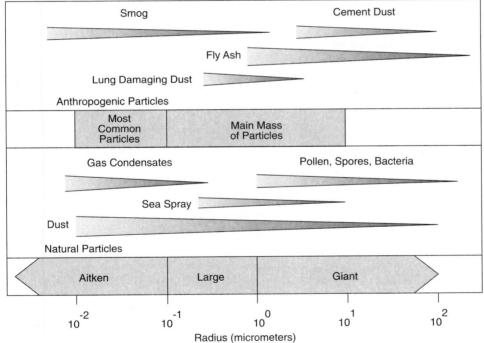

FIGURE 3-5

A. Histogram of temperatures at which common gases have vapor pressure of 1 atmosphere. Inorganic gases, which make up most of the atmosphere, reach this vapor pressure at low temperatures. **B.** Size classifications and sources of some atmospheric aerosol particles (width of bar indicates relative number of particles). Minimum size of the atmospheric aerosol is limited by coagulation of particles with radii less than 0.01 micrometers. Maximum particle size of about 200 micrometers is limited by gravitational settling. The main mass of the atmospheric aerosol is contained in the larger particles. Modified from Horne (1978), Winchester (1980), Fergusson (1982), and Warneck (1988).

centration of microorganisms, allowing the plant to take up elements that would otherwise remain undissolved in the lithosphere (Cloud, 1983; Froth, 1984).

Dead organic material in soils comes largely from plants and consists of *sugars, proteins,* and *cellulose,* which are relatively easily degraded by bacteria, and *fats, waxes,* and *lignin,* materials that do not degrade easily. Degradation of this material yields *humus,* which is divided into two groups. The dominant *humic group,* is an amorphous, black to brown substance, part of which can be dissolved in acidic or basic solutions and is known as *fulvic acid.* Part is soluble only in basic solution and is known as *humic acid.* The remaining material that will not dissolve is called *humin.* The less abundant *nonhumic* group consists of smaller, sugar-like molecules, including *polysaccharides* $[C_nH_2O_m]$ (Brady, 1990). Partly decomposed organic material known as *kerogen* is the raw material from which coal, crude oil, and natural gas form and is also an active ion-exchange medium similar to clay minerals and iron and manganese oxides.

Biomaterial and body fluids incorporate phosphorus, sulfur, and nitrogen, as well as many trace metals, such as fluorine, strontium, and lead. Some of these elements are essential for life, whereas others can be damaging (Table 3-2). Some are beneficial in small amounts, but toxic in large amounts (Fergusson, 1982). Metals combine with organic matter largely in compounds called *chelates,* which are complex ions with one pair of atoms that can bond with a metal ion. The blood molecule, *hemoglobin,* in which each of four Fe^{+2} ions is bonded to four nitrogen atoms, a protein and an oxygen atom, carries oxygen during the respiration process in animals. *Chlorophyll,* which is essential for photosynthesis in plants, has a similar structure involving Mg^{+2}. Synthetic chelate compounds such as EDTA are used to remove heavy metals from the human body because they form unusually stable complex ions involving six separate bonds with a single metal ion. One way in which metals become incorporated into living organic material is *methylization,* in which metals combine with methyl ions (CH_3^+), a reaction that is probably facilitated by microorganisms (Craig, 1980).

Elements that exist in more than one oxidation state under surface conditions are important energy sources for life. Bacteria use many of these elements and can greatly enhance the rate and completion of oxidation and reduction reactions. Sulfur, which has several stable and metastable oxidation states from $+6$ to -2, is widely used by bacteria. *Desulfovibrio, Desulfotomaculum,* and *Desulfuromonas* convert dissolved sulfate (SO_4^{-2}) to H_2S. *Thiobacillus thiooxidans* and related strains convert dissolved H_2S to SO_4^{-2}. *Ferrobacillus* converts Fe^{+2} to Fe^{+3}. These reactions facilitate rock weathering and might be used in the future to recover metals from their ores or to remove sulfur from coal.

TABLE 3-2

Biological importance of selected elements. As indicated in the second part of this table, some elements that are necessary to life are toxic in higher concentrations.

Element	Biological Importance
Elements Necessary for Life	
Cr	Aids insulin, which controls blood sugar
Mn	Necessary for some enzymatic reactions
Fe	Metal in blood molecule hemoglobin
Co	Component in vitamin B_{12}
Ni	Component of some enzymes
Cu	Component of some enzymes, produces pigment
Zn	Component of insulin and enzymes

Elements Inimical to Life (oral toxicity to small animals)

Very Toxic	Moderately Toxic	Slightly Toxic	Harmless
As	Cd	Al	Cs
Pu	Cu	Mo	Na
Se	Hg	Ta	I
Te	Pb	W	Rb
Tl	Sb	Zn	Ca
	V	Zr	K

GEOCHEMICAL CYCLES AND THE DYNAMIC NATURE OF GLOBAL RESERVOIRS

Geochemical Cycles

The four major global reservoirs are a dynamic function of Earth's continuing evolution. Substances are continually moving from one reservoir to another, and between smaller reservoirs within the global ones. The amount of a substance that moves between reservoirs in a given period of time is known as *flux*, its average stay in each reservoir is known as *residence time*, and its overall pattern of movement from reservoir to reservoir is known as its *geochemical cycle*. At any point in the cycle, reservoirs that produce a substance are known as *sources*; those that consume it are known as *sinks*. One task of environmental geochemistry is to distinguish *natural* sources and sinks from those created by humans, which are known as *anthropogenic,* and to determine whether anthropogenic contributions disturb the natural cycle (Broecker, 1985; Holton, 1990).

The global *sulfur cycle* (Figure 3-6) illustrates these principles as well as the challenges facing environmental chemistry and regulation (Zenhder and Zinder, 1980). The most important reservoirs of sulfur are the lithosphere and hydrosphere, both of which contain large amounts of sulfur with long residence times. Sulfur in the lithosphere is largely in the form of the sulfide mineral *pyrite* and the sulfate mineral *gypsum,* some of which form mineral deposits. Sulfur in the hydrosphere is present largely as dissolved sulfate in the oceans, which comes from weathered sulfide and sulfate minerals. Sulfur returns to the lithosphere by the evaporation of seawater to form gypsum and other sulfate minerals or by precipitation of metal sulfides. Sulfur enters the atmosphere as SO_2, H_2S, and other gases through volcanic eruptions, weathering of sulfide minerals, and dispersal of sea spray. It also enters the atmosphere from anthropo-

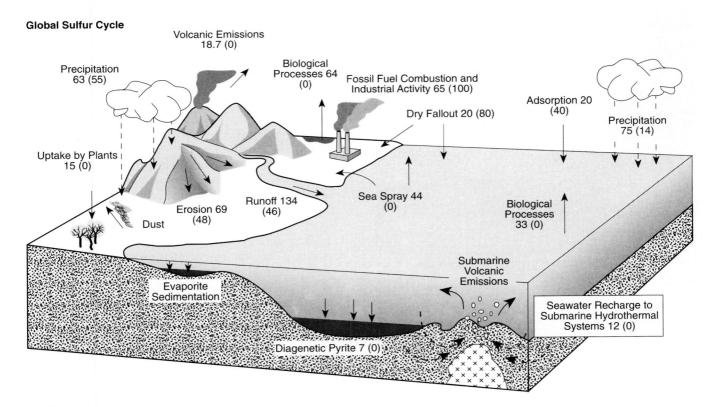

Global Sulfur Cycle

FIGURE 3-6

Generalized global sulfur cycle, showing major sources, sinks, and fluxes (amounts in millions of tonnes annually). Numbers in parentheses are percentage of emissions contributed by anthroprogenic activity. Modified from Zenhder and Zinder, 1980; Adams et al., 1981, and Stoiber et al., 1987.

genic sources including the burning of fossil fuels containing sulfur-bearing minerals and organic compounds, where it is a potential cause of acid rain, as discussed later.

In order to determine the role of anthropogenic sources in the sulfur cycle, we must first recognize and measure its natural sources. This is a major challenge, whether it is done on a global or local scale. Many natural sources (and sinks) are not well understood and emission rates from others are difficult to measure. Volcanoes are a case in point (Malinconico, 1987; Stoiber et al., 1987). For volcanoes on land, such as Mount Pinatubo in the Philippines, activity is inter-

mittent and unpredictable, although we can at least observe and attempt to measure the eruptions (Figure 3-7). Measurements at submarine volcanoes along the mid-ocean ridge are more complicated because they are submerged beneath thousands of meters of water. The entire volume of water in the oceans circulates through the hot, fractured rocks of the mid-ocean ridges within only a few million years and could obviously have a profound effect on the global sulfur cycle. Similar problems beset all environmental studies, making it very important to obtain reliable measurements of geochemical abundances and to understand their relation to geologic processes, global reservoirs, and anthropogenic contributions.

The main goal of these measurements, of course, is the construction of a model that accounts for the geochemical cycle of a substance. With such a model, it is possible to evaluate the potential effects of changes in natural or anthropogenic contributions, thereby identifying factors that require greatest attention. In constructing such models, it is important to recognize that pollution does not result only from anthropogenic contributions. Rocks and mineral deposits were an important source of natural pollution long before we began to exploit them. Pollution is most easily recognized, of course, when it affects a small reservoir in a big way, but modern analytical methods have allowed us to identify more subtle examples of pollution and the list of natural and anthropogenic contamination continues to grow.

Natural Contamination

The chemical relation between rocks and associated soils, water, plants, and even gases is the province of *landscape geochemistry* (Fortescue, 1992). Where these relations affect public health, they become the concern of *geomedicine* (Warren, 1989). Health problems can be caused by deficiencies or enrichments of a natural substance and they can be centered on regional-scale rock units or on smaller features such as mineral deposits. It is widely known that different types of rocks have different average minor and trace-element contents. Ultramafic rocks, for instance, lack potassium and phosphorus that encourage plant growth, and rarely support good crops. Rocks that lack adequate trace metals such as copper, cobalt, molybdenum, and zinc can also cause problems (Allcroft, 1956; Thornton, 1983). Sedimentary rocks that are enriched in selenium cause "blind staggers" in cattle that graze on the land and deformities in birds that live in surface water fed by drainage from the rocks (Anderson, 1961; Presser and Barnes, 1985).

Rocks have their greatest effect on life through their effect on the composition of water. In earlier times, *goi-*

FIGURE 3-7
The main crater at White Island volcano in New Zealand has collapsed partly, allowing this view from the ocean directly into the active vent, where steam, CO_2, SO_2, and HCl escape through fumaroles as hot as 800°C. In its current relatively quiet state, White Island is estimated to release 10^5 tonnes of SO_2 annually, about the same as the Bingham copper smelter (Plate 4-17B). The eruption of Mount Pinatubo, Philippines, in 1991, released about 2,000 times more SO_2.

ter, a thyroid disease caused by iodine deficiency, was a common problem in populations far from the sea where iodine-bearing salt spray did not contaminate streams and lakes, and rocks lacked iodine. Some granitic igneous and metamorphic rocks, as well as phosphate sediment, are enriched in uranium, thorium, and their radioactive decay product radon, which concentrate in groundwater supplies and present health concerns (Aieta et al., 1987; Wanty et al., 1991). Similarly, the absence of selenium in soil and water has been implicated in the high incidence of stroke in the southern United States and the absence of calcium and magnesium in areas of *soft water* has been related to increased incidence of cardiovascular disease (Barringer, 1992; Pocock et al., 1985).

The relation between health and regional geochemical variations can be determined only through geochemical surveys in which the average composition of the soil is determined at closely spaced sites throughout a large area. Studies of this type are just beginning, but early results show surprising patterns, such as the curious barium-enriched zones that cross Finland (Koljonen et al., 1989). Sometimes, the trace-element content of soils is not even a function of the underlying rocks. In areas where soil development is slow, a large part of the soil might be from windborne dust, often from thousands of kilometers away, bringing with it trace elements that are not of local origin. In glaciated areas, soil develops on material dropped when the ice sheets melted (Plate 2-3).

Mineral deposits are important local sources of natural contamination (Plate 2-2), commonly forming patterns that are used in *geochemical exploration,* as discussed in the next chapter. In the vicinity of metal deposits, it is common to observe natural metal concentrations at the surface, such as the copper-rich bogs of Coed-y-Brenin, Wales (Figure 3-8A). The rhizosphere surrounding plant roots facilitates the uptake of metals from mineral deposits into trees, which are very enriched in metals around some deposits. This effect can produce *natural stress,* which limits tree growth and is sometimes manifested by the early loss of leaves in the fall (Canney et al., 1979). Animals also can show variations in composition related to mineral deposits. Fish livers in lakes in British Columbia have been shown to be enriched in metals near mineral deposits.

Drainage and Groundwater Pollution

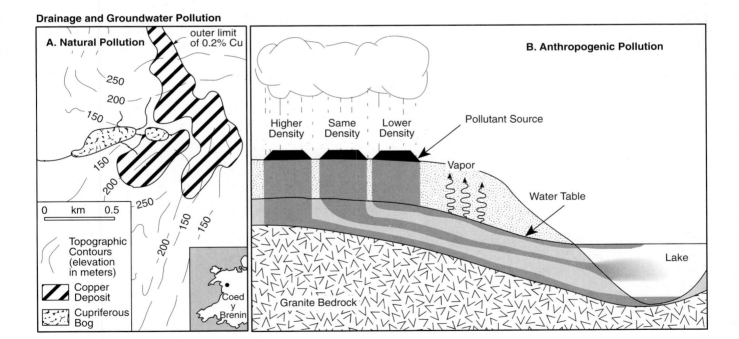

FIGURE 3-8

A. Copper-rich bogs formed by natural weathering of the Coed-y-Brenin copper deposit, Wales, showing outline of area containing soil with more than 0.2% Cu (after Andrews and Fuge, 1986). **B.** Schematic illustration of contaminant plumes in groundwater formed by liquids with different densities.

Anthropogenic Contamination

Anthropogenic contamination of the biosphere has overprinted natural contamination, creating the complex geochemical pattern that affects us today (El-Hinnawi, 1981; Kelly, 1988). We are currently removing metals and other mineral compounds from Earth at a rate equal to or greater than the rate at which they are liberated by the weathering of rocks and other natural processes. Comparison of atmospheric emissions, for instance, shows that anthropogenic sources exceed natural ones by a factor of more than five for cadmium, two to four for vanadium, zinc, and lead, and one to two for antimony, arsenic, copper, mercury, and nickel (Nriagu and Pacyna, 1988; Nriagu, 1990b). Anthropogenic emissions are less than natural emissions only for chromium, manganese, and selenium. Although old mines, oil fields, smelters, and refineries are important sources of some of these anthropogenic emissions, newer facilities are much cleaner (Chander, 1992). More important are sources related to the consumption of minerals, such as the burning of fossil fuels and waste incineration.

Although atmospheric pollution is most obvious, water pollution is a growing threat. Worldwide anthropogenic metal emissions are largest for lead, manganese, chromium, copper, and zinc, with most contributions coming from manufacturing and domestic wastewaters followed by atmospheric fallout (Nriagu, 1990b; Lind and Hem, 1993). Contamination of surface water by fertilizers and road salt is an inevitable result of the requirement that these materials be used in highly soluble forms. Groundwater aquifers have been damaged by drainage from these applications, as well as from point sources such as salt storage areas, sewage treatment plants, and storage facilities for crude oil and refined liquids. Contaminated groundwater commonly forms a *plume* that originates at the point of input and gradually enlarges down the flow path, with its ultimate form determined by the relative densities of contaminated water and groundwater (Figure 3-8B). Because hydrocarbons are lighter than water, they remain close to the surface, invading sewers and basements where they cause explosions, such as the 1992 tragedy in Guadalajara, Mexico, that killed several hundred people. Most modern industrial facilities use impermeable underlayers in order to prevent groundwater contamination, but individuals continue to dump wastes at the surface, creating a layer of contamination that is descending into the water table.

Recent high-sensitivity geochemical studies have shown that anthropogenic contamination patterns are truly global in scale and that they have affected the composition of the biosphere. Global effects are most

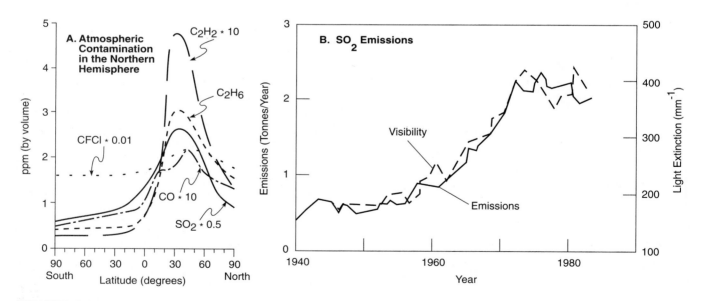

FIGURE 3-9
A. Variation with latitude in some atmospheric trace gases. Note that gas abundances are highest at about 40 to 50 degrees north latitude where largest industrial activity is located (from Eiden, 1990; Warneck, 1988). **B.** Historical relation between sulfur emissions and visibility during the summer in the southeastern United States (from NAPAP, 1991).

obvious in the atmosphere, where mobility is highest, and they include NO_x, CO_2, and SO_2, largely from the combustion of fossil fuels, and *volatile organic compounds (VOC)* such as methane (CH_4), from oil and gas production and industrial applications. Soot, smoke, road dust, and chemical conversion of anthropogenic gases are thought to make up about 12% of global particulate emissions. The effects of these contributions are seen in the concentration of contaminants in the northern hemisphere where population is greatest and in the close correspondence between visibility and SO_2 emissions (Figure 3-9). Two aspects of anthropogenic contamination, acid pollution and global change, have become household terms and require special attention.

Acid Pollution

The most important pollutant of the hydrosphere is acid (H^+), in the form of acid rain and acid mine drainage. *Acid mine drainage* results from the decomposition of pyrite, usually catalyzed by bacteria, to produce iron hydroxide and dissolved H^+ and SO_4^{-2} (Table 3-3). An estimated 93% to 97% of the acid mine drainage in the United States comes from coal mines or waste piles, which contain small amounts of pyrite (Baker, 1975; Powell, 1988). Most of the remaining acid comes from pyrite in metal mines or wastes. Acid water formed by dissolution of pyrite dissolves more pyrite, thus accentuating the effect. As the acid water moves downstream, it mixes with less acid water causing dissolved iron to precipitate as oxides, which produces further acid. As can be seen in Table 3-3, the multiplicative effects of these reactions are so great that oxidation of a single molecule of pyrite can yield at least four H^+ ions. Acid water produced by these reactions can dissolve

other metal sulfides (Plate 2-2) and leach metals that are adsorbed on the surfaces of clays and other poorly crystallized minerals, thus increasing the trace-metal content of streams (Johnson and Thornton, 1987). Water from waste piles recharges groundwater systems, where it can form plumes of acid water (Zaihua and Daoxian, 1991 and Figure 3-8B).

Acid rain is the product of reactions between atmospheric water and CO_2, SO_2, and NO_x (Table 3-4). Although natural levels of CO_2 in the atmosphere would give precipitation a pH of about 5.7, it is much more acid than this in eastern North America and western Europe, reaching levels as low as 2.9 (Figure 3-10). The close correlation of this acidity with areas of sulfur deposition (Figure 3-11) is interpreted to indicate that anthropogenic SO_2 is the main source of acid. The effect of acid rain on surface water and soils is governed in part by the presence of underlying limestone, which consumes acid to produce dissolved calcium and bicarbonate (Table 3-4). Igneous and metamorphic rocks, which consist largely of silicate minerals, also react with acid but at a much slower rate. Thus, lakes in areas underlain by limestone are usually less acid than those underlain by silicate rocks, even when they receive the same acid precipitation.

The distribution and history of acid precipitation has long been a focus of environmental concern. Lakes underlain by silicate rocks in the Adirondacks and

TABLE 3-3

Reactions that decompose sulfide minerals to produce acid mine drainage and natural geochemical dispersion halos. Me = metal.

A. Oxidation of pyrite in contact with air-saturated water to form acid, sulfate, and iron ions in solution:
$$2FeS_2 + 2H_2O + 7O_2 = 4H^+ + 4SO_4^{-2} + 2Fe^{+2}$$
(Fe^{+2} is then oxidized to Fe^{+3} by bacteria)

B. Deposition of hydrated iron oxide to produce more acid:
$$Fe^{+3} + 3H_2O = Fe(OH)_3 + 3H^+$$

C. Dissolution and oxidation of sphalerite:
$$ZnS + 2O_2 = Zn^{+2} + SO_4^{-2}$$

D. Leaching of metals from clay minerals by acid water:
$$Clay\text{-}Me^{+2} + H^+ = Clay\text{-}H^+ + Me^{+2}$$

TABLE 3-4

Generalized reactions that produce (A,B,C) and consume (D,E,F) acid rain. Reactions shown here are *net reactions* that show reactants and final products but omit intermediate products. Because limestone is more soluble, this reaction is much more important than the silicate reaction. Both reactions are essentially the same as reactions that take place during normal rock weathering.

A. Production of carbonic acid:
$$CO_2 + H_2O = H_2CO_3$$

B. Production of sulfuric acid:
$$SO_2 + H_2O + \tfrac{1}{2}O_2 = H_2SO_4$$

C. Production of nitric acid:
$$NO + \tfrac{1}{2}H_2O + \tfrac{3}{4}O_2 = HNO_3$$

D. Limestone dissolution:
$$CaCO_3 + H^+ = Ca^{++} + HCO_3^-$$

E. Silicate dissolution (feldspar):
$$2KAlSi_3O_8 + 2H^+ + H_2O = 2K^+ + Al_2Si_2O_5(OH)_4 + 4SiO_2$$

F. Lime dissolution:
$$CaO + H^+ = Ca^{++} + OH^-$$

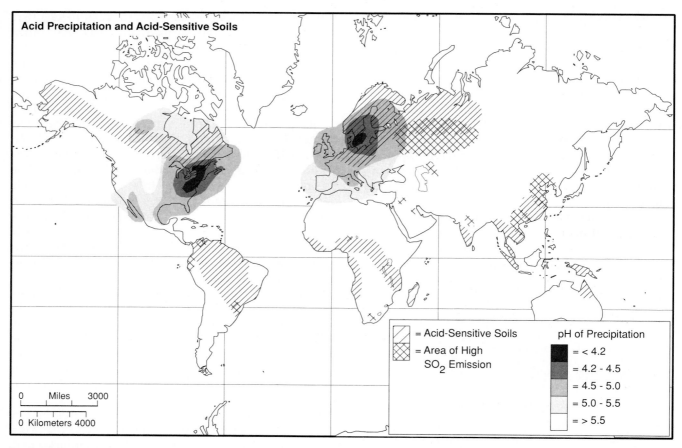

FIGURE 3-10

Distribution of acid precipitation (shown as pH of precipitation for mid-1980s) and acid-sensitive soils. Areas of high current SO$_2$ emissions are shown where pH of precipitation is not available. Compiled from OTA (1984), Rodhe and Herrera (1988), and Irving (1991).

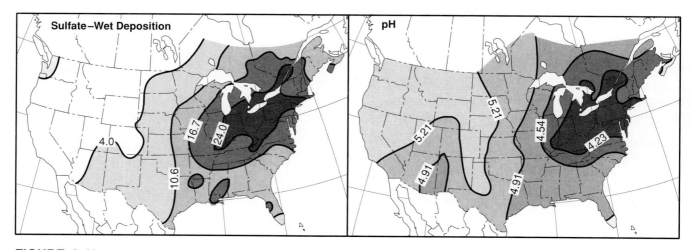

FIGURE 3-11

Close correspondence between pH and dry sulfate deposition (in kilograms/hectare) of precipitation in eastern North America for 1985–1987. Dry deposition of nitrates shows approximately the same pattern (from Irving, 1991).

southern parts of the Canadian and Baltic Shields are surprisingly acid (Figure 3-12A). In the Adirondacks and the Upper Peninsula of Michigan, about 10% of the lakes larger than 4 hectares have a pH of less than 5 (OTA, 1984; Irving, 1991). Diatoms and other fossils show that this acidity has developed within the last hundred years or so, and it appears to have caused large declines in fish populations (Figure 3-12B), probably through increases in dissolved aluminum and other metals that are more soluble at low pH (Longhurst, 1989; NAPAP, 1991). Parallel declines have been observed in the health of forest trees, particularly red spruce.

Although the correlation between anthropogenic emissions and acid precipitation is widely accepted, it is not without complications. Some areas with high SO_2 emissions, such as China, do not have widespread, strongly acid precipitation and some Pacific islands which lack SO_2 emissions, have acid rain. Lake water in the Okefenokee and Everglades swamps is just as acid as that in the Adirondacks, despite the lack of SO_2 emissions or acid rain in the area and the presence of underlying limestone. These relations suggest that naturally acid water might come from the decay of organic matter or from nitrogen gases produced by lightning. Furthermore, acid water might be neutralized during precipitation by reaction with alkaline dust particles in the atmospheric aerosol, which are most abundant in arid regions, just the area where acid rain is least common (Ray, 1987). These concerns in no way diminish the threat of acid rain, but they do point to a need for careful measurements to distinguish anthropogenic and natural acid-related emissions.

Global Change

Global change is our acknowledgment that Earth did not stop changing just because humans showed up. We have always recognized that short-term changes take place, including volcanic eruptions, rising lake levels, and earthquakes, all of which regularly disrupt life. We have also known for many years that Earth underwent longer-term changes, such as the glacial advances that swept over North America and Eurasia intermittently for the last million years, and the unusually hot climate that prevailed about 150 million years ago in Cretaceous time. It now appears that anthropogenic emissions might affect these longer-term changes and the greatest present concern centers on the possibility that they are causing global surface temperatures to increase dramatically (Bolin et al., 1986). Recent analyses suggest that global temperatures have increased by as much as 0.3° to 0.7°C since the early 1900s (GEMS, 1989).

The increase is related to carbon dioxide, methane, water vapor, nitrous oxide, chlorofluorocarbons, and

Acid Lakes

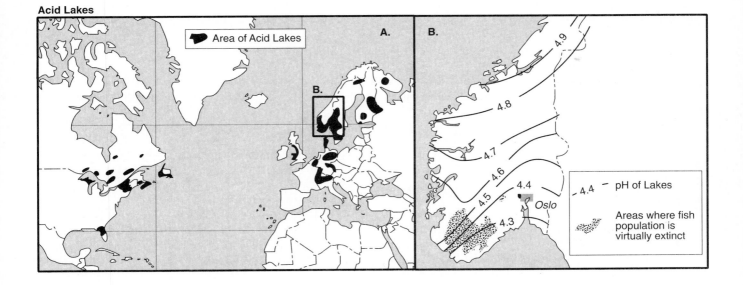

FIGURE 3-12

A. Areas of acidic lakes in North America and northern Europe. **B.** Relation between lake acidity (indicated by pH contours) and fish mortality in Norway. Compiled from GEMS (1989) and Taugbol (1986).

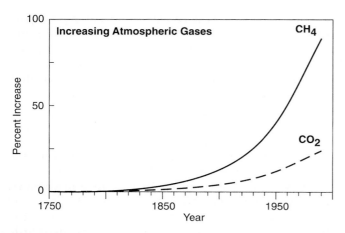

FIGURE 3-13

Increase in CO_2 and CH_4 contents of the atmosphere expressed as percentage change since the period 1750 to 1800 (based on analytical data of Neftel et al., 1985 and Khalil and Rasmussen, 1987).

other gases, which are known as *greenhouse gases* because they absorb infrared radiation that would otherwise radiate from Earth into space, thus heating the surface of the planet. Historical records and analyses of gases trapped in ice caps show that CO_2 and CH_4 concentrations in the atmosphere have increased significantly over the last few centuries (Figure 3-13). The possibility that gas variations control temperature is supported by the close correlation through time between the composition of bubbles of air trapped in the polar ice caps and world ocean temperatures indicated by isotopic measurements (Barnola, 1987).

Although the composition of the atmosphere has clearly evolved through geologic time, the recent increase is thought to be more abrupt than normal, further supporting its suggested relation to anthropogenic emissions. There is no question that we are emitting more of these gases. Compilations of global CO_2 emissions suggest that they tripled between 1950 and 1990 to about 8.5 billion tonnes of CO_2 annually (Cunningham and Saigo, 1992). Most of this comes from fuel combustion, with about 27% from oil, 27% from coal, and 11% from natural gas. The remainder comes from cement production (2%) and land clearing and burning (33%) (Morgernstern and Tirpack, 1990). As can be seen in Figure 3-14, the United States is the largest source of CO_2 emissions, as it is for SO_2, but large LDCs such as China are closing the gap rapidly.

Preliminary calculations using projected anthropogenic greenhouse-gas emission rates suggest that they could cause global temperatures to increase by up to 5.5°C by 2030 (GEMS, 1989). A temperature increase of this magnitude would equal the entire temperature change of the last major glacial advance and could melt enough of the polar ice caps to increase global sea level and flood our large coastal cities. The specter of these problems has caused a surge of interest in anthropogenic contributions to the atmosphere and the factors that control their impact on the planet. Natural geologic and biologic processes might well consume additional CO_2 as it accumulates in the atmosphere, for instance. Pre-

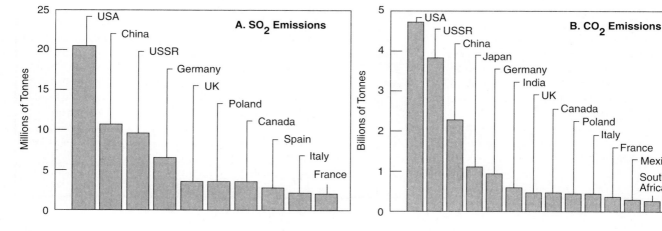

FIGURE 3-14

Anthropogenic atmospheric emissions of SO_2 (**A**) and CO_2 (**B**) in 1989. Compiled from World Resources 1992-1993.

cipitation of calcite in the oceans would have this effect. Increased plant life, which would also generate O_2, would have a similar effect, as suggested by the *Gaia hypothesis* (Lovelock, 1988; Schneider and Boston, 1991). Natural and anthropogenic processes might compensate for global warming, particularly the emission of dust and aerosols into the atmosphere, which shield Earth from incoming solar radiation. Agriculture remains the largest anthropogenic dust emitter, particularly since emissions of particulates from fossil fuel combustion have been stopped. These hardly hold a candle, however, to those of large volcanic eruptions. The eruption of Mount Pinatubo in 1991 cooled the planet for at least a year, and it has been suggested that larger eruptions of this type could cause *volcanic winter*, possibly triggering glacial advances (Rampino, 1988).

Thus, there is no doubt that Earth changes. The only question is whether anthropogenic effects can outweigh natural ones. If global change is placed on the doorstep of modern civilization, it will have an enormous impact on our consumption of minerals. Fossil fuel use will have to be curtailed and substitutes found, possibly including nuclear power. Most of our mineral extraction and processing methods will have to change, as will our current mineral-use patterns, which are based on extensive global trade. In the meantime, we have to test the global-warming hypothesis and make efforts to curtail offending emissions of all types, actions that involve assessment and remediation, as discussed later.

ASSESSING AND REMEDYING THE ENVIRONMENTAL IMPACT OF POLLUTION

Most modern environmental studies begin with chemical surveys of the abundance of the contaminant, which are followed by assessment of the relation between the contaminant and possible health effects. Finally, remediation studies are undertaken, if contaminant levels need to be reduced. The survey of chemical abundances actually involves two separate steps, sample selection and sample analysis. This sequence of efforts provides the only hard facts on which to base environmental assessments, predictions of future conditions, and regulations. Unfortunately, they can be tedious and time consuming, and sometimes they do not bear out the hypothesis that started the survey. They are, however, absolutely essential to the responsible management of our environment.

Sample Selection and Sample Statistics

The nature of the samples used in a study is of critical importance. Regardless of the imaginative nature of a hypothesis or the quality of the analyses carried out to test it, a study is only as good as the samples on which it is based. Samples must faithfully represent the material that is being investigated, and there must be enough samples taken to illustrate any compositional variations in the study group. The selection of proper samples depends on a full knowledge of the science of the material under investigation. To obtain representative samples of lake sediment, it is necessary to understand limnology and geology; to obtain samples from humans a knowledge of biology and anatomy is needed. There are many examples of confusion caused by improper sample material or sampling methods. Early ice cores used for mercury and lead dispersion studies were obtained with equipment that contained traces of lead, giving the mistaken impression that they contained more lead than they actually did (Patterson and Ng, 1981; Boutron and Patterson, 1983). Similarly, many analyses made as recently as the 1980s for mercury were contaminated by air and dust, giving results that were too high (Porcella, 1990). Asbestos samples used in some early studies contained several types of asbestos, rather than the potentially dangerous amphibole asbestos, thus producing a confused relation between exposure and symptom. It can even be important to analyze material from the right part of a sample. As can be seen in Figure 3-15, lead contributed by atmospheric fallout is concentrated in the upper part of the soil horizon, whereas lead from natural soil formation is concentrated in the B horizon of the soil. The fact that this upper lead-rich zone is now being buried beneath cleaner soil and sediment in many parts of the world suggests that it came largely from leaded gasoline, although coal-fired electric generating plants are also an important source.

Attention must also be given to *sample statistics*, which deal with the characteristics of groups of samples called *populations*. The first question posed by statisticians is whether the population of samples used in the study truly represents the group from which it came. We know intuitively that the gold content of any part of Earth cannot be determined accurately with one sample and that even a great number of samples would only give a very good approximation. However, in the interests of cost and convenience, we need to know the minimum number of samples that can be used to make this estimate with the desired degree of certainty. In order to answer this, statisticians usually need to know the type of *distribution* that the sample

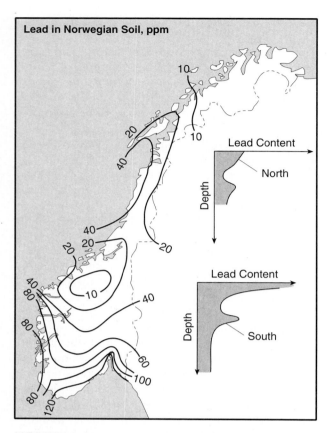

FIGURE 3-15

Lead content of soils in Norway with insets showing the position of lead enrichment in soil profiles. Soil in northern Norway has normal E (leached) and B (enriched) zones, whereas the profile from southern Norway has enrichment of lead in O zone, reflecting atmospheric fallout (from Steinnes, 1987).

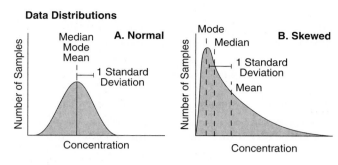

FIGURE 3-16

Frequency diagrams for normal (Gaussian) and skewed populations.

population shows. Populations can be described by their *mean* or average value, *mode* or most common value, *median* or middle value, and *standard deviation,* the square root of the sum of the squares of the deviation between all individual values and the population mean.

These descriptors can be used to distinguish between two distributions that are important in mineral resource and environmental studies. The population labeled *normal* in Figure 3-16 has a mode, median, and mean that are identical, and the values form a bell-shaped distribution. In the other type of population, here called *skewed,* the mode and median values are significantly lower than the mean. Many environmental and geological populations, such as the concentrations of copper in soil, the grades of gold deposits, or the sizes of oil fields, are skewed, with a few large values and many small ones. As can be seen, the mean is much more representative of a normal population than it is of a skewed population. It follows that means are useful only if the type of population they represent is known and if the population truly represents the rocks or soils or lake being investigated. Although the statistical basis for most environmental and geological studies is adequate, conclusions from these studies are sometimes confused in media reports by careless use of these concepts.

Many of the analyses carried out in mineral exploration and environmental studies are designed to determine whether any samples are *anomalous.* There is no single, widely accepted way in which to designate a sample as anomalous, and great care must be taken in using this term. Clearly, anomalous samples are relatively scarce and they differ from most samples in the population under consideration, which are often referred to as *background* samples. Anomalous samples usually differ in having a higher concentration of some element or substance, although samples with unusually low concentrations can also be anomalous. Anomalies for a single survey are often defined as those values that differ from the population mean by two standard deviations. More sophisticated anomaly-recognition techniques include the use of *geostatistics,* which is based on the fact that adjacent samples of Earth are related to each other and can be used as predictors. Comparison to surveys in other areas is also necessary to determine whether all of the samples in a survey are anomalously high (or low) with respect to global averages. Finally, it is important to look for patterns of variation in samples. For instance, otherwise normal samples can cluster in ways that identify underlying causes for the features being studied.

Analytical Preparation and Methods

Once a group of samples of rocks, minerals, water, plants, or air has been collected, the next step is measurement of the parameter of interest, usually by chemical analysis (Marr and Cresser, 1983). These measurements are used by the exploration geologist to search for ore, by the miner to determine that the ore is being recovered effectively, by the environmentalist to determine the dispersal of polluting substances, and by the public health professional to determine the toxicity of a substance. Often, they are compared to allowable amounts that have been specified by government agencies (Table 3-5).

Analytical measurements involve two important steps, preparation of the sample and the actual analysis. For most atmosphere and water samples, the only *sample preparation* needed is filtration, which separates fluid and particle fractions so that they can be analyzed separately. For solid samples such as rocks, soils, or trees, matters are more difficult because elements and compounds can be present in different minerals and other forms. For instance, lead in stream sediments might be found as adsorbed ions on the surface of fine-grained clay minerals, iron oxides, or kerogen, in trace amounts substituting for potassium in silicate minerals, or as clastic grains of the lead mineral *galena*. To determine the total lead content of the sample, a *total analysis* must be performed, such as by completely dissolving the sample

and analyzing the resulting solution. If, on the other hand, it is desired to estimate the amount of lead that was introduced by waters flowing through a sample, some means must be found for removing lead from the appropriate part of the sample, a technique known as a *partial analysis.* One way to do this is to leach the sample with a weak acid that removes only adsorbed lead without decomposing lead-bearing minerals such as galena or feldspar. Analysis of the residue from this leach treatment provides information on the amount of lead in galena and silicate minerals. Many such methods are available, each with its capabilities and weaknesses.

Modern analyses of properly prepared samples are made by an amazing array of methods, and anyone carrying out a geochemical study is immediately faced with the need to select the proper analytical method for the study at hand. Although it is best to compare analytical methods on the basis of the chemical principles they employ, a quick comparison can be based on three measures of their capability. *Detection limit* refers to the minimum concentration that can be measured. Detection limits for different elements or compounds vary tremendously,

TABLE 3-5

Maximum allowable concentrations of pollutants of interest to mineral resources (from Environmental Protection Agency and U.S. Public Health Service Drinking Water Standards).

Substance Dissolved Elements	Maximum Allowable Concentration (mg/liter)	Gases in Atmosphere (micrograms/m^3)
Arsenic	0.05	SO_2 80 (0.030 ppm) - annual mean
Barium	1.0	365 (0.014 ppm) - 24-hr mean
Boron	1.0	1300 (0.500 ppm) - 3-hr mean
Cadmium	0.01	TSP* 75 - annual geometric mean
Chloride	250	260 - 24-hour geometric mean
Chromium	0.05	Pb 1.5 - maximum quarterly mean
Copper	1.0	VOC 160 (0.24 ppm) - 3 hours
Flourine	2.0	
Iron	0.05	
Lead	0.05	
Manganese	0.05	
Mercury	0.002	
Selenium	0.01	
Silver	0.05	
Sulfate	250	
Zinc	5.0	
pH	6.5-8.5	
dissolved solids	500	

*TSP = total suspended particles

even for a single analytical method. Complicating matters is the fact that different elements and compounds are usually present in a single sample in concentrations that differ by many orders of magnitude. Even when you are only analyzing one element with one method, things can be complicated because samples in a large group can have a wider range of concentrations than can be measured by that method. Similarly, it is almost impossible to determine the concentration of all elements in a single sample with one analytical method. Instead, the analytical method used is commonly selected to give the best results for most elements or samples of greatest interest.

Analytical methods also differ in their *precision,* which is the ability to obtain the same result if the same sample is analyzed again. Precision is commonly expressed as the deviation from an average value for a sample that has been analyzed repeatedly throughout the duration of the study. Some methods are considerably more precise than others. *Accuracy,* which is the ability to obtain the correct answer for a sample of known composition, is based on the analysis of a *standard sample,* in which the concentration of the component of interest has been agreed upon by a large number of analytical laboratories. It is possible for analyses to be precise without being accurate and vice versa. Because mineral resource-related samples have so much economic and environmental importance, it is critically important that they be reported along with information on the analytical method used and its precision and accuracy. Ideally, analyses of standard samples should be provided as well.

As the detection limit, precision, and accuracy of analytical methods improve, we are able to see variations that were not detectable previously. One famous instance of this was the development of the electron-capture detector for gas chromatographs, which allowed measurement of low concentrations of chlorofluorocarbons (CFCs) in air and led to the recognition that they were accumulating in the atmosphere (Lovelock, 1971).

Risk and Dosage in Environmental Chemistry

Many environmental studies related to mineral resources seek to determine whether specific elements, compounds, or materials are damaging to life. The term *toxic,* which is frequently used in this context, refers to any substance that causes undesired effects such as alterations in DNA, birth defects, illness, cancer, or death (Doull et al., 1980). In most envi-

ronmental studies, toxicity is measured by exposing a number of individuals of a specific organism to a solution containing a known concentration of the compound of interest, and then observing how many of the organisms die in a specific time period. By convention, the results are reported as the concentration necessary to produce 50% mortality, a factor known as LD_{50}. Toxicity measurements of this type can be made on any living organism, although aquatic organisms are used most commonly. There is no universal scale of toxicity because different organisms respond differently to the same compound. Furthermore, these measurements assess only the mortality that occurs in a short time span, usually 96 hours, and do not take into account the long-term effects such as the increased risk of cancer.

Studies of humans cannot use this approach and commonly rely on *epidemiology,* which is the study of patterns of disease or effects from toxic exposure. Most such studies are based on *mortality ratios,* in which death rates in one population are ratioed against that in another, with appropriate standardization for age, sex, race, and other possible controlling factors (Zopf, 1992). The effects of exposure to material such as chrysotile asbestos can be investigated by determining whether groups that have been exposed to it exhibit a higher mortality ratio (when compared to the general population) than other groups with no exposure. Although this approach is reasonable in theory, it is difficult to identify groups that differ only in exposure to asbestos, without differences in other possible complicating factors, such as smoking. The fact that lower dosages require longer periods of time to manifest symptoms aggravates this problem because it is necessary to continue studies over decades, thus increasing the variation in exposure to other factors.

One method designed to cope with these complications is the *extrapolaiion technique,* which involves the determination of the effect per unit of high dosage and extrapolation of this relation to lower dosages to give the effect typical of natural settings. The simplest assumption, represented by the proportional curve in Figure 3-17, accepts a linear relation between dose and effect. As one decreases, the other does so also, at a constant rate. A frequently quoted pitfall in this method is illustrated by its application to estimate fatalities associated with crossing a river. If 100 out of 1,000 of the people attempting to cross a 10-meter deep river die by drowning, then 10 out of 1,000 people should die trying to cross a river 1 meter deep, and 1 out of 1,000 people should die attempting to cross a river only 0.1 meter deep. This analogy is obviously not correct and it suggests that there is a mini-

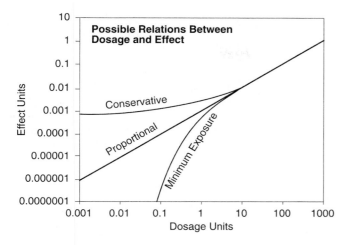

FIGURE 3-17

Three possible relations between dosage and effect. The conservative curve is based on the assumption that a constant amount of damage is caused below some lower limit. The proportional curve assumes that there is a constant relation between dosage and effect over the entire range of dosages. The minimum exposure curve assumes that exposure below a specified minimum has no detectable effect.

mum threshold value below which no damage is experienced (minimum exposure curve, Figure 3-19).

Dose-effect relations are even more complex for some compounds and elements. For instance, underexposure to elements that are essential to life, such as zinc and copper, can be just as debilitating as overexposure. In contrast, it is unclear whether the body can avoid damage from even small levels of exposure to metals such as mercury and lead. Statistical efforts to clarify these questions have not been completely successful (Cohen, 1990), and it is hoped that molecular biology will shed additional light on the problem by making theoretical estimates of dose-effect relations.

It is no surprise, then, that the term *toxic* is often used confusingly. Plutonium, for instance, is less toxic than arsenic when taken orally, but its alpha radiation can cause lung cancer if it is inhaled. In spite of this, a full-page ad against Japanese nuclear policy in *The New York Times* of June 1, 1992, began "Plutonium, the most toxic substance ever created. . .", a misconception perpetuated in subsequent articles on the topic.

Remediation

Once an anthropogenic emission has been identified as damaging, means must be found to stop it or clean it up, efforts known generally as *remediation*. Most regulatory agencies require that pollutants in water

and air leaving mineral and industrial facilities be at or below recommended standards (Table 3-5), even if they did not meet those standards entering the property. Treatment of some type is commonly necessary to accomplish this. The simplest methods involve *physical separation* of particulate material, which can be carried out by filtration, sedimentation, or electrostatic methods, in order of increasing cost and effectiveness. The least expensive and most common method uses a *cyclone* (Figure 3-18). Greater efficiency can be obtained at additional cost by passing the fluid through a filtration system that removes smaller particles. Very small particles must be separated by more expensive electrostatic methods that ionize particles, causing them to be attracted to an electrode (Figure 3-18). Even after they have been recovered, the disposal of particles is a major problem. Although *biological separation* methods have not been widely used yet, genetic engineering promises to open some very exciting possibilities, including plants that take up several percent of their weight in metals and bacteria that consume crude oil.

Chemical separation methods are much more varied and complex. In liquids, they can involve the exchange of some less harmful substance for the pollutant, precipitation of the pollutant in solid form, or adsorption of the pollutant on an active surface. *Ion exchange* is the most commonly used process, with typical water softeners being the prime example. These units operate by exchanging the offending calcium in hard water for sodium in *zeolite* minerals to produce sodium in solution and calcium in the zeolite. Natural and synthetic zeolites are being developed to remove metals and other ions from solution. *Precipitation* and exchange are actually closely related. Acid mine drainage, for instance, can be neutralized by the addition of *calcite, limestone,* or *lime* to the solution (Table 3-4). If the concentrations of dissolved calcium and sulfate in the resulting solution are low enough, this reaction is essentially an exchange of calcium in the rock for hydrogen ions in the water. On the other hand, if the concentrations of calcium and sulfate in the solution become high enough, *gypsum* or *anhydrite* precipitate, forming a sludge that must be removed from solution. Substances can also be removed from solution by adsorption on *activated charcoal,* a form of carbon that has been treated to increase its ion-exchange capacity. For dissolved ions, this is actually the same as exchange because it is necessary to maintain the electrical balance of the solution (that is, equal total charges of cations and anions). Thus, most of the commonly used water-cleaning methods simply replace undesirable dissolved constituents with those

FIGURE 3-18

Methods of separating particles from liquids or air. In filter systems (A), particles are removed when they fail to pass through small holes. In the cyclone (B), the fluid-particle mixture moves down along the walls of the cone where particles fall out and the fluid rises to exit through the top. In the electrostatic separator (C), particles are charged by electrons in a high-voltage field and attracted to one of the poles where the charge is neutralized, causing them to fall to the bottom of the container.

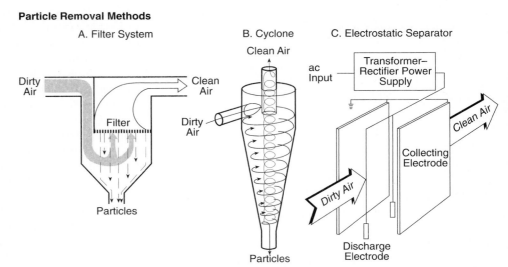

Particle Removal Methods

that are less undesirable; if you want to remove all dissolved material, you must *distill* the sample and dispose of the precipitate, a very costly process.

Chemical separation processes for gases are different. Gases are not electrically charged, but can be adsorbed onto active surfaces such as charcoal, zeolites, and related substances known as molecular sieves. They can also be removed by reaction with other chemicals, either in the gaseous state or after being dissolved in water. At present, air-cleaning systems are used largely to remove SO_2 from exhaust gases, a process known as *flue-gas desulfurization (FGD)*. When the concentration of SO_2 in the exhaust gas is high, as in many modern metal sulfide smelters, it can be reacted with water and oxygen in an *acid plant* to make sulfuric acid (H_2SO_4) (Table 3-6). If the SO_2 concentration is low, it is usually treated in a *scrubber*, where it

reacts with a finely dispersed compound such as calcite or lime to form calcium or sodium sulfite (Table 3-6). In most cases, this is done by spraying a water slurry of the compound through the flue gas, although "dry" scrubbing can be carried out by spraying a finely divided powder of water and a reactive mineral. Concentrations of SO_2 are so low in many coal-burning electric utilities that neither process works very efficiently, and large amounts of SO_2 still escape from these sources, as discussed in the section on coal. Scrubbers can be *regenerative* or *nonregenerative*, depending on whether the chemical process used to trap SO_2 can be reversed. Most nonregenerative scrubbers produce large volumes of waste that must be disposed of and is an environmental contaminant in its own right. Other gases can be removed by similar systems, and the next one on which attention will focus is ni-

TABLE 3-6

Common reactions used to remove SO_2 from flue gases, including those that produce sulfuric acid (acid plant) and those that precipitate calcium or sodium sulfite (scrubber). Both systems are most efficient on gas with a high concentration of SO_2, making them more useful in cleaning gases from smelters rather than those from electric utilities, which have lower SO_2 concentrations (modified from Elliot and Schwieger, 1985 and Manahan, 1990).

Process	*Chemical Reaction*	*Advantages/Disadvantages*
Sulfuric Acid Reaction	$SO_2 + H_2O + \frac{1}{2}O_2 = H_2SO_4$	
Sulfate and Sulfite Precipitating Reactions		
Lime slurry	$Ca(OH)_2 + SO_2 = CaSO_3 + H_2O$	large volume of waste $CaSO_3$
Limestone slurry	$CaCO_3 + SO_2 = CaSO_3 + CO_2$	less efficient than $Ca(OH)_2$
Magnesium hydroxide slurry	$Mg(OH)_2 + SO_2 = MgSO_3 + H_2O$	can regenerate $Mg(OH)_2$

TABLE 3-7

U.S. federal laws related to mineral activity (compiled from Ely, 1964; Flawn, 1966; Culhane, 1981; Vogely, 1985; Cameron, 1986; Foss, 1987; Balzhiser, 1990 and U.S. Bureau of Mines Mineral Commodity Summaries). The main Canadian federal laws regulating mineral activities are the Canadian Environmental Protection Act, Fisheries Act, and Environmental Assessment and Review Process Guidelines Order, although numerous provincial laws are also relevant.

Title	Date
Rivers and Harbors Act	1899
Reclamation Act	1902
Antiquities Act	1906
Migratory Bird Treaty Act	1918
Fish and Wildlife Coordination Act	1934
Multiple Use-Sustained Yield Act	1960
Clean Air Act	1963
Endangered Species Act	1963
Wilderness Act	1964
National Historic Preservation Act	1966
Forest and Rangeland Resources Planning Act	1966
Air Quality Act	1967
Bald Eagle Protection Act	1969
National Environmental Policy Act	1969
Coal Mine Health and Safety Act	1969
Clean Air Act	1970
Mining and Minerals Policy Act	1970
Land Policy Management Act	1970
Hazardous Substances Act	1970
Noise Pollution and Abatement Act	1970
Occupational Health and Safety Act	1970
Coastal Zone Management Act	1972
Ports and Waterways Safety Act	1972
Water Pollution Control Act	1972
Marine Mammal Protection Act	1972
Endangered Species Act	1973
Safe Drinking Water Act	1974
Toxic Substances Control Act	1976
National Forests Management Act	1976
Noise Control Act	1976
Resource Conservation and Recovery Act	1976
Surface Mining Control and Reclamation Act	1976
Clean Water Act	1977
Coal Mine Health and Safety Act	1977
Soil and Water Resources Conservation Act	1977
Clean Air Act Amendments	1977
Hazardous Materials Transportation Act	1978
Mine Safety and Health Act	1978
Comprehensive Environmental Response, Compensation and Liability Act	1980
Nuclear Waste Policy Act	1982
Resource Conservation and Recovery Act Amendments	1984
Safe Drinking Water Amendments	1986
Superfund Amendments and Reorganization Act	1986
Clean Air Act Amendments	1990
Oil Pollution Act	1990

trogen from air that is converted to nitrogen oxides (NO_x) during combustion, a target of the *Clean Air Act Amendments of 1990.*

CONCLUSIONS

Environmental geochemistry is essentially the study of global and local geochemical cycles for substances that move through the biosphere. Although the biosphere is not the major reservoir for any important elements, many substances affect life through it. Hence, environmental geochemistry takes a somewhat distorted view of Earth processes, but one that is critical to our continued survival on the planet. Many of the cycles studied for environmental reasons, such as that of salt, involve natural, geologic materials. Others, such as that for chlorofluorocarbons, involve synthetic materials. Still others, ranging from the global cycle of zirconium to the cosmic cycle of hydrogen, are studied by geochemists and cosmochemists, but are not of major interest to environmental geochemists because they do not impinge significantly on the biosphere.

Studies involving environmental geochemistry have several goals. First, the reservoir of interest must be recognized and its average composition for the element or compound of interest must be determined. Second, the form of the element or compound in related reservoirs must be specified. Third, reactions that transfer the element or compound from reservoir to reservoir must be determined. Finally, the impact of all parts of this cycle on the biosphere must be understood. When this is done completely, it is possible to distinguish between natural and anthropogenic inputs to environmental systems.

The results of these studies are commonly used to formulate legislation and regulations. As can be seen in Table 3-7, environmental legislation has occupied a large part of the United States Congress's time over the last few decades. This list will undoubtedly grow as we recognize new mineral-related environmental problems that require further legislation. It is up to us to be certain that the studies on which this legislation is based are carried out and interpreted in ways that are consistent with geologic and chemical principles. Poorly designed or badly conducted environmental studies can miss the real problem, wasting our energy on regulating phenomena that turn out to be unimportant.

4

Mineral Exploration
and Production

OUR GLOBAL APPETITE FOR MINERAL RESOURCES IS SO GREAT THAT even large mineral deposits are exhausted rapidly, making it necessary to search constantly for new ones. No example of this is more dramatic than the Prudhoe Bay oil field in Alaska. With an original recoverable oil reserve of 9.45 billion barrels (Bbbl), Prudhoe Bay was the largest oil field in North America and the eighteenth largest in the world (Carmalt and St. John, 1986). Before Prudhoe Bay was discovered in 1967, the largest oil field in the United States was East Texas, which ranked fifty-sixth in the world with "only" 5.6 Bbbl. The discovery of Prudhoe Bay increased U.S. oil reserves by about 30%.

Your first reaction to this might be that we can relax now; this huge field will take care of our needs. Sadly, this is not so. Even under normal production, huge deposits like Prudhoe Bay last only a few decades. At present consumption rates of over 6 Bbbl of oil annually, the United States could exhaust Prudhoe Bay in less than two years. This is a distressingly short interval in terms of the thousands to millions of years that nature needs to form a new deposit, confirming that mineral deposits are *nonrenewable resources*. They differ greatly from renewable resources such as trees and fish that can be replenished naturally in periods similar to our lifetimes. It follows that we must think of Earth as having a fixed inventory of minerals for which we must constantly explore.

At this point, some will say that there is no point in looking for deposits if even the large ones will be exhausted so rapidly. Few of those making this argument appear to have given up the comforts of modern life, which come almost entirely from minerals, however. In fact, this argument is no more logical than it would be to stop fixing meals because each one is eaten as soon as it is prepared. So, let's look into what is involved in mineral exploration.

EXPLORATION SUCCESS RATES AND EXPENDITURES

The odds of success in exploration are not good. Every new discovery leaves one less deposit to find. Furthermore, we usually find the most obvious deposits and miss the ones in more remote locations and at greater depths. Things are made worse by the fact that mineral deposits come in a wide range of sizes and form *skewed populations* in which large deposits are a very small proportion of the total (Figure 4-1). These distributions tell us that we have a much better chance of discovering small deposits than we do of finding large ones. Unfortunately, many of the small deposits cannot be extracted at a profit. Thus, most of our "discoveries" are geologic successes but would be economic failures.

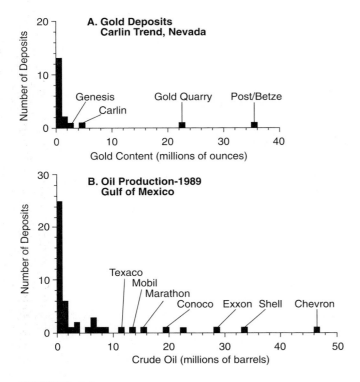

FIGURE 4-1

Skewed size distribution of mineral deposits showing how a few deposits can contain a large fraction of total mineral reserves. **A.** Gold content (in Troy ounces) of deposits in the Carlin area, Nevada (Rendu and Guzman, 1991). **B.** U.S. Gulf of Mexico oil production by operating companies (MMS, 1991), and one hole can discover it.

It should be clear from these factors that mineral exploration is a business in which failure is the norm and success is the exception (Slichter, 1960; Woodall, 1984, 1992). Continued success in mineral exploration requires enough capital to withstand a long series of failures. Recent estimates put the average cost of finding a 100-Mbbl oil field at about $700 million and the cost of a single metal discovery at about $200 million (Cook, 1986; Hamilton, 1990; EIA, 1991). Thus, most oil companies spend hundreds of millions of dollars each year in exploration, and most metal companies spend tens of millions (Table 4-1). The combined effects of inflation and the increasing difficulty of discovery will require that these expenditures increase significantly, if we are to maintain our mineral reserves at present levels.

With all of these complications, success in the mineral business is understandably elusive and anything that can be done to enhance it is most welcome. This is where prospectors, geologists, geochemists, and geophysicists come in. Some of the concepts and methods used in their search are discussed later in this chapter.

MINERAL EXPLORATION, PERSONAL INVESTMENTS, AND YOU

This is a good place to put in a word of caution. Mineral exploration is a fascinating business. Not only are the minerals pretty, but they are valuable and have been the source of many fortunes. Opportunities for individuals to invest in mineral exploration are widespread, and you may have one yourself some day. Hopefully, the information in this chapter will help you evaluate the opportunity. The first step is to determine the "maturity" of the project. Is it simply *moose pasture,* with only a hypothesis that a deposit might be there, or have samples with anomalous mineral contents been collected on the property? If a deposit is actually present, how well has it been delineated? How big is it? Who controls the land on which the deposit is located? Have any tests been made to determine that the mineral can be extracted and processed? Can all of this be carried out economically under current environmental regulations and tax laws?

If these questions look daunting, call on a professional for advice. Be wary of volunteers. Some people are simply not very careful about other people's money. Some are even dishonest. Things were so bad during the early days of the American West that Mark Twain is reported to have said that a mine was a hole in the ground with a liar on top. *Salting,* the practice of scattering ore minerals around a worthless property, still happens (Greeley, 1990; Danielson, 1992). You can do it yourself by rubbing your gold ring onto a rock, leaving a trace of gold that would make the rock seem attractive to an uninformed investor. Fortunately, most explorationists are simply highly enthusiastic rather than sneaky. In fact, enthusiasm is a requirement; no one could survive the many disappointments of exploration without a strong dose of it. And, in spite of all the poor odds, important discoveries are made every year and mineral exploration remains one of the few businesses where a small investment can turn a huge profit.

MINERAL EXPLORATION METHODS

Although the basic function of mineral exploration is to replace deposits that become exhausted, it is not directed evenly at all mineral commodities. This is particularly true in market economies, where exploration is driven by the desire to make a profit and focuses

TABLE 4-1

Annual exploration budgets for publicly-owned mineral companies. The list is representative only and omits many important companies (compiled from annual reports, largely 1992; all amounts in millions of U.S. dollars).

Company (Oil Companies)	Exploration Budget	Company (Hard Mineral Companies)	Exploration Budget
Exxon (USA)	957	CRA (Australia)	102
Amoco (USA)	693	Rio Tinto Zinc (UK)	100
Mobil (USA)	686	Anglo American (South Africa)	85
Arco (USA)	648	Placer Dome (Canada)	69
Chevron (USA)	642	Newmont/Newmont Gold (USA)	62
Conoco (USA)	560		
Texaco (USA)	449	M.I.M. (Australia)	60
BP America (USA)	245	Falconbridge (Canada)	51
Phillips (USA)	229	Phelps Dodge (USA)	47
Marathon (USA)	229	Homestake (USA)	39
Sun Energy (USA)	219	Inco (Canada)	44
Occidental (USA)	130	Noranda (Canada)	38
Kerr-McGee (USA)	95	Battle Mountain (USA)	36
Petro Canada (Canada)	93	Genmin (South Africa)	37
Sunoco (USA)	92	AMAX (USA)	34
Union Texas (USA)	83	Asarco (USA)	27
Imperial Oil (Canada)	79	DeBeers (South Africa)	27
Unocal (USA)	74	Cominco (Canada)	26
Murphy (USA)	63	Hemlo (Canada)	26
Louisiana Land (USA)	62	Gold Fields Mining (UK)	20
Pennzoil (USA)	38	Lac Minerals (Canada)	18
Diamond Shamrock (USA	16.5	Rio Algom (Canada)	15
Santa Fe Energy (USA)	11	Teck Corporation (Canada)	14
Mesa Limited Partners (USA)	9	Renison Goldfields (Australia)	13
Plains Petroleum (USA)	3	Cyprus (USA)	13
		Peñoles (Mexico)	13
		Goldfields of South Africa	12

therefore on commodities with rising demand or price. Figure 4-2, which shows the relation between mineral reserves and recent price increases, provides a good indication of global exploration targets. As can be seen here, natural gas, gold, cobalt, oil, gem diamond, and vermiculite have low reserves and have experienced significant price increases since 1960. In contrast, the large reserves of commodities such as iron ore, aluminum, and soda ash have discouraged price increases and exploration. Whatever the target, however, mineral exploration can be undertaken in several different ways.

Prospecting and Random Exploration

Mineral exploration has not always been a science. Many good deposits were found by *prospectors* with nothing more than hope and determination. The Spindletop oil field in Texas was found by Patillo Higgins, who embodied just these qualities. Spindletop was the first important field in the U.S. Gulf Coast, and at the time it was found there was general agreement that commercial oil was not present in the area. Higgins, a timber expert and land promoter from Beaumont, believed otherwise. He had rights to Spindletop, a small hill around which there were seeps of oil and gas. Rather than giving up after several unsuccessful drilling efforts at Spindletop, Higgins kept trying to find more backers, even against active discouragement from government agencies. In 1901, he drilled to a depth of 300 meters with the backing of one last investor. There, the well blew out to become the famous Spindletop gusher, which lost almost 750,000 bbl of oil before it was

FIGURE 4-2

Global exploration targets. Mineral commodities plotting in the lower right-hand quadrant are the major targets of current mineral exploration. They have low reserves and have had significant price increases since 1960. Prices of mineral commodities plotting in shaded zone have not increased as fast as inflation over this period.

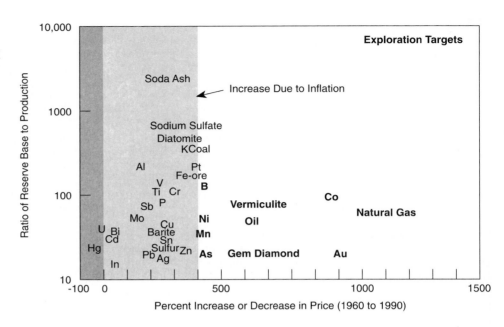

brought under control ten days later (Wheeler and Whited, 1975). Clearly, this discovery did not owe much to geologic insight. *Dowsers,* who profess to be able to sense everything from water to gold in the subsurface, carry on Higgins's tradition today.

Because of the serendipitous nature of mineral exploration, some people suggest that it be carried out by the *random drilling* of enough holes to find anything that is there. Geologic clues to the location of deposits would not be ignored, but they would not be the primary guide in the search. The number of exploration drill holes required by this approach and their distribution would be determined by the nature and size of the deposit being sought and the degree of statistical certainty desired for the search (Figure 4-3). In an early proposal of this type, it was suggested that the entire United States could be explored for oil with holes drilled on a grid-shaped pattern spaced 5.6 kilometers apart (Drew, 1967). According to this estimate, the value of oil that would be found would be about double the cost of the drilling program. Reducing the area of exploration to parts of the United States covered by sedimentary rocks, the type of rock that hosts oil, would have doubled this success rate. This approach is rarely used, however, because it is risky to assume that individual deposits can be recognized with a single hole and because it is impossible to obtain the necessary level of funding and land access (Drew, 1990).

A modified form of random drilling was used with some success in zinc exploration in central Tennessee (Callahan, 1977). Inspiration for the search came from the eastern part of Tennessee, where the largest zinc

deposits in the United States fill parts of a buried cave system (Plate 1-5) in a thick sequence of Ordovician-age carbonate sedimentary rock known as the Knox Group (Figure 4-4). These deposits, which are known as Mississippi Valley-type deposits and will be discussed in more detail later in this book, formed when basinal brines were expelled from nearby sedimentary basins. Study of the Knox Group in the 1960s showed that its buried cave system extended throughout much of the southeastern United States, suggesting that it might be filled with ore in other locations. Evaluation of this possibility was difficult, however, because the Knox Group was buried beneath hundreds of meters of younger sediments in much of this area. It came closest to the surface in the area around Nashville, where the sedimentary rocks formed a broad dome. Younger sedimentary rocks covering the Knox Group here contain small veinlets of zinc ore thought to have leaked upward from larger deposits in the buried cave system. To test this possibility, drilling within the favorable region was carried out in a random sequence determined by the availability of exploration leases signed with land owners (Figure 4-4). After drilling 78 holes in an area of 18,000 km², the seventy-ninth hole intersected what became a major zinc-producing district.

The Knox Group extends northward beneath much of Kentucky (Figure 4-4) and comes closest to the surface in a dome under the Lexington area where there are many veinlets in the overlying, younger rocks. Would you explore for another big deposit in this area? It would not be an easy job. The favorable region measures over 5,000 km² and a large deposit

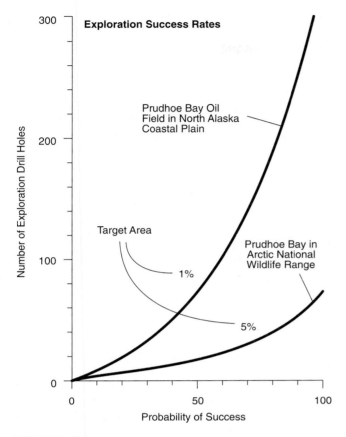

FIGURE 4-3

Odds of using properly spaced exploration drill holes to intersect a field the same size as Prudhoe Bay in the entire northern coastal plain of Alaska and the Arctic National Wildlife Reserve. Prudhoe Bay field has an area equal to about 1% of the coastal plain and 5% of the wildlife reserve. Curves show number of drill holes needed to assure indicated probabilities of success. For instance, 20 holes are needed in the reserve for a 50% chance of discovering a Prudhoe Bay-sized field, *if one is there to be discovered, and one hole can discover it.*

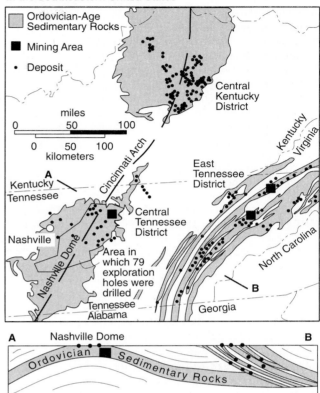

FIGURE 4-4

Geologic map and cross section of Tennessee and adjacent states showing the distribution of Mississippi Valley-type zinc deposits and prospects in Ordovician-age sedimentary rocks (compiled from Callahan, 1977, *Economic Geology,* 1977, v. 72, p. 1382 and Clark, 1989).

might have an area of only 1 km², only about 0.2% of the geologically favorable region. Obviously, it would help to have additional information to guide exploration and that is where geology, geochemistry, and geophysics come in.

Geological Exploration

Geological exploration works because most mineral deposits form in specific geological environments by processes that leave evidence in the rock (Ohle and Bates, 1981). By studying deposits that have already been discovered, geologists learn about these environments and processes and then look for them elsewhere.

Discovery of the Las Torres silver deposits in Guanajuato, Mexico, resulted from extrapolation of a favorable environment (Gross, 1975). Mining of vein deposits at Guanajuato started in 1548 and had produced over 31,000 tonnes of silver and 125 tonnes of gold by the 1960s, when most veins seemed to be exhausted. Reexamination of mined-out areas suggested that the main vein, the Veta Madre, contained ore in upper and lower levels and that only the upper levels had been mined in the southern part of the vein (Figure 4-5). Deep drilling beneath the upper level in this area resulted in the discovery of the Las Torres deposit, which revived mining at Guanajuato.

Geological exploration is often aided by *hydrothermal alteration,* the process by which hot water flowing through fractures reacts with and changes the composition of adjacent rocks (Plate 1-1). Hydrothermal al-

FIGURE 4-5

Geologic maps of the Guanajuato district and longitudinal section (drawn in the plane of the vein) showing the location of ore zones between A and B in the Veta Madre. Ore zones appear to fall into a lower zone (light shading) containing the Valencia, Cata, and Rayas deposits and an upper zone (dark shading) containing the San Rafael and Cedros deposits, which overlie the Torres discovery in the lower zone (after Gross, 1975, *Economic Geology,* 1975, v. 70, p. 1175).

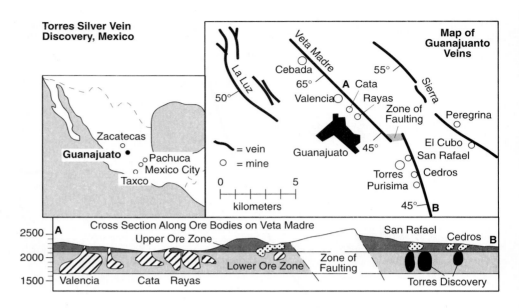

Geochemical Exploration

Geochemical exploration depends on the fact that nature works just as hard to destroy mineral deposits as it does to create them. At Earth's surface, weathering disperses the components of mineral deposits into the surrounding water, soil, vegetation, and air to create chemically enriched zones known as *geochemical anomalies* (Rose et al., 1979; Overstreet and Marsh, 1981; Levinson, 1983). The Montico copper-zinc deposit in the Dominican Republic provides a good example of geochemical dispersion (Figure 4-7). Montico is at the top of a hill in humid, tropical forest and grassland. It consists of copper, zinc, and iron sulfides in a matrix of quartz, all of which have been weathered to depths of at least 30 meters by reactions similar to those that generate acid mine drainage (Table 3-5). Thus, the ore at Montico weathered to form quartz grains, rusty colored limonite, and acidic water containing dissolved copper and zinc (Plate 2-2).

Almost as quickly as they formed, these weathering products were dispersed into the surrounding surface environment. Quartz and limonite grains became part of the regolith, which moved downslope by slump, landslide, and sheetwash into arroyos, where it was carried downstream as sediment. Soluble elements such as zinc were carried away in surface or groundwater. Some dissolved metals were adsorbed on clay and iron oxide minerals, and some were taken up by plants. Thus, the passing geologist sees rusty rock on the hillside at Montico, and the copper and

teration halos usually contain minerals in a zonal arrangement that can be used to focus on the center of the alteration zone, as is the case in porphyry copper deposits, the most important source of world copper (Lowell and Guilbert, 1970). When it was first mined, the large San Manuel porphyry copper deposit in Arizona appeared to consist of only half of a deposit. The other half seemed to have been cut off by a fault (Figure 4-6). Early exploration holes drilled to look for this lost half of San Manuel went to depths of 630 meters, but stopped when good copper values were not intersected. Review of these holes with hydrothermal alteration in mind, however, suggested that they had entered the outer alteration zones around a porphyry copper deposit, and subsequent deepening of the holes resulted in the discovery of the Kalamazoo ore body, San Manuel's other half (Lowell, 1968).

It should be obvious from these descriptions that basic geologic information is a critical part of mineral exploration. This exploration is usually based on geologic maps and reports prepared by government agencies such as the U.S. Geological Survey, the Geological Survey of Canada, and the Consejo de Recursos Minerales (Mexico), along with state and provincial agencies. More detailed maps are then prepared by exploration personnel, who also carry out geochemical and geophysical studies. Hamilton (1990) has estimated that government surveys were important aids in the discovery of more than half of the mines found in the United States, Canada, and Australia by Cominco, a major metal exploration company.

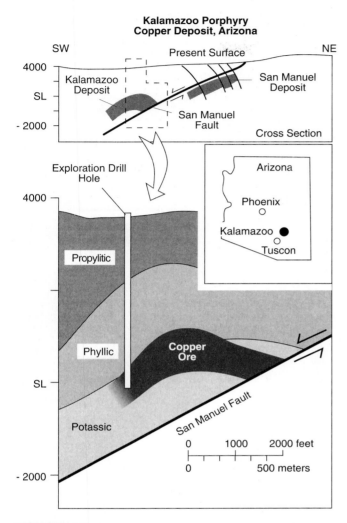

FIGURE 4-6

Kalamazoo-San Manuel area, Arizona, showing how hydrothermal alteration zoning around the Kalamazoo deposit was used to guide exploration drilling; SL=sea level (compiled from Lowell, 1968, *Economic Geology*, 1968, v. 63, p. 646 and Chaffee, 1982).

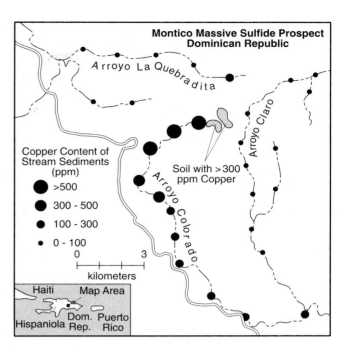

FIGURE 4-7

Copper content of stream sediments and soils around Montico deposit, Dominican Republic. Soil containing more than 300 ppm Cu covers a large area over the deposit and stream sediments draining this area contain over 500 ppm Cu near the deposit and as much as 250 ppm many kilometers downstream.

zinc contents of arroyo sediment around Montico are much higher than normal. In fact, the main arroyo that drains Montico is called Arroyo Colorado, reflecting the reddish color of iron oxides that weathered from the deposit. Names like this can be helpful exploration guides. The famous Rio Tinto, or colored river, drains the large copper deposits near Huelva in southern Spain.

Geochemical exploration surveys also use other natural features, such as plants, whose roots reach deep into the underlying regolith, as well as lakes and swamps (Figure 4-8). In areas where glaciers have covered the bedrock with sand and gravel, these can be sampled for traces of ore scraped from the underly-

ing rock. Surveys are also based on gases, such as methane and ethane, that leak upward from oil and gas fields (Figure 4-8). Radioactive decay of uranium produces helium and radon that can percolate upward through pores in the soil, and the weathering of metal sulfides forms mercury and sulfur gases that can be used in surveys.

Geochemical exploration is living proof that nature is far from pristine. In fact, the natural landscape varies greatly in chemical composition and contains many natural geochemical dispersion patterns that would be classified as highly polluted if they were of anthropogenic origin. Regional geochemical surveys help locate these areas, as well as stimulating exploration and aiding strategic planning. They have been used by the United Nations to encourage mineral exploration in LDCs (Guy-Bray, 1989). MDCs have used geochemical surveys, such as the U.S. National Uranium Resource Evaluation (NURE) program, to locate areas favorable for uranium, and similar studies have long been used to assess areas proposed for wilderness status.

FIGURE 4-8
Geochemical dispersion around mineral deposits. **A.** Zinc content of lake sediments around Daniel's Harbour zinc deposits in northwestern Newfoundland (from Davenport et al., 1975). **B.** Methane and ethane contents of soil gas over the Albion-Scipio oil field, Michigan. Soil gas contents are thought to be highest on edges of field because rocks immediately above field are less permeable (after Burns, 1986).

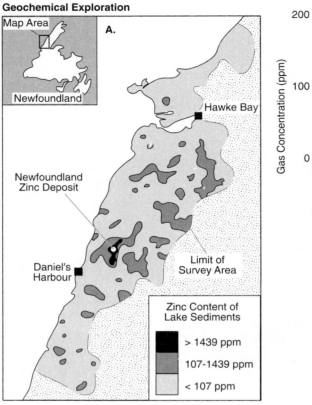

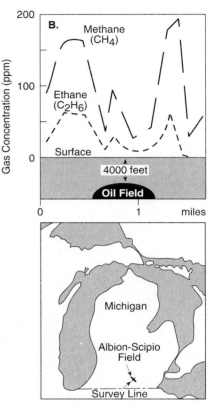

Geophysical Exploration

Geophysical exploration involves the search for deposits by measuring physical properties of rocks, such as magnetic intensity, electrical conductivity, radioactivity, and the speed of shock (seismic) waves passing through them (Dobrin, 1976; Telford et al., 1976). Some of these measurements can actually detect the presence of the element or mineral that is being sought, although most reflect only the general nature of buried rock. Metal and oil exploration programs usually use different geophysical exploration methods because the deposits have different properties and host rock characteristics.

Geophysical methods in metal exploration emphasize magnetic, electrical, and radioactive methods. Under favorable circumstances, measurements made at the surface can detect metal ore bodies at depths of a hundred meters. Thus, they can be used to "see through" younger sediments that cover ore-bearing rocks, a capability that is particularly useful in large areas of Canada and northern Europe where Pleistocene glacial sand and gravel deposits cover most of the bedrock that contains deposits (Plate 2-3). Many of these geophysical measurements can be made

from the air, thus increasing the speed of surveys and cutting down on their environmental impact (Figure 4-9).

Whether a mineral or rock will be detected by geophysical measurements depends on its composition, crystal structure, density, and electrical conductivity. Some minerals, such as magnetite and pyrrhotite, are *magnetic*. Where they are present in sufficient abundance, they distort Earth's natural magnetic field, producing sizeable magnetic anomalies (Figure 4-10). Many metallic minerals conduct electricity better than the surrounding rocks (Hohmann and Ward, 1981), a phenomenon that can be used to carry out a wide array of *electrical* and *electromagnetic* measurements (Plate 2-4). Surveys of this type have been a critical element in exploration of the Canadian Shield and adjacent parts of Minnesota, Wisconsin, and Michigan, where glacial deposits obscure most underlying rock (Figure 4-10). *Radioactivity* can be used to detect uranium, thorium, and potassium, all of which have naturally radioactive isotopes. The original measurements were made with a *geiger counter*, which detected alpha radiation, but modern *scintillometers* and *gamma-ray spectrometers* measure gamma radiation. *Gravity* methods, which measure the

FIGURE 4-9
Geophysical exploration. The DIGHEM helicopter-borne geophysical system shown here contains high-resolution navigational systems, radar, and altimeters, as well as gamma-ray spectrometers to measure radioactivity (courtesy of John Buckle, Dighem Surveys and Processing).

strength of Earth's gravitational field, can reflect differences in the type of rock present in large regions, but are not commonly sensitive enough to be used directly to locate mineral deposits.

Exploration for oil and gas is done largely by *seismic methods*, in which shock waves from an explosion or a vibrating source are sent into the ground from a land-based unit or a ship (Figure 4-11). By determining how long it takes these shock waves to reach detectors around the area, it is possible to reconstruct an image of the layers that are present in the subsurface and to delineate traps that might contain oil or gas. Older seismic surveys produced cross sections show-

ing subsurface geologic relations in two dimensions, but modern methods yield a three-dimensional image of the underlying rock that has significantly improved exploration success. Geophysical measurements can also be made down the length of a drill hole. This method, which is known as *well-logging* (Glenn and Hohmann, 1981), is widely used in oil and gas exploration and, along with seismic surveys, is expected to account for worldwide expenditures of over $6 billion annually during the mid-1990s.

DEPOSIT EVALUATION

Drilling

Evaluation of potentially attractive deposits must be carried out by *drilling,* which is by far the most expensive part of the exploration process (Kennedy, 1983). Drill holes used to explore metal deposits cost up to $200 per meter of hole, and a single deposit can require thousands of meters of drilling. Resulting costs, exclusive of administration and other support, range up to $5 million per deposit. Costs for individual holes are much higher in the oil and gas business, averaging about $400,000. Exploration wells in frontier areas cost as much as $50 million. Keep in mind that these costs apply to both successful and unsuccessful exploration drilling programs. Over 70% of all oil and gas holes in the United States are dry and must be paid for by revenue from the other holes.

Drilling is carried out with a *drill rig,* which consists of a superstructure with pumps and a motor that rotates a "string" of steel *drill pipe* into the ground (Figure 4-12A; see pg. 66). The pipe commonly comes in segments that are screwed together, and on their bottom end is a *drill bit,* impregnated with diamond chips that are hard enough to cut through any type of rock. Rock fragments cut by the drill bit are washed to the surface by air, water, or a suspension of minerals and chemicals known as *drilling mud,* which is pumped down through the hollow center of the pipe. Drill bits can be designed to cut the rock into tiny pieces, called *cuttings,* or to cut a cylindrical sample of rock, called a *core* (Plate 2-5), which provides more detailed information on the nature of the buried rocks. The most famous type of drill bit, consisting of three cone-shaped cutting heads and known as the Tricone, was first marketed by the Hughes Tool Company and formed the basis for the fortune accumulated by Howard Hughes. The purpose of drilling is to obtain samples of buried rock. If only a partial sample is re-

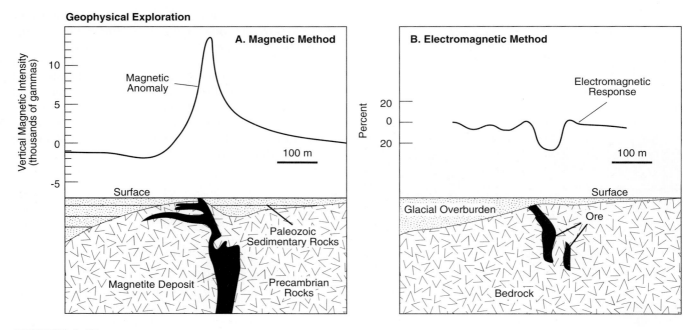

Geophysical Exploration

FIGURE 4-10

Geophysical anomalies associated with mineral deposits. **A.** Magnetic anomaly over iron ore body at Iron Mountain, Missouri. The clarity of magnetic anomalies depends on the orientation of Earth's magnetic field and shape of the ore body and not all anomalies are as obvious as this one (from Leney, 1964). **B.** Electromagnetic anomaly (inphase MaxMin-II horizontal loop at 888 Hz) over Montcalm township nickel discovery, Ontario (from Fraser, 1978 with permission from the Canadian Institute of Mining, Metallurgy and Petroleum).

covered, information is incomplete and important deposits can be missed.

Drilling technology has come a long way. The most impressive improvements are directed drilling and horizontal drilling, both of which offer great promise for diminishing the impact of drilling on the environment. *Directed drilling* involves the use of special wedges and motors to make the drill hole go in a desired direction. Contrary to popular notion, drill holes do not go straight down, even if they started as vertical holes at the surface. Instead, their path is influenced by heterogeneities in the rock and by spinning of the drill rods, causing some holes to wander long distances from their intended targets. Directed drilling, which was developed to prevent this, is now used to direct holes toward locations that are hard to reach from above. This method can be used to drill a large number of holes into many parts of an oil field from a single platform. *Horizontal drilling* is a variation on this procedure in which a vertical hole is turned so that it proceeds horizontally for a thousand meters or more.

Offshore exploration has been one of the most active areas of innovation in oil drilling. It started in 1896,

when rigs were placed on piers to drill the Summerland field in southern California. In 1938, true offshore drilling began with discovery of the Creole field in 4.5 meters of water about 2.4 kilometers off the Louisiana coast. Since then, exploration has reached distances of several hundred kilometers from shore and water depths of 2,400 meters. Drill rigs with jack-up legs can work in water as deep as 130 meters, but special drilling ships or floating platforms are needed for greater depths (Figure 4-12B). During deep-water exploration, the ship must be kept on location above the hole by constant positioning adjustments. Floating ice complicates offshore exploration and production in arctic areas (MMS, 1986). In water up to 15 meters deep along the northern Alaska coast, drilling has been done from islands that are constructed during the summer. In somewhat deeper water, it can be carried out from conical structures that cause encroaching pack ice to bend and fail or from special platforms (Plate 5-1). In really deep water, such as the east coast of Labrador, exploration holes are drilled from ships that must be moved out of the way when the attending army of tugs fails to deflect an approaching iceberg (Munro, 1982).

Geophysical Exploration

Seismic Section

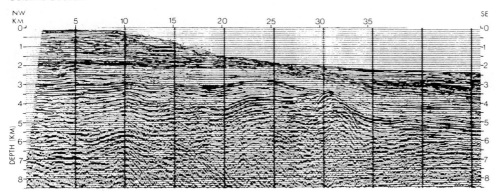

Geologic Interpretation

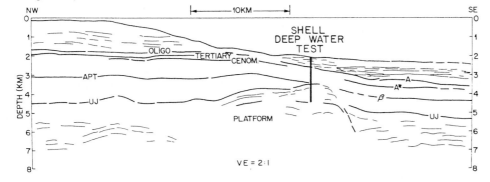

FIGURE 4-11

A. Marine seismic survey ship, which uses collapsing air bubbles to send shock waves (that are harmless to marine life) through the water and into the underlying rocks. Returning waves are detected by measuring devices known as hydrophones, which are towed behind the ship (courtesy of Jack Bull, Western Geophysical). **B.** Seismic survey across the Baltimore Canyon offshore from the eastern United States showing the edge of the carbonate platform that was the target of exploration drilling for oil (from Klitgord and Watkins, 1984).

A.   **B.**

FIGURE 4-12

Drill rigs. **A.** Small "back-pack" drill rig in use on a Mexican hillside. Drill rods are
rotated down the hole cutting a core of rock, while drilling mud is pumped in from the
top to wash away rock cuttings. **B.** Sedco/BP 471 drill ship on location. It can drill in
water depths of up to 8 kilometers and has been used for both oil and gas exploration
and scientific drilling of the ocean crust (courtesy of Ocean Drilling Program, Texas
A&M University).

Reserve Estimations and Feasibility Analyses

If drilling indicates that a deposit is present, it is necessary to undertake a more formal *reserve estimate* using the drill hole data (Cummings and Given, 1973;
Royal Dutch/Shell, 1983). In the simplest approach,
holes are drilled on a regular grid. Where surface features prevent this, they are distributed randomly. The
depth and grade of ore or oil and gas are determined
in each hole and a judgment is made about its continuity from hole to hole. If continuity is assured, the
amount of ore can be calculated. The grade of the ore
is determined by calculating an average of the grades
of ore in all drill holes, giving each hole a weighting
consistent with the volume of ore that it represents.

The uncertainty of these measurements can be minimized by drilling holes closer together, but this is expensive and the most successful drilling programs
prove the most ore for the smallest expenditures. So,
the challenge in reserve estimation is to obtain an estimate that is highly accurate and precise at a reasonable cost. *Geostatistics,* which started in the gold mines
of South Africa, has been most successful in achieving this goal (David, 1977). The approach works
because the grade in two parts of an ore body are
related to each other, with the relation decreasing in
significance as the distance separating the two points
increases. Thus, the grade of one hole can predict the
grade in another. By determining the distance at which
this relation breaks down, it is possible to specify the
minimum spacing of drill holes needed to get a reliable estimate of the mineral reserve in a deposit. This
relation also allows bankers and environmental regulators to judge whether a deposit or area have been
adequately evaluated. When the reserve estimation is

complete, a deposit can be divided into *proven, probable,* and *possible reserves.* These terms are commonly used for ore in a single deposit, whereas the generally similar terms, *measured, indicated,* and *inferred reserves,* are used to discuss estimates for a larger region containing many deposits.

Reserve estimates form the basis for a *feasibility analysis* to determine whether the mineral deposit is an ore deposit (Wellmer, 1989; Harris, 1990). First, engineers determine the rate and cost of extraction, as well as costs for processing, transportation, and administration. Estimates must also be made of costs related to environmental monitoring and reclamation, including the amount and nature of any bonds that are required, and of taxes and royalties and other applicable charges. Finally, it is necessary to estimate future prices for the commodity of interest. Once these estimates are on hand, the cost of extracting the resource and selling it can be compared to estimates of the future prices of the commodity to determine the potential profitability of the operation. If estimated costs are significantly less than the estimated value of future production, the deposit will probably be put into production, a process known as *development.* This involves the construction of drilling and production platforms and pipelines for oil and gas or a beneficiation plant and tailings disposal areas for hard minerals, as discussed later. If the deposit does not look economically attractive at this stage, it will be abandoned or held for possible later reconsideration. Many projects are stalled at this stage, indicating how difficult it is to determine the economic viability of a deposit early in an exploration program.

Most mineral deposits require outside *financing* to cover the cost of development. Although funds are sometimes raised by public stock offerings, most financing is accomplished by bank loans. Some banks and investment houses employ or retain mineral advisors to review the technical quality of data supporting loan requests. During the 1970s, predictions of future mineral prices were unrealistically high, even among bankers who are naturally conservative. Loans made on the basis of these overly optimistic predictions led to several bank failures and reorganizations, including Penn Square in Oklahoma and Continental Illinois in Chicago.

Environmental Effects of Mineral Exploration

Geological, geochemical, and geophysical exploration has limited impact on the environment, usually confined to access roads, survey lines, and sample trenches put in during the evaluation of anomalies, all of which require government permits. More damage is caused in wetlands where canals must be cut for access and boat traffic causes bank erosion, as has occurred along the 20,000 kilometers of canals in the Mississippi delta (Getschow and Petzinger, 1984). In the United States, dewatering of abandoned, flooded mines for further exploration requires a permit from the National Pollutant Discharge Elimination System, as well as relevant state agencies.

Drilling is of more concern environmentally because it requires larger-scale access. It also has the potential to introduce drilling fluids into the ground and release natural fluids, such as brines, natural gas, and H_2S, to the surface. These risks are greatest in oil and gas exploration, where reservoir characteristics are not known and pressurized fluids might be encountered unexpectedly. Fluid escape can be minimized by the use of drilling mud and blow-out preventers. *Drilling mud* consists of a suspension of fine-grained barite and bentonite clay in water. The bulk density of this slurry is high enough to wash rock cuttings made by the drill bit to the surface and to help prevent the escape of pressurized fluids. Thousands of gallons of drilling mud can be used at a single well, all of which is recycled, unless it is lost when the hole enters a cave system or other large void. Little evidence has been found that the small amount of mud that leaks to the surface causes important damage, a result consistent with the fact that barite is chemically inert under oxidizing, surface conditions (Miller et al., 1980; Miller and Pesaran, 1980). Drill rigs used in oil and gas exploration are required to have a *blow-out preventer* (Plate 6-5) mounted below the drill platform to prevent the escape of highly pressurized oil, gas, or water that cannot be stopped by drilling mud. In the event of a sudden increase in pressure, this valve shuts the hole automatically. If this fails, oil or brine can be caught in large dikes that surround wells on land, although offshore wells cannot be protected in this way. Where pressurized oil or gas reaches the surface, it forms a blowout or *gusher.* In the worst case, the oil and gas are ignited by heat or sparks from collapsing drilling equipment to become a tower of fire. Once the fire is extinguished, some blowouts can be stopped by capping the well, whereas others require the injection of mud and cement from new wells drilled to intersect the reservoir at depth.

MINERAL EXTRACTION METHODS

Once a mineral deposit has been found and judged to be economically attractive, it must be extracted if we are to benefit from it in any way. This can be done by mining in which the rock is physically removed from the ground or by the use of well systems that remove only a fluid, leaving the rock behind. Extraction methods of this type are by their very nature damaging to the environment, although damage can be minimized, as discussed later.

Mining

Mining involves removal of rock from the ground (Hartman, 1987). In general, deposits within a hundred meters of the surface are extracted from open-pit mines and those at greater depth come from underground mines (Figure 4-13). Deposits with intimate mixtures of ore and worthless rock are usually mined by *bulk* extraction methods, whereas rich zones of ore are often mined by more *selective* methods. Open-pit mining is almost always less selective than underground methods. In general, the cost and difficulty of mining increase with depth and extremely deep ore deposits are not mineable at a profit by any method.

Curiously, open pits account for over 90% of the ore mined in the United States but only about 50% of the ore mined in the rest of the world. There is no geological reason for this, nor are there tougher environmental regulations in LDCs, which might favor underground mining. Instead, the controlling factor is economic and relates to average wage scales in LDCs. Because underground mining requires more worker-hours per tonne of ore produced, wages are a larger proportion of mining costs. Thus, in areas with lower wages, underground mining and other labor-intensive ore concentrating processes can be used on deposits that could not be mined in the United States and other MDCs (Figure 4-14).

A typical *open-pit mine* removes ore and *overburden,* the worthless rock that overlies ore (Plate 3-1). The ratio between the volumes of overburden and ore, known as the *stripping ratio,* rarely exceeds ten and commonly is less than five. The overall slope of the pit wall must be low enough to prevent failure by landslides and the pit walls have steps known as *benches* that prevent rock from falling to the bottom where people are working. As a result, pits are much wider at the top than at the bottom, and they include large amounts of barren rock near the surface. In most mines, the rock must be broken by blasting before it can be hauled out the top of the pit or through adits exiting at the bottom. One of the largest open-pit mines in the world, the Bingham mine just outside Salt Lake City, Utah, removes over 225,000 tonnes of rock each day, only 20% of which is ore (Figure 4-15).

Strip mining is a special open-pit method that is used on shallow, flat-lying ore bodies such as coal (Atwood, 1975). In spite of the unsavory connotation of its name, strip mining is actually one of the most environmentally acceptable forms of mining because it fills the pit and restores the land surface to its original form (Plates 3-2 and 3; Figure 7-5). *Contour mining* is a special form of strip mining that is applied almost exclusively to coal in areas where the overburden is too thick or the coal layer is too thin to pay for complete stripping (Plate 4-1). *Dredging,* also a special type of

FIGURE 4-13

Cross section showing depth of four actual metal deposits and the way in which they can be mined by open-pit or underground methods.

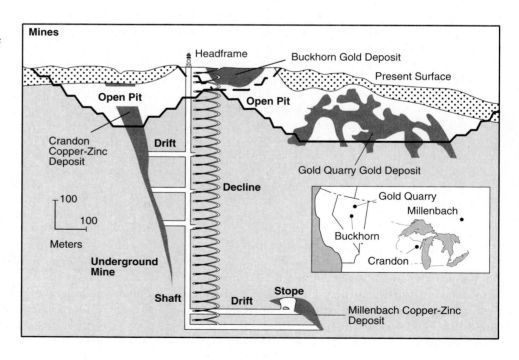

A.

B.

FIGURE 4-14

Mineral production in LDCs. **A.** Miners at San Augustin strontium mine, Coahuila, Mexico. In the early 1980s, when this photo was taken, they earned only a few dollars per day. **B.** First stage of concentrating tungsten-tin ore from the Xihuashan tungsten mine, China, involves hand separation of ore and waste minerals by about 200 people working in three shifts (courtesy of James E. Elliott, U.S. Geological Survey).

open-pit mining, is used where the inflow of water is too high to be pumped out economically and the ore is sufficiently unconsolidated to be dug without blasting (Figure 9-10). It is almost always used on placer deposits. The dredge digs its way forward, bringing along its own lake. Mined material is processed on board and gangue minerals are disposed of behind the advancing dredge. *Hydraulic mining* is a cross between dredging and conventional mining in which a jet of

water is used to disaggregate the rock and wash it into a processing facility.

A typical *underground mine* is entered by a vertical *shaft* or a horizontal *adit* (Hartman, 1987). Tunnels or *drifts* enter the deposit at different levels and ore is extracted from holes called *stopes*. It is taken to the surface in trains or conveyors, or it is lifted to the surface by a large bucket in the shaft called a *skip*. Older underground mines used small equipment on

A.

B.

FIGURE 4-15

Bingham copper mine in the Oquirrh Mountains 40 kilometers west of Salt Lake City has a surface area of 7 km². **A.** Mine in background sends ore to Copperton plant (right foreground) where ore is beneficiated and sent by pipeline to smelter 26 kilometers away. **B.** Electric shovel scoops 55 tonnes of ore into 210 tonne-capacity haulage truck (courtesy of Alexis C. Fernandez, Kennecott Corporation).

rails to move ore, but many modern mines have rubber-tired, diesel equipment that enters the mine via spiral ramps similar to those in parking garages. Most mines use a cyclic system of drilling and

blasting followed by digging out the broken ore and hauling it away. Mines in flat-lying ores such as coal and potash use *continuous mining,* in which a machine advances steadily, breaking up the ore and sending it by conveyor to a collection point. The specific mining method used depends heavily on the shape and geologic setting of the ore body. Tabular and flat-lying ores, such as coal and salt, are mined by *room and pillar* and *longwall* methods (Figure 7-5). Steeply dipping ore bodies are usually mined by highly selective methods such as *cut and fill* or by bulk mining methods such as *sublevel caving* or *block caving* (Figure 10-18). Geologic factors complicating underground mining include weak rock that caves into stopes and dilutes the ore, and groundwater that flows in too rapidly to be pumped away economically.

Depth limits for open-pit mining are caused by the ever increasing volume of overburden that must be mined to permit a cone-shaped pit to reach great depths, by difficulties of maintaining high pit walls, and by the high cost of hauling ore out of the pit. In practice, not many open-pit mines reach depths of more than 200 meters. The depth of underground mines is limited by the *geothermal gradient,* Earth's downward increase in temperature (Figure 4-16). Geothermal gradients are high in areas of active volcanism and relatively low in areas of old, Precambrian crust. For instance, the Limón gold mine along the Central American volcanic arc in western Nicaragua is only a few hundred meters deep but contains 75°C water. At depths of more than 1000 meters, some mines experience *rock bursts,* which are small explosions that take place as rocks in newly opened, underground workings decompress due to the release of confining pressure (Heunis, 1985). Shaft depths are also limited by the maximum length of cable that will not break from its own weight, usually 2,600 meters. Levels deeper than that must be accessed through internal shafts, which greatly increase the time required to get workers to their stations and ore out of the ground. These factors combine to limit underground mining to about 2,250 meters in most areas, although the deepest operating mine in the world, Western Deep Levels gold mine near Carletonville, South Africa, has reached a depth of 3,466 meters.

Pumping and Well Systems

Pumping and well systems are used to recover groundwater, crude oil, natural gas, and some solid minerals that can be dissolved or melted. Most well

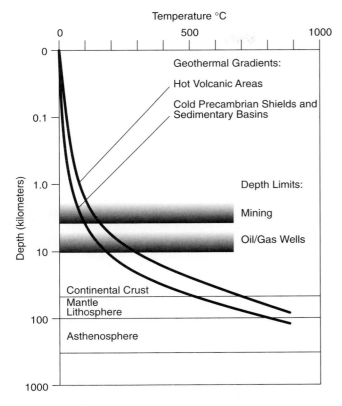

FIGURE 4-16

Geothermal gradient showing how Earth's temperature increases downward in hot volcanic areas and in colder rocks in Precambrian shield areas, as well as the lower limit of mines and oil or gas wells.

systems are devoted to oil and gas extraction and an entire discipline known as *reservoir engineering* involves the determination of the best distribution and type of wells for most efficient extraction (Koederitz et al., 1989). Reservoir rocks for oil and gas are, by definition, porous and permeable, but it is common to enhance their permeability in the area of producing wells by injecting sand or other granular material at high pressure to fracture the rock. *Drill stem tests* are used to determine how much oil and gas can be extracted from a well by flowing it through an opening of specified size. If these tests are successful, pumping and storage facilities are installed, a process known as *well completion*, and the exploration well becomes a producer.

Production from *onshore fields* is achieved simply by connection to a crude oil or natural gas pipeline. Wells in frontier areas must sometimes wait for years for pipelines to reach them. In the meantime, oil can be pumped into storage tanks and trucked or taken by tanker to a pipeline connection or refinery. *Offshore production* takes place from platforms and islands (Figure 4-17; Plate 5-1) that are among the engineering wonders of the world (Crooke and Otteman, 1984). Water depths in which these platforms operate have increased steadily over the last decade and are approaching 700 meters. Production from deeper water comes from subsea wellhead systems, although deeper water platforms are in various stages of development. Fields with more than one owner usually undergo *unitization* in which each owner agrees to a specific share of the production, as discussed further in the next chapter.

Solution mining, the process by which solid minerals are dissolved or melted and then pumped from a well system, is the most environmentally acceptable form of mining because no overlying rock has to be removed. Evaporite minerals, such as halite and sylvite, are uniquely suited to this process because they are easily dissolved in water. A variant of solution mining, known as the *Frasch process*, is used to mine native sulfur, which can be melted by superheated steam and flushed to the surface. Matters are not as simple for other minerals because they require more aggressive solvents. Some metal oxide and carbonate minerals can be dissolved by sulfuric acid, uranium oxide can be dissolved by acidic or alkaline solutions, and gold can be dissolved by cyanide solutions. For these solutions to be effective mining agents, they must react only with the desired mineral. For instance, sulfuric acid cannot be used to leach metals from a limestone because the acid will be consumed by reaction with the limestone first. An additional problem confronted by solution mining is channeling, in which the leach solution finds a few easy paths through the ore, never contacting the rest of it.

It is likely that mining will change greatly in the future, spurred both by the need to reduce labor costs and the desire to produce materials with less environmental damage. In spite of its problems, solution mining will almost certainly expand as better solvents are developed and ways are found to fracture rock more evenly. Microbes offer promise for selective decomposition of metals and organic materials, and *microbiological mining* will probably become more common (Brierly, 1982). In all mineral recovery, the emphasis will be on the conversion of the raw material to a marketable form as close to the deposit as possible. An early start on this will probably be made with *in situ combustion*, in which combustible ores are burned underground to produce a gas or liquid that can be pumped to the surface (Hammond and Baron, 1976). Where physical removal of ore must still be done, it will undoubtedly involve larger-scale, continuous mining equipment that will increase processing volumes but decrease unit costs.

Frontier Exploration and Production

As nearby deposits are exhausted, production must move into frontier areas. In fact, resource-based communities have been the traditional way that many frontier areas were settled. A few decades ago, some Canadians thought that a complex of mining and oil production communities would stretch across the Arctic. Although a few were established, notably Nanisivik and Pine Point in the Northwest Territories (Figure 5-8), most modern production facilities did not turn out that way. For instance, workers at the Lupin gold mine in the Northwest Territories live (Figure 5-8) in Edmonton and other cities far to the south, or in established northern communities and work "fly-in" shifts of several weeks duration followed by extended vacation time. This arrangement produces a happier, more stable work force, as well as eliminating the need for a large infrastructure at the mine, with attendant environmental problems. It also does away with *ghost towns*, the legacy of exhausted mineral deposits.

FIGURE 4-17

Offshore oil production facilities. **A.** The Monopod platform in Cook Inlet in southern Alaska was installed in 1966 at a cost of $21 million. It stands on a single leg that is less vulnerable to floating ice and 32 wells pass through the 8-meter diameter leg into the oil field (courtesy of M. T. Hogelund, Unocal Corporation). **B.** The artificial ice-proof island that produces from Endicott oil field near Prudhoe Bay on the north coast of Alaska was constructed in the mid-1980s for over $1 billion. The 45-acre (18-hectare) island has an unusually close spacing of wells designed to minimize its area and environmental impact (courtesy of M. Mogyordy, BP America).

A.

B.

Environmental Effects of Mineral Extraction

Mines occupy about 3700 km^2 in the United States (about 0.26% of the land area) compared to 2500 km^2 each for airports and railroads, 13,500 km^2 for highways, 100,000 km^2 for national parks and wildlife refuges, and over 385,000 km^2 for the wilderness system (Barney, 1980). Barney has also estimated that cumulative use of land by mining throughout the world between 1976 and 2000 will be about 37,000 km^2 or about 0.2% of Earth's land surface. Because the important construction materials stone, clay, and sand and gravel make up such a large proportion of mined material (Table 4-2), MDCs have more than their share of land disturbed by mining. About 60% of disturbed areas is used for excavation and the remaining 40% is used for the disposal of overburden and similar wastes, which account for about 40% of the solid wastes generated annually in the United States.

The most important statute regulating surface mining in the United States is the *Surface Mining Control and Reclamation Act of 1977 (SMCRA)*, which pertains largely to coal mines and is administered by the Office of Surface Mining and related state agencies. Mining wastes are regulated in the United States by the Environmental Protection Agency (EPA) and related state agencies under provisions of the *Clean Water Act, Clean Air Act, Comprehensive Environmental Response, Compensation and Liability Act,* and *Toxic Substance Control Act,* among others. A small amount of mining wastes is regulated under Subtitle C of the *Resource Conservation and Recovery Act of 1986 (RCRA)*, which addresses hazardous waste disposal, but most are deemed nonhazardous. Finally, the *National Environ-*

mental Policy Act requires that an *environmental impact statement* be prepared before any large mineral operation is started, and federal regulations require that any operation on land that will disturb more than five acres (12.35 hectares) must present and have approved plans for operation and reclamation, and deposit a bond to assure that reclamation will take place.

In the United Kingdom, environmental aspects of mineral production are regulated under the Environment Protection Act of 1990 (IMM, 1992). In Canada, environmental regulation is carried out largely by the provincial governments. The earliest and most comprehensive of these laws were Bill 56 in British Columbia and Bill 71 in Ontario (Champigny and Abbott, 1992).

Reclamation includes removal of buildings, restoration of the land surface to an acceptable contour, and alleviation of acid mine drainage caused by weathering of rock and unmined ore (Johnson and Paone, 1982; Carlson and Swisher, 1987). In the United States, SMCRA requires that strip mines be restored to their original contours, with no slopes greater than 20° (Plates 4-1 and 2). It is not usually possible to fill other types of mines because they proceed deeper and deeper in the same spot (Plate 3-1), although some operations that progress from one pit to another during mining fill older pits with overburden from later pits. Consideration is also being given to using abandoned open-pit mines for disposal of municipal waste. Preliminary tests of this are underway at the Mezquite gold mine in California and have been suggested for the Island Copper mine on Vancouver Island. Even where the pits are not filled, present regulations require that pit walls and overburden dumps be reshaped and revegetated in a manner consistent with local topography. Mine reclamation can be quite successful; Butchart Gardens, one of the main tourist attractions in Victoria, British Columbia, was once a rock quarry, as was Queen Elizabeth Park in Vancouver, and many homes and recreation areas surround flooded gravel pits that have become lakes. Most governments require that a reclamation plan be approved and a bond securing completion of the plan be paid before a mine can begin operation. Where an operator plans to abandon a mine, a bond is also often required to cover future environmental damage such as the failure of the waste confinement system, seepage of mine water into nearby streams, or collapse of the mine.

Although reclamation requirements are part of the law for active mines in most areas, many mines were abandoned before these laws were in place. These old mines themselves are safety hazards, but the real concern centers on mine wastes, which contain metals and

TABLE 4-2

Estimated land area used annually by various types of world mineral production in 1976, 1985, and 2000 (from Barney, 1980).

Mineral Commodity	1976	1985	2000
	(thousands of acres)		
Bituminous Coal	455	520	624
Sand and Gravel	345	525	865
Stone	231	246	444
Copper	101	165	279
Iron Ore	85	136	187
Clays	72	96	132
Phosphate	26	52	130
Uranium	5	21	39
All Other	90	132	210

other potentially toxic chemical compounds that are being dispersed into the surrounding surface and groundwaters. Estimates as high as $70 billion have been made for cleanup of these wastes, but they are essentially meaningless. Each situation has to be judged individually. Some will require complete isolation of wastes in new landfills underlain by impermeable clay barriers. Others, such as those in Park City, Utah, can be covered by layers of soil. The real question is who pays for the reclamation. Whereas considerable federal funding has gone into the cleanup of chemical wastes, mine wastes have received much less government funding. Current practice in some jurisdictions is to require owners of old mines to accept the responsibility for environmental cleanup, even if they did not cause the original problem. For example, prevention of acid drainage from old mines in the Leadville area of Colorado is estimated to cost $17 million, most of which will be paid by the current owners rather than the original miners. Although this approach is easy from the standpoint of enforcement, it discourages the well-established method of looking for elephants in elephant country (e.g., looking for big deposits near large mines) and it ignores society's collective responsibility for the original mess. A more equitable solution would be to reclaim abandoned properties with funds from severance taxes that are already in place, as is done with coal. Proposals have even been made to dedicate any royalty imposed by revision of the 1872 Mining Law to reclamation costs. A final consideration is the need to dispose of mine wastes in a way that allows the next generation to reach them. After all, today's waste is likely to be tomorrow's ore.

Health and safety considerations are also a major aspect of current mining regulations. Although fatalities do occur because of the collapse of rocks and the failure of equipment, underground *mine fires* are the most serious problem. Most fires begin in wood-supporting materials, but they often spread to flammable rock such as coal. Even metal sulfides will burn if the fire is hot enough. Fatalities from fires are usually caused by carbon monoxide (CO), a highly toxic gas. Since the Sunshine silver mine fire in Idaho, U.S. mines have been required to provide all personnel with a portable unit to convert CO to CO_2, although they are still not used in many other countries. Dust is actually of wider concern because it is the source of so many lung and systemic ailments (Table 4-3). The biggest problem is crystalline silica, including quartz, which causes the lung disease *silicosis*. Silicosis was not understood until well into the twentieth century, and many outbreaks were caused by poor dust control in silica-rich mines.

Present practice requires wet drilling and the collection of all dust from crushing and grinding. Dust and other particles are also of concern in uranium mines because they contain adsorbed radioactive isotopes that become lodged in the lung, where radiation damage can cause *lung cancer*. *Black lung disease* or *coal worker pneumoconiosis (CWP)* is a related disease observed in coal miners, as discussed in the section on coal.

The main environmental problems related to oil and gas extraction are the escape of underground fluids and land subsidence. Large-scale fluid escape from wells and local distribution systems is amazingly rare, amounting to only about 0.003% of the approximately 13 Bbbl of oil produced from United States' offshore leases between 1964 and 1991. Only one well of the 30,000 that have been drilled during that time has blown out. *Subsidence* of the land surface, which is a problem only over shallow oil and gas fields and groundwater aquifers, has been minimized by modern reservoir engineering designed to maintain efficient production. Even small amounts of subsidence can be damaging in low-lying areas, however. Major problems have been encountered in the Mississippi delta area where flood-control measures have channeled the river out to the Gulf, cutting off the supply of sediment just when it is needed most to counteract production-induced subsidence.

The most troublesome fluid that comes from oil and gas wells is *brine*. Although some brines contain high enough concentrations of sodium, potassium, bromine, or related elements to be of commercial interest, most are worthless and must be disposed of, usually by injection back into the sedimentary strata from

TABLE 4-3
Lung diseases (pneumoconiosis) associated with mineral dusts and fumes

Disease	Related Mineral Material
Anthracosis	carbon (soot)
Asbestosis	asbestos
Bauxite lung disease	bauxite fumes
Berylliosis	beryllium minerals and fumes
Cadmium pneumonitis	cadmium fumes
Coal worker's pneumoconiosis	coal
Lung cancer	asbestos
Mesothelioma	amphibole asbestos
Silicosis	crystalline silica
Silicosiderosis	hematite

which they came. The early practice of collecting the brine in unlined ponds called oil pits near producing wells is no longer permitted because it contaminates local groundwater. Instead, brines must be held in lined ponds or in tanks, and then reinjected into the ground. Brines have elicited additional concern because of their dissolved metals and organic compounds, an observation consistent with the fact that brines act as hydrothermal solutions. A few brines contain anomalous amounts of radium, a radioactive decay product of uranium and thorium that was probably leached from surrounding rocks. Radium is also radioactive and some old oil pits, pipes, and other facilities around such wells are more radioactive than uranium mines and nuclear power plants. Reinjection of brines has greatly minimized this problem (Schneider, 1990).

PROCESSING OF MINERAL RESOURCES

Processing of Metal Ores

Production of metals from ore is usually a two-step process. In most cobalt, copper, chromium, iron, nickel, lead, tin, and zinc ores, the first step involves production of a *concentrate* consisting largely of the ore mineral. This process is known as *beneficiation* and the gangue (waste) minerals that are rejected are called *tailings.* The concentrate is then decomposed chemically to separate the desired metal from the ore mineral, a process known as *smelting.* In most aluminum ores, as well as some uranium, vanadium, gold, and other ores, concentrates are not produced and metal is recovered directly from the ore (Cummings and Given, 1973).

Beneficiation, which involves crushing and grinding ore to separate grains of ore mineral from grains of gangue minerals (Figure 4-18A), is usually carried out at the mine site in order to avoid the cost of transporting worthless rock. Coarse-grained ore minerals can sometimes be separated by handpicking, but most ores are so fine-grained that they must be pulverized into a fine powder and separated by more complex methods based on the different physical properties of ore and gangue minerals. The simplest method recovers grains of magnetic minerals, such as magnetite and pyrrhotite, with an electromagnet. Density differences can also be used because most ore minerals are heavier than gangue minerals. In one such method, pulverized ore is immersed in a *heavy medium* consisting of fine

A.

B.

FIGURE 4-18
Ore-processing methods, Bingham mine, Utah. **A.** Crushing and grinding mill at Copperton pulverizes ore to the consistency of face powder before ore minerals are concentrated by flotation (worker at left center provides scale). **B.** Smelter on shore of Great Salt Lake produces over 155,000 tonnes of copper and captures 93% of its sulfur emissions through a duct system immediately below the 365-meter stack. This facility is now being replaced by a smelter that will trap 99.9% of emissions (courtesy of Alexis C. Fernandez, Kennecott Corporation).

particles of a high-density mineral suspended in water, similar to drilling mud. Mineral grains float or sink in this suspension depending on whether their density is more or less than the bulk density of the suspension. *Froth flotation,* the most widely applied beneficiation method, uses a hydrophobic organic liquid to coat specific minerals, which then attach to bubbles that are passed through the suspension, causing the grains to float. Unwanted minerals, which are not coated by the organic liquid, do not attach to bubbles and sink (Figure 4-19). An imaginative method, borrowed from the processing of rice to remove stones and black rice grains, uses a sensor to test grains falling off the end of a conveyor and a jet of air to blow away grains that do not meet predetermined criteria.

Separation of metal from the concentrate is usually done in smelters (Figure 4-18B) using *pyrometallurgy,* in which the melted concentrate separates into two immiscible liquids (Gilchrest, 1980). One liquid contains the metal and the other, which is called *slag,* contains the waste elements. The metal-bearing liquid is heavier than slag and sinks to the bottom of the mixture where it can be removed (Table 4-4; Figure 4-20). Smelting of iron ores in a *blast furnace* yields a metal liquid *(pig iron),* which is transferred to a second furnace for further processing to make *steel.* Conventional smelting of base metal sulfide concentrates produces a metal-sulfide liquid known as *matte,* which must be treated in a second furnace known as a *converter* to produce metal. This metal must be purified in a final step known as *refining,* in which traces of gold, silver, and minor metals are separated from the principal

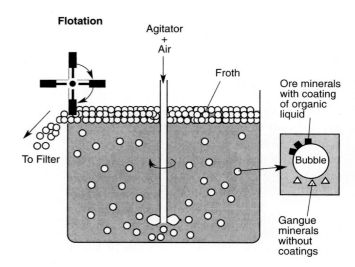

FIGURE 4-19

Cross section of a flotation cell showing how minerals adhere to bubbles, floating to the surface of the cell.

metal. Recently developed smelting methods use a single, continuous process to limit energy consumption and the escape of gases. These include the use of *direct reduction* for producing iron and *continuous smelting* for producing copper and other base metals, as discussed in those sections.

Metals can also be produced by *hydrometallurgy,* which relies on a caustic solution or solvent to leach metal from the ore or concentrate. The metal is recovered from solution by precipitation (usually as a com-

TABLE 4-4

Generalized chemical reactions that take place during smelting of iron and base metal ores.

Iron-smelting reactions

$$Fe_2O_3 \; (\text{hematite}) \; + \; 3CO \; (\text{carbon monoxide}) \; \rightarrow \; 2Fe \; (\text{pig iron}) \; + \; 3CO_2 \; (\text{carbon dioxide})$$

$$Al_2O_3 \; + \; SiO_2 \; (\text{gangue minerals}) \; + \; CaCO_3 \; (\text{limestone}) \; \rightarrow \; \text{Ca-Al-Si-silicate} \; (\text{slag}) \; + \; CO_2$$

Copper-smelting reactions

$$CuFeS_2 \; (\text{chalcopyrite}) \; + \; O_2 \; (\text{air}) \; \rightarrow \; \text{Cu-Fe-S} \; (\text{matte}) \; + \; SO_2 \; (\text{sulfur dioxide})$$

$$Al_2O_3 \; + \; SiO_2 \; (\text{gangue minerals}) \; + \; CaCO_3 \; (\text{limestone}) \; \rightarrow \; \text{Ca-Al-Si-silicate} \; (\text{slag})$$

$$\text{Cu-Fe-S} \; (\text{matte}) \; + \; O_2 \; (\text{air}) \; + \; SiO_2 \; (\text{silica}) \; \rightarrow \; \text{Cu} \; (\text{copper}) \; + \; \text{Fe-silicate} \; (\text{slag}) \; + \; SO_2 \; (\text{sulfur dioxide})$$

Metal Reduction Facilities

A. Blast Furnace

B. Smelter-Noranda Continuous System

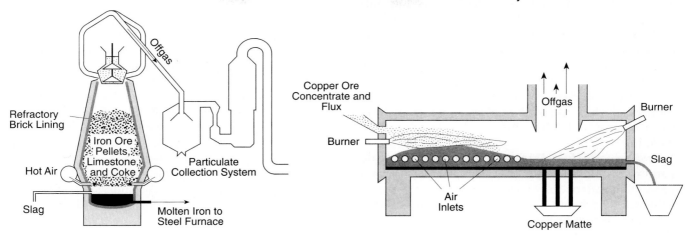

FIGURE 4-20

Metal ore reduction facilities. **A.** Blast furnace used to convert iron ore concentrates (usually pellets) to pig iron, which is then transferred to steel furnaces. **B.** Smelter used to convert metal sulfide concentrates to molten metal sulfide (matte), which is then transferred to converters to remove remaining sulfur as SO_2.

pound containing the metal) or by an electrolytic process known as *electrowinning* in which pure metal is precipitated onto the cathode of an electrochemical cell. This overall process is applied widely in many types of metal recovery systems, where it is often referred to as *solvent extraction-electrowinning (SX-EW)*. Some hydrometallurgical processes require large plants, but others can be carried out on piles or heaps of ore under the open sky. This process, which is known as *heap leaching*, requires less investment and is effective only on some gold, silver, copper, and uranium-vanadium ores. Although present hydrometallurgical processes are based on inorganic chemistry, future methods will probably utilize microbes.

Processing of Fossil Fuels

Fossil fuels consist of carbon and hydrogen with small amounts of sulfur, oxygen, and nitrogen. They form molecules that range from CH_4 with only five atoms, to others with thousands of atoms. Earth has separated these molecules into vapors (natural gas), liquids (crude oil), and solids (coal, asphalt, bitumen, oil shale, tar sands, gilsonite). *Natural gas* is made up of *methane* (CH_4), by far the most abundant component, along with small and variable amounts of other natural gases including *ethane* (C_2H_6), *propane* (C_3H_8), *butane* (C_4H_{10}), *hydrogen sulfide* (H_2S), *helium* (He), *carbon dioxide* (CO_2), and *nitrogen* (N_2). *Pentane* (C_5H_{12}) and

heavier molecules, which can be present as vapors in natural gas at depth in the crust, will condense at the surface and are commonly removed at the wellhead to form *lease condensate* or *natural gas liquids*.

Natural gas is processed to remove ethane, propane, butane, and related gases, which constitute *liquified petroleum gases*, as well as the nonhydrocarbon gases, which can be sufficiently abundant to constitute resources in their own right. Removal of essentially all H_2S is of particular concern because it is toxic and reacts with moisture in pipelines to create highly corrosive sulfuric acid.

Processing of crude oil is carried out in *refineries*, which separate the hydrocarbon molecules by molecular weight and modify them further to produce hundreds of products, including asphalt, fuel oil, gasoline, jet fuel, lubricating oils, naphthas, paraffins, petroleum coke, petroleum jelly, wax, and white spirit, as well as feedstock for petrochemical manufacture. Although individual refineries differ according to the type of oil they process and the types of product that they make, they usually require three steps (Figure 4-21). The first step is *distillation* in which much of the crude oil is vaporized by heating. The vapor is passed to an atmospheric pressure distillation tower where it cools, with lighter molecules, such as gasoline condensing high in the tower and heavier molecules, such as fuel oil, condensing at lower levels. Unvaporized heavy oil is passed to a vacuum distillation tower, where it too is vaporized and condensed into fractions. In the second step, *cracking*, heavy molecules from distillation are heated under pressure and broken down into smaller molecules that can be used in gasoline and other light hydrocarbon products. The third step, *reforming*, is a generally similar process in which the actual molecular structure of each product is changed to make it more acceptable to today's markets (Royal Dutch/Shell, 1983). The most important refinery product, *unleaded gasoline*, contains less than 0.05 grams of lead and 0.005 grams of phosphorus per gallon.

Processing of the solid fossil fuels varies widely depending on the desired product. The simplest form of processing for coal involves cleaning to remove noncombustible ash and sources of sulfur such as pyrite. Coal can also be heated and reacted with steam and other gases in a process known as *coal gasification*, which produces an impure natural gas that can be reformed into a wide range of products including gasoline. Other solid and semisolid hydrocarbons can also be treated to yield oil and gas, largely by heating to drive off the lower molecular weight fraction, which

can undergo further conventional refining (Hammond and Baron, 1976).

Processing of Industrial Mineral Ores

Most industrial minerals require less processing than metals or fossil fuels and most of it is confined to beneficiation. Some commodities, such as sand and gravel, require only washing and sizing. Others, such as diamonds, require complex separation techniques that rival those used on any metal ore. Among the resources that require actual decomposition of the original ore mineral are some potash and phosphate fertilizers, which are purified by dissolution and reprecipitation, and cement, lime, and plaster, which are the result of heating limestone, clay, and gypsum to create new mineral forms.

Environmental Effects of Mineral Processing

The most important environmental aspect of beneficiation relates to water availability and quality. In

FIGURE 4-21
Schematic diagram of an oil refinery showing the relation between distillation, cracking, and reforming.

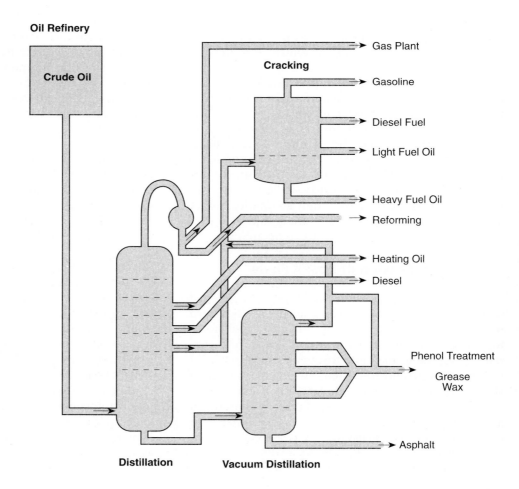

much of western Australia, almost all water is brought from the coast by pipelines and mineral operations have to wait for water or find ways to work without it, even though they are important sources of income in the region. In the western United States, water availability might limit oil shale processing and coal slurry pipelines. The main concern, however, is the quality of water leaving mineral processing facilities and its interaction with local ground and surface water. Almost all beneficiation plants at mines in MDCs must recycle their water, even in areas as remote as the Red Dog mine in northwestern Alaska. Surface water flowing through the property must meet local water quality standards when it leaves the property, even if it did not meet them when it came onto the property and was not used in the process. The escape of water into groundwater aquifers must also be limited, and leach pads, overburden dumps, and tailings piles containing cyanide and other caustic or toxic solutions must be shielded by impermeable barriers to prevent the downward movement of these waters. As discussed in the section on gold, cyanide solutions require special handling.

Although modern mining and beneficiation operations are very clean, older plants and wastes are major sources of environmental pollution. Butte, Montana, for example, was the largest copper producer in the United States from the 1880s until the 1940s and has produced about 450 million tonnes of copper ore. Prior to the installation of tailings ponds in the 1950s, about 100 million tonnes of tailings and smelter wastes were dumped into tributaries of the Clark Fork River, which runs 550 km into Lake Pend Oreille (Figure 4-22A). Stream sediment near Butte contains over 100 times more copper than average background values for this part of the world, and anomalous copper values are found all the way to Lake Pend Oreille (Axtmann and Luoma, 1991), considerably farther than most metal-rich dispersion trains from natural deposits. Metal from these sediments is now finding its way into aquatic plants and animals. Each rain or increase in the flow of the streams stirs up the sediment, making new metal available to the system. There is enough metal in the streams for this process to continue for hundreds of years. The only real solution is the complete removal of the contaminated sediment, a huge task that is being actively considered. In a strange twist of fate, this problem is sometimes less acute in LDCs because small miners set up operations along rivers to recover ore minerals from the sediment. In Chile, many such operations were set up along 100 kilometers of river draining tailings from copper deposits at El Salvador. Even so, beach deposits at Chañaral, where these tailings reached the sea, contain 30 million tonnes of sands grading 0.25% copper. Even where tailings and smelter wastes were disposed of

Pollution Related to Mineral Production

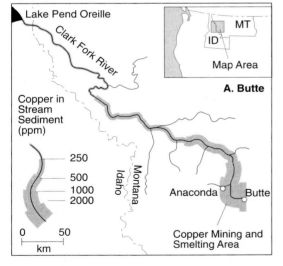
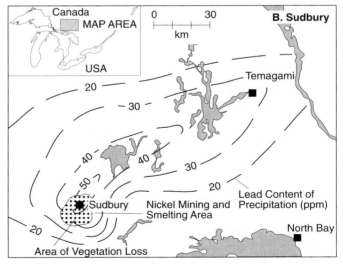

FIGURE 4-22

Pollution around old mineral operations. **A.** Copper content of stream sediment in Clark Fork River draining Butte, Montana (after Axtmann and Luoma, 1991). **B.** Lead content of precipitation around Sudbury, Ontario (from Semkin and Kramer, 1976). Asymmetric pattern reflects prevailing winds. Pollution of this type is not allowed at modern operations in MDCs.

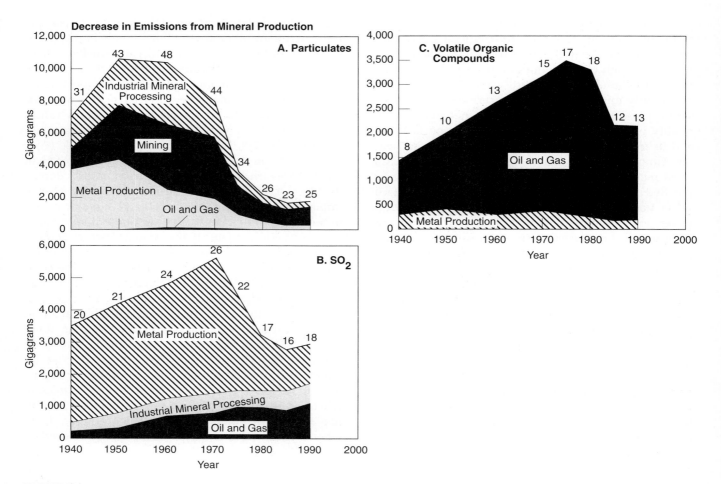

FIGURE 4-23

Decrease in air pollution caused by mineral production in the United States; numbers refer to the percentage of total anthropogenic emissions caused by mineral activities (from Environmental Protection Agency, EPA-450/4-91-004).

in piles or dumps, they remain as eyesores, especially because their metal content discouraged the return of natural vegetation. In the Old Lead Belt south of St. Louis and the Witwatersrand near Johannesburg, tailings piles are a source of windblown dust as well as metal-rich recharge waters that flow down into aquifers. Unlike metal-rich drainage pollution, however, many of these tailings piles are being removed by industry because they provide a ready source of construction sand, after suitable processing to remove metal-rich minerals. With the advent of better environmental regulations, mine wastes are rarely left in an unreclaimed state, although it will take many years to remove the pollution made before these regulations came into effect.

Smelting is of great environmental concern because it produces gases and dust. Early concern in Pittsburgh and other steelmaking cities focused on dust

emissions, a problem that was largely eliminated by compacting the ore mineral concentrate into pellets. SO_2 from metal sulfide smelters has been of concern because of its contribution to acid rain, as discussed in the previous chapter. The effects of smelter-induced acid precipitation are most obvious in humid areas because there is more vegetation to kill and because the soil is naturally more acid (in contrast to the alkaline soils of many arid areas). The Sudbury area of Ontario, the world's largest source of nickel, is a case in point. A century of smelting left Sudbury surrounded by 100 km^2 of barren land and another 360 km^2 of stunted birch-maple woodland (Winterhalder, 1988). Soil near the smelters had a pH of almost 3 and that of nearby lakes was as low as 4. Metals such as mercury, arsenic, and cadmium, which vaporize at the high temperatures used to smelt ore, were dispersed over an enormous area encircling the smelters (Figure 4-22B).

All this has changed, however. In response to the Clean Air Act, U.S. smelters now recover almost all of their SO_2 and other emissions (Figure 4-23). Although progress in Canada has not been as good, emissions at Sudbury have declined 50% since the mid-1970s and are scheduled to decrease another 50% by 1994. Some natural recovery has taken place around Sudbury in response to this decrease, but the dramatic improvement has been caused by seeding, tree planting, and lake liming, which is discussed in the section on lime and limestone. About 40% of the barren ground has been revegetated since 1978 and the pH of lakes in the area has risen an average of 0.5 units (Gunn and Keller, 1990). The decline in U.S. smelter emissions is particularly impressive when compared to the steady trend of coal-fired power plant emissions, the major anthropogenic source of SO_2 (Figure 9-14).

The effort to clean smelter emissions has centered on the SO_2 content of flue gas. Gases from reverberatory furnaces, the most common first step in most old copper smelters, have low SO_2 concentrations that were difficult to clean. Modern smelters, which have been designed to emit gases with higher SO_2 concentrations, have recovery rates exceeding 90%, and 1.4 million tonnes of SO_2 that used to escape into the environment are now recovered each year from smelters. Smelters under construction now will recover almost 99% of their sulfur emissions, effectively removing smelters from the list of anthropogenic sources of SO_2. Facilities that melt scrap metal for recycling, which are known as *secondary smelters* to distinguish them from *primary smelters* that process ore concentrates, are not sources of important sulfur emissions. Without proper exhaust gas cleaning, they can emit metals that vaporize at low temperatures, such as lead, however.

The most important environmental effects of fossil fuel processing are caused by the escape of hydrocarbons from refineries. Hydrocarbons can be lost as volatile organic compounds, including important greenhouse gases such as methane, or in wastewater, crude oil and synthetic organic liquids. Most environmental attention has been given to waste waters and refinery sludge, which contain organic compounds and heavy metals that can be toxic (Akhter et al., 1988). Escape of these liquids from early refineries has created contaminated groundwater plumes that vary in form depending on the degree to which the organic pollutant dissolves in water. Modern refineries now recycle wastewater and are isolated from groundwater aquifers by impermeable underlayers (Kapoor et al., 1992), but many older installations are still being cleaned (Chestnut, 1990).

CONCLUSIONS

Estimates of environmental impact are made by many regulatory agencies and they provide useful mileposts to assess the performance of the mineral industry. Recent data for atmospheric pollution in the United States, show that mineral production accounts for about 30% of anthropogenic lead emissions, 25% of particulate emissions, 18% of SO_x emissions, 13% of volatile organic compound (VOC) emissions, 3% of CO emissions, and 2% of NO_x emissions. For the three pollutants to which mineral production makes a major contribution, mining is the most important particulate source, smelting the most important SO_x source, and crude oil and natural gas processing the most important NO_x source. As can be seen in Figure 4-23, the proportion of these emissions generated by mineral resource extraction has declined significantly during the last few decades. Decreases were first seen in particulate emissions, then in SO_x and more recently in NO_x. When viewed in the context of the effort that all industries have made to clean up their emissions, these data indicate that cleanup within the mineral industry has been more successful than in other segments of industry.

It is particularly important for the average citizen to get this news. In the same way that small investors learn about a rising stock market just in time to buy at the top, average citizens often get the news just as the problem is being solved. We are only now becoming fully aware of the incredible messes that were made during mining and oil and gas production in the past. We are also just becoming aware of the tremendous cost that we must bear to clean up these polluted areas. As enormously important as these problems are, we must not let them cause us to lose site of the fact that modern mineral extraction is much cleaner. This does not mean that things are perfect, that mistakes will not happen, or that emissions will not reach unacceptable levels locally. However, with continued vigilance on the part of operating personnel, regulatory agencies, and the general public, we can hope to extract Earth's remaining resources in a manner more consistent with our obligations to preserve the planet.

5

Mineral Law and Land Access

LAND IS THE CURRENCY OF MINERAL EXPLORATION AND PRODUC-
tion. Without land there is nowhere to look and nothing to produce. This simple constraint is frequently
forgotten in the growing concern over mineral supplies, and it is one of the most pressing aspects of any
country's mineral policy. At the bottom line of the controversy is the need for a balanced, multiple-use approach to land, with the recognition that we must explore for mineral deposits and they must be extracted
where they are found. In this chapter, we review forms
of land tenure and mineral law, as well as the policy
that controls the implementation of these laws. Although the focus of this discussion is U.S. federal law,
the generalizations apply equally to mineral access in
other jurisdictions, for which some specific information is also given.

TYPES OF LAND AND MINERAL OWNERSHIP

The simplest form of landownership involves control
of the entire parcel, including the surface and everything on or below it. It is possible, however, to divide
landownership into separate parts, which are referred
to as rights. Common divisions include *surface rights,*
water rights, timber rights, and *mineral rights,* which
constitute ownership of these parts of the land. These
rights reflect the variety of interests that people and
businesses take in land, as well as their willingness to
permit multiple use of the same parcel of land, an aspect of land tenure that will undoubtedly become
more common as population pressures increase (Figure 5-1).

The possibility of divided landownership might
come as a surprise. Some landowners have found that
they do not control the mineral rights beneath their
property. Split ownership of this type is common in
older mineral-producing regions such as the northern
Appalachian coal fields where farsighted companies
bought mineral rights early in the 1900s but did not
mine the ground until much later. The reverse situation, in which several owners claim a single part of
the land, can also occur. For instance, who owns methane that contaminates some coal mines? This type of
methane is one of several unconventional sources of
natural gas and some farsighted companies have attempted to acquire rights to extract it. But, is it owned
by the owner of the coal, by the property owner or by
some third party? Similarly, what is the basis for separating mineral and surface rights? Most jurisdictions
consider surface rights to stop below the immediate
land surface. However, Texas courts have allowed the
owner of surface rights to claim minerals down to a
depth of 200 feet if their removal would result in
destruction of the present surface (Aston, 1993). This
curious twist of logic reflects the long history of oil
production in Texas, which does not usually result in
significant disturbance of the surface.

Laws governing mineral ownership take two approaches (Ely, 1964). One is based on the concept

FIGURE 5-1

Multiple land use in the Black Hills, South Dakota. The Terry Peak ski area looks down on three gold mines that have produced more than 60 million ounces of gold and are a major factor in the local economy. At the base of the hill is the Golden Reward open-pit mine, in the far right distance is the Gilt Edge open-pit mine, and barely visible in the far left distance is the headframe for the giant Homestake underground mine (courtesy of Paul A. Bailly, MinVen Gold Corporation).

that minerals belong to the crown or government, whereas the other holds that private citizens and corporations can own minerals. These contrasting approaches to mineral ownership evolved into the two major types of mineral law that prevail in the world today. *English mining law,* which held that minerals were owned by the owner of the surface, became the basis for mineral law in the United States and Canada. In contrast, the early German and *regalian legal systems* held that minerals were owned by the state, regardless of surface landownership. Because Spain operated under this system at the time of exploration of the New World, most of the rest of the Western Hemisphere adopted variations of this system (Prieto, 1973).

The rights of overlapping governments, such as states and federal governments, to ownership of public land has been handled in different ways. In Canada, public lands and mineral rights were given to the provinces at the time of confederation. In the United States, in contrast, public lands and their mineral rights reverted to federal control as states joined the union. The only important exception made by Congress was Texas, which was permitted to retain title to its public lands when it was annexed in 1845. This moment of weakness cost the federal Treasury millions of dollars in oil and gas royalty payments that went to the state and private landowners.

Ownership of *water and water rights* is a separate issue in many countries, particularly arid lands (Walston, 1986; Williams et al., 1990). U.S. water law

reflects this experience during the settlement of the West. Most western states apply the doctrine of *appropriation rights,* which grants access to water on the basis of the time it was first used and the purpose for which it was used. Other states apply the doctrine of *riparian rights,* which holds that the right to use water comes with ownership of the land and is maintained even if the water is not used. Groundwater law developed separately from surface water law because of the mistaken impression that the two water reservoirs were not related. Although most states accept that groundwater ownership accompanies landownership, at least ten states have declared that they own the groundwater. This claim is usually manifested in agencies that control groundwater use and does not mean that groundwater is reserved for state use.

The difference between the English and regalian legal systems has had a big impact on economic development in the New World. Most of the early personal fortunes in the United States, for instance, were based on the extraction and processing of natural resources, especially minerals. Included in these were the Rockefellers, Mellons, Carnegies, and Guggenheims, all of whom created important corporations that continue in one form or another today. These families made major philanthropic donations to cultural and historic institutions that have enriched American life. In countries with government ownership of minerals, it was more difficult to develop such fortunes and related philanthropic activity has been more limited.

LANDOWNERSHIP AND LAW
IN THE UNITED STATES

In most countries, the government is responsible for surveying the land and providing a regional location system on which ownership is based. In the United States, locations are based on three survey systems (Figure 5-2). Land in the original U.S. colonies and some of the older states was divided into irregular tracts based on the English system of metes and bounds (Muhn and Stuart, 1988; PLS, 1988). Divisions between parcels were indicated by landmarks such as trees and fences, many of which have long since disappeared. Land in the southern and southwestern states, which was originally under Spanish influence, was divided even less systematically. Present land boundaries in many of these states have been relocated by modern surveys but persist by tradition and agreement in a few areas. Land west of Pennsylvania was surveyed into 1-mi^2 *sections* (640 acres; 2.59 km^2) under the provisions of the *Land Ordinance Act of 1875*. These sections were part of a N-S, E-W grid of townships and ranges into which almost all of the central and western United States was divided. The resulting system includes *townships* that measure 6 miles on a side and contain 36 sections.

Control over the 9.3 million km^2 of land that makes up the United States is about equally divided between the federal government, state and local governments, and private groups and individuals. Earlier in U.S. history, the federal government controlled another 4.6 million km^2 that have since been transferred to state and local governments, private groups, and individuals. The *Homestead Act of 1862*, the *Desert Land Act of 1877*, and related laws, which encouraged settling of the West, accounted for 1.16 million km^2 of these transfers. Larger blocks of land, totaling 1.33 million km^2, were granted to the states for construction of schools, railroads, canals, and other improvements. Between 1850 and 1870, another 562,000 km^2 were transferred to companies and interested states as incentives to build railroads into the West. These grants commonly took the form of odd-numbered sections of land in a strip 20 miles (32 kilometers) wide centered on the rail line. Known mineral lands were specifically excluded from these grants, but many deposits have since been discovered on them.

In spite of all of these transfers, the federal government controls and manages most of the land in the western states, ranging from highs of 87% in Alaska and 85% in Nevada to a low of 29% in Washington (Table 5-1). This land includes national parks, monuments, and other scenic and recreational areas, as well as wilderness areas, where mineral exploration is largely prohibited. It also includes vast areas of national forest, grasslands, and other land, where min-

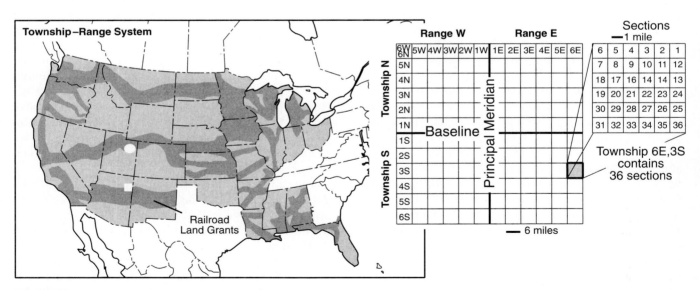

FIGURE 5-2

Areas of the United States surveyed by the township-range system; inset shows the relation between townships, ranges, and sections. Note that the term *township* refers to both the east-west tiers of divisions and the divisions themselves. Also shown are areas covered by railroad land grants, which consisted of alternate sections that created a "checkerboard" landownership pattern that persists today.

eral entry is permitted and which supply the bulk of domestic mineral production.

U.S. Federal Land Laws

Access to U.S. federal land for mineral exploration and production is governed by a host of laws and regulations (Table 5-2). Of most importance to exploration are the laws that deal with the discovery-claim and leasing systems, as discussed next.

Discovery-Claim System

U.S. federal mineral law is based on the experience of prospectors, such as those in the California gold rush of 1849. In the system that they developed, the discoverer of a deposit had the right to claim ownership in it. This concept was put into federal law by the *General Mining Law of 1872,* which is a combination and amplification of the earlier *Lode Mining Law of 1866* and *Placer Mining Law of 1870* (Ely, 1964). These laws divided mineral deposits into two geological types;

those in actual bedrock were called *lodes* and those in river and stream gravels were called *placers.* The 1872 law allows any citizen or domestic corporation to stake claims and no limit is set on the number of claims. Claims must measure 600 by 1,500 ft (183 by 457 meters) for lode deposits and 20 acres (49.4 hectares) for placer deposits. Their location must be marked on the ground and they must be registered with the government through the Bureau of Land Management, which is discussed later.

According to the original law, claims could be held in perpetuity by doing $100 worth of exploration, known as *assessment work,* annually on each claim, a stipulation that was changed by the *1993 Appropriations Act* to an annual rental of $100 per claim. The law also permitted the holder of a claim to purchase title to its surface and minerals rights from the federal government, a process known as *patenting.* Original patenting requirements included the completion of $500 worth of improvements on the claim, public notification of intent, and payment of $2.50/acre for a placer claim and $5/acre for a lode claim. Early abuses of the patent option to obtain land for other purposes,

TABLE 5-1

States with important federal landownership, showing value of nonfuel mineral production. Top 20 states listed here account for 90% of federal land holdings and 67% of U.S. nonfuel mineral production. (From *Public Land Statistics,* 1988, v. 173, Department of the Interior, Bureau of Land Management and U.S. Bureau of Mines, State Mineral Summaries, 1989).

State	Federally Owned Land Total Area (km²)	Percent of Area	Nonfuel Mineral Production (1989, $ million)
Alaska	1,288,229	87.1	174
Nevada	240,986	85.1	1867
California	188,038	46.4	2851
Idaho	136,444	63.7	339
Utah	135,710	63.6	990
Arizona	126,681	43.1	2829
Wyoming	124,957	49.5	755
Oregon	121,288	48.7	169
Colorado	97,308	36.2	375
New Mexico	94,461	31.3	1007
Washington	50,503	29.2	387
Florida	17,244	12.4	1524
Michigan	14,277	9.7	1549
Minnesota	14,001	6.8	1391
Arkansas	13,453	9.8	321
South Dakota	11,055	5.6	298
Virginia	9,978	9.6	473
North Carolina	8,942	7.0	523
Missouri	8,371	4.7	969
Georgia	8,213	5.4	1345

TABLE 5-2
Laws controlling U.S. mineral land access and ownership.

Law	Date	Topic
Mining Act	1866	Superseded by 1872 law
Placer Mining Act	1870	Superseded by 1872 law
General Mining Law	1872	Discovery-location system for hard-rock minerals
Coal Lands Law	1873	Established rules for purchase of federal land containing coal
Building Stone Act	1892	Applied placer claim procedures to building stone
Pickett Act	1910	Permitted withdrawal of public land; created petroleum reserves
Alaska Coal Lands Leasing Act	1914	Established leasing system for coal in Alaska
Mineral Leasing Act	1920	Established leasing system for land with oil, oil shale, natural gas, coal, sodium minerals, and phosphate rock
Taylor Grazing Act	1934	Permitted classification of public lands
Acquired Minerals Leasing Act	1947	Extended leasing to acquired lands
Surface Resources Act	1947	Established multiple-use management for surface of public lands
Outer Continental Shelf Lands Act	1953	Extended leasing system to shelf more than 3 miles offshore
Multiple Mineral Development Act	1954	Clarified relation between mineral leases and claims on same land
Multiple Surface Development Act	1955	Prevented use of mining claims for nonmineral purposes
Multiple Surface Use Act	1955	Removed sand, stone, gravel, pumice from 1872 Act
National Wilderness Act	1964	Created wilderness areas and restricted mineral access
National Historic Preservation Act	1966	Protected prehistoric and historic properties
Wild and Scenic Rivers Act	1968	Provided for preservation of free-flowing rivers
National Trails System Act	1968	Provided for a national trail system
National Environmental Policy Act	1970	Required assessment of environmental impact of all activities on federal land
Geothermal Steam Act	1970	Provided for leasing of geothermal resources on federal land
Alaska Native Claim Settlements Act	1971	Resolved land claims of Alaskan native groups
Forest and Rangeland Renewable Resources Planning Act	1974	Established environmental regulation of Forest Service land
Federal Land Policy and Management Act	1976	Empowered Bureau of Land Management to regulate use of public lands
Federal Coal Leasing Amendments Act	1976	Established environmental and economic guidelines for leasing
Surface Mining Control and Reclamation Act	1977	Regulated surface mining and reclamation
Alaska National Interest Lands Conservation Act	1980	Divided Alaskan land into ownership and use categories; established parks and wildlife refuges
Federal Tar Sands Leasing Act	1981	Established leasing program for tar sands
Federal Oil and Gas Royal Management Act	1982	Required proper and timely accounting for onshore production revenue
Federal Onshore Oil and Gas Leasing Reform Act	1987	Changed leasing of oil and gas to all competitive
Appropriations Act	1993	Set rental of $100/claim

including vacation homes, were prevented by stipulating that claims could be patented only if they contain deposits that are good enough to merit development by a "prudent" person and that the minerals must be "marketable." Mining claims could be located under this law in Alaska, Arkansas, Arizona, California, Colorado, Florida, Idaho, Louisiana, Mississippi, Montana, Nebraska, Nevada, New Mexico, North Dakota, Oregon, South Dakota, Utah, Washington, and Wyoming. Mineral access on federal land in other states is administered under the Mineral Leasing Act of 1920, which is discussed later.

Periodic efforts have been made to clear the books of old claims. The most recent of these was part of the *Federal Land Policy and Management Act of 1976 (FLPMA)*, which required that all active mining claims be registered with the Bureau of Land Management by October 1979. This requirement is estimated to have removed from the books inactive claims covering at least 1.2 million acres. The newly established rental fee will probably have a similar effect.

The General Mining Law of 1872 is not without faults. It was intended to protect small deposits that could be covered by a few claims, and even included the concept of *extralateral rights* for lode claims (Figure 5-3) permitting the miner to follow the vein if it extended downward off the claim. Modern exploration for larger, deeply buried deposits requires many claims covering geologically favorable ground, where regional and detailed studies are carried out until a deposit is found or the project is abandoned. Because the law was designed to encourage settlement of the land and development of its mineral resources, the law made access to the public land for minerals relatively easy. It gave miners priority over other land uses and it required only work, not payment of fees, to hold the land. It also made no provision for environmental safeguards. Recent efforts to revise the law have focused on the need to charge annual rental on claims, to impose a royalty on production, and to control mineral access by more stringent environmental

regulations. These changes will almost certainly discourage domestic mineral exploration and production, and it is important that they be made with careful comparison of their benefits, including increased tax revenues and less environmental disruption, and their costs, including decreased employment, increased import reliance and declining balance of trade.

Mineral Leasing System

The 1872 mining law was not well suited to exploration for oil, coal, gravel, and other minerals that formed large or concealed deposits. This led to laws that established leasing as a second system of mineral entry on federal land. The first of these, the *Mineral Leasing Act of 1920*, originally applied to coal, oil, oil shale, natural gas, sodium minerals, and phosphate rock. It was later amended to include sulfur in New Mexico and Louisiana, as well as potassium minerals and geothermal resources. Under the 1920 law, leases can be acquired by either competitive or noncompetitive bidding, with competitive bidding required on land with known mineral deposits.

The leasing system was plagued from the start by leaseholders making big discoveries or quick profits on land obtained through noncompetitive bidding. Although this was not illegal, many felt that the federal government was missing out on important income. In one infamous case in the 1980s, a tract of land in the Amos Draw area of Wyoming sold for $1 million shortly after it was obtained in a noncompetitive lease.

FIGURE 5-3

Lode mining claim. Note that the miner has the right to pursue veins off the claim (extralateral rights) if their uppermost part or vertex is on the original claim. This practice is not consistent with the anastamosing nature of many veins and led to major legal battles in Butte, Montana, and other western mining districts (from U.S. Forest Service General Technical Report INT-35; courtesy Intermountain Research Station, USDA).

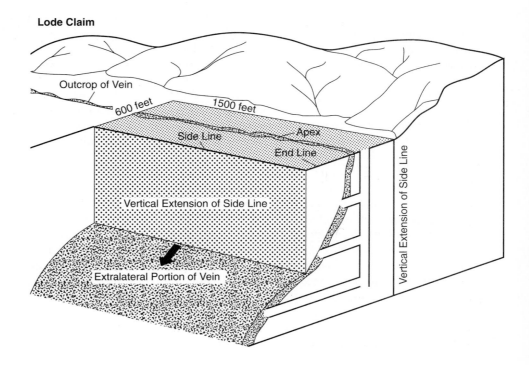

Because these problems became apparent first for coal, noncompetitive coal leasing was eliminated by the *Federal Coal Leasing Amendment Act of 1976.* The *Federal Onshore Oil and Gas Leasing Reform Act of 1987* provided that noncompetitive leases of oil and gas land could be made only on tracts that had been offered unsuccessfully for leasing in an earlier competitive sale. The leases require the payment of royalties and rentals to the federal government, and the title to all leased land is retained by the government. Although the law works well now, it created a long-lived political football in the form of claims that were staked prior to 1920 for minerals that must now be leased. In particular, efforts to patent oil shale claims under the 1872 law have gone on for decades, amid debate about whether the claims had commercial value to a prudent person when they were originally filed, or even now.

The leasing system was modified by the *Materials Act of 1947,* which pertains to most of the construction and industrial minerals. This law permits operators to sell materials from federal land under permits obtained by noncompetitive or competitive bids, depending on the volume of planned production. The *Acquired Lands Act of 1947* authorizes leasing of acquired lands, which are those obtained by purchase, condemnation, donation, or exchange.

Administration of U.S. Mineral Laws

Management of mineral resources on U.S. federal lands is the responsibility of the *Bureau of Land Management (BLM)* (OTA, 1978; Muhn and Stewart, 1984). The *National Forest Management Act of 1976* provided that the U.S. Forest Service manage the surface impact of mining and other activities on national forest lands. The BLM monitors the location and status of all claims on federal land and is responsible for all subsurface mineral activities. The BLM claims to have under its supervision 1.4 Bbbl of oil, 12.5 Tcf of natural gas, 80% of the nation's oil shale, all of its major tar sand, 35% of its uranium, 33% of the coal in the western states, world-class deposits of phosphate, sodium, lead, zinc, and potash, and most of our undiscovered but geologically predictable aluminum, antimony, beryllium, bismuth, cadmium, chromium, cobalt, copper, fluorite, lead, manganese, mercury, molybdenum, nickel, platinum-group metals, silver, tungsten, and vanadium deposits. With the collapse of Soviet government mineral organizations, the BLM administers the largest range and volume of known minerals in the world, and its approach to land management is a major factor in productive mineral exploration on U.S. federal lands.

The BLM is required by the *National Environmental Policy Act of 1970* to determine the environmental effect of all activities, including mineral exploration and production, on federal land. The first step in BLM implementation of this requirement is an *environmental assessment,* which is a generalized evaluation of the nature of the land and how it might be affected by various potential activities. This is used to guide decisions on all types of land use, including mineral entry. If the land is opened for mineral entry, and part of it is leased or claimed, exploration plans must be reviewed and approved. If the activities have potential for significant disturbance to the land, further permitting is required. If the actions described in the notice of intent are judged to be sufficiently serious, a more comprehensive *environmental impact statement (EIS)* will be required.

Leasing of onshore land is administered by the BLM and leasing of offshore mineral land is administered by the Minerals Management Service (Figure 5-4). These agencies identify tracts of land and offer them for competitive bids, which must contain a work plan, including an environmental assessment that has to be approved by the leasing agency. Annual rental for onshore oil and gas leases is $1.50/acre for the first five years and $2/acre thereafter, with a royalty of 12½% of the value of any production. Offshore oil and gas leases require royalties of 16⅔% in shallow water and 12½% in deep water. Terms for other minerals differ, but have approximately the same economic effect. Bids are compared on the basis of a work plan and the amount of a nonrefundable *bonus bid* that is offered for the right to explore the lease. This payment as well as the first year's rent are due on signing the lease.

About 70% of federal leases are for oil and gas, with coal and other minerals accounting for about 15% each. About 10% of the leases are on Native American land, most of which is also administered by the federal government. Federal leased land increased by about 50%, from 14 million acres in 1980 to about 21 million acres in 1990. Although the amount of land leased for all mineral types is up over that period, oil and gas rose steadily, whereas land leased for coal and other minerals rose and fell along with changing economic expectations.

Mineral production from leased lands is a major proportion of U.S. national production and an enormously important source of funds for the U.S. Treasury. Between 1950 and 1989 leased lands produced 16 Bbbl of oil, 123 Bcf of natural gas, 2 billion tonnes of coal, and about 100 million tonnes of phosphate, potash, and sodium. In 1990, offshore leasing was the fourth largest source of federal revenue, with contributions of about $3 billion. Bonus bids are a very important part of these revenues because of their large size and the fact that they are paid regardless of whether the tract ever produces minerals (Figure 5-5).

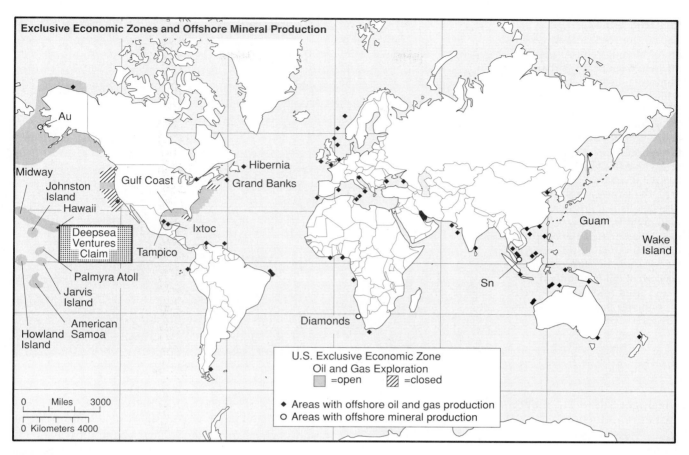

FIGURE 5-4

Offshore areas claimed by the United States as Exclusive Economic Zones (EEZ) and the claim filed by Deepsea Ventures for mining of manganese nodules. Shaded areas in the EEZ were closed to offshore leasing for oil and gas exploration in 1990. Regions of the world with offshore oil, gas, or mineral production are also shown (compiled from *U.S. Geological Survey Circular 1018*).

In fact, the nature of the bidding process seems to encourage large overbids, with winning bids averaging 45% higher than the next highest competing bid (Megill and Wightman, 1984). Between 1954 and 1992, bonus bids paid on U.S. federal leases totaled over $57 billion dollars, almost all of that going for offshore oil and gas (Figure 5-6A). Interestingly, royalty income from offshore oil and gas production totaled only about $40 billion over the same period. The relative importance of royalty income and bonus bids has changed with time, as oil has been discovered in previously leased areas, although it is clear that the U.S. federal leasing system depends strongly on bonus bids for revenue (Figure 5-6B).

Revenues from the leasing system are distributed to the states, the general fund of the U.S. Treasury, and several specific funds. The *Land and Water Conservation Fund Act of 1965* commits a portion of offshore lease revenue to federal and state purchases of park and recreation lands. Additional offshore lease revenues support state purchase of land for historic preservation under the *National Historic Preservation Act of 1986*. The Reclamation Fund supports reclamation of mined land by state governments. In addition, states receive a proportion of revenues from onshore mineral production on federal land within their boundaries, and funds are distributed directly to Native American tribes and organizations with leased land. For 1991, about $45 million of offshore revenues were distributed to the states, with $2.8 billion remaining in the U.S. Treasury, $873 million going the Land and Water Conservation Fund, and $150 million going to the Historic Preservation Fund. Another $483 million was paid directly to the states, largely from onshore production. Largest payments from offshore production went to Texas, Louisiana, and California, each of

FIGURE 5-5

Zapata Concord drilling vessel on location at Point Arguello, offshore from Santa Barbara, California, in 1982. A $333.6 million bonus bid was paid to the U.S. government for rights to explore this 5,131-acre (2,076-hectare) tract, which turned out to host part of a reserve of 300 Mbbl of oil and 500 Bcf of natural gas in this area, the largest western U.S. discovery since Prudhoe Bay. Although the field was developed at a cost of about $2 billion, production was delayed for a decade by environmental concern about how oil would be transported from offshore wells to refineries (courtesy of Vivian Edwards, Chevron Corp.).

which got $10 to $13 million. Largest payments from onshore production went to Wyoming ($189 million), New Mexico ($103 million), and Colorado ($53.7 million).

It is not clear how much longer we will be able to enjoy the largesse of the federal leasing system. Its main source of funds, bonus bids from offshore leasing, is shrinking. Good tracts have already been leased and large areas of offshore land have been closed to leasing, as discussed later. The Gulf of Mexico is essentially the only area of active, large-scale leasing, and most shallow areas of the Gulf have been leased, leaving only the more deeply submerged areas for future leasing. These lands will be more expensive to explore and might not attract the generous bonus bids to which we have become accustomed.

State and Local Mineral Laws in the United States

Discovery-location systems are not applied by the states, and access to any state land opened for exploration is by some form of leasing. Private land of all types is also leased, with most attention focused on large landholders such as ranchers and railroad, timber, or paper companies. Lease agreements vary greatly according to location and commodity. The situation is simplest for oil and gas exploration, where a more or less standard lease form is widely accepted. Leases for other types of minerals are not standard and must be negotiated separately for almost all cases. The main sticking point in these negotiations is the royalty, where many landowners have unrealistic expectations with respect to prevailing practice. Several U.S. states have "mineral lapse" laws that terminate mineral interests that have been undeveloped for 20 years.

States have made important contributions to mineral law in two main areas, both of which relate to oil and gas (Ely, 1964). Interest in *tidewater land* was generated by the need to apportion royalty and tax revenue from offshore production. The *Submerged Lands Act of 1953* divided land between the states and the federal government. States bordering the Atlantic and Pacific Oceans, as well as Louisiana, Mississippi, and Alabama, obtained land within 3 miles (4.8 km) of shore. For Texas and Florida, the boundary was set at 3 marine leagues (16.65 km) from shore to reflect original Spanish conventions. Leasing regulations for these federal offshore lands were spelled out in the accompanying *Outer Continental Shelf Lands Act of 1953*. The key word in these statutes was coast, a term that turned out to require clarification. In southern Louisiana, the ocean shallowed and freshened gradually into swamp and in many other areas, it was clear that the shoreline was eroding landward or growing seaward. Offshore islands further confused the situation. Attempts to make these dynamic geologic processes conform to acceptable legal terminology have occupied years of court time.

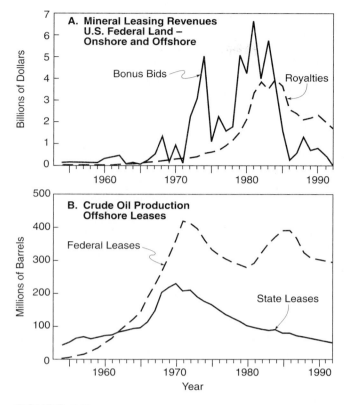

FIGURE 5-6

Oil and gas revenues and production from federal land in the United States **A.** Relation between bonus bid and royalty revenue from federal offshore oil and gas leases. **B.** Change in oil production from state and federal offshore leases. Decline in production from state leases reflects the exhaustion of fields in shallow state waters (from Federal Offshore Statistics, Minerals Management Service).

Unitization of oil and gas fields was also largely a state legal issue, mostly because much early oil production was on private and state land. Unitization refers to the process by which production from an oil or gas field is administered as a single unit, regardless of the number of land parcels and owners involved in the field. Before unitization was applied, landowners drilled as many oil wells as possible on their land in order to prevent their oil from being pumped out of wells on adjacent land. This practice, which placed a premium on closely spaced wells and rapid pumping, greatly decreased the amount of recoverable oil in a field and was soon recognized as wasteful. In its place, each owner was assigned a proportion of the production that was anticipated from the entire field. The field was then produced with the well spacing and pumping rate that was judged most favorable to maximize recovery. With the exception of

California, states took on the job of enforcing unitization, beginning with the mis-named Texas Railroad Commission (Prindle, 1981). The interest that these groups had in production rates gradually grew to take in the relation between production rates and prices, especially since oil prices had a strong effect on state royalty revenues. For many years, oil production rates determined by the Texas Railroad Commission had a strong impact on world oil prices.

Most states and many local jurisdictions also have environmental agencies, resulting in overlapping jurisdictions that can cause considerable difficulty in determining which agency has the appropriate authority. Most counties or municipalities zone land, and some of these zoning regulations deal with exploration activities such as drilling. Almost all projects that can be carried out under the zoning laws must be assessed for environmental impact. These agencies require reclamation, although many restrict such requirements to operations that move more than 1,000 yd^3 of material per acre. An area of growing legal interest is the degree of control that state and local agencies can exert on the use of federal land in their jurisdictions. Recent Supreme Court decisions indicate that states may control federally approved mineral activities on federal land if operators do not comply with reasonable state environmental regulations.

MINERAL OWNERSHIP AND LAND LAWS OUTSIDE THE UNITED STATES

Land Laws in Other Countries

In *Canada*, private citizens can own mineral deposits. From the standpoint of mineral exploration, however, this is not very important because most land in Canada is government controlled. Under provisions of the Constitution Act (originally known as the British North American Act of 1867), provincial governments own and regulate mineral resources (Peeling et al., 1992). The Canadian federal government controls public lands and minerals in the Yukon and Northwest Territories, although some control has been ceded to aboriginal groups, as discussed later. Individual provinces retain title to their public lands and set their own mineral laws. Claims on federal and provincial land can be staked by any person over the age of 18 or by a domestic corporation. They range in size from 1,500 by 1,500 feet (457 by 457 meters) in the Northwest Territories, Yukon, British Columbia, and Manitoba to 40 acres (98.84 hectares) in Saskatchewan, Ontario, Quebec, Nova Scotia, New Brunswick, and Newfoundland. Claims can be staked for most minerals, annual assessment work must be carried out, and patenting is allowed in some provinces. Unlike U.S. mining law, royalties are payable on mineral production from most claimed ground. Oil and gas exploration and production offshore from most provinces is controlled by joint federal-provincial agreements such as the *Canada-Newfoundland Atlantic Accord Implementation Act.* Competition for most leases has not been strong enough to justify bonus bidding, and most lease applications are judged on the amount and type of work that is proposed.

In *Mexico* all minerals are owned by the federal government, regardless of surface ownership. Oil and natural gas exploration and production are operated as a monopoly by the Mexican state corporation, Petroleos Mexicanos (Pemex). Uranium and other nuclear fuel minerals are also reserved for the government. Most other minerals can be produced by individuals and Mexican corporations. The stipulation that Mexican mineral deposits must be controlled by Mexican citizens or corporations was a major impediment to foreign participation in nonfuel exploration in Mexico for many years. Mexican joint-venture partners lacked adequate capitalization and foreign investors were unwilling to forfeit control over projects. Relaxed enforcement of this requirement has led to renewed exploration by foreign groups and the opening of several new mines. Ground control for exploration and mining is accomplished by staking claims, which are called concessions, with no limit on the number of concessions held. Annual rental is charged on claims and assessment work is required.

In *Australia* and *New Zealand*, title to most minerals is also held by the federal government. Exploration access is by way of *concessions*, which are large tracts of land obtained by application to the government. Controversy in New Zealand centers on just who owns their extensive geothermal resources. When the Maori people signed the *Treaty of Waitangi* with the British in 1840, they retained ownership of fishing and sacred places. As it turns out, many of these sacred places were geothermal areas. The significance of this situation to ownership of New Zealand's geothermal power plants and deposits must be clarified by litigation. It could become even more important if the Maori claim ownership of extinct geothermal systems, many of which contain gold deposits. In *England* and *Ireland,* land control is generally similar, although some older landowners hold *grandfathered* mineral rights retained from before the present laws were written. Irish exploration concessions convey the right of access to the land, with appropriate compensation paid for damage to anything on the surface. In England, the owner of the surface can stop exploration

for a wide range of reasons, which has greatly discouraged activity.

In *Japan,* where as much as 70% of the land is used only for timber and is thus potential exploration country, surface rights are privately owned but all mineral rights are owned by the federal government. The federal government grants claims or concessions, although exploration on them is complicated by the need to negotiate land access with surface owners. In *South Africa,* most land is privately owned and exploration is carried out largely on privately negotiated leases. The lack of large tracts of government land is, in fact, an impediment to exploration because it takes so much time to arrange for land access.

In an interesting twist to the land access problem, the *Brown Coal Act of 1950* in *Germany* allows companies to use the right of *eminent domain* for access to coal on private land. In the United States, this right, which allows purchase of the land at a fair price for a specific use, is commonly available only to railroad, pipeline, powerline, and hydroelectric companies. This law, which reflects a strong motivation to minimize dependence on imported oil, could become a model for legislation in other countries as domestic resources dwindle, requiring more systematic exploration.

Mineral land access in LDCs is usually by concessions that must be negotiated with the government, although the process is not always straightforward. Some countries have mineral laws on the books, but incumbent governments are often not in agreement with the laws and do not facilitate operation under them. Reasons for these disagreements include valid environmental and economic concerns and the lack of technical and clerical staff to process applications, but they can also reflect the desire for extra payments for approving applications. The incidence of graft and the misuse of bonus bids in concession applications in some LDCs has led to the practice of offering funds for public improvement projects that are carried out or monitored by independent groups, in lieu of bonus bids. Where concessions are obtained, they are often set up with low rentals in the first few years and increased payments after that. The increased payments are often linked to *"drop-out" clauses* requiring that a proportion of the land be released each year. This approach stimulates concession holders to search diligently and prevents hoarding of mineral concessions.

A final element of concern to explorationists and those who wish to maintain mineral supplies is the legal link between exploration and extraction. Although this link is a fundamental part of the law in many countries, giving the discoverer the right to extract a mineral deposit that he or she has found, some countries require completely new negotiations. In many

LDCs, these negotiations focus on economic factors, especially the division of profits. In Panama, for instance, several large porphyry copper deposits have never been developed because the government required such a large share of the profits that no one would invest in the operation. In some MDCs, notably the United States, these negotiations focus on environmental issues. Because of inadequate preparation on the part of regulatory agencies and extremely high levels of public participation and conflicting interests, it has become difficult to agree on all environmental aspects of proposed mineral projects and some have failed on the basis of these complications, particularly in Wisconsin and California. Although the high level of concern about environmental issues is necessary and desirable, the failure to reach a compromise permitting production is not encouraging for domestic mineral supplies in MDCs.

Law of the Sea

A world mineral law would be an ideal solution to our vexing mineral land access problems. Although such a law for land-based minerals is unlikely, it has been drafted for marine mineral resources as part of the 1982 United Nations *Law of the Sea (LOS)* (Morell, 1992; Schmidt, 1989). The Law of the Sea holds that ocean mineral resources are the "common heritage of mankind." It puts them under an *International Seabed Authority (ISA),* which would license mining projects, receive royalties from production, and disburse them to all members of the United Nations. Royalties from production are proposed to be 5% to 12% of the gross value of production or 2% to 4% of gross plus 35% to 70% tax on net profits. The law also stipulates that the ISA receive all technical data and that it can choose to develop areas being explored by private groups.

Concern over these provisions has prevented wide acceptance of the LOS. Between 1982 and 1993, it had been ratified by only 45 countries, all of which are LDCs except Iceland. At least 60 countries must ratify the treaty for it to become international law and none of the MDCs appear likely to do so. In response to the failure of the LOS to address interests of MDCs, the United States proclaimed in March 1983 that the ocean area between 3 and 200 miles (4.8 and 320 km) offshore from its territory was part of an *Exclusive Economic Zone (EEZ)* under U.S. control (Figure 5-4) and many other coastal nations followed suit. The EEZ has an area of 15.8 million km^2, considerably larger than the land surface of the United States, and contains a wide variety of mineral deposits (Cronan, 1992). Even earlier, in 1974, a cooperative venture known as Deepsea Ventures, consisting of U.S. Steel, Tenneco, Union Miniere, and several Japanese firms, filed a claim with the U.S. government for the right to mine manganese nodules in a 60,000-km^2 area of the Pacific Ocean floor (Figure 5-4). Lower demand for manganese during the 1980s diverted attention from the still incompleted determination of who owns marine mineral resources, but the question will undoubtedly return to prominence as land-based reserves dwindle and marine minerals assume center stage in our thinking (Borgese, 1985; OTA, 1987; Earney, 1990).

LAND ACCESS AND LAND POLICY

Many jurisdictions have mineral laws on their books but control access to mineral lands and production by other means, which constitute their *land policies* (Culhane, 1981; Foss, 1987; Hargrove, 1989). These policies have come to rival geology as controls on world mineral reserves. The most important policy matters of current interest are land withdrawals and land classification.

Land Withdrawal

Land withdrawal, the prohibition of mineral, grazing, agricultural, or related activities on public land, has become an important part of U.S. land policy. It began in 1872 with the creation of Yellowstone Park, an event that marked the end of a century in which federal land policy consisted largely of giving away land to settle the country. As the population grew and no more land was acquired, however, the need to protect some land became obvious. In 1906, Theodore Roosevelt withdrew more than 267,000 km^2 from homesteading and leasing under existing agricultural laws. Passage of the *Pickett Act or General Withdrawal Act of 1910* empowered the President to withdraw federal lands for classification and to stop mineral entry to these lands if it was in the public interest. These laws were not motivated by environmental concerns. Rather, they reflected a desire to set aside land with oil, coal, and fertilizer minerals thought to be vital to national security. (The major source of U.S. fertilizer minerals at that time was Germany.) The *Naval Petroleum Reserves,* areas set aside to preserve oil for the U.S. fleet, were started at this time, as was federal control over oil shale lands.

Federal land withdrawals continued intermittently after passage of the Pickett Act, but reached a new level with passage of the *Wilderness Act of 1964,* which set aside 36,800 km^2 of federal land to be preserved as wilderness. Through this act and the FLPMA, the Department of Interior was charged with evaluating all federally owned roadless areas larger than 5,000 acres (20.2 km^2) for possible inclusion in a *National*

Wilderness Preservation System. The wilderness system has grown at a rapid pace since 1964 (Figure 5-7). By 1991, the system contained about 385,000 km², more than ten times larger than in 1964. Evaluation of land for inclusion in the wilderness system is continuing and the ultimate size of the system could reach 500,000 to 600,000 km². These lands come from areas controlled by the Bureau of Land Management, Forest Service, Fish and Wildlife Service, and Park Service, which have been responsible for making recommendations to Congress on which lands to include in the wilderness system. Over the life of the system, Congress has actually approved about 25% *more* wilderness land than the agencies have recommended. Including federal land withdrawn for other purposes, conservative estimates indicate that 50% to 60% of federal land is or will be closed to mineral exploration and production by the year 2000.

Unfortunately, most of the land that has been withdrawn from exploration has not been adequately explored. The Wilderness Act recognized the possibility that unknown minerals might be present on this land and called for geologic studies as part of this evaluation. These studies have been largely the responsibility of the U.S. Geological Survey, Bureau of Land Management, and U.S Bureau of Mines, whose work has been of a reconnaissance nature, involving largely geological mapping, and geochemical and geophysical surveys. For instance, the Bureau of Land Management spent about $100 million to evaluate about 105,000 km² of their land that was proposed

for the wilderness system. That amounts to about $1,000/km², considerably less than it would cost to mow a lawn of that size. Full evaluation requires drilling, which was not done for cost and policy reasons. Thus, there is little doubt that potentially economic deposits have been included in withdrawn areas, including the wilderness system. Present administration of the wilderness system gives essentially no attention to the mineral potential of land after it is included in the system. In proceedings volumes from 1989 and 1990 conferences on the status of the wilderness system, for instance, only one paper out of about 100 mentioned minerals, and it was only in the context of protecting the lands from development (Lime, 1989; Reed, 1990).

Even offshore areas are subject to withdrawals. A 1990 presidential decision withdrew most of the Pacific offshore, as well as parts of Florida and the Grand Banks off New England (Shabecoff, 1990). This move, which postponed leasing in most of these areas until the year 2000, was considered moderate in view of environmental demands that exploration be outlawed forever. The withdrawal also contained provisions for the return of payments made for existing leases in these areas, although all claims so far have been rejected. Pressure is also on for a permanent withdrawal of the *Arctic National Wildlife Refuge (ANWR)* from oil exploration. ANWR, which has an area of 76,900 km², is about 100 kilometers east of the Prudhoe Bay oil field, the largest oil field in the United States (Figure 5-8). The geology of ANWR is very similar to that of Prudhoe Bay, suggesting that it could contain similar oil fields (USGS, 1989). Based on 1,300 kilometers of seismic surveys carried out in a 6,000 km² part of ANWR, the U.S. Geological Survey and Bureau of Land Management have estimated that there is a 50% chance that this area could contain 3.2 Bbbl of economically recoverable oil, giving ANWR the fourth largest field in the United States. This oil would be welcome because the Alyeska Pipeline, which carries oil from Prudhoe Bay to the port of Valdez on the south coast of Alaska (Figure 7-16), must have a minimum daily throughput of 300,000 bbl to operate. Prudhoe Bay fields are estimated to fall below that level by 2009, which would cause the pipeline to be shut down, leaving as much as 1 Bbbl of oil in the fields. Once the pipeline is shut, it would be prohibitively expensive to rehabilitate and reopen. In spite of these factors and widespread exploration in the Alaskan offshore arctic (Figure 5-8), exploration of ANWR will likely be delayed for years.

Most other countries have not been as aggressive in withdrawing land. Canada has no national wilderness law, although it does have an extensive system of national and provincial parks (McNamee, 1990). Although wilderness laws have been enacted by some

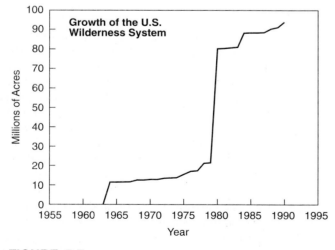

FIGURE 5-7

Historical growth of acreage placed in the U.S. Wilderness Preservation System. Large increase in 1978–80 consists largely of Alaskan land (compiled from information supplied by the Bureau of Land Management).

FIGURE 5-8

Northern North America showing Canadian aboriginal territory of *Nunavut* and areas of major mineral interest, as well as oil and gas exploration wells. Note the extensive offshore exploration throughout arctic waters. Detailed map shows relation between Arctic National Wildlife Refuge (ANWR) and oil and gas fields in the Prudhoe Bay-MacKenzie Delta area, along with possible oil and gas traps indicated by seismic surveys in ANWR. Exploration well KIC#1 was drilled on Native American land (compiled from *U.S. Geological Survey Professional Paper 1850;* Standard Oil Scene, Spring, 1987; *International Petroleum Encyclopedia,* 1991).

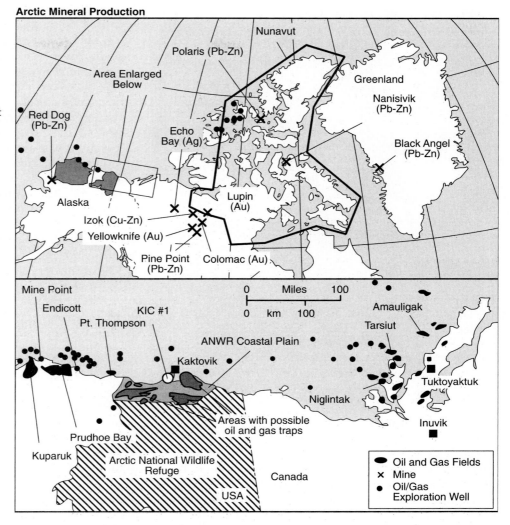

provinces, only British Columbia and Newfoundland have proposed significant withdrawals. Nevertheless, almost 260,000 km² of Canada's 10 million km² land surface is protected in some way, and grassroots campaigns are underway to enlarge this. The impact of such movements on Canadian economic activity will be considerably greater than that in the United States, in view of the greater proportion of its GNP that is contributed by mineral production. Norway, another country with large wilderness areas, has taken a different approach. Even though Norwegians take more time to enjoy their countryside, they have not established formal wilderness systems and, in fact, have promoted the multiple-use concept for most land (Kaltenborn, 1990).

Antarctica has become the ultimate land withdrawal (Zumberge, 1979; Riding, 1991). This continent is jointly administered by members of the *Antarctic Treaty,* which consists of Argentina, Australia, Bel-

gium, Brazil, Britain, Chile, China, France, Germany, India, Italy, Japan, New Zealand, Norway, Poland, Russia, South Africa, United States, and Uruguay. Antarctica contains many potentially important mineral deposits (de Wit, 1985), including the Dufek Intrusion, the closest possible analogue to South Africa's huge Bushveld Igneous Complex, which hosts the world's largest reserves of platinum, chromium, and vanadium (Figure 5-9). Pressure to explore these deposits has been dampened by distance and harsh operating conditions, although it should increase as more readily accessible deposits are exhausted. Rather than allowing this to occur, the Antarctic Treaty group agreed in 1991 to ban mineral and oil exploration on the continent for the next 50 years. This agreement must still be ratified by the signatory governments, but it is likely to become international law, thus confining us to our home continents for the next half century.

FIGURE 5-9
The Dufek Igneous Complex in Antarctica, as seen from a passing aircraft, consists of layered mafic igneous rocks similar to those in the Bushveld Igneous Complex and could host important mineral deposits (courtesy of S.B. Mukasa, University of Michigan).

Land Classification

Land withdrawals are a first big step toward national *land classification*. Whether we like it or not, all land is likely to be classified according to preferred use in the near future. To do this intelligently, maps must be prepared for all areas showing how the land is used presently and the distribution of special factors or attributes that might affect classification decisions. This is not a new idea. Most municipalities consider minerals in their zoning regulations, usually focusing on the preservation of conveniently located deposits of sand, gravel, and crushed stone for local construction. Land classification in larger regions requires attention to all types of potential mineral resources. Studies done by the U.S. Geological Survey, Bureau of Land Management, and Bureau of Mines set an important standard for such efforts (Shawe, 1981). Their work provides information on the type of mineral concentrations that are present in an area, the geologic and geochemical environments in which these deposits are found, and the potential for the occurrence of undiscovered deposits. This information is presented in maps (Figure 5-10) that can be compared by nontechnical people to maps for other potential land uses, thus facilitating informed land classification.

The most important land classification effort in history has taken place in Alaska since the late 1960s (Singer and Ovenshine, 1979). When Alaska joined the Union in 1959, it was given the right to select about 413,000 km² of federal land by 1984. Fortunately for the state, it went about this selection process slowly, focusing primarily on agricultural land. Then came the 1967 discovery of the Prudhoe Bay oil field on land that the state had selected. A lease sale in 1969 for oil exploration land around Prudhoe Bay yielded over $900 million in bonus bids to the state. Suddenly, mineral potential became the focus of state land classification efforts. In addition to claims by the state, the federal government was faced with long-standing claims by Alaskan native people. These claims were recognized by the *Alaska Native Claims Settlements Act of 1971*, which provided 178,000 km² to Native Americans along with $962.5 million, of which $462.5 million was paid immediately. The act also provided for the formation of regional corporations controlled by natives, with shares in the corporations awarded to all native residents of each region. In the face of competing claims from these groups, land distribution in Alaska was stopped by the federal government in 1978 and an enormous program of land classification was undertaken to guide further decisions. By 1992, about 146,000 km² had been conveyed to native groups and another 348,000 km² had gone to the state. Remaining native land claims were some 109,000 km² greater than permitted under the law, and state claims were about 22,000 km² over their allowable limit, thus assuring another decade or so of negotiation and land distribution.

The Alaskan land distribution effort shows the growing importance of *aboriginal land claims* and their emergence as a factor in land classification. After a century of quiet, many aboriginal groups are asking the courts to support their claims to land. Some claims are based simply on prior occupancy, others on religious significance, and still others on treaties that were signed and forgotten by settlers. In a recent settlement of this type, the Canadian federal government granted to the Gwich'in tribes 6,162 km² of land and mineral rights in the Mackenzie River Delta, an area of oil exploration interest. A much larger agreement reached in 1991 converts an area of 2 million km² of the eastern Northwest Territories into a new Inuit-administered territory known as *Nunavut* (Figure 5-8). This agreement also awarded Nunavut's 17,500 Inuit residents $1.4 billion over a 14-year period and ownership of about 20% of the territory, with the rest remaining in Canadian federal hands (Farnesworth, 1992).

Not everyone approaches land classification in the same way. A few simply ignore geologic or other information. Some conservation groups actually target areas with mineral potential for withdrawal. The idea

FIGURE 5-10

The system of geologic land classification advocated by the U.S. Geological Survey begins with the preparation of a map (top) showing geologic features related to potential mineral deposits, in this case Mississippi Valley-type (MVT) lead-zinc deposits in the Viburnum Trend of Southern Missouri. As tabulated below the map, these deposits are found in the Bonneterre Formation where it has been altered to dolomite, usually around older, Precambrian rock or along faults. Deposits are also surrounded by traces of base metals in the rock. Second map (middle) shows the distribution of rocks with these favorable characteristics, and third map (bottom) uses these zones to classify the area in terms of the probability of containing MVT deposits (from *U.S. Geological Survey Circular 845, 1981*).

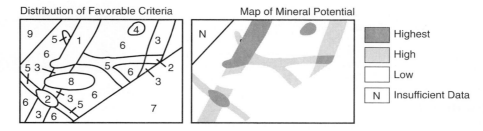

Ranking Exploration Potential of an Area

Modified Geologic Map

- ⊥⊥⊥⊥ Bonneterre Formation (dolomite) is present northwest of this line
- Major faults
- Bonneterre is limestone
- Buried Precambrian rocks (from aeromagnetic map)
- Zones of abnormally high content of base metals in drill samples
- N No subsurface information is available

Criteria for mineral deposits · Areas numbered on map

Criteria for mineral deposits	1	2	3	4	5	6	7	8	9
Bonneterre Formation	■	■	■	■	■	■	■	■	■
Dolomite	■	■	■	■	□	■	■	⊠	
Buried Precambrian knobs	■	□	■	■	□	□	■		
Major faults	□	■	■	□	■	□	□	⊠	
Base metals	■	■	■	□	□	□	□	⊠	

- ■ Listed criterion is present
- □ Listed criterion is absent
- ⊠ Insufficient data to infer presence or absence

Distribution of Favorable Criteria

Map of Mineral Potential

- ▨ Highest
- ▨ High
- □ Low
- N Insufficient Data

here is that withdrawal of ground with high mineral potential will cut down future environmental problems by locking up minerals in wilderness areas, scenic rivers, or wildlife refuges. A case in point is the Irish Wilderness area in Missouri. This 16,500-acre (42,750-hectare) area is near the southern end of the Viburnum Trend (Figure 5-11), which contains one of the largest reserves of lead and zinc in the United States. Mining in the Viburnum Trend, which has been a major employer in the area since 1970, takes place along an almost continuous line of Mississippi Valley-type lead-zinc deposits, at least 70 kilometers long and 1 kilometer wide, buried at a depth of about 300 meters. Almost no evidence of these deposits can be detected at the surface, and they were discovered by risky deep drilling based on geologic concepts.

Limited drilling around the Irish Wilderness showed that it appeared to be the southern continuation of the Viburnum Trend and had high potential for mineable ore. Although this information was provided to all interested parties during classification decisions, the area was included in the wilderness system. Ironically, the Irish Wilderness is not wilderness or even unique countryside. It was logged extensively early in the 1900s, and was described then as . . . "The pines

were gone. Wildlife, deprived of food and cover, also was gone. The soil, stripped of its protective trees and even of much of its grasses, eroded and washed into the streams and choked them with gravel" (Barnes, 1983, p. D-1). By the time it entered the wilderness system, the land was covered by scrub forest similar to that found throughout the area. In this case and in many others, wilderness classification was assigned with little consideration of local employment or long-term mineral requirements.

CONCLUSIONS

In the preceding chapters, we have established that land access is essential if we are to maintain our mineral supplies. We must explore large areas to find deposits and we must produce minerals where the deposits are found. It is interesting to speculate on how a visitor from Mars might view our present land laws and policies in this regard. Our Martian might ask where it is that we expect to get minerals for the next generation if we withdraw all distant and wilderness land from exploration. The only land left will be in more settled areas of the world, where the *NIMBY*

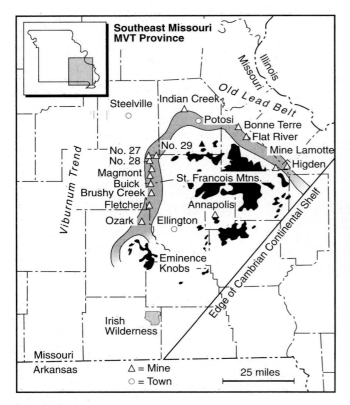

FIGURE 5-11
Schematic geologic map of southeastern Missouri showing the location of important Mississippi Valley-type lead-zinc mines of the Old Lead Belt and Viburnum Trend. Old Lead Belt mines are exhausted and flooded, and explored only by local scuba divers. Lead-zinc reserves remain only in the Viburnum Trend and possibly in its southern extension (stippled zone), which continues into the Irish Wilderness (compiled, in part, from Larsen, 1977, *Economic Geology,* v. 72, p. 411).

FIGURE 5-12
Oil production island immediately offshore from Long Beach, California. This and several other islands had yielded over 700 Mbbl of oil through 1990, paying over $3 billion in taxes to the city and state. As can be seen here, drill rigs are enclosed and the island has structures along its front that make it look like a residential complex, complete with waterfalls (courtesy of S.J. Marsh, THUMS Companies).

(Not In My Back Yard) syndrome is most strongly entrenched. This visitor might note that most of us lack information on the distribution of mineral operations in our own backyard. Oil exploration and production has been going on in the Los Angeles area since the early 1900s with little attention from the general public, often from wells housed in large structures that look like office buildings (Figure 5-12). Sand and gravel, crushed stone, and other construction materials are mined near most towns from operations that are similar in size to large metal mines. Landfills for municipal waste are even more common and make up the largest "mining" activity in most MDCs.

Hopefully, our Martian visitor would conclude from this that we have coexisted with mineral extraction for centuries and that we must continue to do so. There is no question that early operations made terrible messes, but we know how to prevent this now and with more of us interested, we can assure that operations are environmentally as well as economically acceptable. But we must accept the concept of multiple use for land and defer to the natural law, which dictates that minerals can be extracted only where deposits are found. Thus, mineral production will probably be granted precedence in some areas over other uses that can be relocated more easily.

6

Mineral Economics

STRUCTURE OF THE MINERAL INDUSTRY

As we have said repeatedly, a mineral deposit must be an ore deposit before it can be extracted. In other words, it must have the potential to yield a *profit*. But what is a profit? Some define it as any improvement or advancement. Others define it more narrowly as income remaining after deduction of expenses and taxes. The definition that prevails depends in large part on the organization that is extracting the resource and on its goals. As it turns out, many types of organizations produce minerals around the world and each type defines profit differently.

At the beginning of the twentieth century, most minerals were produced from mines and wells that were operated by individuals or small groups. Funds to start these operations were usually provided directly by the owners or wealthy backers. As the scale of operations grew, the size of producers also grew until they became sizeable *corporations*. These obtained funds from *loans* provided by banks and institutions or, where this was not possible, from proceeds of *stock offerings*, in which *shares* in the corporation (also known as stock) were sold. Where these shares were sold to a few people, the operation was a *privately-owned corporation*. Where shares were sold to the general public, a *publicly-owned corporation* was created. In some areas, an intermediate-sized organization known as a *partnership* could be created if share-hold-

ers exceeded a certain minimum number but ownership was not open to the general public. Shares of publicly-owned corporations are usually traded on a *stock exchange*, whereas shares in privately-owned corporations and partnerships are not.

The transition from small operations to corporations was hastened by *amalgamation* or *consolidation*. This became necessary when continued extraction of what first appeared to be isolated deposits showed that they were, in fact, parts of larger deposits. The classical example of this took place in the diamond mines of Kimberley, South Africa, which were discovered in 1871 (Wheatcroft, 1985). Early arrivals to Kimberley simply staked a claim and began digging. At the height of activity, as many as 30,000 people were working, almost shoulder to shoulder, as the mines slowly progressed downward (Figure 6-1). Gradually, adjacent claims were joined to form groups, which combined with neighboring groups to form small syndicates. By the mid-1870s, the original 3,600 claims had been condensed into 98 groups and by 1889 two groups, De Beers Mines headed by Cecil Rhodes and Kimberley Mines headed by Barney Barnato, vied for total control. With payment of over 5 million pounds, control was gained by De Beers, which went on to become the most powerful corporation in the diamond business (van Zyl, 1988). The fortune that Rhodes earned from this venture and its sequels funded many philanthropies, including Rhodes Scholarships to Oxford University, an institution that had the foresight to let Cecil Rhodes attend university intermittently for sev-

FIGURE 6-1

Kimberley mine, South Africa, began as small diggings that gradually grew to form this deep pit, which extended downward into an underground operation as shown in Figure 10-18. The dark, high-rise building in the background is Harry Oppenheimer House, where diamonds from South Africa and Botswana are sorted and graded.

eral years, as he commuted to and from South Africa building his fortune.

As the scale of world mineral production grew, corporations expanded their activities into other parts of the mineral business. Producers of crude oil moved into refining, and then into marketing of gasoline and other products. Metal-ore producers set up smelters and began to fabricate bars, plates, wires, and other metal products. As raw material needs grew and deposits were exhausted more quickly, it became necessary to explore for new deposits on a continuous basis. This led corporations into countries where geologic conditions were favorable for mineral discover-

ies even though there was no significant domestic demand for the minerals. This, in turn, produced a new type of corporation, the *multinational,* which extracted minerals in one country, moved them to other countries for processing, and sold them wherever the demand was greatest (Jacoby, 1974; Sampson, 1975; Thoburn, 1981; Mikdashi, 1986; Mikesell and Whitney, 1987). Long before electronic, food, or drug manufacturers were setting up plants in the Third World, the oil and mining multinationals were major factors in the economies of these countries. In fact, they have been called the pioneers of international business because they operated in countries where economic, political, and legal conditions discouraged other multinationals. The most famous of these, the *Seven Sisters* (British Petroleum, Chevron, Exxon, Gulf, Mobil, Royal Dutch/Shell, and Texaco), controlled more than half of world oil production until the late 1970s.

With the growth in mineral production came an increased awareness of its importance to national welfare and the advent of government-owned *national corporations* (Grayson, 1981; Radetsky, 1985; Mikesell, 1987). Driving forces for government involvement included security of mineral supplies and a desire for more of the profits from domestic economic activity. The first such company was created in 1909 when discovery of oil in the Persian Gulf resulted in the establishment of Anglo-Persian Oil, the predecessor of British Petroleum. It was designed to prevent control of the oil from falling into the hands of corporations from other countries, a situation that would have made the Royal Navy dependent on foreign firms for its oil supplies. Some oil companies, such as Compagnie Francaise des Petroles in France and Ente Nationale Idrocarburi (ENI) in Italy, grew from small government investments in oil exploration and production organizations. Others, such as Statoil, the national oil company of Norway, were created by government decree and assigned an interest in all domestic concessions that were granted to other companies. Still others, such as Petro-Canada, were started by government purchase of publicly-held oil companies. Note that all government-owned corporations are not publicly-owned, a distinction that is frequently missed.

The trend toward government participation was strongest in the LDCs, where the definition of profit and its distribution between the operating company and the host government became a major sticking point (Nwoke, 1987; Dasgupta and Heal, 1980). During this period, which reached its peak in the 1970s, it was common to hear mineral resources referred to as part of the national heritage of a country, in which all citizens should have an interest. Lack of funds prevented most LDCs from purchasing an equity interest in mineral operations by conventional means. This led to the use of *nationalization* or *expropriation*, the forced transfer of all or part of the operation to the government. The pioneer in such efforts was Mexico, which nationalized its oil industry in March 1938 to form the giant government-owned Pemex corporation (Grayson, 1980). This was followed in 1951 by the nationalization of British oil interests in Iran, a move that was reversed when the government of Iran was overthrown, placing the Shah in power. By the early 1960s, however, nationalization was widespread in areas as far apart as Cuba, Zambia, Zaire, and many Arab and South American countries (Baklanoff, 1975; Ingram, 1974; Akinsanya, 1980). Although many such takeovers were carried out with fair compensation, some were not and many led to long legal and political battles (KCC, 1971; Girvan, 1972).

The rise in oil and other mineral prices, along with publication of *Limits to Growth*, brought mineral companies to the attention of the global financial community during the mid-1970s and early 1980s. Anticipation of quick profits generated by growing mineral shortages and rising prices made the mineral companies targets of *takeover* efforts in which their publicly-owned stock was bought on the open market by other groups, which thereby obtained control of the compa-

nies. Although some takeovers led to improved production, most were motivated financially and resulted in the closing of major operations and the loss of important exploration expertise, particularly when mineral prices did not rocket upward as first anticipated. But these changes were minor compared to those caused by government takeovers. In 1970, multinationals controlled about 70% of world oil and gas production. By 1980, that share had declined to about 50%. At present, 11 of the top 20 oil producers in the world are government-owned and the top 20 government-owned producers control over half of world oil production (Table 6-1). The potential longevity of the national oil companies is even more impressive. On the basis of 1992 reserves, the seven largest multinationals can continue to produce at current rates for 11 years, whereas the seven largest national oil companies will last over 120 years.

Ironically, many national oil and mineral companies that were set up to assure domestic control of domestic resources have now become multinationals themselves with exploration and production interests outside their own borders. This is particularly obvious in the oil industry, where national oil companies have almost doubled their control of world refining capacity

TABLE 6-1

Leading world oil producers in 1990. Companies are divided into publicly-owned and government-controlled organizations. Top 20 government-controlled corporations (not all of which are listed here) control about 50% of world oil production and 85% of world oil reserves.

Organization	Production (million bbl)	Reserves (billion bbl)	Type of Organization
Saudi Arabian Oil Co.	1823.2	255	Government
National Iranian Oil Co.	1075.7	93	Government
Iraq National Oil Co.	1051.4	100	Government
Petroleos Mexicanos	917.4	51	Government
Petroleos de Venezuela	693.3	59	Government
Royal Dutch/Shell	676.0	9.5	Public
Exxon	621.0	6.3	Public
Nigerian National Petroleum Co.	610.3	16	Government
Kuwait Petroleum Corp.	581.4	95	Government
Abu Dhabi National Oil Co.	536.6	92	Government
British Petroleum Co.	515.4	4.9	Public
Petramina (Indonesia)	514.3	11	Government
National Oil Corp. (Libya)	412.5	23	Government
Chevron Corp.	312.0	2.9	Public
Amoco Corp.	296.0	2.7	Public
Texaco Inc.	269.0	2.3	Public
ARCO (Atlantic Richfield Corp.)	267.0	3.0	Public
Sonatrach (Algeria)	260.6	9.2	Government
Ste. Nat. Elf Aquitaine (France)	249.7	NA	Government
Mobil Corp.	249.0	2.6	Public

during the last decade, as well as becoming major marketers of gasoline and other products in North America and Europe.

The transition to government ownership of nonfuel minerals was also strong in the 1970s and early 1980s, with large fractions of world iron ore, bauxite, copper, tin, and phosphate production assumed by government corporations (Table 6-2). In the 1950s, 75% of world copper production outside the centrally planned countries was controlled by eight publicly-owned corporations; now more than 50% is controlled by government-owned corporations. CODELCO, the Chilean copper mining company that took over operations of the Kennecott and Anaconda mines in that country, is the largest copper producer in the world.

Interestingly, a growing trend toward *privatization,* the sale of government corporations to private investors, has surfaced since the late 1980s. British Coal Corporation announced in 1992 that it would privatize after 45 years of declining results under government control, and parts of ENI are for sale in Italy. In an unusual privatization attempt, Empresa Minera del Hierro del Perú, a government-owned iron mining corporation, was sold at the end of 1992 to a state-owned Chinese company, Capital Steel (Shougang Corp.) for $312 million. (The purchase price, which was more than three times the assessed value of Hierroperú, reflected the urgent need in China for iron ore to support growing demand for steel.) The largest privatization, if it takes place, will involve state-owned mineral operations in the former Soviet Union.

PROFITS IN THE MINERAL INDUSTRY

Differing Views of Mineral Profits

With its complex structure, it is no surprise that the definition of profit is not simple for the world mineral industry. Consider the Real del Monte silver-gold mines near Pachuca, Mexico, for example. These veins have been mined since the 1500s (Randall, 1972) and by 1973 had produced more than 20 billion ounces of silver, over 6% of total world silver supplies to that time. By 1974, however, the mines were not profitable and they were sold to Fomento Minero, a branch of

TABLE 6-2
Commodities with important production by government-controlled corporations (other than those in the original centrally planned economies). Compiled from Mikesell, 1987 and U.S. Bureau of Mines.

Commodity	*Country*	*Major Producer*
Iron Ore	Brazil	Companhia Vale do Rio Doce (CRVD)
	India	National Mineral Development Corp.
	Liberia	Bong Mining Co.
	Mauritania	Societe Nationale Industrielle et Miniere
	Venezuela	C.V.G. Ferrominera del Orinoco
Bauxite	Ghana	Ghana Bauxite Co.
	Guyana	Guyana Mining Enterprise (Guymine)
	Jamaica	Bauxite and Alumina Trading Co. of Jamaica
	Indonesia	INALUM
	India	NALCO
	Guinea	Compagnie de Bauxites de Guinee
Copper	Chile	CODELCO-Chile
	Peru	Empresa Minera del Centro del Peru (Centromin Peru)[a]
		Empresa Minera del Peru (Minero Peru)[a]
		Empresa Minera Especial Tintaya[a]
	Zaire	La Generale de Carrieres et des Mines du Zaire (Gecamines)
	Zambia	Zambian Consolidated Copper Mines
Nickel	Finland	Outokumpu Oy
Gold	Dominican Republic	Rosario Dominicana
Tin	Bolivia	COMIBOL[b]
	Indonesia	P.T. Timah

[a] Government decided to privatize in 1991 and is seeking investors.

[b] Government decided to privatize in 1989 and is seeking investors.

the Mexican government, for $3.5 million. A major reason for Fomento's purchase was the Mexican government's need to provide continued employment to thousands of miners, rather than to have them drift around the country looking for work, perhaps contributing to social unrest. The decision was probably a good one; when the price of precious metals rose in the late 1970s new veins were found, good profits were made, and the deposits were rejuvenated.

Even where the government does not buy the mineral property, it can still have a strong effect by *subsidizing* the operation. During the 1960s, gold producers were especially in need of help because the world price had been held constant at $35 per ounce since 1934 by government action, as discussed in the section on gold. As costs rose, many mines were forced out of business, a problem that was most acute in the big gold mining areas of Canada and South Africa. In Canada, the federal government subsidized operations between 1948 and 1973 through the *Emergency Gold Mining Assistance Act.* Through 1972, over $C302 million was paid on production of 61.8 million ounces of gold, a subsidy of almost $5/ounce at a time when gold sold for only $35 per ounce (EMR, 1973). A similar subsidy offered by the South African government was particularly necessary because many of the mines are joined together underground. Thus, groundwater filling abandoned gold mines would have increased pumping costs at adjacent operating mines, dragging them into bankruptcy.

Several other considerations might allow mineral deposits to be operated under conditions that are probably not profitable in the strict sense of the word. *Strategic minerals,* which are usually defined as minerals that are essential for national defense, often receive special treatment from governments (van Rensburg, 1986). U.S. government purchases of uranium for weapons manufacture were the main support for the domestic uranium industry until the advent of nuclear power plants. SASOL in South Africa, the world's largest facility for conversion of coal to oil, was begun as a government effort to limit dependence on imported oil. Some governments maintain *stockpiles* of strategic minerals. Although the original intent of most such stockpiles was defense, purchases for them can be used to support mineral producers, or sales from them can be used to depress commodity prices to levels that encourage consumption. Sales from stockpiles can even be used as a source of revenue, particularly in periods of budget deficit. Sales from the Russian gold stockpile have served this purpose. The United States maintains an extensive stockpile, including oil, as discussed further in the final chapter.

The need for *foreign exchange* is another consideration affecting government attitudes toward mineral production. The sale of minerals on international markets provides funds for other foreign purchases. Zambian Consolidated Copper Mines, for instance, which is 60% controlled by the Zambian government, accounts for 85% of the foreign-exchange earnings of the entire country. Some countries barter minerals for other commodities. For many decades, Cuba traded nickel for Soviet oil. Even in publicly-owned or privately-owned corporations, the extraction of a mineral deposit might not be controlled exclusively by its immediate profitability. The most obvious case occurs when the price of a commodity decreases to a level at which production is unprofitable, but management is convinced that the price will rise. Under those conditions, it can be less expensive to continue operations at a loss for a short period rather than incurring care and maintenance costs after shut-down without any offsetting revenues.

Accounting Methods and Mineral Profits

In free-market economies, it is necessary to calculate the profit of any business operation. The exact steps taken to determine profit depend on prevailing accounting principles, but they boil down to a comparison of income and expenses directly attributable to the operation (Gentry and O'Neil, 1984; Vogely, 1985; Wellmer, 1989). This statement is useful not only as a measure of the benefits derived from the operation but also as the basis for calculation of any taxes or royalties that it might owe, as discussed later in this section.

Publicly-owned companies publish an *annual report* outlining their activities and financial results, including a *balance sheet* that compares assets, liabilities, and owners' equity, as well as a *statement of income* that records income and expenses of the company. Lornex Mining Corporation provides an example of this type of information and shows the critical role of taxes in the economic viability of mineral properties. Lornex owns and operates a large copper mine in the Highland Valley of southern British Columbia (Figure 6-2). During the late 1970s, the company was partly owned by Rio Algom, a Canadian mining company, which was partly owned, in turn, by RTZ Corporation, a British multinational. Shares of Lornex not owned by Rio Algom were available for purchase on the Vancouver stock exchange, shares of Rio Algom were available on the Toronto stock exchange, and shares of RTZ, which started in the Rio Tinto area of Spain, were traded on the London stock exchange. (In later transactions, Lornex was acquired by Rio Algom and another Canadian mining company and RTZ sold its interest in Rio Algom.) In its 1975 report, Lornex showed that it produced 12,893,000 tons of ore, which

FIGURE 6-2
Lornex mine in British
Columbia, Canada, showing
concentrator in foreground,
dumps for overburden in
middle ground and mine in
right background (courtesy of
Rio Algom, Ltd.).

were beneficiated to yield concentrates containing 107,160,000 pounds of copper and 3,084,000 pounds of molybdenum. Sale of these concentrates to smelters in Japan yielded *revenues* of $51,043,000 to the company.

In order to determine how much of this $51,043,000 was profit, Lornex accountants deducted expenses related to the production of the ore and concentrate. The largest such deduction, amounting to $30,818,000, consisted of *operating expenses.* These included salaries of employees, heat and light for the buildings, fuel for the trucks, paper for the report itself, and many other items, all of which were recognized as deductible expenses by the Canadian federal tax code administered by Revenue Canada. *Amortization and depreciation* deductions in 1975 amounted to $5,917,000. These included a fraction of the cost of all depreciable assets such as trucks, ore beneficiation equipment, and computers. By allowing such a deduction, tax codes permit operators to set aside funds to purchase replacements as the original equipment wears out (although the funds do not have to be used for this purpose). Finally, a deduction of $6,146,000 was taken on net *interest payments* made on a debt of $78,174,000 of which $66,952,000 was *long-term debt,* most of which had been borrowed to set up the mining operation. These deductions totaled $42,881,000, leaving an operating profit (or *earnings before taxes)* of $8,162,000. *Taxes* on this income amounted to $7,536,000, leaving *net earnings after taxes* of $626,000.

The statement of earnings shows that the Lornex mine earned a profit, although you might have been surprised by the size of the tax bill. In fact, you might even ask whether it is fair for taxes to consume 87% of the profit from an operation, leaving only 13% for the owners. This is a hard question to answer, but it is at the heart of our efforts to assure long-term mineral supplies, and we must try. To do it, we must review the concepts of distribution of profits and return on investment.

Distribution of Profits and Return on Investment

Recall that in 1975 Lornex was owned partly by Rio Algom and partly by public shareholders. Even if you do not own shares in a corporation personally, other groups that you depend on such as your bank, credit union, or pension fund do own shares. Thus, we all have an interest in the profitability of corporations. This interest derives from the fact that some of the profits made by corporations such as Lornex are paid as *dividends* to the owners, on the basis of their share of ownership. Most companies hold back some of the profits, known as *retained earnings,* to pay for future projects. For instance, in 1990, the Amoco Corporation had net income after taxes of $1.913 billion but paid dividends of only $1.038 billion, retaining the rest for corporate projects.

Shareholders who invest in any corporation do so with the hope of making their money grow. This growth, which is referred to as *return on investment,* is the most important parameter controlling whether an investment will be made. It is usually expressed as the percentage of the investment that is returned to the investor each year. For a bank account yielding 5% interest annually, the return on a $100 deposit is $5. For most corporations, a comparable number can be obtained by dividing *operating profit* by *shareholders' equity,* which is the amount of money that the shareholders have invested in the corporation. Shareholders' equity is essentially the difference between the *assets* (cash, funds owed to the corporation, goods that have not yet been sold [inventories], and other supplies) and *liabilities* (debt or taxes that have not yet been paid). Although shareholders' equity is a positive number in most cases, it can be negative, if the corporation has borrowed lots of money and lacks cash in the bank or inventories. In the case of Lornex, shareholders' equity at the end of 1974 was $64,538,000. Thus, the operating profit of $8,162,000 was 12.6% of shareholders' equity.

This percentage is a very useful index of the profitability of the Lornex operation and it allows us to assess whether the profits were fair. The simplest comparison to make is between the return that Lornex shareholders made on their investment and the return that they would have gotten at the bank. Bank interest rates vary, so we will have to take a representative rate for this comparison, such as the 5% mentioned earlier. If the Lornex shareholders had put their $64,538,000 in a bank paying that interest rate on deposits at the beginning of 1975, they would have earned about $3,227,000 compared to the $8,162,000 that they earned from the mine. However, we must remember taxes, which apply to both of these profits. At present U.S. and Canadian tax rates, about one third to one half of the bank interest would be taken as tax. For the sake of argument, let's use one half, which would make the bank deposit yield an after-tax profit of about $1,613,000 or 2.5%. In contrast, Lornex yielded $626,000 or less than 1% after tax.

Something is wrong here. It is common knowledge that people put money into banks partly for safety and ease of access, and that they accept a lower return on their investment for the peace of mind. We also know that people occasionally risk money on an investment that has only a poor chance of success, but which would pay a really big return if it did succeed. In other words, investors look for a relation between the risk that they take and the reward that they expect. The higher the risk, the higher the reward that must be paid to attract investment. The mineral business is risky, as you might expect from the many technical challenges associated with exploration and production. So, how could Lornex's investors have gotten into a situation where they were making an after-tax profit smaller than they would get in a bank? As you can see, the answer has nothing to do with the technical aspects of mineral exploration and production. Instead, it is related to the two main factors that control mineral profits, mineral commodity prices and taxes, which are reviewed next. In the course of this review, we will see how Lornex's profits were decimated by a peculiar coincidence of these two factors.

MINERAL COMMODITY PRICES AND MINERAL PROFITS

Mineral Commodity Prices

Over the long term, most mineral prices have not kept up with inflation. As can be seen in Figure 4-2, prices for only natural gas, gold, cobalt, oil, gem diamond, vermiculite, and possibly nickel, manganese, and arsenic outpaced inflation between 1960 and 1990. In fact, most mineral commodities are considerably cheaper today than they have been at any time in the past.

These long-term trends disguise the fact that mineral commodity prices respond to short-term changes in supply and demand, often referred to as *marginal demand* (MacAvoy, 1988). Short-term effects can be caused by accidents that limit production at a major deposit, technological developments that permit substitutions in an important application, or governmental changes in a major producing country. These and other factors exert a powerful force, as you can imagine by estimating the price of air in a sealed room holding 100 people but containing only enough air for 99 of them. If supply or demand are delicately balanced, as they are in most cases, only small perturbations in either one can cause large changes in price (Figure 6-3).

To minimize the impact of short-term price fluctuations, *commodity futures markets* provide a forum in which buyers and sellers establish future prices. These markets trade *contracts* for the delivery of specified quantities of commodities at a certain date and price. Copper contracts on the New York Commodity Exchange are for 25,000 pounds of the metal for delivery on any of the next eight months into the future. Contracts can be closed before they expire by selling those that were originally bought or vice versa. Thus, no one has to deliver or take delivery of the commodity. This permits speculators with no use for the commodity to participate in the market, thereby in-

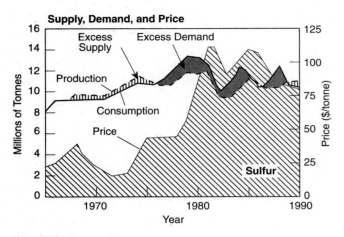

FIGURE 6-3

Changes in price and supply of sulfur in the United States since 1965 showing how small changes in production and consumption can cause major price changes (compiled from *U.S. Bureau of Mines Minerals Yearbook*).

creasing the number of potential buyers and sellers and improving the *liquidity* of the market. To further increase liquidity, most exchanges permit contracts to be bought and sold with a small down-payment, often 10% of the value of the commodity in the contract, a practice referred to as *trading on margin.*

Markets of this type permit anyone to attempt to profit from their best guess of the future price of a commodity. Producers who feel that the price will decline might sell a contract to deliver their future production in six months at a fixed price. On the other end of the trade might be a consumer or speculator who is convinced that the price will rise in that same period. By examining the futures markets, producers and consumers can determine the consensus regarding future prices for mineral commodities. Commodities exhibiting *cotango* have future contracts that are more expensive than the cash, or present, price. Those with cheaper future contracts are said to exhibit *backwardation.*

Price Controls and Cartels

The important control that mineral prices exert on profits has led to numerous efforts at *price control* by individuals, corporations, and governments. These efforts have been motivated by a desire for increased profits, protection from high prices, and prevention of losses in a declining market (Prain, 1975). Most price control efforts have tried to control the supply of the material, although a few efforts have been made to control demand. When control involves a single individual or organization, it is known as a *monopoly;* when a group is involved, it is known as a *cartel.*

Modern mineral monopolies are rare simply because of the large funding necessary to control world markets. One recent monopoly effort involved the Hunt brothers, who tried to corner the world silver market in 1979 (Hurt, 1981). The Hunt group purchased more than $1 billion in silver in the form of futures and other contracts, helping the price rise from $9/ounce to almost $50/ounce. When the group could not meet a margin call, that is, put up additional funds as collateral for the funds borrowed to buy the silver, they had to sell some of their contracts, causing prices to collapse and driving the Hunts to declare bankruptcy. In a similarly ill-fated venture, the government of Malaysia lost hundreds of millions of dollars in 1981 and 1982 during a clandestine attempt to control the price of tin on world markets (Pura, 1986).

Cartels, which are more common, include both corporations and governments, with government cartels usually being more powerful. There is a wide variety of opinion on cartel activity among governments of the world, with the United States at one pole and centrally planned economies at the other. Cartel activity in the United States was limited by antitrust legislation of the *Sherman and Clayton Acts of 1890,* which had a big effect on the structure of the domestic mineral industry. The first major action against a mineral corporation under the law came in 1911, when the Standard Oil Trust was broken up to form Amoco, Chevron, Exxon, Mobil, and Standard of Indiana. Legal action was also brought by the U.S. federal government against U.S. Steel and Alcoa, the dominant producers in the steel and aluminum industries, although with less effect.

Government efforts at price control by edict have usually been given palliative names such as *mineral price stabilization.* The U.S. National Bituminous Coal Commission, which operated between 1937 and 1943, had the power to set coal prices and was intended to stabilize the weakened coal industry. The Texas Railroad Commission, which has controlled the production of oil in Texas since the mid-1930s, was set up ostensibly to prevent damage to the natural drive of oil fields by overproduction. In fact, it was able to control world oil supplies because of the large percentage of production coming from Texas during this time. In some cases, government price stabilization groups in the same country have come into conflict with one another. During the 1970s in the United Kingdom, the National Coal Board and British Gas Council, which were in charge of commodity production and prices in their respective industries, engaged in direct competition for energy expenditures by the public.

One of the largest price stabilization efforts ever undertaken has been the quixotic attempt by the U.S. government to control domestic oil prices. The modern chapter in this continuing saga began in 1971 when President Nixon froze all prices and wages in the economy under the auspices of the *Economic Stabilization Act of 1970*. Controls on oil were replaced briefly by voluntary guidelines in 1973, when the *Emergency Petroleum Allocation Act* set the price for "old" oil, defined as an amount equal to the oil that had been produced annually before 1973, at $5.25/bbl. Oil production in excess of 1972 levels was termed "new" oil and was allowed to sell at an uncontrolled price, as was imported oil. This approach failed almost immediately in the face of the OAPEC oil embargo and resulting price rises that began in late 1973 (Figure 7-1C), as oil companies increased the price of new oil so that the average price of oil reflected real demand. To combat this, buyer-seller agreements were frozen in December 1, 1973, and the Federal Energy Office was given the responsibility of controlling oil prices. In the ensuing confusion, prices were determined by endless bickering over byzantine formulae that required producers of "old," cheap oil to pay *entitlements* to rival companies with higher-priced crude, simply for the right to refine their own crude oil. By late 1975, payments totaled over $1 billion and the system collapsed.

The *Energy Policy and Conservation Act of 1975* then divided oil into three price-control tiers, but failed to cope with the strain of price changes associated with the Iranian revolution of 1979. As an acknowledgment of the failure of price controls, Congress replaced price controls with the Windfall Profits Tax in 1980, which allowed prices to move without significant restraint but attempted to capture most of the profits resulting from the price rise, as discussed later (Chester, 1983; Ghosh, 1983).

Although the U.S. government has discouraged cartels (other than its own efforts at price control), governments in other countries have been more accommodating. During the last century, cartels or monopolies have been organized with varying success in aluminum, bauxite, coal, copper, diamonds, lead, mercury, nickel, nitrate, crude oil, potash, silver, steel, sulfur, tin, tungsten, uranium, and zinc. Several of these were active as recently as the last few decades (Table 6-3), but only the Central Selling Organization (CSO) of De Beers, which deals in diamonds, has maintained long-term control over its market. Even OPEC, which was the most powerful economic force in the world during the late 1970s and early 1980s, failed to control oil prices through the crude oil glut of the late 1980s. The rise of OPEC was due largely to an incredible concentration of production in the Middle East at the time. In response to the rapid increase in oil prices brought about by the oil embargo of 1973, however, new supplies came into production in non-OPEC nations including the United Kingdom, Canada, Norway, and Mexico, breaking OPEC's influence on world oil prices (Johany, 1980; Al-Chalabi, 1986; Kohl, 1991).

One of the strangest market control efforts in recent times concerns the *Producer's Club*, a group of uranium-producing companies that was formed in the early 1970s (Table 6-3). This group, which was encouraged by their host governments, attempted to place a floor on uranium prices as government purchases for nuclear weapons declined. Uranium markets were made even smaller for non-U.S. producers when imports to the U.S. market were curtailed to support domestic producers. At about the same time, nuclear reactor manufacturers began to compete for the electric power generating market. In order to induce utilities to use their reactors, some companies offered to supply reactor fuel at low prices. By late 1974, the Westinghouse Corporation had contracted to deliver about 80 million pounds of low-priced uranium to its reactor customers, but held contracts to buy only about 15 million pounds. This short position, which amounted to about two full years of U.S. uranium production, caused upward pressure on uranium prices at the same time that the overall uranium market picked up because of forecast demand for reactor fuel. The resulting price rise quadrupled the price of U_3O_8 to about $40/pound, leaving Westinghouse with a potential loss of $1 to $2 billion if it lived up to its contracts. Instead, it defaulted and began a protracted legal war against the Producer's Club, charging that they fixed the price of uranium, and the utilities countersued to get their contracted cheap uranium (Taylor and Yokell, 1979; Gray, 1982).

The history of cartels is littered with failures. To succeed, a cartel must control the supply of the commodity and the commodity itself must be essential and nonsubstitutable. In addition, members of the cartel must be well financed and willing to withstand periods of low demand. Perhaps most importantly, they must have similar internal cost structures so that low commodity prices do not cause competition among members that are still making a profit and others that are already losing money. A mutual belief or value system is also helpful to hold groups of this type together during trying times. Most recent cartels have lacked at least one of the key elements, and the future of cartels appears dim, indeed. In fact, present efforts to stabilize prices focus more on *market transparency*, in which producing countries share information on production, consumption, and other aspects of a commodity, making it easier for producers and con-

TABLE 6-3
Mineral resource cartels and market study groups active since 1970. Most were formed
when commodity prices were rising and have since been disbanded or lost power to
market study groups (compiled from Cammarotta, 1992, and *U.S. Bureau of Mines
Minerals Yearbooks*).

Cartel	*Members*
Association of Iron Ore Exporting Countries	Algeria, Australia, Chile, India, Mauritania, Peru, Philippines, Sierra Leone, Sweden, Tunisia, Venezuela
Association of Tin Producing Countries[a] (ATPC)	Australia, Bolivia, Indonesia, Malaysia, Nigeria, Thailand, Zaire
Central Selling Organization (CSO)	De Beers Centenary AG as agent for most diamond-producing countries in the world
International Association of Mercury Producers (1976)	Algeria, Italy, Peru, Spain, Turkey, Yugoslavia
International Bauxite Association	Australia, Guinea, Jamaica, Suriname
International Council of Copper Exporting Countries (CIPEC)	Australia, Chile, Peru, Zaire, Zambia
Organization of Petroleum Exporting Countries (OPEC)	Algeria, Ecuador, Gabon, Indonesia, Iran, Iraq, Kuwait, Libya, Nigeria, Qatar, Saudi Arabia, United Arab Emirates, Venezuela
Tungsten Producers Association[b] (1976)	Australia, Bolivia, Peru, Portugal, Thailand
Uranium Producer's Club[c]	Denison Mines (C), Eldorado Nuclear (C), Electrolytic Zinc (A), Getty Oil (US), Gulf Oil (US), Imetal (F), Noranda (C), Nuclear Fuels (SA), Pancontinental (A), Pechiney (F), Peko-Wallsend (A), Queensland Mines (A), Ranger Mines (A), Rio Tinto Zinc (UK), Uranex (F), Uranium Canada (C), Western Mining (A)
Study Group	*Membership, Production Represented*
International Copper Study Group	16 countries, 59% of world trade
International Lead and Zinc Study Group	32 countries, 90% of production
International Nickel Study Group	14 countries, 80% of production

[a]Cooperates with International Tin Council, which includes producing and consuming nations.
[b]Largely superseded by International Tungsten Industry Association with 12 mining companies, 16
  trading companies, and 27 processors in 19 countries.
[c]This group did not function as a public cartel and its membership has not been listed publicly;
  this list includes companies said to have had ties to the Uranium Producer's Club (Taylor and
  Yokell, 1979, pp. 71-72) along with their national affiliation (A = Australia, C = Canada, F =
  France, SA = South Africa, UK = United Kingdom, US = United States).

sumers to make plans (Cammarota, 1992). Most such organizations have started life under the wing of the United Nations Conference on Trade and Development (UNCTAD), although many have opted to operate autonomously (Table 6-3).

Relation of Mineral Profits to Mineral Prices

Returning to Lornex for a moment, its 1975 earnings fell victim to a severe decline in the price of copper. In 1975, the average world copper price was only $0.54/lb, a big decline from the $0.90/lb that prevailed in 1974 (Figure 9-1). The 1974 price was an abrupt increase from much lower levels in 1973, when the price averaged only $0.46/lb. These wide swings in the copper price had an enormous impact on the revenues of Lornex, which went from $85,421,000 in 1974 to $51,043,000 in 1975, a decrease of about 40%. Some of this decrease was caused by a reduction of about 16% in production, but most of it reflected declining copper prices.

The profits of most mineral producers are similarly *cyclical* and rise and fall with changes in the price of the mineral commodities that they sell. Although producers might lose money in some years, the cyclical highs and lows must average to an acceptable profit level if the operation is to continue (Figure 6-4A). As the Lornex situation shows, the pendulum swings both ways. When copper prices were high in 1974, for

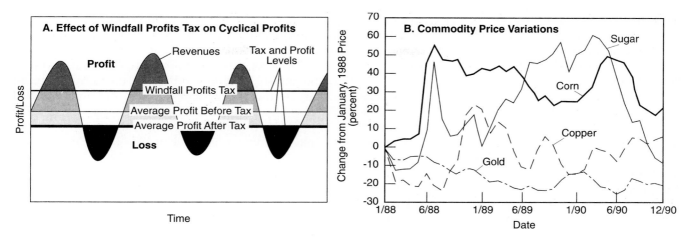

FIGURE 6-4

A. Cyclical nature of mineral profits showing how high and low profits average to an acceptable level. **B.** Variation in futures cash price of copper, gold, corn, and sugar from 1988 through 1990 showing that mineral price variations are not greater than those of other commodities.

instance, the company had after-tax profit of $22,681,000, which yielded a return on investment (as a percentage of shareholders' equity at the time) of 54%, a level that would be considered excellent by anyone.

The huge price rise that occurred in 1974 affected most metals and resulted in similarly large profits for most producers. Thus, governments throughout the world were faced with high commodity prices, large corporate profits, and a generally disgruntled citizenry that did not like what had happened to the prices they were paying for mineral products and resented the profits that were being reported to them, usually in a rather sensational way, by companies and newscasters. In British Columbia, the government responded by levying extra taxes, in the form of royalties, that were retroactive to the start of 1974. These increased provincial taxes, which were originally deductible from Canadian federal taxable income, caused an decrease in federal tax revenues and led to a dispute between the two governments over entitlement to revenues from taxation of mineral production (a direct outgrowth of the way government mineral ownership was originally defined in Canada). As a result, the federal government disallowed deduction of the provincial taxes and royalties in the calculation of federal tax. This resulted in an extremely high tax burden, which essentially nullified profit in 1975. So, to understand mineral profits, one must also understand mineral taxes and the factors that motivate them.

MINERAL TAXATION AND MINERAL PROFITS

The link between mineral profits and taxes is inescapable, and taxation is strongly influenced by mineral prices. Although, the fluctuations in mineral prices are not really different from those seen in prices of agricultural products (Figure 6-4B), they affect the prices of larger expenditures such as a tank of gasoline or an entire car, rather than a cup of coffee or a can of orange juice. In addition, producers of minerals tend to be large and visible parts of the economy and they are correspondingly large sources of revenues to government. Thus, when mineral profits rise, government has a natural tendency to raise taxes to obtain some of this new income, making taxes the single largest factor affecting the economic viability of mineral operations.

Although the original purpose of taxation was to generate income to support government activities, lawmakers have not been able to resist using taxes as instruments of policy. One rationale for this approach is the assumption that taxing less desirable activities will promote more desirable ones. The many conflicting interests in conservation, exploration, inexpensive energy, and other features of our mineral supplies make them obvious targets of policy-oriented taxes.

Income Tax and Depletion Allowances

The most common tax is *income tax*. It is usually expressed as a rate, or percentage of operating profit, often with a sliding scale such that the percentage of tax increases as profit increases. In addition to the income tax rate, the amount of income tax paid by any organization is controlled by the deductions that can be taken in determining taxable income.

Because it involves the extraction of nonrenewable resources, the mineral industry is sometimes afforded a special deduction known as the *depletion allowance* (O'Neil, 1974; Gentry and O'Neil, 1984). Unlike clothing manufacturers, for instance, which can operate indefinitely by buying cloth and replacing sewing equipment, a mine or oil well must stop production when the reserve of ore or oil is exhausted. Thus, as soon as it begins operation, a mine or well is putting itself out of business. In a way, the depletion allowance is a special form of depreciation, which recognizes that the mineral deposit wears out with time. It permits mineral producers to deduct a portion of income from tax each year to compensate it for the loss of this asset.

The problem comes, of course, in determining the magnitude of the depletion allowance deduction. Present practice in the United States uses two methods, *cost depletion* and *percentage depletion* (Table 6-4). Producers are required to calculate both types of depletion and to use the larger of them in calculating taxable income. Although percentage depletion is usually the larger of the two, cost depletion provides a deduction even when no profit has been made. Per-

centage depletion was introduced in response to excessive litigation resulting from disagreement between producers and the Internal Revenue Service over the calculation of cost depletion. Actual rates used range from 5% to 22%, depending on the commodity. In many accounting systems, profits, depreciation, and depletion in any year are totaled to provide a value of *cash flow,* the funds actually available to the corporation for operations (exclusive of any new funds raised by borrowing or sale of stock).

Depletion is one of the most contentious aspects of the tax code. Supporters argue that it is a logical extension of the depreciation concept and that it is required to make mineral production competitive with other investment opportunities. Critics say that it is a huge loophole through which important amounts of revenue escape taxation. Opponents are particularly disturbed when these deductions are used for purposes other than replacement of exhausted mineral deposits, a practice that is legal (recall that depreciation deductions taken by a business need not be used to buy new equipment). The acquisitions of Montgomery Ward by Mobil and of Reliance Electric by Exxon caused great consternation, for example. Adverse opinion about the depletion allowance reached a crescendo in the United States during the OAPEC oil embargo of 1973 and the Iranian revolution of 1979, when the rapid rise in oil prices greatly increased oil company profits. In response to public concern about these

TABLE 6-4

Comparison of cost and percentage depletion deductions in the U.S. federal tax code for a typical small porphyry copper deposit that produced 11 million tonnes of ore in the tax year, with revenues of $52 million and taxable income of $4.5 million. The cost of the deposit was $1.5 million, preproduction expenses were $3 million, annual exploration costs were $1.5 million, and remaining reserve is 100 million tonnes.

Cost Depletion	= *prorated value of each ton of ore times number of tons produced that year.*
Deposit Value	= cost of deposit + preproduction exploration expenditures
	= $1,500,000 + $3,000,000 = $4,500,000
Value of Ton	= deposit value/number of tonnes
	= $4,500,000/100,000,000 tonnes
	= $0.045/tonne
Cost Depletion	= $0.045/tonne * 11,000,000 tonnes
	= $495,000
Percentage Depletion	= *allowable percentage multiplied by revenue, not to exceed 50% of taxable income.*
	= $52,000,000 * 15%
	= $7,800,000
But this is larger than half of taxable income (excluding depletion allowance),	
	= $4,500,000 * 50%
	= $2,250,000
Applicable Depletion Allowance	= *larger of cost and percentage depletion*
	= $2,250,000

profits, the depletion allowance was disallowed and the Windfall Profits Tax was established. (In spite of its misleading name, the Windfall Profits Tax is an excise tax and is more appropriately discussed in a later section.) Metal mining companies also made large profits during the early 1980s in response to rapid increases in metal prices, but their depletion allowances were not cancelled.

Other governments approach the depletion deduction differently. Until 1989, the Canadian federal tax code included the concept of *earned depletion* (Parsons, 1981), in which corporations were able to deduct four thirds of exploration and other qualifying expenditures. Although it has been suggested that this deduction is an improvement over a pure depletion allowance because it focuses the deduction on mineral-related expenditures, earned depletion was discontinued in Canada after 1989. The Canadian tax code retains a deduction of 25% of resource profits, which is very similar to percentage depletion as applied in the United States. Because of the importance of mineral production to the Canadian economy, other deductions have been allowed from time to time, including *flow-through financing*, in which individuals can deduct the cost of exploration ventures directly from current income. This approach, which has been discontinued, pushed almost $2 billion into exploration between 1983 and 1987, creating many jobs, some discoveries, and criticism that the regulations should have been structured to encourage production of minerals rather than to reward those who spend the most in exploration (Woodall, 1988).

Tax holidays have been used as a means to encourage mineral exploration and production in areas with high unemployment. When Ireland passed legislation in 1956, exempting mineral production from income and corporate taxes, an exploration boom resulted in several important discoveries, including Navan, the largest zinc producer in Europe. By 1973, mining was well established in Ireland and taxes had been raised to world norms. The same approach was taken in Australia where gold mining, the most important productive activity in the arid, western part of the country, was granted freedom from taxation. Gold mining increased tremendously in the area, due in part to this incentive and in part to the increased world price of gold during the late 1970s and early 1980s, and the law was later repealed. Both of these successful tax holidays were originally passed without a termination date. In contrast, tax holidays with specific terms have been notably unsuccessful. Prior to 1973, the Canadian federal tax code permitted a three-year tax-free production period for new mines. Unfortunately, this led to wasteful extraction practices such as *high-grading*, the removal of only the highest-grade ore, during the

tax holiday in order to maximize profit. This made it harder to mine remaining low-grade ore and caused many mines to close prematurely.

Ad Valorem Taxes

In addition to taxing income, it is possible to tax the value of assets and the commodity that is produced. These taxes must be paid whether the operation makes a profit or not, and are often regarded as regressive for that reason. Nevertheless, they are a common part of mineral industry taxes in most countries.

Reserve taxes are a special form of *ad valorem* or *property tax*, which is a levy on physical property. They are usually expressed as a percentage of the value of the property. Property taxes are applied almost everywhere to the value of the office and physical plant at mineral operations, but some jurisdictions place an additional levy on the value of the mineral reserve itself. Proponents of the reserve tax view it as a logical extension of the property tax concept. Opponents counter that minerals in the ground have no value. They say that value is realized only when minerals are produced and sold, and that changing economic conditions might convert ore to waste before it is mined. It is also argued that reserve taxes limit recoverable reserves because they apply equally to both high-grade and low-grade ores, making it proportionally less economic to produce low-grade ores. Sticklers even point out that it is not logical to disallow the depletion allowance on the one hand, thereby indicating that the asset has no value, and impose a property tax on the other hand, which accepts that a deposit has value. This logical conundrum has not daunted legislators in the United States largely because property taxes are levied by state and municipal governments, whereas the depletion allowance is part of the federal tax code.

Severance taxes are also specific to the resource industries and are applied to each unit of production, as are sales taxes. Severance taxes are similar to *royalties* (Bourne, 1989; Olson and Kleckley, 1989), which are commonly granted to the owner of land from which minerals are produced. In the case of production from private land in a state or province with a severance tax, the producer would pay both levies. In theory, these taxes are supposed to compensate the owner of the land for any loss in value caused by removal of the minerals. Arguments against severance taxes note that the minerals had no value until someone risked funds to set up an extraction operation and that groups not willing to accept the risk by way of ownership in the mineral production organization have no right to the rewards. This argument has not been very convincing, however, and severance taxes currently yield about $4.5 billion to the 28 states that levy this tax. As can be seen in Table 6-5, essentially all revenues come from taxes on oil and gas production, with 47% coming from Alaska and Texas. Royalties are also important sources of income to private individuals and corporations. Although royalties differ greatly for most hard minerals, they are almost always the same, one eigth of production, for oil and gas. As discussed earlier, anticipated changes in the Mining Law of 1872 could well incorporate a royalty for production of minerals from federal land.

Two aspects of severance taxes merit special attention. The first is their application to reimburse citizens for a loss of what is sometimes referred to as their *natural (or national) heritage*. The idea here is that removal of the minerals diminishes future employment and economic opportunities in the area. Some governments put part of their severance tax revenues into funds that will be used to relieve these problems in the future. The largest such effort is the Alberta Heritage Fund, which was set up by the Alberta government to invest in businesses that will provide employment in the future when oil revenues are no longer important. Minnesota allots 50% of its severance tax revenue to a fund for use in the northern part of the state where iron ore is produced, and other states use some of the revenues for environmental cleanup.

Concern has been expressed recently about whether severance taxes restrain *interstate trade* (Stinson, 1982). The issue was brought to a head by the very high severance taxes imposed on coal by Wyoming and Montana (Table 6-5). Low-sulfur coal in these states is in high demand to cut SO_2 emissions from power plants in the eastern United States. Because there are no other domestic sources of low-sulfur coal, these states can impose large severance taxes without fear of losing sales. The issue of just how large these taxes can be without imposing an unfair burden on consumers in other states has been brought to the courts, which have ruled that it should be decided by legislative action. Unless it is decided, resource-related severance tax wars between states or provinces are a distinct possibility.

Excise taxes are also essentially sales taxes in that they are levied on the value of production with little or no regard to the profitability of the operation. The most famous excise tax in recent history is the inaptly named *Windfall Profits Tax*. This tax, which was signed into law on April 2, 1980, resulted from the decision of the federal government to discontinue efforts to control crude oil prices and, instead, to let prices rise and capture most of the higher prices in taxes. The national preoccupation over large profits made by oil companies in the late 1970s when the price of oil rose from about $12/bbl to over $30/bbl added support for this move (Figure 7-1). In spite of its name, the Windfall Profits Tax was a pure excise tax. Its main thrust was to place a 70% tax on the sales price of oil above a *base price* of $12.81/bbl (Table 6-6). Formulation of this tax occupied Congress for several months, and negotiations over the provisions of the tax were based largely on economic projections of income that it would generate. The final version of the tax that was hammered out in conference between House and Senate committees was supposed to bring in $227 billion by 1990. Not surprisingly, this economic projection was wildly wrong. By the late 1980s, the price of oil had fallen to $12/bbl and the tax was generating essentially no income. In August 1988, it was repealed,

TABLE 6-5

Ten largest state severance tax revenues for 1988, showing major source of revenues and percentage of revenues from oil and gas. Only Arizona, Utah, and Nevada obtain significant severance tax revenue from nonenergy minerals (State Fiscal Capacity and Effort, 1990, Congressional Information Service).

State	Total Revenues	Principal Sources	Percentage from Oil and Gas
Alaska	$1,072,000,000	Oil, gas	100
Texas	1,058,800,000	Oil, gas	98
Louisiana	465,700,000	Oil, gas, sulfur	84
Oklahoma	386,700,000	Oil, gas	92
New Mexico	293,400,000	Oil, gas, coal	57
Wyoming	230,500,000	Oil, gas, coal	69
Kentucky	210,000,000	Coal	7
West Virginia	148,000,000	Coal	20
Montana	112,800,000	Coal, copper	17
North Dakota	90,900,000	Oil, gas, coal	78

TABLE 6-6

Provisions of the Windfall Profits Tax. Oil was divided into three tiers, each of which had a base price above which all income was taxed at the rate specified (*Energy Policy Congressional Quarterly*, 1980, pp. 223–224).

Base Price Tax ($/bbl)	Rate	Comments
Tier 1 Oil — Oil in production before 1979		
12.81	70%	50% tax on first 1,000 bbl/day of production by crude producers with no refining capacity
Tier 2 Oil — Stripper Oil (oil produced from old wells in amounts of less that 10 bbl/day)		
15.20	60%	30% tax on first 1,000 bbl/day of production by crude producers with no refining capacity
Tier 3 Oil — Oil discovered after 1978, heavy oil, and oil recovered by tertiary extraction methods		
16.55	30%	base price adjusted 2% annually in addition to inflation adjustment

Additional specifications:
1. Total tax not to exceed 90% of new income from property.
2. Tax deductible as business expense for income tax purposes.
3. Amount subject to tax could be reduced by most state severance tax payments.
4. Tax to phase out over 33-month period beginning in January 1988, if it raised $227.3 billion, or by January 1991 otherwise.

Exemptions:
1. Alaskan oil except that from Sadlerochit field, which was taxed as pre-1979 oil.
2. Oil from properties owned by Native Americans and state and local governments.
3. Oil from properties owned by nonprofit medical or educational institutions or by churches that dedicated proceeds to these purposes.

after having generated $77 billion in taxes (Crow, 1991).

In retrospect, most people agree that the Windfall Profits Tax was a failure. It was not very successful as a revenue generating vehicle. Much worse, it consumed important national energies at a time when they were needed to address the real problem of how to lessen dependence on foreign oil. The obvious solution was to find oil closer to home, a challenge that had to be met by increased exploration, which required money. It was ironic that the main move taken by Congress in response to this national emergency was to decrease oil company profits, thereby cutting funds available for exploration (Figure 6-5). In the end, this move simply weakened the multinationals and strengthened the national oil companies (Table 6-1), which control oil reserves and oil prices today.

Present interest in mineral resource excise taxes focuses on the so-called *carbon tax*. This tax would impose a levy on the carbon content of fossil fuels to compensate the world for the CO_2 that they generate and the global warming that they cause. It is hoped that the tax would encourage development of alternate energy sources. The size of the tax has not been established, but estimates range up to $13/bbl of oil

and $60/ton of coal. Opponents point out that the tax will generate large government revenues, but that no plans have been made to spend these to encourage alternate energy sources. Excise taxes on oil are already a major source of government revenues. For instance, taxes of this type generate over $200 billion for the 12 European community governments, about three times the value of the oil that they import. Consideration has also been given to an *energy tax*, particularly in the United States, where it is regarded as one possible way to raise government revenues without resorting to an income tax. Such a tax might be applied equally to all forms of energy, thus having limited impact on efforts to develop alternate energy sources.

Tariffs and Import Quotas

Governments can also affect mineral profits through *tariffs* or import quotas, which are often used to protect a weak or inefficient domestic industry from foreign competition (Peck, 1992). The motivation for such protection has commonly been the desire to maintain employment levels in a large industry such as coal or steel, or the need to preserve domestic productive capacity for a strategic mineral commodity such as oil.

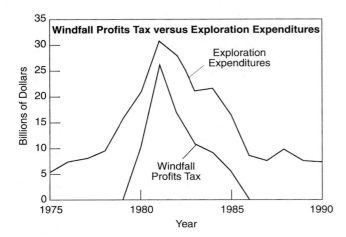

FIGURE 6-5

Oil exploration expenditures and payments to the Windfall Profits Tax in the United States during the 1970s and 1980s compiled from data of American Petroleum Institute. Exploration expenditures rose when the tax was in effect in response to anticipated high future oil prices.

Tariffs imposed by the U.S. government are divided into lower most-favored-nation levies and higher non-most-favored-nation levies. The latter group, which included Afghanistan, Albania, Azerbaijan, Byelarus, Cuba, Georgia, Kampuchea, Kazakhstan, Laos, North Korea, Romania, Tajikistan, Turkmenistan, Uzbekistan, and Vietnam in 1993, is almost certainly destined to change as new alliances are forged. Specific tariffs against individual mineral products vary widely in the United States and reflect attempts to protect domestic copper, lead, and zinc production.

Import quotas have been employed to protect the U.S. steel and oil industries. In steel, the U.S. government met with other governments and foreign producers to set up *voluntary restraint agreements (VRAs)* that specified the fraction of the U.S. market that could be supplied by each country or economic region. For instance, in 1990 these agreements allowed imported steel to supply 20% of the U.S. market, with 7% coming from the European Community, 5.3% from Japan, and the remaining 7.7% from South Korea, Brazil, Mexico, Australia, and other countries. Although VRAs provided some shelter for the U.S. steel industry, they were replaced in 1993 by steep tariffs on steel imports from 19 countries, most of which were alleged to have provided unfair support to their steel industries (Greenhouse, 1993).

The largest import quota effort in recent history was the attempt by the U.S. government to control oil imports. In contrast to efforts on behalf of steel, which involved cooperation among nations, the oil import effort was largely a unilateral move. It began in 1955 when oil companies were asked to limit their growing imports of cheap foreign oil to 1954 levels, in order to protect the domestic industry. This failed, and in 1957 the *Voluntary Oil Import Program* was proclaimed because oil importation was deemed to be a threat to national security. The Secretary of the Interior was authorized to establish permissible import levels and allocate quotas. Lack of cooperation led the government to establish the *Mandatory Oil Import Program* in 1959 under the authority of the National Security Provisions of the *Trade Agreements Extension Act of 1958*. To make a long story short, this also failed to control imports in a way that furthered the interests of the country, the consumer, or the industry. Refineries in the northeastern United States, where cheap oil was usually imported, were without adequate supplies of crude. Foreign producers attempted to sneak oil into the United States via Canada as "strategically secure" Canadian oil. Consumers were denied the cheaper gasoline that could be refined from imported oil. Just as the system was about to collapse on itself, the world oil situation changed with the emergence of OPEC and the start of oil embargoes.

Relation Between Mineral Revenues and Taxation

The general objective that most governments have in taxing the mineral industry is to capture as much of the profit as possible, leaving only enough to attract new investment. The amount or fraction of profit that meets these criteria is known as *economic rent,* and it is indeed an elusive quantity. Any combination of income tax, severance tax, royalty, or even government equity can be used to approximate economic rent. In general, corporations are most interested in repaying their investment and achieving an acceptable return on their investment in as short a period as possible. Some governments allow this by way of depreciation and depletion deductions and attempt to maximize their return over a longer period. This is sometimes done with a *resource rent tax,* which is similar to the Windfall Profits Tax, and which is imposed after corporate goals have been attained. Australia imposed a tax of this type on oil production in 1990.

In some cases, taxes and other payments can be structured to provide special support to an industry. This is usually done to make exports more competitive. Where such assistance is obvious, countries are accused of *dumping* the commodity. Less commonly, companies will dump commodities to penetrate international markets. The most common mineral product to be cited in dumping charges is steel, although aluminum and cement also figure prominently because of the high cost of energy in their manufacture. China has been a frequent focus of dumping complaints related to raw ores, largely because of its extremely low labor costs and large production.

VALUATION OF MINERAL DEPOSITS

The economic factors discussed previously are critically important in determining the value of a mineral deposit. No matter how rich a deposit is, or how favorably located it is, it will not be extracted unless the economic regime is favorable. To be sure, environmental concerns also limit mineral development, and deposits in many MDCs cannot be economic until they are approved by local environmental authorities. But this is often just another cost that must be integrated into the overall cost of the operation. The pivotal factor that always remains after all of the real costs of doing business are counted is the division of spoils between the owner and the host government.

One of the most important numbers that is used in negotiations between government and industry on tax rates and other economic parameters is the *discounted present value* of a deposit. Various figures of this type are also widely used by industry to decide whether a deposit can be extracted profitably. The discounted present value of a deposit is obtained by summing the present values of after-tax cash flow for the deposit for each future year of operation using the relation:

$$P = S/(1 + i)^n$$

where P is the present value of the deposit, S is the after-tax cash flow from operation of the deposit in the nth year, and i is the *discount rate*, which is the expected return on the investment.

This calculation is needed because a dollar received today is worth more than one received tomorrow. It requires that certain estimates be made about the future of the deposit. The value of S for any year can be determined from information on the cost of production and the price of the commodity, which are gathered in a feasibility study. This also involves an estimate of the period over which the deposit will be operated, which depends largely on the scale of extraction, the size of the deposit, and the amount of money to be invested in the first place. Obviously, it makes no sense to spend huge amounts of money to build a large plant and buy a hundred trucks to mine a gold deposit in one year, when a smaller operation can do it in ten years. The value of i to be used in the calculation depends on the risk involved in the business. For money to be invested in a savings account

at the bank, prevailing and projected future interest rates would be used. But the goal for any business is to make its investment grow more rapidly than inflation by taking risks. The risks might be technical (perhaps the oil will not flow from the reservoir at the predicted rate, or maybe the reserve estimates are wrong), economic (maybe the price of oil will change), or governmental (the tax rate could be raised or lowered, or the oil field might even be nationalized). Arriving at a figure to represent these factors is obviously difficult, but it must be done.

As its name implies, the present value calculation provides an estimate of the present cash value of anticipated future profits from a mineral deposit. Figure 6-6 shows just how important the choice of a discount rate can be to calculation of the present value. The higher the discount rate, the smaller the present value of the deposit. In other words, knowledgeable buyers will pay more for a project with a low anticipated risk than they will for one with a high anticipated risk. It is also apparent from the calculation that the earliest years in the operation of a deposit have the greatest effect on present value. Thus, the present value of a large deposit that will last for 100 years might be only twice as great as one that will last for ten years.

Present value calculations are most useful in comparisons among different deposits, whether it is a corporation attempting to decide which deposit to extract, or a government trying to decide how large a tax or environmental cost to impose on a group of deposits. As shown in Figure 6-6, the time necessary to repay an initial investment of $1 million required to put a hypothetical deposit into operation, known as the *payback period,* would be about five years at a discount rate of 10% and seven years at a discount rate of 20%. At a 30% discount rate, however, the deposit would not recoup its initial investment even within ten years. Any similar deposit with lower payback periods would obviously be more attractive unless other factors intervened.

The reverse of this relation concerns *lead time,* the period between discovery and production (Wellmer, 1992). Obviously, a deposit must be put into production as soon as possible in order to minimize the payback period. An exploration investment of $1 million can be paid back the next year for only $1.1 million at a discount rate of 10%. If development of the deposit is delayed for ten years after the exploration expenditure, the amount required to repay the original investment would be $2.7 million. This is a very real amount for the investor whose money is tied up in the deposit because this money could otherwise be earning interest in a bank. Although geological and engineering complications often delay the development of mineral properties, the most common culprits are disagree-

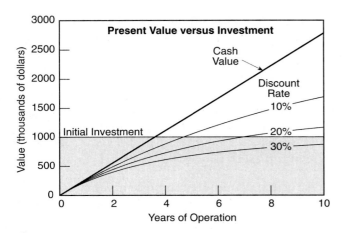

FIGURE 6-6

Present value of a hypothetical mineral deposit with an initial investment of $1 million and annual after-tax profit of $275,000, showing the time needed to pay back the initial investment for different discount rates. If no discount rate is used (that is, the prevailing interest rate is assumed to be 0 percent), cash value reaches the initial investment in slightly less than four years.

ments about economic rent or environmental control. For example, construction of a mine and mill at the Flambeau copper deposit in Wisconsin was stopped recently when an unprecedented second environmental impact study was required to quiet public concern. This study found that the deposit could be developed without significant damage, as had the first one, but it increased the payback cost of the operation by delaying it for a year. Thus, we have an economic responsibility to ourselves as the customers for these products to carry out evaluations of proposed mineral operations as carefully and expeditiously as possible to avoid imposing on them (and ourselves) unnecessary interest costs.

CONCLUSIONS

Mineral economics may be the great, final battleground between modern civilization and the natural order. We have not been following Mother Nature's system and it is unclear just how much longer we will be able to flaunt her authority. As originally planned, Earth concentrated minerals into deposits, we explored for them and then extracted the best ones, perhaps moving on to poorer ones as time passed. By erecting an economic structure that renders some rich deposits worthless through high taxes and encourages extraction of poor ones through subsidies and tariffs, we have confused the system.

As minerals become more difficult to find and produce, we need to establish *national mineral policies* that recognize the strong linkage between mineral supplies and economic policies. These policies must provide mineral producers with a secure and reasonable return on investment, governments with a fair share of profits derived from their soil, and citizens with a clean environment. While we wait for these miracles to occur, we should work to encourage more accurate use of economic terms in discussions of mineral resources. We have shown here that there is a huge difference between the revenues of a mineral operation and any profits that might be obtained from it. Both are very different from the gross value of minerals in the deposit. In the end, it is only the profit that counts. In spite of this, mineral companies persist in describing their discoveries in terms of gross value, perhaps because that makes them seem more important, and newscasters and politicians perpetuate the error. This exaggerates the perception of profits that are available from deposits, which encourages debate over the division of fictional spoils between government and owners, thus delaying decisions to develop mineral deposits. It would be naive to think that such debates penalize only the corporation; higher costs are passed along to the consumer. Thus, we must strive to make the period of time between discovery and development of mineral deposits as short as possible without jeopardizing our important economic and environmental goals. This requires an informed public; one that does not have to spend years catching up on basic geologic, economic, and environmental concepts in order to express an informed opinion.

7

Energy Resources

THE VALUE OF ANNUAL WORLD ENERGY MINERAL PRODUCTION has ranged between $500 and $1,500 billion since the mid-1970s. It makes up about 2% of U.S. GNP and is one of the largest components of world productive capacity. Its importance extends far beyond these numbers, however, because most other activities depend on energy, as is shown by the strong correlation between energy production and GNP for most countries (Starr, 1971). Energy production, particularly oil, has also formed the basis for many of the world's great fortunes. Some, such as those of the Rockefeller (Standard Oil), Mellon (Gulf Oil), and Getty (Getty Oil) families, have supported philanthropic activities. Others, such as those administered by some heads of state, have been used to further territorial, religious, and philosophical ideals (Yergin, 1990).

World energy has been derived from minerals since the late 1800s, when coal began to surpass wood as the main source (Figure 7-1). Coal gave way to oil in the mid-1900s and natural gas and nuclear energy began to grow in importance in the latter half of the century. Since 1960, world natural gas use has grown at a relatively high rate, whereas uranium has stalled.

FOSSIL FUELS

Coal, oil, natural gas, oil shale, and tar sand are called *fossil fuels* because they consist of plant and animal remains, known as *organic matter*, that have been preserved in rocks (Fulkerson et al., 1990). Organic mat-

ter is changed easily by geologic processes and its many forms are known by a large number of names (Figure 7-2). Most organic matter decomposes by oxidation to form CO_2 and H_2O, which are recycled into the hydrosphere and atmosphere. A small amount of organic matter is shielded from the atmosphere by burial beneath other sediments where it is preserved to become fossil fuels, an important endowment of solar energy that Earth has set aside for later use. This accident of geologic history was probably the most important single factor in the development of our present industrial civilization. Had it not occurred, we would have had no widely available fuel to bridge the gap between wood and more advanced, nonfossil fuels, and the Industrial Revolution might have been delayed indefinitely.

Coal

Coal is both the fuel of the past and the potential fuel of the future. Annual world coal production is worth over $100 billion now and could grow tremendously, if we fall back on it as the replacement for our dwindling oil supplies. Coal is plentiful because it forms so simply, as discussed next.

Geology and Global Distribution of Coal

Coal is a sedimentary rock made up largely of altered *cellulose, lignin,* and other plant remains. Some coals contain fossil bark, leaves, and wood, largely from ter-

117

Energy Minerals

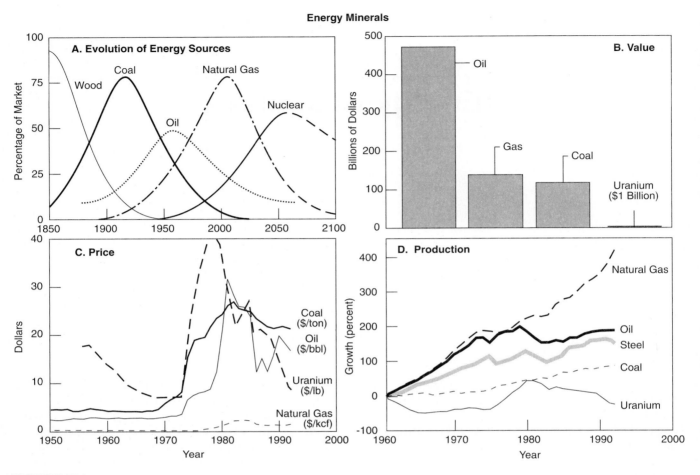

FIGURE 7-1

A. Evolution of energy use since 1850; height of curve represents the fraction of the total energy market supplied by that source. **B.** Value of world energy mineral production in 1991. **C.** Change in prices of energy minerals since 1960; reference to Figure 4-2 shows that only oil and natural gas prices have increased faster that inflation over this period. **D.** Growth in production of energy minerals since 1960, compared to growth in steel production (compiled from Marchetti and Nakicenovic, 1979, U.S. Bureau of Mines and Department of Energy; uranium production does not include CIS countries).

restrial *vascular plants,* but the fossils in most coal have been obliterated, leaving a hard, black rock that will burn. Under the microscope, coals can be seen to consist of grains called *macerals,* which include *vitrinite,* the remains of wood or bark, *liptinite* or *exinite,* the remains of spores, algae, resins, and needles and leaf cuticles, and *inertinite,* a complex mixture of fungal remains, oxidized wood or bark, and other altered plant materials (Hessley et al., 1986; Schobert, 1987).

Coal begins as *peat,* an accumulation of partly decomposed, brownish plant remains. As peat is buried beneath new sediment, the increasing temperature and pressure cause it to release water and other gases, gradually increasing the proportion of carbon, a pro-

cess referred to as *coalification* (Table 7-1). As coalification proceeds coal increases in *rank* from *lignite,* through *subbituminous* and *bituminous,* to *anthracite.* The thermal stability of lignite and high-rank bituminous coal indicates that the maximum temperatures that they experienced were about 200°C and 300°C, respectively. Anthracite is found in rocks that have been folded and subjected to even higher temperatures, such as those typical of low-grade metamorphism (Ward, 1984).

Peat accumulates in swamps, which form in flood plains and deltas of rivers, coastal barrier island systems, and poorly drained regions underlain by glacial deposits. Although glaciated muskeg regions of

Classification of Organic Material

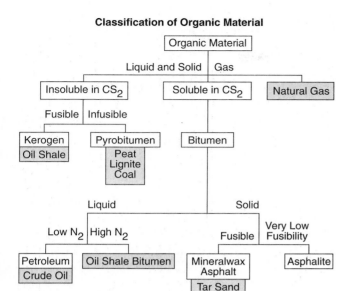

FIGURE 7-2

Classification of organic material (modified from Chilingarian and Yen, 1978).

fication. With good coal layers, or *seams* as the miners call them, ranging from 2 m to 10 m in thickness, and some reaching 100 m, this means that original peat thicknesses must have been enormous. They also covered large areas. The famous Pittsburgh seam in the Appalachian basin covers about 52,000 km², somewhat more than the present Okefenokee or Everglades Swamps (Figure 7-3). It is only one of more than 50 coal seams that are mined in the Appalachian basin.

The distribution of coal-forming swamps through geologic time has been a function of Earth's plate tectonic and climatic history. Temperate conditions apparently extended to higher latitudes when global climates were warmer, enlarging potential coal-forming regions. None of this mattered, of course, before Late Silurian time when vascular land plants developed. Thus, pre-Silurian sedimentary basins, notably in South America and Africa, lack coal (Figure 7-4). Even after the development of vascular land plants, coal formation was intermittent, and took place largely during the Late Carboniferous (Pennsylvanian) and Permian Periods and during the Jurassic through Tertiary Periods (Fettweiss, 1979; Thomas, 1992).

Canada and Russia contain over 12 million km² of peat, most of the world's large coal deposits formed from peat that accumulated in temperate swamps that contained fresh or, at most, brackish water. The swamps were largely along flat coastal plains and continental interiors, where surrounding land had been leveled by erosion and could not contribute much clastic sediment. Modern equivalents of this environment might include the Dismal Swamp of Virginia and North Carolina, and the coastal swamps of Sumatra (Figure 7-3).

Large coal-forming swamps were remarkable features that required stable sedimentary conditions over large areas and for long periods of time. Peat layers decrease in thickness by almost ten times during coali-

Coal Production and Related Environmental Concerns

Coal can be mined from both underground and open-pit mines (Meyers, 1981; Hartman, 1987). Most open-pit coal mines are *strip mines*. Although strip mines have a poor reputation among most people, operations of this type are actually the most environmentally acceptable form of mining because they fill the pit and reclaim the land surface. Coal is amenable to strip mining where it forms flat layers that are covered by only a few tens of meters of overburden. In the first step of strip mining, overburden is removed; in the second step, the underlying coal layer is removed, and in the final step the overburden is replaced (Figure 7-5; see pg. 122). This process is re-

TABLE 7-1

Composition and energy content of coals of different rank (compiled from Schobert, 1987 and Meyers, 1981).

Rank of Coal	C	H	Volatile	Fixed Carbon	Calorific Value (Btu/lb)	(MJ/kg)
		(weight percent)				
Lignite	73.0-78.0	5.2-5.6	45-50	50-55	<8,300	<19.31
Subbituminous	78.0-82.5	5.2-5.6	40-45	55-60	8,300-11,500	19.31-26.75
Bituminous						
High-volatile	82.5-87.0	5.0-5.6	31-40	60-70	11,500-14,000	26.75-32.56
Medium-volatile	87.0-92.0	4.6-5.2	22-31	70-80	>14,000	>32.56
Low-volatile	91.0-92.0	4.2-4.6	14-22	80-85	>14,000	>32.56
Anthracite	92.0-98.0	2.9-3.8	2-14	85-98	>14,000	>32.56

Coal Basins

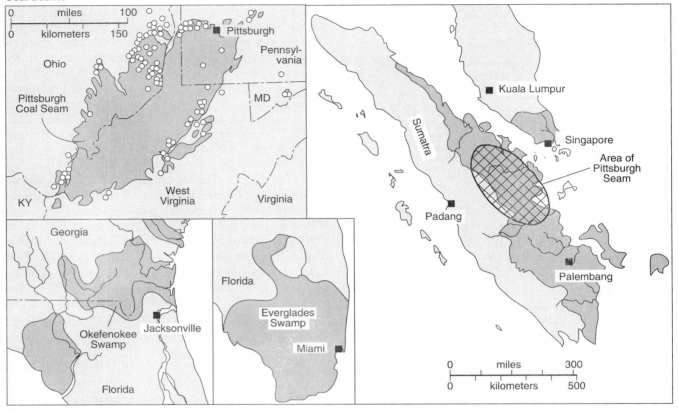

FIGURE 7-3

Comparison of the Pittsburgh coal seam with modern swamps containing peat.
Distribution of test holes in the Pittsburgh seam shows the level of information about
the quality of this seam (modified from *U.S. Geological Survey Professional Paper 979*,
p. 170).

peated back and forth across strips of land, giving the method its name. Most strip mines are large-scale, expensive operations (Figure 7-6A; see pg. 123; Plate 3-2). A single dragline, the large shovel commonly used to remove overburden, can cost tens of millions of dollars. *Contour mining* is a special form of strip mining that is used around the sides of hills where flat-lying coal layers are too thin to merit the removal of overburden from the entire hill. Instead, the outer part of the layer is mined from a cut or *bench* that is dug around the hill to the economic limit of overburden removal (Figure 7-5; Plate 4-1). An auger is sometimes used to drill out parts of the layer under deeper cover (Figure 7-6B). Contour mining requires smaller investments in equipment and permits recovery of coal seams that cannot be extracted economically by strip mining.

Early strip and contour mines were major environmental offenders. In old strip mines, overburden was not graded flat after it was replaced, leaving an ugly,

hummocky topography and no vegetation. Old contour mines were simply left as open cuts along the hillsides that continued to erode long after they were abandoned. Pyrite, a common contaminant in coal, was left at the surface in these old mines, and it oxidized (Table 3-3) to create sulfuric acid that contaminated surrounding drainages (Figure 7-7; see pg. 124). The *Surface Mining Control and Reclamation Act of 1977 (SMCRA)* was a response to these abuses in the United States. This regulation is vigorously enforced; over 352 km^2 of old mines, including 8,202 mine portals and 5,281 shafts, were reclaimed between 1978 and 1991 (OSM, 1992). Nevertheless, some small contour mines continue to operate without proper reclamation, largely where local politics impede enforcement, much of which is carried out by state agencies. *Groundwater quality* is an important concern in reclaimed strip mines. Although original aquifers are destroyed by mining, broken rock that is used to fill the mine is commonly more porous and permeable,

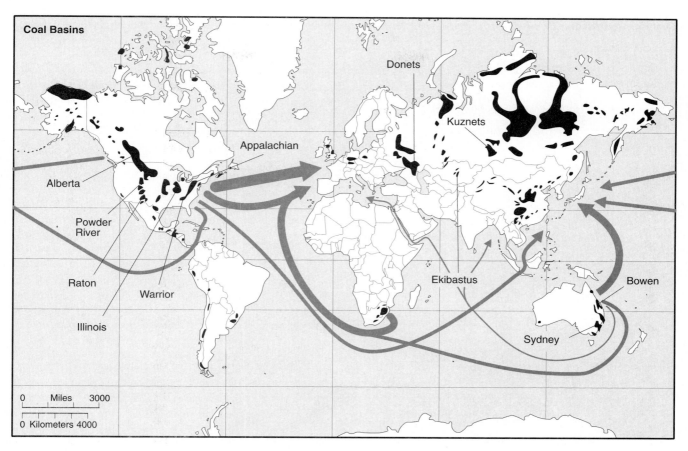

FIGURE 7-4

Location of coal basins of the world (modified from Fettweis, 1979), including names
for important producing basins. Major lines of marine coal trade are also shown (from
International Energy Annual, U.S. Department of Energy).

and actually increases water storage capacity. However, exposure of this rock to weathering during mining allows groundwaters to dissolve elements such as calcium, sulfur (as sulfate), and some metals, which are enriched in coals. Although this can be minimized by the addition of lime to reduce the pH of groundwater, other elements such as selenium are soluble in alkaline solutions. Uranium is also concentrated in some coals and disturbances related to mining can release radon (Johnson and Paone, 1982; Hossner, 1988; NRC, 1990).

Underground coal mines are used to mine coal that is too deep for surface mining. Most modern underground mines compete with open-pit mines through the use of *continuous miners*, which are large rotating cutters that break coal away from the layer and drop it onto a conveyer belt that carries it out of the mine (Figure 7-8; see pg. 125). *Room and pillar mines* remove rooms of ore, leaving walls or pillars between them to support the overlying rock (Figure

7-5). Although some pillars are eventually removed, most are left to prevent collapse that would damage the mine. *Longwall mines* recover the entire ore layer with a continuous miner, which advances beneath a moveable artificial roof that protects it as the mined-out area collapses (Figure 7-5; Plate 3-3). Longwall mining began in Europe and was adopted slowly in North America because it requires expensive equipment and extensive excavation prior to coal extraction with consequent delayed cash flow. It increases mine productivity and is clearly the way of the future, however (Chen, 1983; Peng, 1985).

Coal mining is one of the most rigorous lines of work in the world and underground mines are the worst. *Black lung disease* or coal worker pneumoconiosis (CWP), which results from the inhalation of coal dust, is a major health problem in coal-producing countries. As many as 4,000 retired miners in the United States are thought to die each year from the

Coal Mining Methods

A. Strip Mining

B. Contour Mining

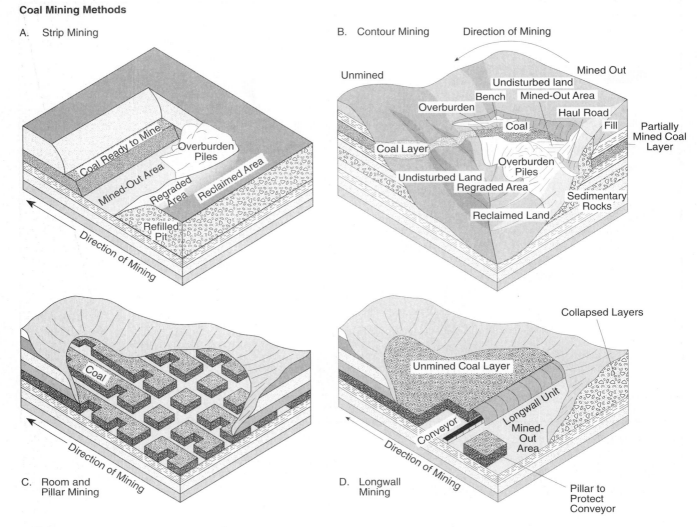

FIGURE 7-5
Coal mining methods, with unmined coal layer on left and mined area on right in all
diagrams. **A.** *Strip Mining.* Unmined coal seam has been stripped of overburden, which
has been put in spoil piles that are regraded and revegetated to produce reclaimed land
after coal is removed. **B.** *Contour Mining.* Unmined coal seam on hillside is mined from
bench (center), which is then reclaimed. **C.** *Room and Pillar Mining.* Unmined coal seam
has been cut into a series of rooms and pillars, which are left to support the roof.
D. *Longwall Mining.* Unmined coal seam is mined by the longwall method, which
removes the entire layer, allowing complete collapse of overlying rock.

illness or related complications, and 5% to 10% of active coal miners show evidence of the disease in
X-rays (Hilts, 1990). These statistics reflect past practice, however. Present U.S. federal regulations require
that dust levels in the mines be kept below 2 mg/m^3
of air, and dust levels are monitored in more than
2,000 mines across the country. U.S. coal producers
pay $1.10 per ton of underground-mined coal and 55
cents per ton of surface-mined coal into the Black
Lung Disability Trust Fund, which provides benefits

to eligible miners. *Fires and explosions* caused by methane and coal dust are also of concern. Methane-rich
gas intersected during mining can flow into mine ventilation systems, where it builds up so rapidly that explosions occur before the mine can be evacuated. In
addition to the cost in lives, explosions start fires that
can destroy a mine by burning wooden supports and
the coal itself. The famous Centralia mine fire in Pennsylvania has burned for decades and may not go out
until the supply of air or coal is exhausted. In the west-

ern United States, outcropping coal seams have been-struck by lightning and burned to great depths, destroying significant coal reserves. Despite these problems, coal mine safety continues to improve. In 1968, 222 mine deaths were recorded in the United States, versus only 54 in 1992 (both of these numbers are huge in comparison to uranium mining, however, as noted below).

Underground mines also have *subsidence* problems, and they are of greatest concern where surface and mineral rights have been divided (Bise, 1981). In gen-

eral, mining that causes subsidence will be restricted to areas without buildings, although mining has extended under farm dwellings and subdivisions have been built over abandoned mines. Most governments specify minimum roof support that must be left to prevent the collapse of abandoned room and pillar mines and insurance is available to cover repair costs if subsidence does occur. Longwall mines are designed to collapse within a few months, however, requiring that operators and landowners agree on compensation caused by this subsidence.

A.

B.

FIGURE 7-6

Surface coal mining methods. **A.** Coteau strip mine, North Dakota, showing dragline that cuts trench by removing overburden from right side and putting it on left side, exposing coal layer to be removed by smaller equipment. **B.** Close-up view of auger operation on bench of contour mine (courtesy of Bill Suter, Coteau Properties and Charles Meyers, Office of Surface Mining).

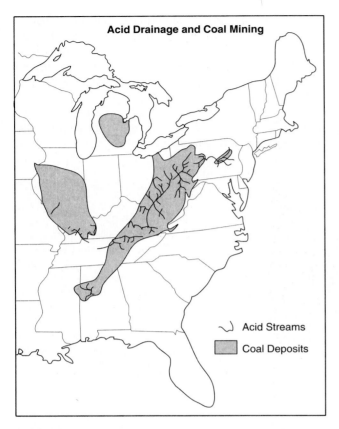

Acid Drainage and Coal Mining

~ Acid Streams

■ Coal Deposits

FIGURE 7-7

Distribution of streams made acid by coal mining prior to present mining regulations (from CEQ, 1989).

The cost of reclamation in U.S. coal mines that were started after 1977 is secured by a bond put up by the mine operator. This bond is not released until reclamation has been approved, commonly in five to ten years. Mines abandoned before 1977 are reclaimed by the *Abandoned Mine Land Reclamation Fund (AMLRF)*, which was set up by SMCRA (St. Aubin and Massi, 1987). Funding comes from production fees of 10 to 35 cents/ton of coal, which are paid by all active mines. As of 1990, $2.6 billion had been paid into this fund. Half of the funds are returned to their state of origin, where they must be applied to reclamation, and the other half goes to high-priority reclamation, regardless of location (Table 7-2). The *Energy Policy Act of 1992* extended the term of payments to AMLRF until 2004 and provided funds to support retired coal-workers and pay death and orphan benefits for them.

Coal mining has been a leader in mechanization, a trend that correlates with an increase in the importance of surface mining, a decrease in coal-related employment, and an increase in productivity (Figure 7-9A). The slow increase in productivity that characterized early mechanization in the United States was reversed by efforts to comply with the *Coal Mine Health and Safety Act of 1969*, a pattern observed throughout the developed world. As these changes were implemented and surface mining increased in importance, productivity again increased to reach present levels of about 3.6 tonnes/miner/hour. Over the same period, union labor decreased over 50% without significant reduction in wages. The story is similar in Canada and Australia but differs in much of western Europe, where mechanization is also widespread. In spite of significant improvements, average hourly productivity in the United Kingdom and Germany is only about 0.6 tonnes/miner/hour (Hessling, 1992). During the 1980s, the Soviet Union required seven times as many people to match U.S. production. The situation is worse in LDCs that lack mechanized coal production. Even though India produces most of its coal from open-pit mines, average hourly productivity is only 0.125 tonnes/worker. Low productivities are associated in almost all cases with state-run mining operations, which will probably be reorganized in the future. A first step in this direction is the debate over the closure of many British coal mines (Gladstone, 1992).

Coal Use and Quality

Coal started the 1900s as a fuel for home heating, manufacturing, and rail transport, and the main feedstock for the dye and chemicals industry. Now, its markets in the United States and other MDCs are *thermal coal* for electric power generation and *coking coal* to make coke for steel production (Figure 7-9B). Sales of coking coal in the United States have actually decreased, paralleling the decline of the domestic steel industry, but increasing electric power generation has made up for this and overall coal consumption has increased steadily. The pattern differs in LDCs such as China and Poland, where large amounts of coal are still used as household fuel.

Concerns over *coal quality* are growing in response to recognition that it is the dirtiest of the fossil fuels (Table 7-3). The most important factors affecting coal quality are ash, sulfur, and trace elements (Krishnan and Hellwig, 1982; Shephard et al., 1985; Cecil and Dulong, 1986). *Ash,* the residue that remains after burning, can make up as much as 30% of the coal. It consists of clastic material such as clay minerals and quartz that were part of the original sediment and were accidentally included with the coal during mining. During combustion, heavy *bottom-ash* grains settle out and lighter *fly-ash* grains go out with flue gases, where about 90% is recovered by particulate recovery systems. Some coal ash is used in construction materials, although most is disposed of in landfills.

FIGURE 7-8
Underground longwall mine using continuous miner. Note the cutting head at left and the protective cover above the operator at right (courtesy of J.D. Pyle, Eickhoff Corp.).

Sulfur occurs in coal in three main forms (Arora et al., 1980). Most obvious are *sulfide minerals* such as *pyrite,* which are often visible in coal as small, shiny grains. In contrast, *organically bound sulfur* is invisible because sulfur is part of the complex organic molecules that make up coal. *Sulfate minerals* such as *gypsum* are a third, considerably less important, type of sulfur. The distinction between high-sulfur and low-sulfur coals is made at sulfur contents of 1.5% (by weight), although coal can contain as much as 12% sulfur. The main source of this sulfur is sulfate-rich water, such as seawater, that enters coal-forming environments. Coals in eastern and central North America, which are high in sulfur (Figure 7-10),

TABLE 7-2
Disposition of Abandoned Mine Land Reclamation Funds (in millions of dollars) in important coal-producing states and Native American lands, 1978–1986 (Office of Surface Mining, 1987). Most reclamation funds are spent on mines abandoned before 1977. Lower amounts spent in some western states reflect extensive reclamation undertaken during present coal mining.

State	Amount Collected	Amount Awarded
Alabama	$ 54.7	$ 36.4
Colorado	$ 42.5	$ 20.4
Illinois	$119.1	$ 55.0
Indiana	$ 89.2	$ 43.3
Kentucky	$299.1	$174.7
Montana	$ 82.9	$ 29.3
Navajo	$ 58.9	$ 2.6
Ohio	$ 93.3	$ 76.3
Pennsylvania	$182.0	$271.8
Virginia	$ 59.5	$ 45.5
West Virginia	$189.9	$140.4
Wyoming	$311.7	$113.3

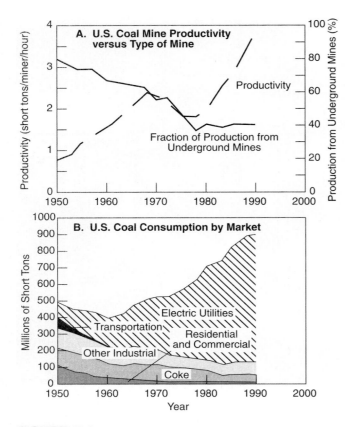

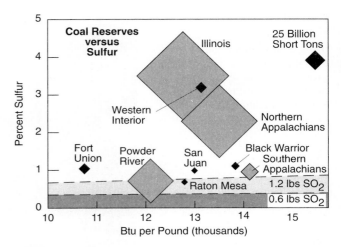

FIGURE 7-10

Sulfur content and reserves in U.S. coal basins. Reserve of each basin is shown by the size of the symbol. Coal falling below the line labeled 1.2 meets emission standards for SO_2 (expressed in pounds of SO_2 per million Btu) for the year 2000. Note that the Powder River Basin is the only large, low-sulfur coal reserve (from *U.S. Geological Survey Circular 979* and data of U.S. Department of Energy).

FIGURE 7-9

Patterns of coal production and use. **A.** Increase in productivity of U.S. coal mining with decrease in the number of underground coal mines. Between 1970 and 1990, productivity in German coal mines increased by 25% (Hessling, 1992) compared to the 35% increase in the United States. **B.** Change in U.S. coal markets since 1950 showing growth in the consumption of thermal coal for electric power generation (data from Annual Energy Review, U.S. Department of Energy).

formed in swamps that suffered periodic incursions of seawater, whereas those in western North America, which are low in sulfur, formed largely in fresh water. During combustion, sulfur forms SO_2, which reacts with atmospheric moisture to create acid rain (Table 3-4). *Nitrogen* from coal combustion also forms gases

(NO_x) that contribute to acid rain. Some of this ni-nitrogen is in the coal molecules, although most of it comes from air in which the coal burns.

Coals are also enriched in some *trace elements,* most of which are combined at the atomic scale as chelated organic molecules (Smith et al., 1980; Vourvopoulos, 1987). When coal is combusted, at temperatures of about 1,500°C, many of its trace metals vaporize. Although some condense on fly ash or form aerosol particles that are caught on their way out the exhaust systems, others escape to form fallout around power plants. This metal-bearing fallout is easily leached by acid rain, suggesting that a thin film of ash and aerosol particles around coal-burning plants could release trace elements to groundwater and crops (Grisafe et al., 1988). Problems of this type are especially severe in LDCs where many coal-fired plants lack even par-

TABLE 7-3

Relative contributions to atmospheric pollution in 1989 by coal, oil, and natural gas electric power generation (Department of Energy, Annual Energy Review, 1990, Table 98).

Fuel Type	Carbon Dioxide	Sulfur Dioxide	Nitrogen Oxides
Coal	1090	10.22	4.04
Oil	1205	6.67	2.62
Natural Gas	601	0	2.20

All amounts in tons of gas per million kw-hr of electric power.

ticulate collection systems. As noted later, sizeable numbers of fatalities in China have been attributed to arsenic poisoning of soils and foods caused by emissions from plants of this type.

In the United States, modifications to the *Clean Air Act* limit power plant emissions to 2.5 pounds of SO_2/MBtu (1,080kg/Mjoule) of heat generated by 1995. By 2000, emissions must not exceed 1.2 pounds/MBtu (650kg/Mjoule). Most other MDCs face similar constraints. *Clean coal technology*, the response of industry to these challenges, involves the removal of sulfur from the coal before combustion, the removal of SO_2 from the exhaust gas after combustion, or special combustion methods (Elliott, 1985; Osborne, 1988). Coal can be cleaned before burning by standard beneficiation methods, involving grinding, washing, and flotation, although these processes do not remove organically bound sulfur, nitrogen, or other trace elements. Cleaning after burning centers on particulate recovery systems and scrubbers. Most scrubbers in use on power plants today are nonregenerative and use a spray of powdered lime or limestone (containing Ca) to react with SO_2 gas to form calcium sulfite, calcium sulfate (anhydrite), or gypsum. These units cost as much as $500 million for a single power plant, consume up to 10% of the power from the plant, occupy about half of the space on the power plant grounds, and still permit 10% of the sulfur to escape. They also produce about three tonnes of sulfate sludge for every tonne of SO_2 that they remove from flue gas (Luxbacher et al., 1992). In some cases, gypsum can be sold; otherwise, the sludge must be disposed of in a landfill.

As of 1990, only about 22% of U.S. electric generating capacity used scrubbers, none of which could remove NO_x gases. Sulfur emissions from U.S. power plants have been reduced by about 15% between 1970 and 1990, even though coal use increased by about 50%. Even so, coal-fired power plants remain the major source of atmospheric sulfur pollution. Although it is widely anticipated that most power plants will have to install scrubbers to meet the 2000 emission standards, other technologies will also be tried, including cleaning coal *during* combustion with a scrubber-like process that adds Ca to *fluidized-bed* combustion chambers where the sulfur combines to form anhydrite even before it leaves the furnace (Balzhiser, 1989). To encourage efficiency in these efforts, the *Clean Air Act Amendments of 1990* permit industries or utilities with SO_2 emissions lower than the required limits to sell the right to emit the SO_2 that they are not emitting. This more flexible approach allows polluters who cannot meet existing limits to invest in new facilities where they are likely to be a technical and economic success and to buy exceptions for plants where remediation is not likely to succeed, either for

technical reasons or because the plants are to be put out of service shortly. In the first sale, which took place in March 1993, over $20 million worth of allowances were sold. The approach has been criticized, however, because the market is not currently regulated and therefore no one has oversight over just where pollution will be concentrated.

A final way that coal can be "cleaned" is by *in situ coal gasification,* in which coal is burned while still underground to produce combustible gases (Berkowitz and Brown, 1977). Tests of this process have been hampered by incomplete combustion caused by the channeling of gas flow underground, but the attraction of coal-based energy with little or no mining is probably too great to discount the possibility of future breakthroughs. In view of all these difficulties, however, it is not surprising that low-sulfur coals are of great interest, as discussed later.

Coal Reserves, Resources, Trade, and Transport

World recoverable coal reserves of about 1.1 trillion tonnes are huge in relation to other fossil fuel reserves and global energy needs. This should not be surprising. After all, coal is a common sedimentary rock that forms by a relatively simple process, and sedimentary basins are widespread. Coal reserve and resource estimates depend strongly on the quality of available data, however. Many remote areas are not well explored and legal and geologic factors in well-explored areas constrain the amount of recoverable coal. For instance, only about 20% of the 500 million tonnes of unmined coal known to be present in a coal-rich area of Pike County, Kentucky, can actually be mined. Many coal seams have not been well characterized with respect to variations in ash, sulfur, or trace-element contents (Figure 7-3), factors that will probably further constrain reserves. In some cases, mines have had to stop production because of a decrease in the quality of coal that was originally included in the reserve. Although these considerations are likely to lower present coal reserves, they are probably not sufficient to topple coal from its position as the most abundant fossil fuel.

Exports and imports account for only about 10% of total world coal production, with the flow to Europe and western Asia and from Australia, the United States, South Africa, Poland, Canada, and the CIS countries (largely Russia) (Figure 7-11A). Over the last 40 years, China has emerged as a major coal producer (though not exporter), production from western Europe has declined steadily in the face of declining reserves, and Russian production has stagnated in spite of large reserves (Figure 7-11B). Coal makes an important contribution to the trade balance in Australia,

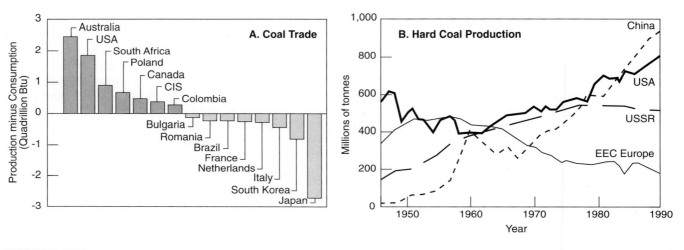

FIGURE 7-11

A. World coal trade for 1990. Amount shown for each country is the difference between coal consumed and coal produced. Exporters plot above the line and importers plot below the line. **B.** World coal production (excluding lignite) since 1945 showing the decline in Europe, increases in the United States and China, and stagnation in the Soviet Union (largely Russia) (from Gordon, 1987; U.S. Bureau of Mines and Department of Energy).

where it accounts for 15% of foreign earnings (Lyday, 1988).

Significant coal trade is also driven by coal quality. The most important quality-based trade at present is inside North America, where large amounts of low-sulfur coal from the Powder River Basin in Wyoming are used in eastern power plants. Coal producers in the eastern and midwestern United States worry that western coal will become the favored way to meet new Clean Air requirements, and their host states see the specter of increasing unemployment in high-sulfur, coal-mining areas. While some states offer tax incentives and other credits to domestic utilities that burn local coal, the coal industry and unions remain divided on the issue of sulfur in coal. Those producing high-sulfur coal advocate mandatory scrubbers and those producing low-sulfur coal favor permitting consumers to meet emissions any way they choose, including the purchase of low-sulfur coal.

The next generation of quality-based coal trade is likely to be international in scale. Large coal deposits recently opened in the Tanjung area of Borneo, for instance, contain only one tenth of the sulfur in Powder River basin coal. The coal is so pure, in fact, that it has been called "solid natural gas" (Wald, 1992). Coal layers that are being mined in Borneo are up to 66 meters thick, under only 3 to 12 meters of overburden, and they are close to ocean transportation. Two factors have prevented coals such as these from tak-

ing over the world market. First, their relatively low-energy content requires that large amounts of coal be burned for an equivalent amount of steam in a generating plant, and many plants lack the throughput capacity. Second, the coal is shipped in large ocean-going vessels that cannot be accommodated at the port facilities of many generating plants. As installations are changed to accept coal of this type, production from U.S. sources is likely to be hurt.

Present coal transport is largely by train, which accounts for about 75% of shipments. In many areas of the United States, large *unit trains* with 100 cars carrying nearly 100 tonnes each move directly from western mines to eastern power plants. Barges or ships and trucks account for about 10% each. Coal can also be transported by *slurry pipelines*, in which a mixture of pulverized coal and water is pumped through a pipeline. This method is particularly suited to coal, which is less dense than most minerals. Coal pipelines were used in England over 100 years ago, and pipelines measuring 173 and 437 kilometers have been used recently in Ohio and Arizona, respectively. Efforts to build larger pipelines from low-sulfur western coal fields to eastern markets have been thwarted by railroads, which do not want to lose their coal transport business and have refused to allow pipelines to cross their right-of-ways. Agricultural interests have expressed concern that water for the coal slurry would deplete already water-poor areas of the upper Colorado River valley (Karr, 1983). Efforts to replace water

Formation of Oil and Gas

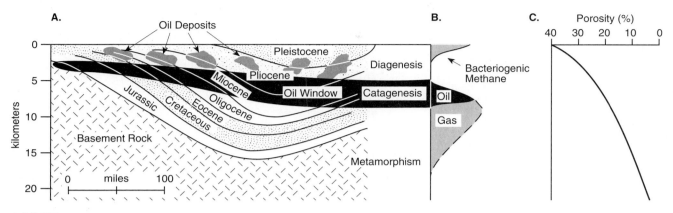

FIGURE 7-12

Formation of oil in a typical sedimentary basin. **A.** Distribution of sedimentary formations, showing the oil window in which temperatures are high enough to form oil and gas from kerogen in source rocks. **B.** Diagram showing the relation between natural gas and oil formation during burial. **C.** Decrease in the porosity of sediments due to compaction and mineralogical changes caused by burial (with permission from Dow, 1978).

with liquid CO_2 and other fluids have been impeded by price.

Crude Oil and Natural Gas

Crude oil has ruled late twentieth-century global economics (Yergin, 1990). With annual world production of about $450 billion at 1993 prices and almost twice that at the higher prices of the early 1980s, oil is the most expensive basic commodity in world history. Natural gas, with current world production valued at about $125 billion, has not been as important simply because it is not as convenient to transport, particularly on the ocean. It is waiting in the wings for the rapidly approaching decline in world oil production, however, when it will compete with coal for world energy markets. In the meantime, the growth of world oil reserves has slowed or stopped, forcing us to press the search to more remote and challenging areas in a constant effort to provide a smooth transition to the next fuel, whatever that will be, and encountering significant environmental challenges along the way. Success in that search depends on an understanding of the processes that form oil and gas.

Geology and Origin of Oil and Gas

Crude oil and natural gas consist largely of chains and other forms of carbon and hydrogen atoms, known as *hydrocarbons*. In natural gas, which consists largely of *methane* (CH_4) with smaller amounts of ethane (C_2H_6)

and propane (C_3H_8), the chains are very short. In crude oil, the chains contain four to 30 carbon atoms and are more complex. Crude oil and natural gas are derived from fats and other *lipids* in marine algae and other aquatic plants that were buried with sediment (Hunt, 1979; Link, 1982). Conversion of lipid-rich organic matter to hydrocarbon fluids begins immediately after sedimentation, when microbes begin to alter organic matter, producing methane (Figure 7-12). As organic matter is buried beneath younger sediments, it is leached and altered by groundwaters until it becomes *kerogen*, a relatively insoluble form of organic matter that consists of molecules much larger than those in oil or gas. With deeper burial, temperature increases and kerogen decomposes to form crude oil and natural gas, leaving a complex, solid residue. Oil generation usually begins when organic matter reaches temperatures of 50° to 60°C, the lower limit of conditions known as the *oil window*. At temperatures of about 100°C, kerogen has released most of its oil, and further reactions produce largely methane gas (Figure 7-12). The conditions under which oil and gas form are commonly known as *catagenesis*.

Sediments with abundant organic matter that might form oil and gas are called *source rocks*. Although the best source rocks are marine shales and other fine-grained clastic sediments, oil and gas are not usually recovered from these rocks. Instead, they come from *reservoir rocks*, which have sufficiently high *porosity* and *permeability* (Figure 2-7) for oil and gas to be

FIGURE 7-13

Oil and gas traps. Two sedimentary sequences shown here are separated by an unconformity which formed when the earlier sequence was folded and eroded. The lower sedimentary sequence contains anticline and fault traps, as well as traps beside salt domes and beneath unconformities. The upper, flat-lying sedimentary sequence contains stratigraphic traps, including a carbonate reef with reef talus, and a sand lens. All of the traps are surrounded or overlain by impermeable shale.

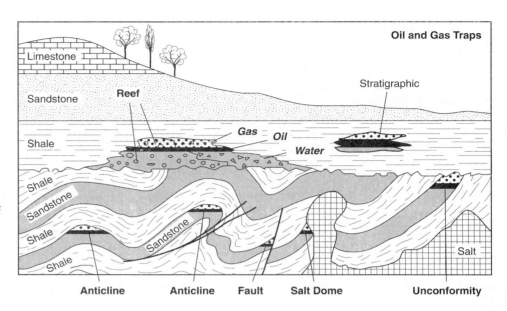

pumped from them. Source rocks such as shale lack permeability and it is not normally possible to produce oil or gas from them. Over geological time, hydrocarbon fluids seep out of shale and migrate into reservoir rocks where oil and gas accumulate in deposits that we produce.

When oil and gas escape from source rock, they flow toward the surface because they are less dense than water, which fills most porosity in the upper crust. Some reaches the surface where natural leakage of oil and gas is widespread (Figure 4-8B), including the famous La Brea tar pits of Los Angeles. Oil and gas deposits exist because some hydrocarbons are trapped before reaching the surface. *Hydrocarbon traps* are essentially reservoirs with an impermeable cap that impedes upward migration. They can be divided into two types, structural and stratigraphic. *Structural traps* form where the shape of the trap or its ability to hold fluid is due to folding or faulting of the host rocks. Structural traps consist of *anticlines,* with alternating layers of reservoir rock and impermeable sediment, or more complex features related to faults or the margins of impermeable salt domes (Figures 7-13 and 14). *Stratigraphic traps* form where variations in porosity or permeability of the sedimentary sequence create isolated reservoirs. If oil and gas are present in the same trap, they tend to separate, with gas floating on oil, which floats on water.

Of the almost 800 sedimentary basins on Earth, only 32% produce oil and gas (Figure 7-15). Another 26% are thought to have some chance of future production, and fully 42% will probably never produce (Klemme, 1980; Nehring, 1982). Age of sediments is not as great a limiting factor as it was for

coal. Marine aquatic plants from which oil and gas could be generated formed long before Silurian time, and Precambrian oil is known. The hydrocarbon endowment of a sedimentary basin is determined by its geologic history. In some basins, such as the delta of the Brahma-Putra River, source rocks are scarce or absent. In others, such as the delta of the Amazon River, source rocks are present but have not been buried deeply enough to enter the oil window. In highly productive basins such as the North Sea and Middle East, all factors converged to maximize hydrocarbon generation and preservation. In the North Sea, important factors were the presence of a rich Jurassic-age source rock and a late stage of deep burial related to rifting, which increased heat flow into the basin facilitating the maturation of the organic material (Prather, 1991). In the Middle East, source rocks were unusually rich, releasing large amounts of oil, and they were close to potential traps, thus preventing the loss of oil during migration. In addition, the reservoirs consisted of highly porous, fractured limestones, many of which had been folded into long, continuous anticlinal traps, and they were covered by impermeable evaporites and shales. Seepage, along with uplift and the erosion of oil fields, probably limit the hydrocarbon endowment of most basins, as shown by the fact that younger basins contain more oil (Tissot and Welte, 1978; Hunt, 1979). Finally, if meteoric water invades a basin during uplift, it can flow through the reservoirs removing much of the oil. Szatmari (1992) has suggested that the preservation of large oil fields has been greatest in arid areas where this could not occur.

FIGURE 7-14
Anticlinal structure at Elk Basin, Wyoming, as delineated by sedimentary layers dipping
away in both directions from center of photo. Oil is produced from the pumping units
at the crest of the anticline.

Oil and Gas Production and Related Environmental Concerns

Hydrocarbons are commonly produced from drill
holes that penetrate traps (Koederitz et al., 1989). In
the simplest reservoir, one with layers of water, oil,
and gas (Figure 7-16), pressure from the expansion of
gas and buoyancy provide a natural *drive* forcing fluids up the well. Where fluids do not reach the surface
naturally, they must be pumped. If oil is produced too
rapidly, water and gas, which flow more easily
through porous rock, will stream into the reservoir
cutting off the flow of oil. *Reservoir engineering* involves the determination of the optimum type and
distribution of wells needed to maximize production
without causing this type of damage (Royal Dutch/
Shell, 1983). Although most producing wells are vertical, *horizontal wells* are coming into increasing use because they intersect vertical fractures and permeabil-

ity that might otherwise be missed. These wells can
even be used to recover oil and gas from shales
(Hayes, 1990; Schmoker, 1992). Because oil and gas extraction rates govern royalty returns, governments
sometimes regulate pumping rates, as in the case of
the Texas Railroad Commission.

Secondary production methods, which are used to
increase oil recovery, were originally applied late in
the life of a field but they are now used from the start
of production. The most common methods, *gas
injection* and *water flooding* (Figure 7-16), use gas and
water from earlier production in the same field to
maintain reservoir pressure. In the early days of oil
production, accompanying natural gas was burned
(flared) rather than being injected. Today, only about
1% of North American gas is flared, although 8.5% of
Middle East gas and 20% of African gas is still flared.
Whatever the production method, it is necessary to

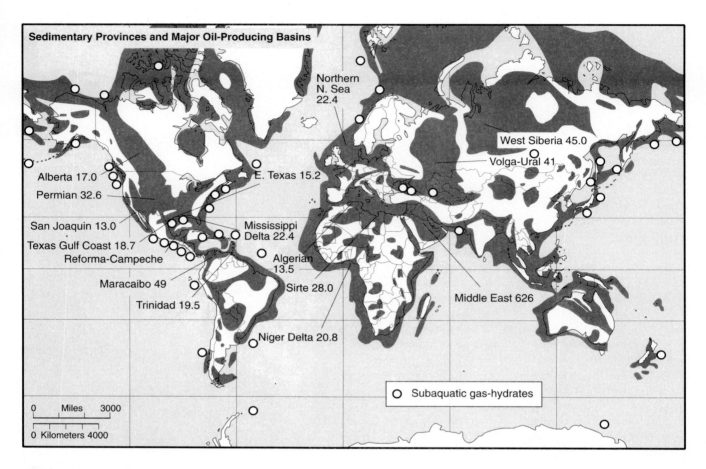

FIGURE 7-15

Distribution of sedimentary basins that contain oil and gas, including major basins
with known oil reserves or basins that have geologic characteristics favorable for its
discovery. Note that many basins are offshore on the continental shelf and slope.
Known areas of gas hydrate are also shown (after Klemme, 1980; Nehring, 1982;
Kvenvolden et al., 1993).

make adjustments to existing wells and drill new ones
to maintain efficient production through the life of a
field. Failure to do this, as has been the case in the
CIS countries, causes a serious decrease in productive
capacity.

Even after the best conventional primary and secondary production methods, as much as 50% of the oil
remains in the ground. Some of this oil can be recovered by *tertiary production methods* or, as they are more
commonly known today, *enhanced oil recovery (EOR)*
(OGJ, 1991). EOR uses methods such as the combustion of oil and gas on the margin of a field or injection
of CO_2 or alkaline solutions to decrease the viscosity
of oil and increase the pressure forcing it to flow from
the reservoir. The largest EOR project in North America, at Cold Lake, Alberta, uses cycles of steam injection followed by the pumping of fluidized oil from
the same well to produce almost 100,000 bbl/day.

Some tax codes encourage EOR to maximize recovery
and increase employment. The only way to improve
on EOR is *oil mining,* in which reservoir rock is removed, disaggregated, and washed clean of oil
(Herkenhoff, 1972). As discussed later, mining is already done for some heavy oil and tar sands. It has
been carried out on some shallow oil fields, but widespread oil mining will be difficult because so many
fields amenable to the method are in populated areas
of Pennsylvania, Ohio, and California.

Production in most onshore fields in populated areas requires simple pumps and other facilities, and hydrocarbons reach markets through pipelines (Figure
7-17). In frontier wells that lack pipeline connections,
oil is trucked or taken by tanker to a pipeline connection or a refinery. *Offshore production* takes place from
very expensive platforms that are being improved
constantly. In 1982, the deepest water platform in the

FIGURE 7-16

Efficient (top), wasteful (middle), and enhanced (lower) oil production (EOR) methods. Pumping rate for efficient production must not exceed rate at which oil can flow into well. EOR methods include water flooding, gas injection, and CO_2 injection (modified from Imperial Oil, Ltd.).

Oil Production Methods

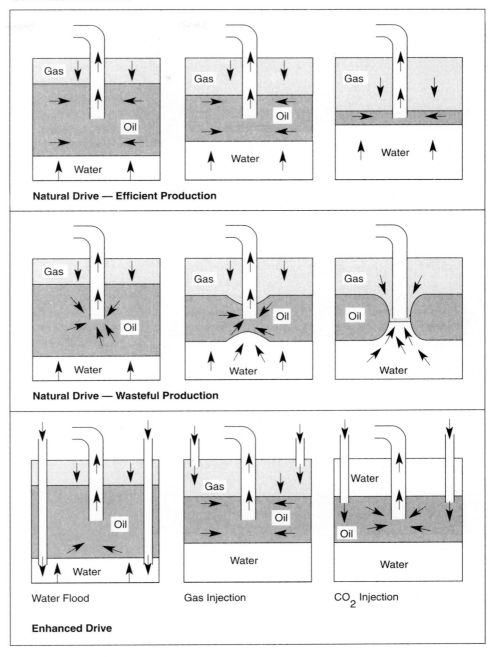

Natural Drive — Efficient Production

Natural Drive — Wasteful Production

Water Flood · Gas Injection · CO_2 Injection

Enhanced Drive

United States, Shell Cognac in the Gulf of Mexico, weighed 56,000 tonnes and sat on rigid legs. It had slots for 62 wells and stood in 311 meters of water. A decade later, the deepest platform was Conoco's Jolliet, also in the Gulf, which weighed only 17,000 tonnes and had only 24 well slots. However, it was in 533 meters of water and was held in place by metal legs that were under tension rather than supporting the platform (Figure 7-18). At present, wells in deeper

water are produced from subsea wellhead systems, although platforms are planned for work in almost 1,000 meters of water in the Gulf of Mexico by the mid-1990s.

Environmental problems related to oil exploration and production include subsidence, as well as spills of water and oil (Williams, 1991). *Subsidence,* which is caused by the collapse of the rock as fluids are withdrawn from the pore space, was more significant in

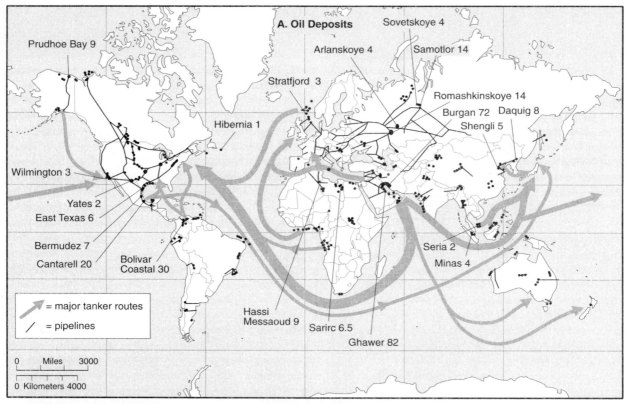

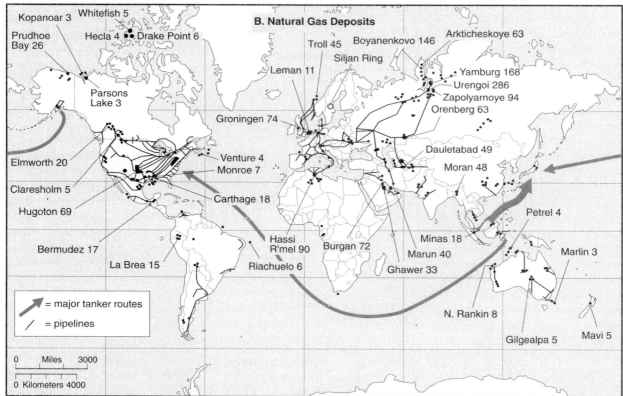

FIGURE 7-17

Distribution of major oil (A) and gas (B) fields showing pipelines and tanker routes (reserves shown for oil in Bbbl and for gas in Tcf; after *Oil and Gas Journal; International Petroleum Encyclopedia,* copyright PennWell Publishing Company, 1990; Carmalt and St. John, 1986). Detail on parts of California and Middle East in Figures 7-19 and 7-20.

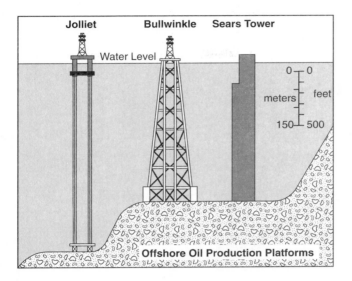

FIGURE 7-18

Offshore production platforms for work in deep water. The Bullwinkle platform stands on rigid legs, whereas the Jolliet platform uses tension legs, which are attached to the floating platform and to a base that is fixed to the sea floor.

early oil fields. Between 1926 and 1967, for instance, production of 1.2 Bbbl of oil from the Wilmington field in Long Beach, California, formed a conical depression at the surface that had an area of 52 km² and a maximum depth of 8.8 meters (Figure 7-19). Water injection, begun in the late 1960s, stopped the subsidence and is widely practiced in most fields now (Mayuga, 1970). Natural and production-related subsidence in deltas, such as the Louisiana Gulf Coast, is aggravated by levies constructed to limit flooding, which prevent river sediment from replenishing the land (Getschow and Petzinger, 1984). *Water spills* can release brines that contaminate local surface and groundwater. As noted earlier, radium in some brines causes radioactive contamination of well equipment (Schneider, 1990).

The risk of *oil spills* is greatest during exploration, when a drilling rig might intersect an unknown, highly pressurized reservoir that would cause a blowout. Oil spills related to production are very rare. Of the approximately 13 Bbbl of oil produced from U.S. offshore leases between 1964 and 1992, only about 0.003% (450,000 bbl), were lost by spills (FOS, 1990; MMS, 1986). Only one of the more than 30,000 wells drilled during that time has blown out. This well, which was drilled off Santa Barbara, California, in 1969, was actually shut off by its blow-out preventer, but about 77,000 bbl were lost through faults and fractures in the adjacent sea floor. It is no coincidence that

this well was in a zone of about 2,000 natural oil seeps, which extends along the Santa Barbara Channel into the La Brea tar pits of Los Angeles (Link, 1952). These seeps bleed naturally from the underlying fields at an estimated rate of 14,600 to 244,500 bbl/yr, and have been a source of beach pollution for centuries (Estes et al., 1985). Natural hydrocarbon leaks of this type are widespread and could be an important part of global hydrocarbon pollution (Prior et al., 1989). If anything, production from a field should lower reservoir pressures, slowing the flow from natural seeps. When they do occur, however, blowouts can release large amounts of oil because it takes a long time to contain them. The largest blowout in recent times, which took place in 1979 on the Ixtoc I offshore exploration platform, near Yucatan in the southern Gulf of Mexico, released more oil than most tanker spills, as discussed later.

Reclamation of abandoned oil fields is a matter of ongoing concern. Surface installations must be removed and wells need to be plugged so that they do not leak to the surface or to other fluid-bearing zones. A typical oil well penetrates numerous layers of sediment, including some with salty brines or drinkable water. Mixing of these fluids is prevented during operation of the well by lining the hole with a *casing* of steel pipe that is perforated only at the level of the oil and gas-bearing layers. After the well is abandoned, it must be filled with cement. Some older wells remain open, however, and programs to shut them down properly are in place in most states. In Texas, a fund established for this purpose receives $100 to $200 in permit fees for each new well, as well as 0.3125 cent/bbl and 0.0333 cent/Mcf of natural gas produced in the state.

Oil and Gas Uses and Quality

Oil is used largely to make gasoline and other fuels, whereas natural gas is used largely for home and industrial heating. As with coal, the key quality factor for both crude oil and natural gas is sulfur, which occurs as atoms in crude oil molecules and as H_2S in natural gas. Sulfur contents can range up to several percent and are most serious in natural gas. Natural gas with more than 1 ppm (0.0001%) H_2S is called *sour gas*; that with less than 1 ppm is called *sweet gas*. The permissible amount is so low because H_2S reacts with air and moisture to form sulfuric acid, which corrodes metal pipes causing leaks. Thus, long before acid rain became a concern, sulfur was removed from natural gas, and major sour-gas producers such as those in Alberta were important sources of by-product sulfur. The main source for sulfur in oil and gas is seawater sulfate, as it is in coal. Although many geologic variables control the presence of sour gas in a reservoir,

FIGURE 7-19

A. Oil fields in the Los Angeles basin. **B.** Contours showing location of maximum land subsidence over the Wilmington field between 1927 and 1969 (from Mayuga, 1970, with permission).

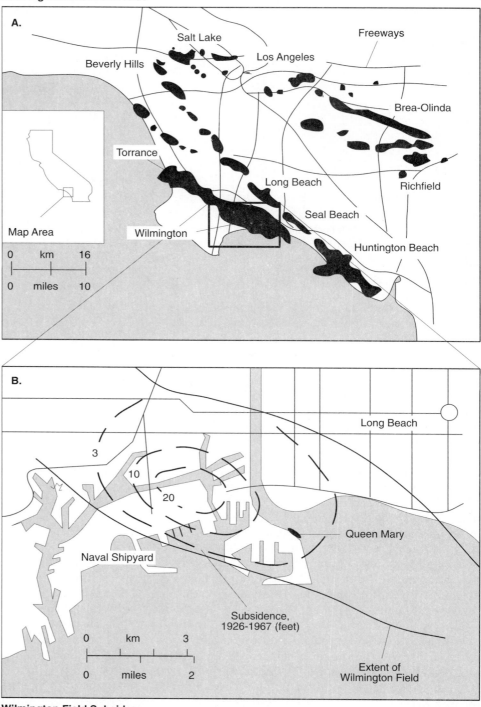

Wilmington Field Subsidence

an important one is the scarcity of metals in the reservoir. Where metals are present, H_2S reacts with them to produce metal sulfides such as pyrite, thereby removing sulfur from the natural gas. Where metals are absent, as in many limestone reservoirs, H_2S content of natural gas can reach high levels. In addition to H_2S, natural gas can contain important amounts of CO_2, N_2, Ar, and He. *Helium* found in some gas fields in the midcontinent and Rocky Mountain regions of the United States is thought to be derived from the radioactive decay of uranium in underlying Precambrian granitic rocks (Leachman, 1988).

The dominant consideration in crude oil quality is its suitability for making gasoline, the most important oil product. Light oil, which yields more gasoline, commands higher prices than does heavy oil. Refining costs for gasoline have long been a driving force in the economics of oil, with a constant battle between cost and environmental considerations (Marshall, 1989). Early concern centered on *tetraethyl lead* [$Pb(C_2H_5)_4$], introduced in the 1920s as the inexpensive way to minimize premature combustion, which created a knock in engines. Tetraethyl lead use has declined since the mid-1970s because it is a major contributor to global lead pollution, as discussed in the section on lead. It also coats the surface of catalytic converters used to clean automobile exhaust from unleaded gasoline. It has been replaced by lead-free gasoline formulations that do the same thing but cost more to produce. Costs are about to rise further in response to concern over the incomplete combustion of gasoline, which produces CO and NO_x that contribute to smog. Incomplete combustion occurs because of the scarcity of oxygen in the combustion chamber, a problem that can be alleviated by increasing the oxygen content of the gasoline. The *Clean Air Act Amendments of 1990* recognized this by requiring that all gasoline contain at least 2% oxygen by 1995. Because conventional gasoline contains essentially no oxygen, organic molecules containing oxygen will have to be added during refining. The best known of the possible additives are *methanol* (CH_3OH), which can be synthesized from other fossil fuels, or *ethanol* (C_2H_5OH), which can be synthesized from sugar cane, corn, and other plants. But methanol has only half as much energy as gasoline per unit volume and would, therefore, require larger fuel tanks. Although methanol evaporates more slowly than gasoline, ethanol evaporates more rapidly and would aggravate atmospheric pollution related to fuel evaporation (Wald, 1992). Worse yet, both are toxic and soluble in water, whereas gasoline is not, increasing the possibility that leakage would damage surface and groundwater supplies. Ether, another organic molecule with somewhat less oxygen, has been proposed as an alternative to avoid some of these complications.

Oil and Gas Reserves

World oil reserves are estimated to be almost 1 trillion barrels (Tbbl) and world natural gas reserves are about 4,208 trillion cubic feet (Tcf). Although these figures are adequate for about 50 years at present production rates, they mask important local complications. Reflecting their more complex origin, oil and gas fields have a wider range of sizes than coal deposits, and more of the world's reserves are found in fewer large deposits (Dreyfus and Ashby, 1989; Masters et al., 1991, 1992). The 500 or so *giant fields*, those containing more than 500 Mbbl of oil or an (energy) equivalent amount of gas, account for two thirds of recoverable global oil and gas reserves (Carmalt and St. John, 1986). The two largest fields, Ghawer in Saudi Arabia and Burgan in Kuwait (Figure 7-20), contain more than 10% of world reserves. By far the largest reserves are in the Middle East, which contain an estimated 600 Bbbl. The largest western hemisphere reserves, in the Maracaibo area of western Venezuela and the Mexican Gulf Coast, contain only a tenth as much oil as the Middle East, and the largest provinces in the United States, the Permian Basin and the Mississippi Delta, are even smaller (Figure 7-15). In fact, total U.S. reserves are only a little more than ten times U.S. annual production, and most other oil-producing MDCs have even lower ratios (Fisher et al., 1992).

World gas reserves are not really any better. Russia has the largest gas reserves, in the Urengoy, Yamburg, Bovanenkovo, and other fields of the West Siberian province (Table 7-4) and large reserves are found in deeper parts of the Middle East basin (Permian Khuff Formation). Smaller but very important reserves are present in the United States, Canada, Algeria, and the Netherlands. As luck would have it, early exploration in the United States found what appears to be its largest field, the giant Hugoton field in Kansas (Table 7-4). Nothing to match this field has been found since. Gas reserves in the United States amount to only eight times annual production, even less than for oil.

Just how good these reserve figures are is difficult to say (Rice, 1986). Estimates for the CIS countries are based on limited data. Recent studies have discounted these reserves significantly. Even in the United States, reserve data depend strongly on the price of oil and gas and on the accuracy of long-term projections. Take the Jolliet platform (Figure 7-18), for instance. Jolliet produces from the Green Canyon field, which is at an average depth of 576 meters of water off Louisiana. The field, which was discovered in 1981, contains 45 Mbbl of oil and 110 Bcf of gas. Although this would be an attractive, medium-sized field on land, it cost about $400 million to put the Jolliet platform in place. After provisions for bonus bids, rental, and royalties on the lease, the profit margin on this field leaves little room for error. Whether similar investments will be made in the future to bring on additional reserve depends both on the price of oil and the economic and regulatory climate.

Reserve estimates for oil and gas are also affected by seemingly magic *oil field growth*, by which total recoverable oil and gas actually grow with time. This

FIGURE 7-20

Distribution and size (Bbbl of oil and oil equivalent) of oil and gas fields in the Persian Gulf showing location of the oil spill created during the Gulf War of 1991. Heavy lines connecting oil and gas fields are pipelines (modified from *International Petroleum Encyclopedia*, copyright PennWell Publishing Company, 1990).

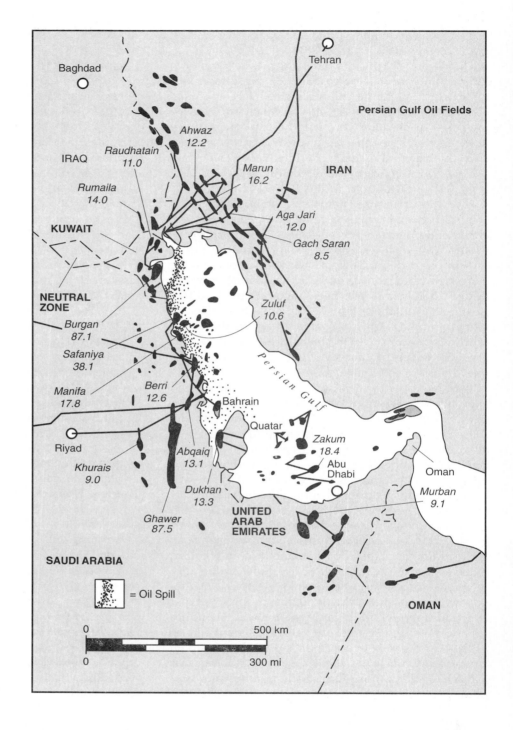

phenomenon reflects an improved definition of recoverable oil in a field as it is drilled and studied more completely, but is also due to the flow of oil into the reservoir from adjacent areas during production. It is estimated to increase the ultimate recoverable reserve of fields in the Gulf of Mexico by about 10% in the first decade of production, declining to about 2% in the fourth decade (Drew and Lore, 1992). According to Root and Attanasi (1992), oil field growth more than doubled the size of oil reserves in the lower 48 U.S. states between 1960 and 1989.

A critical feature of any reserve is the rate at which it is replaced by new discoveries. On that count, the MDCs are in deep trouble; reserves in the United States, Canada, Norway, and the United Kingdom declined by 30% between 1975 and 1992. The United Kingdom, for instance, is nearing exhaustion of North Sea oil that fueled its national economy for decades.

TABLE 7-4
Ten largest oil and gas fields of the world in relation to ten largest fields in major
MDCs. Fields are ranked in order of declining size with gas reserves converted to
oil equivalents. Numbers show their rank with respect to all fields in the world.
Reserves shown here are original and have been depleted by production (from
Carmalt and St. John, 1986, with permission; Bbbl = billions of barrels, Tcf = trillions
of cubic feet).

Field Name	Country	Oil (Bbbl)	Gas (Tcf)	Oil Equivalent (Bbbl)
Largest Fields in the World				
1 - Ghawer	Saudi Arabia	82.000	33.008	87.500
2 - Burgan	Kuwait	75.000	72.50	87.083
3 - Urengoy	Russia	.002	285.59	47.602
4 - Safaniya	Saudi Arabia	36.100	11.79	38.066
5 - Bolivar	Venezuela	30.100	.00	30.100
6 - Yamburg	Russia	.000	167.89	27.983
7 - Bovanenkovo	Russia	.000	146.50	24.416
8 - Cantarell	Mexico	20.000	.00	20.000
9 - Zakum	Abu Dhabi	18.400	.00	18.400
10 - Manifa	Saudi Arabia	17.000	4.79	17.000
Largest Fields in Importing Countries				
18 - Prudhoe Bay	Alaska, USA	9.450	26.00	13.783
21 - Hugoton	Kansas, USA	1.412	69.50	12.995
23 - Groningen	Netherlands	.000	74.00	12.333
36 - Troll	Norway	1.400	45.39	8.966
56 - East Texas	Texas, USA	5.600	.00	5.600
77 - Statfjord	Norway	3.000	2.29	3.383
79 - Elmworth	Alberta, Canada	3.333	.00	3.333
83 - Carthage	Texas, USA	.150	18.00	3.150
91 - Wilmington	California, USA	2.758	1.31	2.977
92 - Brent	United Kingdom	2.163	3.19	2.696

The same pattern prevails for natural gas. Over the same time period, reserves in the Middle East increased by 67% and those in Mexico and Venezuela increased by 300%. Only the MDCs are producing more oil than they are discovering. Thus, world oil trade can only expand as the MDCs increase consumption and imports in the face of declining domestic production.

Trade and Transport and Related Concerns

Oil and gas are among the most important commodities of international trade. Imports (to all countries), which make up about 40% of world oil production, had a 1992 value of almost $200 billion and natural gas imports were worth about $20 billion. These numbers are large enough to have an impact on any national budget, and energy imports account for trade imbalances in many countries. For instance, the United States spends about $500 million annually on imported oil and has accumulated a net deficit of about $800 billion in its oil trade since 1974. As can be seen in Figure 7-21, world oil exports are dominated by Persian Gulf and CIS countries, with shipments going largely to Europe, Japan, and the United States. Natural gas exporters must be adjacent to major consumers to avoid costly tanker transport of liquified natural gas (LNG), as discussed later.

World oil trade for the 1990s depends very much on Russia and other CIS states, where the abysmal state of oil and gas fields became obvious during the collapse of the Soviet Union (Stanglin, 1992). The problems appear to reflect unrealistic requirements by the Ministry of Oil and Gas, which was in charge of the domestic industry. With oil selling for only about $0.60/bbl (even when lifting [pumping] costs were as high as $2/bbl) and no rewards for conservation, production was very inefficient. Recovery rates were less

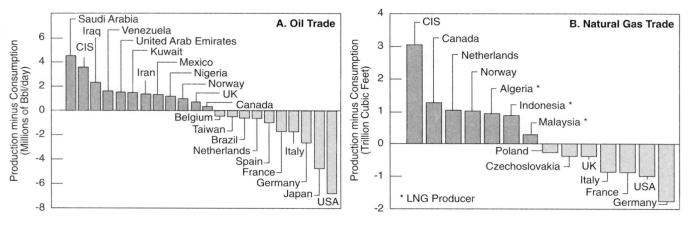

FIGURE 7-21
A. World trade in crude oil for 1991. **B.** World trade in natural gas for 1991. Countries plotting above the line are net exporters and those plotting below the line are net importers (compiled from *International Energy Annual,* Department of Energy).

than half of those in western countries, and distribution systems are estimated to have spilled as much as 20% of the oil they carried. The result is crippled fields, which are major sources of pollution, and which are estimated to require $130 billion in repairs and modernization. These funds are not available today in the CIS, particularly since the countries gained most of their foreign exchange from oil exports, which have dropped steadily. The possibility of foreign loans or equity investment seems remote until governments are more stable and committed to long-term profit sharing. Although this could cause global supply shortages, its greatest effect will probably be on the states themselves, which will be short of income just when they need it most.

Oil is moved by pipelines and tankers, with tankers having the advantage on the longer hauls. Crude oil *pipelines* (Figure 7-17A) extend from producing regions to refinery centers. Natural gas pipelines extend from fields to industrial and residential markets (Kennedy, 1984). Most important oil fields in the Middle East and other producing areas have short pipelines to tanker ports. Long-distance oil and gas pipelines are found largely in Canada, Russia, and the United States, where overland transport is unavoidable (Figure 7-22A). Most pipelines operate at capacity all year. Because of the highly seasonal residential heating market, natural gas pipelines would be largely idle in the summer if gas were not pumped into underground chambers near markets during the summer to supplement pipeline supplies during the winter.

Tankers (Figure 7-22B) are more cost effective for really long distances, in terms of both original investment and operating cost, and they can be directed to new destinations as demand changes. In practice,

world oil tanker shipments, which are about 160 Bbbl-kilometers annually, are dominated by shipments from the Middle East. Tanker transport of natural gas is more difficult because it must first be liquified at temperatures of about −160°C. *Liquified natural gas (LNG)* terminals are expensive facilities that require long-term commitments between supplier and consumer, making LNG shipments more like pipeline transport. Although LNG shipments account for less than 5% of world gas sales at present, shipments currently come from Abu Dhabi, Alaska, Algeria, Brunei, Indonesia, Libya, and Malaysia. The list will undoubtedly grow, although attention will continue to focus on safety issues related to possible explosions in these facilities.

Several decades of operation show that tankers merit more environmental attention than pipelines. Pipelines have closely spaced shut-off valves, that limit the size of any spill. In arctic terranes they have been elevated to prevent the blockage of caribou and other migration routes and to prevent the melting of *permafrost,* which is permanently frozen ground that might become unstable when melted by hot oil moving through the pipeline (Figure 7-22). Early concerns about pipelines were assuaged by the record of the Alyeska Pipeline, which extends for 1,265 kilometers from Prudhoe Bay to the tanker port at Valdez (Figure 7-17A). The $7.7 billion line started service in 1977 and carries 1.5 Mbbl/day, considerably above its original capacity. As of 1992, it had experienced only 14 spills, the largest of which was 15,000 bbl lost when a saboteur blew a hole in the line, although corrosion damage along a 15-kilometer section of the line required repairs estimated to take five years and cost about $1 billion. *Tanker oil spills* (Table 7-5) range

A.

FIGURE 7-22
Crude oil transportation.
A. Alyeska pipeline near Fairbanks, Alaska, where it has been elevated to prevent heated oil from melting permafrost.
B. Patriot double-hulled tanker, shown here under construction in Korea, is 250 meters long with a capacity of 692,000 bbl. The outer hull protects the inner cargo hull in case of collision. Tankers of this type cost about 20% more than single-hulled vessels (courtesy of Carlton Adams, Conoco Inc.).

B.

from small losses from all types of ocean-going vessels, which have polluted many shipping lanes and coastal areas, to occasional large spills from oil tankers, which attract a disproportionate share of attention. There is no question that tanker oil spills are fatal to much marine life, but it is not clear that their long-term effects are particularly serious. Unlike metals, which are not biodegradable, most oil residues decompose with time and form H_2O and CO_2. Studies made in coastal areas where oil was spilled 10 to 20 years ago find little or no evidence for the event in the sediment record (Frey et al., 1989). Detailed studies in Prince William Sound following the spill there in 1989 showed that the tanker oil made up only a small fraction of the total hydrocarbon in the sediments and that it declined rapidly in abundance during the next few years. Most of the

hydrocarbon in sediments of the Sound came from natural seeps to the east and other marine fuels, and their abundance did not change over this period (Page et al., 1993). These results do not mean that we can continue to spill oil into the oceans; there will obviously be a limit to Earth's capacity to disperse spilled oil. In fact, they raise the fascinating question of just how bad a spill would be necessary to leave a recognizable trace in the sedimentary record. It has been suggested, for instance, that large offshore oil fields might have leaked rapidly to the surface along faults formed during earthquakes, covering large parts of the ocean. This could have changed the solar reflectivity of Earth, altering local and possibly global climates. Until we find better ways of recognizing natural and anthropogenic oil spills, this possibility remains a matter of speculation.

TABLE 7-5
Oil spills from tankers (global data) and production/exploration facilities (U.S. data)
since the mid-1960s (from Federal Offshore Statistics, Minerals Management Service,
U.S. Department of Interior). Spills listed in order of declining size (in bbl). For
perspective, daily crude oil consumption in the United States is about 17 Mbbl.

Production/Exploration Spills

Spill/Cause	Date	Location	Amount of Oil
Well blowout	1979	Ixtoc #1, Mexico	3,500,000
Anchor damage to offshore pipeline	1967	West Delta, LA	160,638
Well blowout	1969	Santa Barbara Channel, CA	80,000
Well blowout	1970	South Timbalier, LA	53,000
Well blowout	1970	Main Pass, LA	30,000
Anchor damage to offshore pipeline	1974	Eugene Island, LA	19,833

Tanker Spills

Spill	Date	Location	Amount of Oil
Atlantic Empress/Aegean Captain	1979	off Trinidad	2,100,000
Castillor de Bellver	1983	off South Africa	1,750,000
Amoco Cadiz	1976	off France	1,561,000
Torrey Canyon	1967	off England	833,000
Sea Star	1972	off Oman	805,000
Urquiola	1976	off Spain	700,000
Othello	1970	off Sweden	up to 700,000
Hawaii Patriot	1977	North Pacific	693,000
Braer	1993	Shetland Islands	619,000
Exxon Valdez	1989	Prince William Sound, Alaska	240,000

Security of oil and gas supplies is of great concern, particularly the high *import reliance* of many MDCs. The oil embargo of the 1970s, which crippled these countries, led to the formation of oil stockpiles, such as the 586-Mbbl U.S. Strategic Petroleum Reserve. Although there was widespread interest in the development of a *national energy policy* to decrease future vulnerability, many MDCs, including the United States, are more dependent on imported oil now than in 1970. Interestingly, some countries are developing a parallel import dependence for natural gas. Pipeline and LNG systems built since 1975 have formed regional natural gas links between producers and users. Indonesia and Brunei supply LNG to Japan and other countries. Canada and the United States are linked and it is only a matter of time before Mexico and perhaps Venezuela become part of the system. Much of Europe is supplied by pipelines from the Netherlands, United Kingdom, and Norway, with important additions from Siberia, a connection that was bitterly opposed by the U.S. government during the early 1980s

(Jentleson, 1986). Industrial and residential users in most MDCs now depend heavily on natural gas. Fully 88% of U.S. households and industries heat with natural gas. Obviously, an interruption in natural gas shipments would be just as catastrophic as earlier oil embargoes.

With a high import dependency among developed nations, it is likely that the 1991 Iraq conflict was not the last war we will fight over oil. In these conflicts, oil and gas installations are major targets of hostile efforts. During the Gulf war, sabotage of the Sea Island tanker terminal spilled 6 Mbbl of oil into the Persian Gulf. The resulting oil slick extended 400 kilometers southward along the western side of the Gulf (Figure 7-20). In addition, 752 wells in Kuwait were damaged, spilling out as much as 55 Mbbl of oil into 200 lakes, and burning 5 to 6 Mbbl of oil per day, about the same as prewar oil production from these fields (Ibrahim, 1992). In a grotesque way, these events provide a test of the environmental impact of truly gigantic oil spills and blowouts. We could not have found a less forgiving place for such a test. The Gulf has such a small opening to the ocean that full exchange of water requires about five years and it was already receiving about 250,000 bbl/year of oil from natural seeps and production spills in the area. The results of this "experiment" have been more encouraging than many thought. In contrast to estimates of up to five years to extinguish all of the fires, they were put out in less than a year. Local climate changes predicted from soot and smoke pollution also failed to occur and damage to marine life appears to have been less severe than anticipated.

Frontier and Unconventional Sources of Oil and Gas

Future oil and gas supplies will depend on discoveries of conventional oil and gas in remote, *frontier* areas, as well as unconventional sources. *Frontier exploration areas* in North America include the western overthrust belt, the Arctic slope and islands, and the East Coast (Figure 7-15). The *overthrust belt,* which runs along the eastern margin of the Rocky Mountains, consists of possible reservoir rocks that are buried beneath other rocks that were thrust over them from the west. Some of the overlying thrust sheets contain hard igneous and metamorphic rocks, making drilling through them costly and difficult. Interest in the *Arctic slope and islands* stems from their proximity to the Prudhoe Bay field, the largest field in North America. Two decades of exploration in the Canadian arctic have found two main hydrocarbon areas, the McKenzie Delta, with as much as 4 Bbbl of oil and 56 Tcf of gas, and the Arctic Islands, which are estimated to contain 10 Bbbl of oil (*International Petroleum Encyclopedia*, 1991). Pipeline production from Arctic island fields will be complicated by the fact that they are separated from the mainland by channels in which sonar studies show grooves formed by icebergs dragging across the bottom (Lewis and Benedict, 1981; Barnes and Lien, 1988). To the west in Alaska, the Arctic National Wildlife Refuge, which is directly adjacent to Prudhoe Bay, has become a focus of debate over the environmental effects of oil exploration (Stevenson, 1990). Significantly, industry operations at Prudhoe Bay have not caused severe environmental damage so far and the local caribou herd has actually increased in size. The continental shelf of the *East Coast* is of interest because of its general geologic similarity to the prolific Gulf Coast area. Exploration of the East Coast has found producible reserves in Canadian waters, including the large Hibernia field almost 100 km east of St. John's, Newfoundland. Exploration farther south has concentrated in the George's Bank Embayment, the Baltimore Canyon, and the Southeast Georgia Em-

bayment areas. About 50 exploration wells have been drilled into some large traps (detected by seismic surveys) in these areas since 1975, but commercial reserves of oil and gas have not been found.

Unconventional resources that might add significantly to our oil and gas supplies include heavy oil, tar sands, and oil shale, which are discussed separately later. Another possibility is *coalbed methane* (NPC, 1980). Coalbed methane recovery systems have been developed in the San Juan basin of New Mexico and gas from this source now accounts for about 3% of U.S. consumption. These efforts were stimulated by tax credits provided by the U.S. government in the *Fuel Use Act of 1980*. Another possibility, *tight gas*, refers to gas from reservoirs that has porosity and permeability too small to permit conventional production. Many sands and shales contain this type of gas, and experimentation continues with ways to recover it, including horizontal drilling and explosions to enhance permeability.

Farther out on the horizon are two other possible sources. *Geopressure gas* is gas that is dissolved in waters in the deeper parts of the world's sedimentary basins. This gas attracted attention in the Gulf of Mexico where it is widespread at depths of 4 to 5 km. Although the concentration of dissolved gas is low, there is so much water that the total amounts of gas are huge. *Gas hydrates* are crystalline compounds containing methane and water, similar to a methane-bearing ice. These compounds form under the low-temperature, high-pressure conditions that prevail in the shallow parts of some marine sedimentary basins. Widespread reflective horizons seen in seismic surveys of these basins indicate that methane hydrates may be the most important global reservoir of organic carbon and, as such, potentially important both as a fossil fuel and a source of methane contamination to the atmosphere (Kvenvolden, 1988; Kvenvolden et al., 1993).

All of these possibilities pale in comparison to *abiogenic gas* (Gold and Soter, 1980). The term alone implies trouble. Our entire discussion of oil and gas has been based on the premise that they are derived from the decomposition of organic matter. The abiogenic hypothesis says gas might also form by inorganic chemical reactions, perhaps in Earth's mantle, and be trapped as it rises through the crust. If this were true, gas could occur in the igneous and metamorphic rocks that underlie sedimentary basins, which would greatly enlarge the area that might contain gas reserves. Although very few igneous and metamorphic rocks are porous and permeable enough to contain large amounts of gas, a meteorite impact crater might be an exception. This possibility was actually tested by the Swedish government and other groups, including the U.S. Gas Research Institute, which is funded by government-mandated contributions from gas producers. They spent $25 million to drill a 6,779-m hole into the Siljan ring meteorite-impact feature in central Sweden (Figure 7-17B). The hole found only traces of methane, most of which was biogenic (Juhlin, 1991), and the concept has lost support.

Heavy Oil and Tar Sands

Heavy oil and semisolid bitumen are the wild cards of global hydrocarbon resources (Figure 7-23). *Heavy oil* is oil that will flow under normal reservoir conditions but requires EOR methods for economic production. Oil that will not flow is commonly known as *tar*. Although tar can be found in all types of rocks, most attention has been given to the large accumulations in sandstones, which are known as *tar sands*. In the few places that tar has been recovered from tar sands, it has been necessary to mine the reservoir, physically separate the tar from the rock, and process it to make

it more like oil. Heavy oil and tar are less desirable than crude oil because they cannot be converted to gasoline as easily and they yield a larger fraction of heavy oil products. They also contain more sulfur and nitrogen and have locally high metal concentrations, notably nickel and vanadium (Chilingarian and Yen, 1979; Stauffer, 1981).

Most heavy oils and tar sands appear to be the remains of typical oils that were altered by reaction with groundwater and bacteria at relatively shallow depths. In these environments, oxygen-bearing groundwater dissolves and oxidizes part of the oil by a process known as *water washing*. *Bacterial degradation* also takes place as bacteria in the water use the oil as food, decomposing lighter molecules first. These processes remove hydrogen and increase the fraction of heavy molecules, making the oil viscous. This is an essentially irreversible process in nature; studies show that heavy oil and tar are not easily converted to oil even if they are buried to greater depths at which kerogen would form oil.

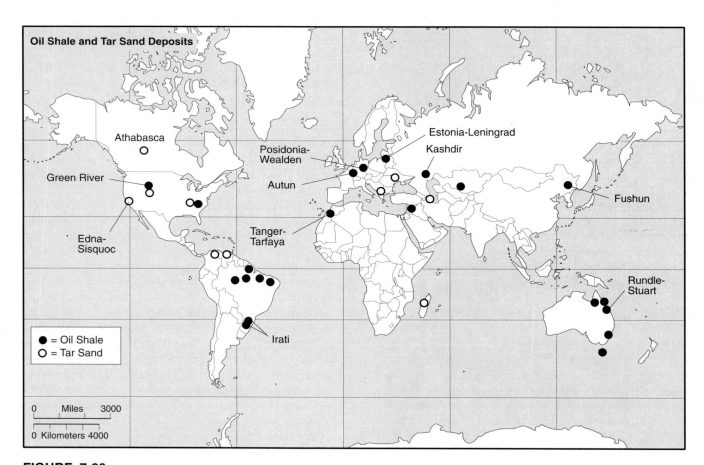

FIGURE 7-23
Location of world tar sand and oil shale deposits. Deposits that are named have had significant production.

The largest deposits of tar sand in the world are in the Athabasca area of northeastern Alberta (McConville, 1975), where two operations produce over 73 Mbbl of oil per year (Figure 7-24). The largest of these, the $3.6 billion Syncrude operation, mines 120 million tonnes of tar sand each year and accounts for about 11% of Canadian oil production, as well as important amounts of coke and fuel gas (Plate 6-1). The sand which has about 8% to 14% bitumen, is removed by strip-mining methods that are among the largest in the world (Singhal and Kolada, 1988). Tar is then removed from the sand by washing in very hot water and is converted to *syncrude* by the addition of hydrogen and other chemical processing. About 83% of the bitumen is converted to syncrude, a much better recovery rate than obtained from most oil wells. The tar contains up to 5% sulfur, over 98% of which is recovered during processing.

Although in situ reserves of tar sand in the Athabasca area might be as high as 1 Tbbl, a much smaller amount is extractable by present methods, possibly 35 Bbbl (Chilingarian and Yen, 1979). The Oligocene Oficina-Temblador belt in the eastern part of Venezuela could contain 200 Bbbl of tar sand and large heavy oil deposits are present in the Maracaibo basin of western Venezuela. Most other deposits are considerably smaller, including the 900 Mbbl deposit at Asphalt Ridge, Utah (Campbell, 1975).

These tar sand reserves cannot be compared directly to crude oil reserves for several reasons. First, they represent total oil in place without consideration of losses during recovery. Second, the actual extent and quality of tar in these fields are not known with certainty because exploration has not been as complete as it has for oil. Third, only a small part of the tar sands can be mined from an open pit, the only economic method at this time (Figure 7-24). Heavy oil in the Cold Lake and Peace River areas is being extracted by well-based EOR methods discussed in the section on crude oil, but efforts to do this on the more viscous tar sands have not been encouraging. Finally, and most important, the economic feasibility of these reserves is not assured. Present estimates suggest that new operations to mine and process high-quality tar sands would require oil prices of $24/bbl to break even, including debt service. These prices do not compare favorably to costs of only a few dollars per barrel to produce Middle Eastern oil and would make large investments of the type needed to develop these reserves extremely risky without support from host governments.

Oil Shale

Oil shale is shale from which oil can be obtained by processing. It is a type of source rock that has not yet given up its oil. In the 1970s, oil shale was thought to be the answer to U.S. energy self-sufficiency. "Realistic" estimates suggested that oil shale operations being developed in the Piceance Creek basin around Grand Junction, Colorado, would be producing 400,000 to 500,000 bbl of oil/day by the mid-1990s (Russell, 1981). By the mid-1980s, however, after a total investment of $4 to $5 billion dollars, all but one of these operations had closed, most without achieving routine production (EMJ, 1982). The last one, Parachute Creek (Figure 7-25), was closed in 1991 after an

Alberta Tar Sands

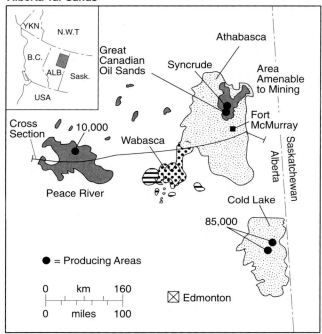

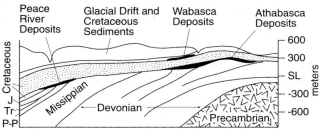

FIGURE 7-24

Alberta tar sands in the Athabasca, Peace River and Wabasca areas and heavy oil-bearing sands in the Cold Lake area. Syncrude and Great Canadian Oil Sands operations are open-pit mines, whereas the other operations produce by EOR from wells (daily production in bbl is also shown). Also shown is the smaller area of tar sands that is amenable to open-pit mining (from *Mining Engineering; Oil and Gas Journal*).

Green River Oil Shale

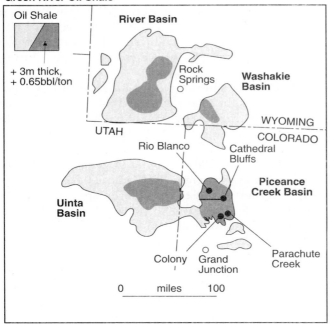

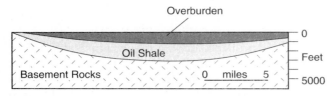

FIGURE 7-25

Geologic map showing distribution of Green River Formation and areas of high-grade, thick oil shale [>3 m thick and >0.65 bbl/ton (0.6 bbl/tonne)]. Cross section shows the relation of oil and overburden along the line extending west from Cathedral Bluffs. Colony and Parachute Creek projects used conventional mining and processing and Cathedral Bluffs and Rio Blanco projects used modified in-situ (MIS) shale combustion (from *Mining Engineering*; Lewis, 1985).

investment of about $1 billion (NYT, 1991). It produced 4.5 Mbbl of synthetic crude oil during its lifetime but never reached design capacity of 10,000 bbl/day and its production costs were above market prices for oil, even with government price supports. Shale oil projects in other parts of the world, including the Rundle Oil Shale Project in Queensland, Australia, suffered similar fates (EMJ, 1981).

Oil shale certainly did not fail for geologic reasons. World reserves of oil shale are enormous (Hook and Russell, 1982). In the United States, the most important reserves are in the Eocene-age Green River Formation, which contains an estimated 2 to 4 Tbbl of shale oil. Huge volumes of Green River oil shale are

present in the Piceance Creek basin, where the mining efforts were centered (Figure 7-25). Much of the eastern United States is underlain by Late Devonian to Early Mississippian-age shales that are thinner and contain smaller amounts of oil, but that have resources of up to 3 Tbbl. Smaller deposits containing billions of barrels of shale oil are found in Alaska and California, and similar deposits are found in Brazil, Australia, and Russia (Figure 7-23). All together, an estimated 17 Tbbl of shale oil are present in rocks that would yield at least 0.65 bbl of oil per tonne, the minimum grade for commercial interest.

Most oil shales, such as the Chattanooga Shale in the eastern United States, are relatively typical organic-rich, marine shales that were deposited along the margins of larger basins where they were not buried by later sediment to reach the oil window. Others, however, formed in very different environments. The strangest of all is the Eocene-age Green River Formation, a sequence of lake, or *lacustrine*, deposits that accumulated in a huge, shallow lake (Lewis, 1985). Water in the lake was alkaline and became saturated with halite and sodium carbonate minerals including *trona*. The Rundle and Stuart oil shales in Queensland, which are sometimes called *torbanite*, also formed in a similar lacustrine environment.

The sad fate of oil shale appears to be largely a matter of difficult engineering and unexpected economics (Ratten and Eaton, 1976). Optimistic predictions for the future of shale oil were based on assumptions of an increasing price for crude oil. Instead, the price weakened in the 1980s, making shale oil production costs of about $45/bbl unacceptable when crude oil was selling for $15 to $20/bbl. Thus, just as oil shale projects all over the world reached a stage that required major capital investment to upgrade them from pilot-scale test plants to large-scale production facilities, adverse prices caused the projects to be canceled. Engineering problems also played a part. To recover oil from shale, it must be pulverized and heated to temperatures of 500 to 1,000°C, where the shale undergoes *pyrolysis,* a process that releases hydrocarbon gases and liquids. Heat for the process can come from the combustion of the shale or the injection of hot gases or solids. Most pyrolysis uses containers at the surface in which mined shale is heated, although *modified in situ systems (MIS)* can be excavated in the shale layer itself. Efforts have also been made to carry out the pyrolysis process in fractured oil shale layers without actually mining it. In addition to their inherent technical difficulties, oil shale projects require water in volumes two to five times as large as the volume of oil produced. To produce 500,000 bbl/day a plant would require most of the precipitation falling on an area of 200 to 500 km². Water is already a big problem in the drainage of the Colorado River where the best oil shale is located, suggesting that large-scale oil shale conversion would require water imports into the area (Saulnier and Goddard, 1982).

Although the latter half of the twentieth century has not been kind to oil shale, it will probably not disappear. Oil has been produced from shale whenever economics made it an attractive option, ever since the first patent on a shale oil process was granted in England in 1694. When oil prices rise again, oil shale will probably return. This time, most of the engineering problems will be solved and the mines and equipment will be waiting.

OTHER ENERGY RESOURCES

Nuclear Energy

Nuclear energy serves us in the form of heat produced by *fission reactions* (Hafele, 1990). In present applications, the naturally occurring isotope, ^{235}U, is bombarded by slow neutrons, which increases the rate at which it splits into smaller *fission products* such as ^{90}Sr and ^{137}Cs. During this reaction, some matter is converted to energy, which we see partly in the form of heat. If the concentration of ^{235}U is high enough and the flux of neutrons is large enough, a self-perpetuating series of ^{235}U fission reactions known as a *chain reaction* will result. If the rate of this reaction is controlled, it becomes a nuclear reactor. If not, it is a nuclear weapon (Duderstadt and Kikuchi, 1979).

Use of nuclear energy is complicated because ^{235}U is the only naturally occurring, readily fissile isotope and it makes up only 0.711% of natural uranium. Most modern reactors require uranium in which the proportion of ^{235}U has been increased by a process known as *enrichment*. The rest of natural uranium is largely ^{238}U, which is not directly fissile, but which can be converted to fissile ^{239}Pu by absorbing a neutron. The main isotope of thorium, ^{232}Th, is converted to fissile ^{233}U by a more complex series of reactions that begin by absorbing a neutron. These isotopes are the basis for possible future nuclear applications involving breeder reactors, which are discussed later.

The history of world uranium production reflects the change from weapons to energy production. Following World War II, uranium was used largely for the manufacture of weapons. As this market declined, its place was taken by peaceful uses for nuclear energy as uranium came into demand to fuel reactors to generate electric power. As safety concerns about nuclear power increased in the 1970s, the growth of nuclear power slowed, causing a second downturn in production (Figure 7-1D), as discussed further later

(Cohen, 1984). At present, world uranium markets amount to about $1 billion annually, far less than the fossil fuels.

Geology and Reserves of Uranium

Much of what we know about the geology of uranium was learned during the exploration boom of the 1970s. This boom started when people noted that uranium reserves known at that time were only a fraction of the amounts needed to fuel the 1,000,000 MWe of nuclear generating capacity estimated to be in place in 1990. This imbalance catalyzed a big rise in uranium prices (Figure 7-1C) and massive exploration programs, involving expenditures of $6 to $8 billion. The results were amazing. New-deposit types were recognized, new districts were found, and world uranium reserves more than doubled to present levels of about 1.55 million tonnes of contained U_3O_8 (Red Book, 1992). Unfortunately, it all happened just as the growth of nuclear power slowed and the price dropped, putting many newly discovered mines out of business.

Uranium forms an unusually complex range of deposit types that reflect its tendency to occur in nature in two oxidation states (Harris, 1979; Nash et al., 1981; Dahlkamp, 1989). The *uranic ion* (U^{+6}) is stable in oxidized environments, and the *uranous ion* (U^{+4}) is stable in reduced environments where oxygen is scarce. The uranic ion dissolves readily, especially in the presence of dissolved carbonate ions (CO_3^{-2}), with which it combines to form soluble complex ions; uranic ions precipitate as minerals such as *carnotite* only if K, Ca, P, or V are relatively abundant. In contrast, the uranous ion is relatively insoluble and forms the common uranium mineral *uraninite* (UO_2), which is called *pitchblende* when it is in a microcrystalline form. Obviously, then, the formation of uranium deposits is favored when waters in oxidizing environments scavenge uranium from surrounding rocks and deposit it where they are reduced.

Three types of uranium deposits contribute most to global uranium production (Figures 7-26 and 7-27). *Unconformity-type deposits* are found at and near an unconformity separating Middle and Upper Proterozoic sediments, particularly in the Canadian Athabasca basin and Australian Pine Creek basin (Clark and Burrill, 1981). The ore, which forms veins and disseminations in the sediments, is thought to have been deposited from oxidizing basinal brines that passed through reducing zones rich in carbon or iron. These deposits have unusually high-uranium grades, with important gold and nickel by-products. The generally similar *sandstone-type deposits* are most abundant in sedimentary rocks of the Colorado Plateau, parts of

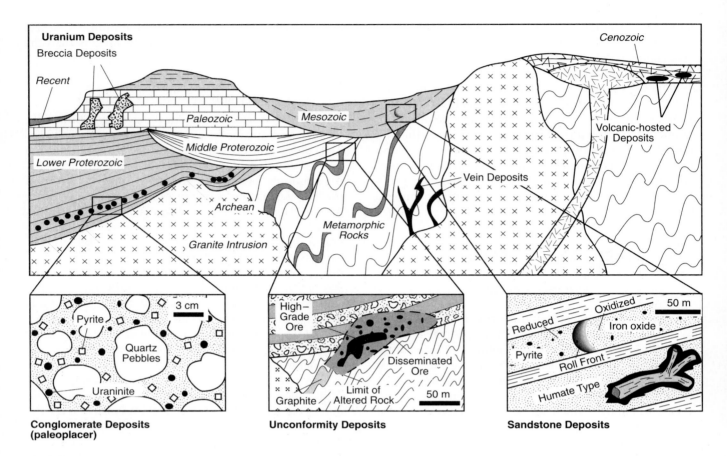

FIGURE 7-26
Geologic setting of important types of uranium deposits. Note that each type of deposit
is shown in host rocks of the most common age.

Wyoming, the Texas coastal plain, and in west Africa. These deposits formed when oxidized groundwater that had leached uranium from surface rocks flowed down into aquifers, where it was reduced to precipitate uraninite. In some deposits, reduction took place along curved zones known as *roll-fronts,* which represent the transition from oxidized to reduced conditions in the aquifer. Elsewhere, it took place around accumulations of organic material, including old logs to form the related *humate-uranium deposits.* Both deposits are also enriched in vanadium, selenium, and molybdenum (Plate 6-2). In some areas, similar waters came close enough to the surface to deposit uranium by evaporation or reaction with other groundwaters. The largest deposit formed by these processes is Yeelirrie, which is in arid Western Australia.

The final important type of uranium deposit, the paleoplacer, is quite different. *Paleoplacers* (Pretorius, 1981) are quartz-pebble conglomerates containing small grains of uraninite (Plate 6-3). Because uraninite is so heavy and occurs as rounded, clastic grains, these deposits are thought to have formed as placers; because they are preserved in ancient rocks, they are called paleoplacers. The largest deposits of this type are at Elliot Lake, Ontario, where several separate conglomerate layers contain mineable ores. Similar deposits, in which gold is the dominant mineral and uraninite is a by-product, are found in the Witwatersrand of South Africa and parts of Brazil (Figure 7-27). All of these deposits are pre-middle-Proterozoic in age (pre-2.2 Ga) and all formed as stream gravels in thick sedimentary sequences around the early continents. The quartz pebbles and uraninite were eroded from veins and pegmatites in granite intrusive rocks in these continents. Because uraninite dissolves fairly easily in modern oxygen-rich stream waters, the abundance of these placer deposits in early Proterozoic time is thought to reflect lower oxygen contents in the atmosphere of the early Earth.

Uranium is also found in a wide range of veins, stockworks, and breccia pipes (Rich et al., 1977; Dahlkamp, 1989) and it has been produced as a by-product

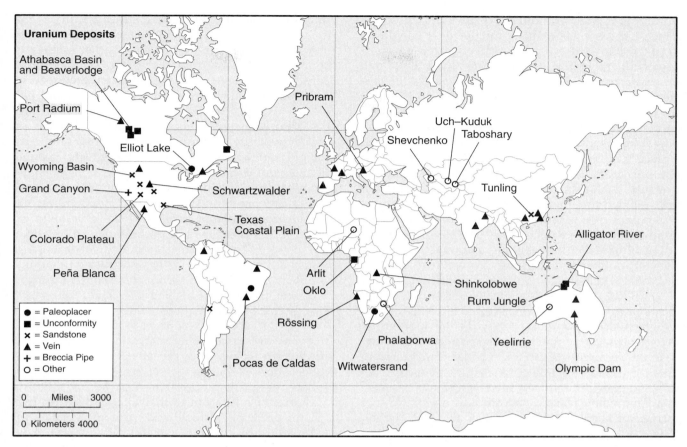

FIGURE 7-27

Distribution of major types of uranium deposits in the world.

from solutions used to leach low-grade porphyry copper ores and those used in processing phosphate deposits in Idaho and Florida. Several countries have tried, largely in vain, to recover uranium from seawater, the most recent effort being made by Japan.

Mining and Processing of Uranium

Uranium has been produced from open-pit and underground mines, and solution mining (in situ leach, ISL) is used in Texas, Wyoming, and Nebraska. The most important environmental hazard of uranium mining is the risk of cancer caused by the inhalation of radioactive aerosol particles that become lodged in the lungs. This problem is worst in underground mines, where particles of rock dust and diesel fumes are more abundant. The radioactivity of these particles is due largely to *radon*, a gas produced by the radioactive decay of uranium and thorium. Radon is liberated during blasting and, within days, most of it decays to other isotopes that are adsorbed onto the aerosol particles (George and Hinchliffe, 1972). When

the particles are inhaled, they are adsorbed onto the lung lining, causing lung cancer and other respiratory diseases. Lung cancer was widespread among workers in the early uranium mines of the western United States during the 1940s and 1950s, especially in Native American miners. In 1990, Congressional action provided for payments of $100,000 each to uranium miners who worked in Arizona, Colorado, New Mexico, Utah, and Wyoming from 1947 to 1971 and who developed lung cancer or respiratory diseases. Increased mine ventilation, the use of air filters and respirators, and fewer smokers among the worker population has greatly mitigated this problem in modern uranium mines. Direct radiation from ore in the walls of uranium mines is not usually a problem, although some very high-grade unconformity deposits require special precautions (Thompkins, 1982).

Tailings from uranium mines are also considered a radiation hazard; only about 20% of the radiation is actually removed from the ore during beneficiation. About 127 million tonnes of these wastes are at abandoned and active mining sites in the United States

(Krauskopf, 1988). Although overall radioactivity of mine wastes is lower than natural ore, their powdered form facilitates the escape of radon and caused problems with early efforts to use waste rock and tailings in local construction projects. The *Uranium Mill Tailings Control Act of 1978* called for cleanup of these wastes at abandoned mines in the United States. Most current environmental regulations require that the wastes be disposed of underground or in piles on the surface, surrounded and covered by impermeable clay (Gershey et al., 1990). In spite of these relatively simple precautions, jurisdictions such as British Columbia have banned uranium mining.

Uranium processing involves three main steps (Duderstadt and Kikuchi, 1979). The first step, separation of uranium from the ore, is carried out at the mine by a chemical leaching process that yields a precipitate known as *yellowcake,* containing about 70% to 90% U_3O_8. Yellowcake is shipped to other locations for the second processing step, enrichment, in which the proportion of ^{235}U in the uranium is increased. Most commercial nuclear reactors require uranium fuel with 2% to 3% ^{235}U, an enrichment of about three to five times the natural abundance of 0.7%. To enrich uranium, it is converted to a gas such as UF_6, which is ionized and diffused under pressure through a barrier. Because ^{235}U is about 1% lighter than ^{238}U, it diffuses slightly more rapidly, thus becoming concentrated. Energy-intensive enrichment plants of this type commonly depend on cheap hydroelectric power, such as at Oak Ridge, Tennessee and Hanford, Washington, although newer enrichment techniques involving gas centrifuges and lasers require less energy. Until recently, uranium enrichment in the United States was carried out at federal laboratories and the cost of enrichment was supposed to have been passed on to the consumer. It was estimated in 1988, however, that up to $9 billion of these costs had not been billed since 1950, apparently in an effort to encourage the use of nuclear power (Franklin, 1988). In 1993, the Department of Energy, which administers United States uranium enrichment facilities, formed the United States Uranium Enrichment Corporation, a semiprivate, profit-making organization that will take over the government's role in uranium enrichment. After uranium is enriched, the third and final process is to fabricate it into reactor fuel.

Uranium Production, Resources, and Trade

Canada and the old Soviet Union were by far the largest historical producers of uranium, followed by the United States, Australia, Namibia, and France. Reasonably assured reserves of uranium, available at a price of $30/lb of U_3O_8 or less, are largest in Australia and South Africa, although much of the uranium in both countries is a by-product of other mining. In South Africa, uranium is in the Witwatersrand gold paleoplacers, where most operations have stopped recovering uranium because of low prices. About half of the Australian reserve is in the Olympic Dam copper deposit, where uranium production depends on copper production. Furthermore, the remaining Australian uranium deposits cannot be put into production without permission of the Australian federal government. Thus, world reserves, in terms of rapid availability, are largely in Canada, the United States, and West Africa. The appearance in 1991 of uranium exports from Russia, Ukraine, Uzbekistan, and Kazakhstan changed this picture, improving global uranium availability but putting further pressure on marginal suppliers in the West. Estimates of uranium resources available at prices as high as $50/lb do not greatly change this pattern (*Red Book,* 1992).

Although uranium resources remain high, production has decreased significantly in response to the precipitous price drop caused by the widespread cancellation of nuclear power plant projects (Figure 7-1C). The greatest effect has been felt by mines in MDCs, particularly those that operated on relatively low-grade sandstone and paleoplacer deposits (VIA, 1991). By 1993, all conventional uranium mines in the United States were closed. Remaining production was from ISL operations or a by-product of phosphate fertilizer production, and imports supplied 85% of U.S. uranium needs. Canadian uranium mining centers were also hurt, particularly the low-grade paleoplacer mines of Elliot Lake, Ontario. This area, once the largest uranium-producing district in the world, cannot compete with high-grade unconformity ores or other deposits that are mined by open-pit methods. In an effort to avert the loss of hundreds of jobs in the area and the virtual abandonment of Elliot Lake, the provincial electric utility company, Ontario Hydro, agreed to pay $160 million above market prices for uranium to be supplied between 1992 and 1996 (*Northern Miner,* June 24, 1991). Pressure on these suppliers is likely to be increased by recent efforts to provide economic aid to Russia through purchases of uranium from decommissioned Soviet weapons, which would be used as reactor fuel in the West. About 300 million pounds of U_3O_8 are estimated to be available from this source through 2012. Another 150 million pounds of weapons-grade uranium that remains in U.S. and Russian government stocks might become available later.

A high import reliance for uranium is common in MDCs. The United Kingdom, Germany, and Japan are

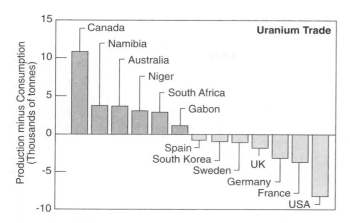

FIGURE 7-28

World uranium trade for 1992 compiled from data of the *Organization of Economic Cooperation and Development Red Book*. Uranium demand (largely for reactor fuel) has been used for consumption.

almost completely dependent on imported fuel, and only Canada, among consuming nations, supplies all of its needs (Figure 7-28). Although this situation is similar to that of the fossil fuels, it does not elicit much comment because of the lower value of uranium production and because imports come, in part, from MDCs rather than LDCs. As discussed earlier, many non-U.S. uranium producers formed a cartel similar to OPEC, known as the Uranium Producer's Club, which almost caused the bankruptcy of Westinghouse Electric in the 1970s, when prices rose so much that

Westinghouse could not supply uranium fuel that they had promised as an incentive for the sale of reactors (Taylor and Yokell, 1979).

Uranium Markets, Reactor Design, and Uranium Reserves

Uranium enters commerce in the form of depleted uranium from which most of the daughter products of radioactive decay have been removed. Small amounts of depleted uranium are used in some chemicals and pigments, where it produces a brilliant yellow color. It is also valued for its extremely heavy atomic weight and consequent high density, which are needed for applications as different as high-performance sailboat keels and armor-piercing ammunition. About 40 tonnes of the metal remain strewn about the surface after the Gulf War, for instance. These markets are very small in comparison to those in nuclear weapons and reactors, however. Of these, reactor fuel is the only market on which uranium can depend if it is to remain an important mineral commodity (CNP, 1991). As of 1992, there were about 400 nuclear power reactors operating in the world, with another 90 under construction (Table 7-6). Whereas nuclear power supplies only 14% and 19% of Canadian and U.S. electrical needs, respectively, it is much more important in Europe and parts of Asia (Table 7-6). Whether nuclear energy continues to grow, however, depends on public perception of the safety of nuclear reactors. An understanding of the controversy requires a brief discussion of reactor design.

TABLE 7-6

Nuclear power plants operating and under construction (data for January 1991 from U.S. Department of Energy Commercial Nuclear Power Report and Uranium Institute).

Country	Operating Reactors	Generating Capacity (MWe)	Percent of Generating Capacity	Reactors Under Construction	Uranium Enrichment Facilities
United States	111	99,588	19.1	3	Yes
France	56	55,778	74.5	6	Yes
Japan	39	30,896	26.5	24	Yes
United Kingdom	39	11,506	20.1	1	Yes
Russia	25	17,385	12.2[a]	21[a]	Yes
Germany	22	22,337	28.2	0	Yes
Canada	18	12,799	14.4	0	No
Ukraine	16	13,030	NA	NA	Yes
Sweden	12	9,817	46.0	0	No
Spain	9	7,607	38.4	6	No
WORLD	397	321,005		90 (91,396 MWe)	

Other countries with large dependency on nuclear energy (percentage dependence given in parentheses) include Belgium (60.1), South Korea (49.1), Switzerland (42.6), Taiwan (38.0), Bulgaria (35.7), and Finland (35.0).
[a] For Soviet Union; separate data not available for Ukraine.

In a nuclear power plant (Figure 7-29), the reactor provides heat, which boils a *working fluid,* driving an electric turbine (Duderstadt and Kikuchi, 1979; Kessler, 1987). In some power plants, the working fluid is separated by a steam generator or heat exchanger from another fluid that circulates through the reactor core. The nuclear reactor itself consists of fuel in some form and a *moderator,* which controls the rate of neutron flux through the fuel and often acts as a *coolant.* All reactors produce some fissile material by neutron bombardment of ^{238}U and ^{232}Th in the fuel or around it. In practice, reactors are divided into *converters,* which produce less fissile material than they consume, and *breeders,* which produce more than they consume. Most commercial reactors in use today are converters, which are further divided into water-cooled and gas-cooled types. These factors control the life span of world uranium reserves because each reactor type has different uranium requirements per kilowatt of electricity generated.

Water-cooled converter reactors come in four main types. In the United States, the most common types are *boiling water reactors (BWR)* and *pressurized water reactors (PWR),* both of which use enriched uranium fuel and a light-water coolant and moderator. Because they have only one fluid cycle, BWRs are more efficient, although they require larger shielding because water expands to steam in a single cycle, going directly from the reactor core to the turbines. PWRs employ two fluid cycles in which the reactor core is separated from the turbine by a heat exchanger, but they require containers and buildings with thicker walls to protect against possible steam explosions. PWRs account for about 60% of U.S. nuclear power capacity, with the rest in BWRs (Figure 7-30). *Heavy-water–cooled reactors (HGR),* such as the Canadian *CANDU,* use heavy water as the moderator and coolant, which permits the use of unenriched uranium fuel, a major advantage for a country without uranium enrichment facilities (Figure 7-31). *Light-water, graphite-moderated reactors (LGR)* are used only in Russia and the Ukraine, where they make up about half of generating capacity, including the ill-fated Chernobyl reactor. They are relatively inexpensive to build and operate but have a higher safety risk because the graphite core can burn in the presence of the water coolant, producing explosive hydrogen gas. *Gas-cooled converter reactors,* which are used principally in the United Kingdom, include the *Magnox reactor,* which uses unenriched fuel and a solid graphite moderator, and the *advanced gas-cooled reactor (AGR),* which uses enriched fuel. The *high-temperature, gas-cooled reactor (HTGR),* with a He coolant, has not been used widely in the present generation of commercial reactors.

Breeder reactors are based on the conversion of ^{238}U to ^{239}Pu, although naturally occurring ^{232}Th can be converted to fissile ^{233}U, and ^{240}Pu can be converted to ^{241}Pu. Breeders use enriched fuel surrounded by a shell of ^{238}U, which allows about 1.5 ^{239}Pu atoms to be produced by each fission event, more than enough to fuel another reactor. Because breeders obtain so much more energy from natural uranium, they could greatly extend the useful life of uranium reserves. Unfortunately, the outlook for breeders is poor. They require concentrated fuel and use liquid sodium coolant, which is highly explosive. Breeders have become lightning rods for economic and environmental concern. The Clinch River project in the United States and the Kalkar project in Germany have been canceled, as has the Dounreay reactor in Scotland. The Phenix and Superphenix reactors in France have experienced numerous operating problems and downtime, leaving the Japanese breeder program as the world's most aggressive (Sanger, 1993). It is not likely that breeder reactors will be able to compete economically with converter reactors in the foreseeable future.

Concerns about reactor safety center around emissions during normal reactor operation and those that might occur when the system fails in some way. It has been demonstrated repeatedly that properly operating reactors are not important sources of radiation, so the problem boils down to what might happen during failure of the system. The principal risk is that of coolant failure, which would cause the fuel to become hot

Nuclear Power Plant

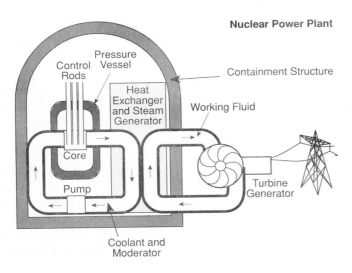

FIGURE 7-29

Schematic illustration of a nuclear power plant. As discussed in the text, some reactors do not use a heat exchanger and pass coolant fluid directly to the generator turbine. In gas-cooled systems and most breeders, the reactor core is surrounded by graphite rather than water as shown here.

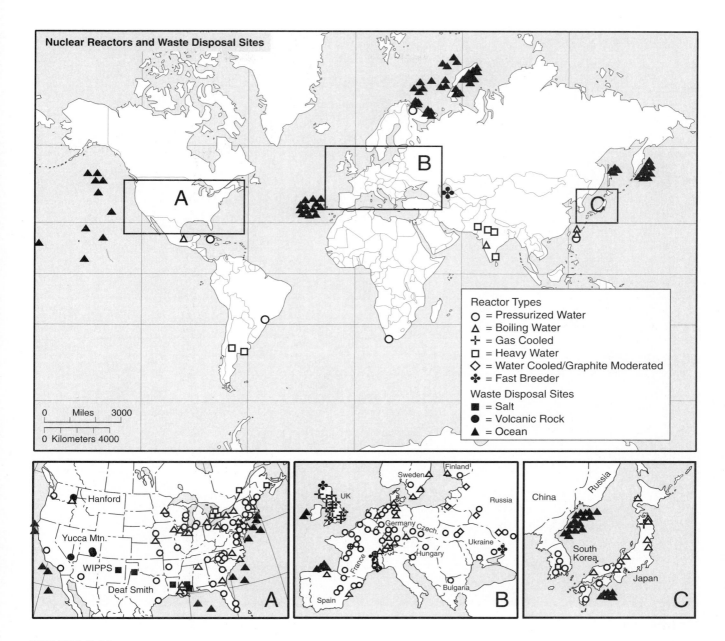

FIGURE 7-30

Location of operating nuclear power plants in 1992. Also shown are sites being used or tested for disposal of nuclear wastes, as well as sites where radioactive materials have been dumped in the ocean. The Asse site in Germany has been used so far only for low-level wastes, whereas the Yucca Mountain site is being considered for high-level wastes (from Department of Energy and International Atomic Energy Agency).

enough to melt. Failures of this type have been mercifully rare but reassuring. In the 1979 accident at Three Mile Island, Pennsylvania, 27 tonnes of fuel melted and ponded in the base of a PWR containment vessel without significant radiation leaks to the environment (Toth, 1986). The tragedy at Chernobyl, Ukraine, which resulted in hundreds of fatalities, was

a special, hopefully unique, case. The Chernobyl reactor was an LGR with a graphite core and no shielding structure, a configuration that would not be allowed in the West. It was operating outside its own safety regulations at the time of the accident, when hot fuel caused the graphite to burn, dispersing radioactive wastes throughout northern Europe (Haynes, 1988,

FIGURE 7-31
Pickering Nuclear Generating Station east of Toronto, which uses CANDU reactors and
has an installed capacity of 4.3 MW. Eight reactor buildings are joined to the vacuum
building by a pressure-relief duct. The turbine hall, auxiliary bay, and administration
and service buildings are to the left of the reactor buildings (courtesy of Kimberley
Beam, Ontario Hydro).

Zhores, 1990). With the exception of this anomalous situation, nuclear power appears to be safer than other types of power, particularly when all aspects of the production process are considered. For instance, coal-fueled power generation in the United States causes seven times more fatalities per unit of electricity generated than does nuclear power, ignoring the potential future problems with black lung disease and glo-

bal warming (Cohen, 1984). Oil causes about twice as many fatalities, and natural gas causes about the same number of fatalities as nuclear power.

In spite of these statistics, public concern about nuclear power has grown since the late 1970s. Sweden held a national referendum in 1980, which might lead to the closure of its reactors by 2010. In 1990, Italy closed its nuclear power plants, even at the cost of 80%

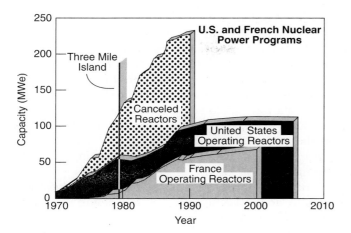

FIGURE 7-32

Growth of nuclear capacity in the United States and France. U.S. power plant cancellations are also shown; France has had no significant cancellations (compiled from data in DOE/EIA0438(91)).

dependence on imported electricity. Although some of this concern resulted from nuclear accidents, over half of U.S. generating capacity had already been canceled before the Three Mile Island accident (Figure 7-32). These early cancellations reflected escalating reactor construction costs caused by government-mandated, safety-related design changes. Costs for many nuclear projects exceeded original estimates by five to ten times (Cook, 1985). Outside the United States, experience with nuclear power has varied from good to bad. The French program, which started later than the United States program, avoided many of its delays, has had very few cancellations, and has not had a significantly worse safety record. The situation in the old Soviet Union has been quite different. There, nuclear power was preferred over oil or gas-fired power plants for domestic supplies because the fossil fuels could be exported. However, reactor design and safety were poor and many Chernobyl-type LGR reactors remain in service. A recent report by the World Bank and International Energy Agency identifies 25 high-risk reactors in Russia, Ukraine, Armenia, Bulgaria, Slovakia, and Lithuania, all of which need to be upgraded or replaced at costs estimated to be as high as $24 billion (Simons, 1993).

Global uranium reserves were actually saved by the decline in uranium demand. Present reserves are barely adequate to supply converter reactors in service or under construction through 2010. The situation would have been much worse if early projections for nuclear power had proved correct and U.S. capacity had been double its present level (Figure 7-32). As discussed next, the problem of nuclear waste disposal also contributed to declining uranium demand.

Nuclear Waste Disposal

Nuclear reactors produce radioactive wastes that give off *alpha particles,* which can be stopped by a sheet of paper or a few centimeters of air, *beta particles,* which are stopped by thin metal or a few meters of air, and *gamma rays,* which require centimeters of lead or even more rock shielding. Alpha particles do the most damage because they are the largest, but gamma rays have greater penetration. Wastes are commonly classified according to origin (Table 7-7) and include gases, dilute solutions, and solids (LWV, 1985; Murray, 1989). Although civilian nuclear wastes have a surprisingly small volume compared to wastes produced by other forms of energy, they contain isotopes such as ^{90}Sr, which substitutes for calcium in skeletal material, localizing radiation damage. Half-lives of many waste isotopes are so long that they must be isolated for

TABLE 7-7

Classification of radioactive waste in the United States (after Gershey et al., 1990). Note that the volume of all high-level wastes would make a cube only a little more than 33 meters on a side.

Type of Waste	Sources/Important Isotopes	Volume (m^3)	Radioactivity (10^6 curies)
High Level	Reprocessing of spent fuel; ^{137}Cs, ^{60}Co, ^{235}U, ^{238}U; $^{239\text{-}242}$Pu	38,200	1,300
Spent Fuel	Power plants, defense; ^{137}Cs, ^{90}Sr, ^{144}Ce, $^{239\text{-}242}$Pu	6,800	18,000
Transuranic	Reprocessing waste with Pu; ^{241}Am, ^{244}Cm, $^{239\text{-}242}$Pu	28,000	4.1
Low Level	Power plants, laboratories fission products, ^{235}U, ^{230}Th	250,000	15

thousands of years before they have decayed to a safe level.

In general, the isolation problem breaks down into two questions: what is the best form for the waste and where should it be put? It was originally thought that *spent fuel reprocessing* would be used to extract plutonium for use in breeder reactors, and that all waste would therefore be put into solution to facilitate reprocessing. This did not occur for U.S. electric power reactors, where concerns about the toxicity and security of plutonium led to the 1972 cancellation of the U.S. fuel recycling program. The wisdom of this decision is questionable, at best. Plutonium is less toxic than cadmium, lead, or arsenic, which do not disappear by radioactive decay. It is of greatest concern when it is inhaled where alpha decay particles will damage lung tissue, possibly causing cancer (NEA, 1981). Furthermore, large volumes of plutonium in U.S. defense wastes are recycled, as are wastes from commercial reactor fuel in most European countries. Nevertheless, spent fuel from commercial reactors in the United States and Canada continues to be stored, largely in water-filled tanks at reactor sites, awaiting decisions on its final disposal form and location, and great public concern accompanied the 1992 shipment of plutonium-rich, recycled fuel from France to Japan.

The *disposal form* for high-level wastes will almost certainly be some type of solid because it is more compact and isolates the isotopes from the hydrosphere and biosphere (Tang and Saling, 1990). Most consideration has focused on sealing wastes in a container made of cement, glass, ceramic, or synthetic mineral or rock (known as *synroc*). The French disposal system uses a borosilicate glass. Some Swedish wastes are put in copper drums because archaeological artifacts made of native copper are preserved for millennia, although concern has been expressed that the copper might be an attractive target for future, resource-starved populations.

Disposal locations for radioactive wastes have been studied for years, with attention shifting from less likely disposal in the ocean or beneath polar ice caps to the need to put the waste back into the rock from which it came (Chapman and McKinley, 1987; Krauskopf, 1988). Rock disposal locations are favored because they have a better chance of remaining undisturbed for thousands of years, the time necessary to isolate waste so that it can decay to an acceptably low level of radioactivity. Rock storage sites must have low porosity and permeability, be free from earthquakes or other natural disturbances, and permit reentry to recover the wastes if necessary. Studies of the *Oklo uranium deposit* in Gabon have been cited as support for rock storage. The proportion of ^{235}U in some uranium from Oklo is much lower than in normal uranium, ap-parently because it was consumed in natural fission reactions that occurred about 2 billion years ago, shortly after the deposit formed and while it was still deeply buried. This provided a natural laboratory to study the dispersion of fission products in rock. Although the analogy to commercial reactors is not perfect, data from Oklo suggest that nongaseous fission products are adsorbed onto clays and other minerals in the surrounding rocks and do not travel far from their source, a result that is encouraging for rock storage.

Rocks evaluated so far as disposal sites include salt, granite, and volcanic rock. *Salt* layers and domes received attention early because their existence proves that abundant water was not present to dissolve them. Project Salt Vault, an early attempt to use salt layers in Kansas for disposal of low-level wastes, was abandoned when it was found that the layers had been penetrated by oil drilling. More recently, attention has centered on the thick Permian salt layer in Deaf Smith County, Texas, and southeastern New Mexico (the Waste Isolation Pilot Project), part of the evaporite sequence that contains important potash deposits, and the Asse salt dome in Germany. *Granite* and other impermeable intrusive igneous rocks in old, stable Precambrian shields have been considered in Canada and the Stripa area of Sweden.

The "winner," as far as high-level wastes in the United States is concerned, is *volcanic rock* at Yucca Mountain in the Nevada Test Site, the location of early nuclear weapons testing. It was selected in 1987 by an amendment to the *Nuclear Waste Policy Act of 1982,* which assessed $0.001/kw-hr of nuclear power to fund studies that would identify the best high-level waste disposal site in the country. Of the nine sites chosen for original study, three remained for further testing in 1987, including Yucca Mountain, basalt lava flows at Hanford in Washington, and the Deaf Smith County salt layers. Congress stopped these studies in 1987 as a cost-cutting measure and Yucca Mountain became the nation's disposal site by default. Fortunately, it has some very attractive characteristics. Unlike other disposal sites, where the water table is near the surface and isolation of the waste depends on the impermeability of the host rock, the water table is 500 m below the surface at Yucca Mountain and likely to stay there for thousands of years. In addition, volcanic tuffs that underlie Yucca Mountain contain minerals that react with and adsorb dissolved ions, making migration of waste products less likely. Present plans call for an intensive study, including extensive drilling (Figure 7-33) and 24 kilometers of tunnels at a cost of $6 billion, to determine whether a repository can be constructed at Yucca Mountain to isolate about 70 tonnes of nuclear waste for 10,000 years.

FIGURE 7-33
Overview of probable U.S. nuclear waste repository at Yucca Flat, Nevada, showing
drill rig in position to test probable level for disposal system in dry rock at 500 to 800
meter depths. Note lack of drilling mud system around the rig, which must drill a dry
hole in order to test the level of saturation of the ground (courtesy of Jerry J. Lorenz,
Reynolds Electrical and Engineering Co.).

Geothermal Energy

Geothermal energy is the small fraction of Earth's
thermal heat that we are able to use, largely in the
form of natural hot water and steam in porous, permeable reservoirs. Where the water in these reservoirs
has a temperature of at least 150°C, enough of it will
flash to steam when it is pumped upward toward the
surface to run turbines that generate electricity (Plate
6-4). Where the water is not that hot, it can still be used
for residential and industrial heating, as is done in
China, Italy, Iceland, and areas as widely separated as
Klamath Falls, Oregon, and Ebino, Japan (Gupta, 1980;

Bowen, 1989). As of 1990, geothermal electric generating facilities throughout the world had a total power
of about 5,827 MW_e, and geothermal heating accounted for another 11,385 MW_t, roughly equivalent
to the annual consumption of 55 Mbbl of oil with a
total value of over $1 billion at 1993 oil prices
(Freeston, 1990; Huttrer, 1990).

Hot groundwater reservoirs are simply active analogues of the *hydrothermal systems* that form many
mineral deposits, and they have been given the special name *geothermal systems* (Rybach and Muffler,
1981; White, 1981). Geothermal systems are *liquid
dominated* if most of the pores in the rock are filled by

FIGURE 7-34
Recent volcanic rocks (foreground) provide heat for geothermal power plants (background) in the Salton Sea, California.

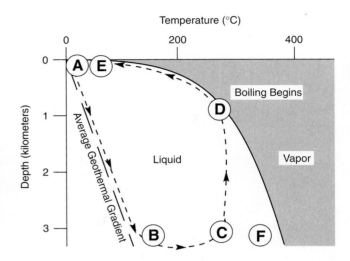

FIGURE 7-35
Boiling curve for water showing the path of geothermal fluid in the liquid-dominated system of Figure 7-37. Pressure is represented by depth of an equivalent column of water (hydrostatic pressure). Recharge water begins at surface (A) and flows downward in an area with a normal geothermal gradient to reach a temperature of about 100°C at a depth of 3 kilometers (B) where it becomes buoyant enough to rise (C). At (D) (whether in the rock or a well), rising water intersects the boiling curve, forming vapor that could be used to drive a turbine. By the time water reaches the surface, it has cooled almost back to conditions represented by (A) (modified from White, 1967).

water, or vapor dominated if the pores are largely filled by water vapor (steam). Vapor-dominated systems tend to have temperatures of about 240°C, whereas water-dominated systems can be as hot as 360°C, although actual temperatures depend on the depth of the reservoir and the resulting pressure, as shown schematically in Figure 7-34. Note in this figure that the *boiling curve* for water, which separates liquid wa-

ter from water vapor, is at higher temperatures than most *geothermal gradients,* the natural rate at which temperature increases downward in the crust. In fact, natural gradients reach typical geothermal temperatures of 250°C only at depths of over five kilometers, considerably deeper than most geothermal fields (Figure 4-13). It follows that hot geothermal fields must be associated with unusually high geothermal gradients, usually where magmas have come near the surface (Figure 7-35). Thus, land-based geothermal systems are most common in areas of recent volcanic activity above subduction zones, as in New Zealand, Philippines, Mexico, Japan, Italy, and the western United States (Figure 7-36, Table 7-8). Similar systems are probably present along the mid-ocean ridges, but they are accessible to us only on Iceland where the ridge reaches above sea level.

The simplest geothermal system consists schematically of a heat source and a porous system through which water can reach the heat (Economides and Ungemach, 1987). The dominant type of water is *meteoric water,* which moves downward along fractures, until it is heated enough to become buoyant and begins to rise (Figure 7-37). Many geothermal reservoirs

TABLE 7-8
World geothermal electric and thermal (heating) capacity.

Country	Capacity (megawatts)
Electric Generating Capacity	
United States	2770
Philippines	891
Mexico	700
Italy	545
New Zealand	283
Japan	215
Indonesia	142
El Salvador	95
Iceland	45
Kenya	45
Thermal Capacity	
Japan	3321
China	2143
Hungary	1276
Russia	1133
Iceland	889

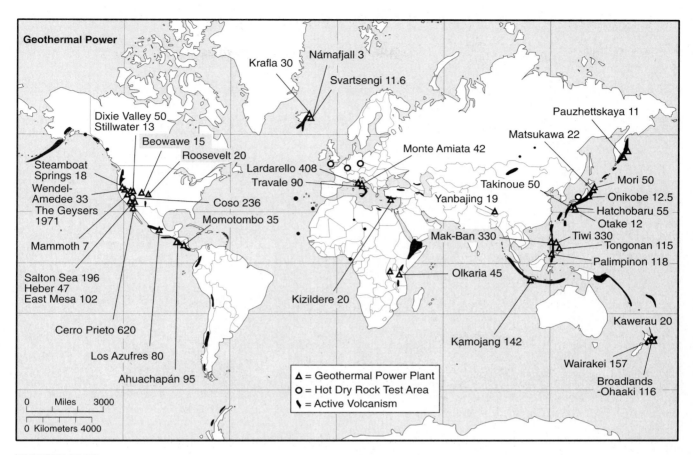

FIGURE 7-36

Distribution of producing geothermal areas in 1990 (capacity in megawatts), areas of hot dry rock tests, and active volcanoes (compiled from information supplied by L.J.P. Muffler and Ellen Lougee, U.S. Geological Survey).

are overlain by a *low-permeability seal* that prevents the hot water from leaking out too rapidly, although small amounts will reach the surface as *hot springs, geysers,* or even steam vents known as *fumaroles.* This seal usually consists of quartz or some other form of silica that has been precipitated in pores in the upper part of the reservoir, usually because of the cooling of the geothermal fluid. In some cases, the seal is lacking, and fluid density and temperature control the structure of the system.

Geothermal power comes close to being a "free lunch," but does not make it. After all, a magma should remain hot enough to power a geothermal system for hundreds or thousands of years, long after we will have found alternate energy sources. In practice, however, it is hard to balance the rate of steam or water production to the rate at which water is recharged to the system, and the drilling and pumping perturb the chemical balance of the system. These effects can cause precipitation of calcite or silica, known as *scale,*

which clogs piping in the systems and possibly also the natural porosity of the reservoir. This complication is most common for geothermal systems containing large amounts of gas, such as Ohaaki, New Zealand, and Ribiera Grande in the Azores, because gas release triggers calcite precipitation. These complications, along with local overproduction of reservoir fluid at fields such as The Geysers, have required that new wells be drilled more frequently to maintain productivity and have raised questions about the ultimate life of geothermal systems.

Geothermal systems emit ten to a hundred times less CO_2 and SO_2 than fossil-fuel power plants per megawatt of electricity generated, but are not without environmental concerns (Axtman, 1975; Bowen, 1989). Some systems contain highly saline water that corrodes pipes and turbines. Some also deposit scale locally, including metal sulfides or precious metals, in amounts that are too small to be of economic interest but large enough to require constant cleaning of the

Geothermal Systems

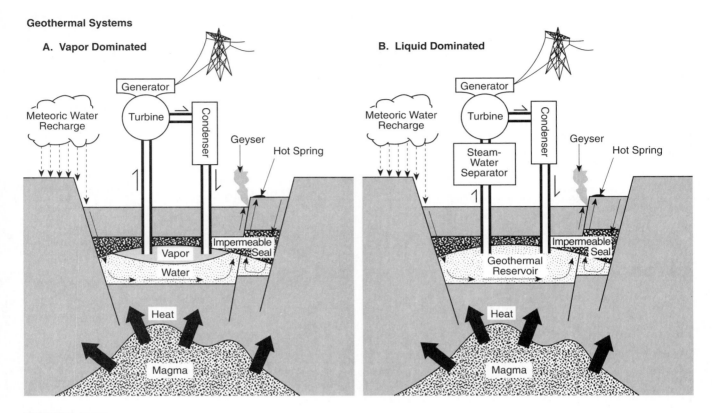

FIGURE 7-37

(A) Vapor-dominated and (B) liquid-dominated geothermal systems showing how meteoric water recharge is heated until it rises to form hot springs and geysers. Conditions within the reservoir are shown by A to E in Figure 7-35. Note that power generation from vapor-dominated systems does not require a liquid-vapor separator (modified from White, 1967).

plumbing system. Most geothermal systems have no significant water emissions because waters are reinjected to recharge the system. This procedure, which is simply prudent operating practice, is also required by law because the waters have locally high dissolved solid contents, including toxic elements such as arsenic, antimony, and boron, which are relatively soluble in low-temperature hydrothermal water and which can enter the vapor phase in small but important amounts (Smith et al., 1987). Gaseous emissions are more difficult to contain and steam loss is common. The main hazard of gaseous emissions is H_2S, a highly toxic gas that is common in some geothermal systems, although minor metal emissions are also observed.

Geothermal power is very important locally, but it is not a major, global source of energy. Even in the United States, which has the largest installations in the world (Table 7-8), geothermal electric power generating capacity is only 2,770 MW_e, about 0.2% of total electric generating capacity. Large, hot, vapor-

dominated geothermal systems, such as The Geysers in California and Lardarello in Italy, produce enough electricity to power a large city, but most geothermal systems are smaller. Existing geothermal plants barely scrape the surface of global geothermal potential, however. Many attractive geothermal areas are known but development of them has been discouraged by the same factor that killed the oil shale industry, that is, low oil prices.

Even beyond obvious geothermal prospects, the potential is great. It has been estimated, for instance, that geothermal energy in the outer 10 km of the crust is 2,000 times greater than the energy in the coal resources of the world. We can never harness all of this energy, but some parts of it might be made available. Lower-temperature systems that will not power a turbine directly can generate electricity by use of liquid with a lower boiling point than water. Waters in sedimentary basins have attracted attention, and low-level heat might be obtained from brines in unconfined aquifers. Warm, confined aquifers, which are known

TABLE 7-9

Comparison of energy content of world fossil fuel resources [in units of 10^{21} joules; from Fettweis (1979) for coal and Masters et al. (1987, 1991) for all other fuels]. Oil, gas, heavy-oil, and tar-sand resources represent recoverable amounts. Oil shale is in situ resources. Both figures are given for coal, because some might be recovered by in situ gasification.

Fossil Fuel	Resources	Energy Content
Coal — recoverable	1076 billion tonnes	21.5
— in situ	2152 billion tonnes	45.4
Oil	1469 million bbl	8.99
Natural gas	9258 trillion ft^3	9.77
Heavy Oil	571 million bbl	3.49
Tar Sand	436 million bbl	2.66
Oil Shale	13883 billion bbl	84.9

as *geopressured zones* because they exceed hydrostatic pressure, are common in the Gulf of Mexico basin and elsewhere, where they are an unconventional source of natural gas, as noted earlier (Wallace et al., 1979). Attention has also been given to *hot, dry rock systems,* where drilling can be used to enhance permeability allowing a fluid to be circulated through the rock to remove its heat. Systems of this type are most common above shallow magmas, such as in the Rio Grande Rift around Los Alamos, New Mexico, or in felsic intrusions with high uranium contents, such as in the Cornwall area of England (Garnish, 1987; Richards et al., 1992). Most of these possibilities have been evaluated by drilling and pilot plants, and none of them have been shown to compete effectively with cheap fossil fuels, although they will be resurrected if fossil-fuel prices climb.

THE FUTURE OF ENERGY MINERALS

The future of energy minerals is inextricably linked to international relations and environmental concerns. Figure 7-38 shows the dramatic difference in energy self-sufficiency between major energy-producing and consuming nations of the world. Canada is the only MDC that is a major energy exporter. These relations are controlled by oil, which is the dominant world energy source, and they are likely to prevail for the next decade or so. But we cannot go on as if nothing were happening. U.S. oil reserves are staggeringly bad (Figure 13-2) and the world is on the verge of a similar problem, with the decreasing success of exploratory wells and greater depth for production wells (Lee, 1989). As discussed in the final chapter, U.S. oil production has declined for years and world production may have begun to decline.

There are two ways that we might respond to the loss of oil as our major fuel. We might continue to use fossil fuels, which provide simple, portable energy. If so, we will have to shift our dependence to natural gas and coal, which have larger reserves (Table 7-9). Coal will bear the brunt of our demand, simply because it is more abundant. But it is unlikely that we will use coal in its crude form for both environmental and technical reasons. Instead, we will place more emphasis on clean coal technologies, particularly *coal gasification and liquification.* This is not a new idea. *Town gas* and *coal oil,* both of which came from coal, were widely used in the 1800s and early 1900s (Perry, 1974). Conversion of coal to gas or liquids requires an increase in H:C ratio by the removal of carbon or the addition of hydrogen, or by decomposing the coal and

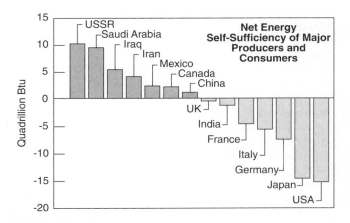

FIGURE 7-38

Net energy self-sufficiency of major producers and consumers (data from U.S. Department of Energy, International Energy Review).

making a new product. The last option allows the removal of sulfur and nitrogen but is more expensive.

Most modern coal conversion plants are based on the *Lurgi gasification process* and their history has been only partly successful. One disappointment was the *Great Plains Coal Gasification Plant* in Beulah, North Dakota, which was designed to use 12,700 tonnes of lignite, 12,700 tonnes of steam, and 2,700 tonnes of oxygen to produce 137 Mcf of gas each day (along with 82 tonnes of sulfur, 95 tonnes of ammonia, and 818 tonnes of ash). The plant was delayed for almost a decade by environmental, engineering, and financial concerns, but was completed in the mid-1980s at a cost of $2.1 billion, considerably higher than original estimates. Its gas production costs turned out to be considerably above market prices, and it cannot be a strong economic performer unless prices rise significantly. The only success story has been the *SASOL* operation in South Africa, a system of three plants built to convert domestic coal into gasoline, diesel fuel, jet fuel, motor oil, and petrochemical products (Figure 7-39). SASOL was begun in 1955 in response to the worldwide embargo of apartheid South Africa, although most of its capacity was constructed after the energy crisis. It has given South Africa, which lacks significant domestic oil production, a measure of freedom from international oil suppliers. The second generation of coal treatment will build on this success. For instance, a plant was constructed recently in the Powder River area of Wyoming to convert low-sulfur coal into a liquid similar to fuel oil and a coal-like residue, both of which are lighter and contain more energy than the original coal. Thus, processed coal is likely to be a major factor in future energy supplies.

Our second possible response might be caused by concern over global warming. Fossil-fuel combustion currently contributes 66% of anthropogenic CO_2 to the atmosphere, as well as 80% of anthropogenic SO_2 and almost all of the NO_x. It is also a major source of airborne particulates (Morgenstern and Tirpak, 1990). We could find temporary relief from this problem by increasing the use of natural gas, which contains no ash and essentially no sulfur and nitrogen (Burnett and Ban, 1989). It has the highest H:C ratio of any of the fossil fuels and yields about 70% more energy for each unit of CO_2 produced. It can also be converted to liquid fuels. The bad news, however, is that methane itself is a stronger greenhouse gas than CO_2 and small leaks in extensive production and delivery systems could be major contributors to atmospheric pollution. Furthermore, unless the unconventional gas resources mentioned earlier turn out to be larger than expected, reserves are not adequate to support us for long (Table 7-8). Oil shale or tar sand might be substituted for natural gas in some places, but their distribution is too spotty to become dominant world fossil-fuel sources without global cooperation in investment and production.

Thus, our alternative will be to return to nuclear energy, or to find a substitute among the nonmineral sources such as wind, wave, or solar, all of which lack portability, the dominant advantage of fossil fuels. Although many early studies predicted that nuclear power would dominate our energy future (Figure 7-1A), the public does not trust it, with most concern focused on reactor safety. The second generation of reactors, which is waiting in the wings for testing funds, offers great hope in this regard (Taylor, 1989). These reactors include high-temperature gas and sodium

FIGURE 7-39

SASOL 2 coal liquefaction plant, half of the huge petrochemical complex at Secunda, South Africa. These plants are supplied by the largest underground coal mining system in the world. They employ 32,000 people and produce 130 different products, all from coal. SASOL is estimated to save $1 to $2 billion annually in oil import costs for South Africa.

metal-cooled types that incorporate innovative new designs to maximize reactor safety. In one design, fuel is encased in small balls that will not melt even if an uncontrolled chain reaction occurs. If these reactors can be perfected, adequate disposal methods found for nuclear waste, and the public educated about these factors, nuclear energy might become a viable alternative, provided we have not consumed all of our uranium resources in converter reactors by that time.

Whatever the outcome, it is clear that we must have a *national energy policy,* something that has not been developed by any MDC outside France and South Africa. Programs were started by many countries during the 1970s, but most of them were discontinued in the face of low world oil prices and public concern over nuclear power. The U.S. Synfuels Corporation, the Brazilian gasohol project, and many others are largely matters of history now. The only aspect of energy policy on which action continues is *conservation,* which is relatively painless, both economically and politically. In fact, most recent energy-related activities in MDCs have been dictated by environmental concerns rather than by the impending shortages. There is even talk of a *carbon tax* to be paid for energy produced from fossil fuels, which pollute the atmosphere with CO_2, or an *energy tax* on all types of energy production regardless of its source. Although these measures will help balance government budgets, they will not contribute to the solution of the growing world energy demands, which remain the principal challenge for the next generation of geologists, politicians, and citizens.

8

Iron, Steel, and the Ferroalloy Metals

IRON AND STEEL FORM THE FRAMEWORK AROUND WHICH WE have built our civilization. Steel is a major component of cars, cans, ships, bridges, high-rise buildings, appliances, and armament. Even energy minerals are useless without furnaces, pipelines, engines, and reactors, which are usually made of steel. This wide range of uses reflects the abundance of iron ore and the relative ease with which it can be converted to steel. The simplest form, *carbon steel*, usually contains less than 1% carbon and 0.5% manganese. *Alloy steel*, in which other metals have been mixed with iron (Table 8-1), accounts for about 15% of world production. These metals are known as *ferroalloy metals* and include *chromium, manganese, nickel, silicon, cobalt, molybdenum, vanadium, tungsten, columbium*, and *tellurium*. They permit steel to be used in a wide variety of applications, thus expanding its market at the expense of aluminum and other metals (Schottman, 1985).

World iron ore and steel production are worth $350 to $400 billion annually (Figure 8-1), slightly less than world oil production. The value of ferroalloy metal production can be quoted in several ways because they are traded as both ores and intermediate alloys such as ferromanganese, ferrochromium, and ferrosilicon. In Figure 8-1A, which is based largely on intermediate processed forms, it can be seen that manganese, nickel, and chromium dominate world production. The total value of primary world ferroalloy metals in

these intermediate forms is slightly more than $30 billion, almost 10% that of iron and steel. The prices of all ferroalloy metals are higher than that of steel (Figure 8-1B, D, F). Thus, they increase the cost of steel and must be used sparingly. Even so, the properties that they impart are so important that consumption for most of them has increased more rapidly than steel (Figure 8-1E, F).

IRON AND STEEL

About 73 countries produce steel, making it the most widely produced metal and one of the most widely produced mineral commodities (Peters, 1990). Fewer countries, about 52, produce iron ore, reflecting the more limited distribution of iron deposits large enough to support steel-producing facilities (Kuck, 1988). Although small iron deposits occur in almost all countries of the world, large deposits have a more restricted distribution.

In view of the enormous value of world steel production, one could well ask why the steel industry has little of the glamor of the oil business. The answer lies in profits. Although steel produced some fortunes in the early 1900s, including that of Andrew Carnegie of U.S. Steel, since then it has not been as profitable as oil production. This has been particularly true for the

164

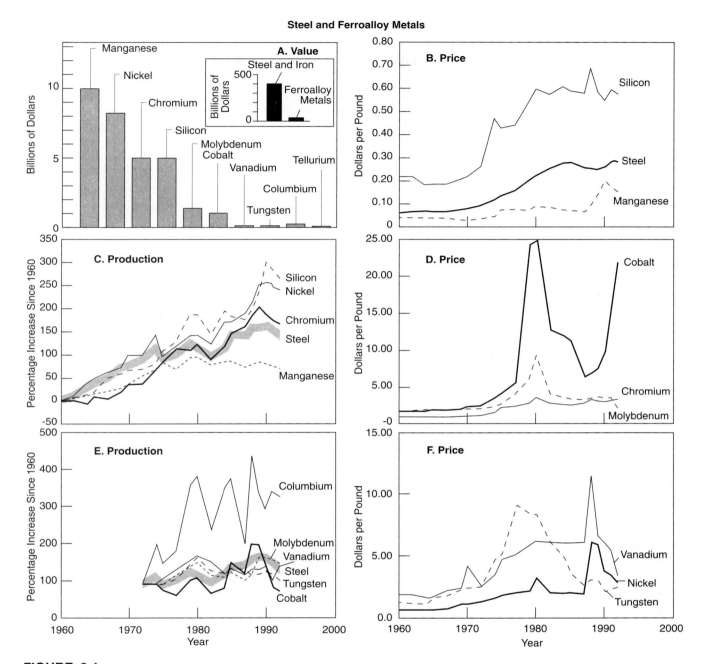

Steel and Ferroalloy Metals

FIGURE 8-1

A. Values of 1992 world iron, steel, and ferroalloy metal production, not including recycled metal or metal from other sources. **B,D,E.** Prices of steel and major ferroalloy metals since 1960. Reference to Figure 4-2 shows that only manganese, nickel, and cobalt prices have matched the rate of inflation over this period. These prices are not strictly comparable because the metals are sold in so many different forms, as discussed in the text. In particular, manganese prices are for metal in metallurgical-grade ore. **C,F.** Growth of world steel and ferroalloy metal production (primary mine production) since 1960 (from *U.S. Bureau of Mines Minerals Yearbooks* with additional information from L.D. Cunningham, H.E. Hilliard, T.S. Jones, and J.F. Papp).

TABLE 8-1
Alloy elements used in steel (*U.S. Bureau of Mines Mineral Facts and Problems*).

Element	Function in Steel
Aluminum	Deoxidation and grain-size control
Chromium	Hardenability, high-temperature strength, corrosion resistance
Cobalt	High-temperature hardness
Columbium	As-rolled strength
Copper	Corrosion resistance and precipitation hardening
Lead	Machinability
Manganese	Deoxidation, sulfur control, hardenability
Molybdenum	Hardenability, high-temperature hardness, temper brittleness control
Nickel	Hardenability and low-temperature toughness
Rare earths	Inclusion control, ductility, toughness
Silicon	Deoxidation, electric properties
Sulfur	Machinability
Tungsten	High-temperature hardness and hardenability
Vanadium	Grain-size control, hardenability, high-temperature hardness

U.S. steel industry, which has lagged behind its competitors, as discussed later.

Iron Deposits

Average crust contains about 5% iron, its fourth most abundant element, whereas iron deposits contain 25% to 65%. Thus, nature does not have to work particularly hard to make iron ore deposits, and makes important deposits by sedimentary, hydrothermal, and igneous processes. The concentration of iron in these deposits depends, in part, on the fact that iron exists in three oxidation states in nature. In some meteorites and Earth's core, it is in the *native (Fe^o)* state, but most iron in the crust is either *ferrous (Fe^{+2})* or *ferric (Fe^{+3})*. Common iron ore minerals such as *hematite* and *goethite* contain ferric iron and are stable in the presence of abundant oxygen, whereas ore minerals such as *magnetite* and *siderite* contain ferrous iron and are stable in more reducing, oxygen-poor environments. The common iron mineral *pyrite* also requires oxygen-poor environments, but is not usually mined for iron because of difficulties in disposing of its sulfur. In general, when an iron-bearing mineral undergoes weathering, the iron will be transported in solution if it is in the ferrous state, but will precipitate as goethite, if it is in the ferric state. Thus, iron minerals that dissolve during weathering in the oxygen-rich environment near Earth's surface form rusty outcrops rich in ferric iron oxides and hydroxides. In contrast, reduced solutions below the surface, including many hydrothermal solutions and some well waters, contain dissolved ferrous iron.

Sedimentary deposits, our largest and most important iron deposits, are chemical sediments. Most of them consist of thinly layered, alternating iron-rich and silica-rich layers that have earned them the name *banded iron formation (BIF)* (Plate 7-1B). It is also known as *taconite* in North America, *itabirite* in Brazil, *jaspilite* in Australia, and *banded ironstone* in South Africa. Iron-rich BIF layers contain iron oxides, sulfides, carbonates, or silicates, and silica-rich layers usually consist largely of fine-grained quartz called chert (James and Sims, 1973; Kimberley, 1978).

There are two main types of BIF deposits. *Algoma-type* BIF deposits are several tens of meters in thickness and a few kilometers in lateral extent. They are closely associated with volcanic rocks and appear to have accumulated where volcanic hot springs released iron-rich hydrothermal solutions into sedimentary basins (Figure 8-2). Algoma-type BIF has been mined at Temagami and Steep Rock in the Algoma region of Ontario, and at Thabazimbi, South Africa (Figure 8-3). It is the larger *Superior-type BIF* on which the world iron industry is based, however. These get their name from Lake Superior, where deposits in Minnesota and Michigan have supplied 80% to 90% of U.S. production annually since about 1900. Superior-type BIF deposits can be several hundred meters thick and many kilometers in regional extent. In the Hamersley Range of Western Australia, a BIF 1,000 meters thick covers over 10,000 km^2, and similarly large deposits are known around Lake Superior, the Labrador Trough of Canada, the Carajás and Minas Gerais districts of Brazil, and the Krivoi Rog and Kursk districts of Russia (Figure 8-3).

Most BIF deposits are late Archean to early Proterozoic in age and their formation is thought to be related to the increase in atmospheric oxygen that took place during that period (Maynard, 1983; Holland,

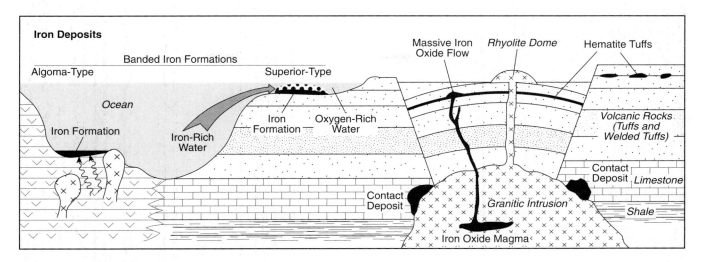

FIGURE 8-2
Schematic illustration of geologic processes that form iron deposits.

1984). This theory holds that the iron that accumulated in the BIFs originally was dissolved as Fe^{+2} in the early Archean oceans. This took place because abundant volcanism and related hot-spring activity fed dissolved Fe^{+2} into the oceans at that time, and a scarcity of atmospheric oxygen allowed most of it to remain dissolved. By late Archean time, however, the oxygen content of the atmosphere had increased enough to precipitate dissolved iron, particularly where iron-rich seawater fed by upwelling currents from the deep ocean came into closer contact with the atmosphere on shallow shelves. The alternating layers of iron and silica minerals might be the result of seasonal sedimentation or ocean current patterns.

The largest BIF deposits did not form after middle Proterozoic time, apparently because the increase in atmospheric oxygen after that time made the oceans too oxidizing to retain dissolved iron. A few deposits formed in late Proterozoic time and it has been suggested that they are the result of an unusual period known as "snowball Earth," when floating ice from glaciers covered the entire oceans limiting the input of atmospheric oxygen (Klein and Beukes, 1992). Or perhaps these BIFs formed in isolated basins. Sedimentary iron deposits that formed during Phanerozoic time are known as *ironstones* and are relatively rare. They lack fine layering and appear to have formed when iron-rich groundwater replaced carbonate sediments. Deposits of this type, including the Clinton Formation of the Appalachians and the minette ores of Alsace-Lorraine in France (Figure 8-3), are not as thick or extensive as BIFs.

Iron is also found in hydrothermal and magmatic deposits. The most important hydrothermal deposits, known as *contact metasomatic deposits,* are found at the contact between igneous intrusions and limestone wallrock (Figure 8-2). These deposits form where hydrothermal solutions containing dissolved iron flow out of intrusive rocks into limestone, causing iron minerals to precipitate. Some hydrothermal solutions are magmatic in origin, but others are simply heated meteoric waters that circulated through the intrusive rock and dissolved iron. The Bethlehem deposits at Lehigh, Pennsylvania, which were the basis for the Bethlehem Steel Company, are of this type. They formed at the contact of mafic intrusions that rose along rifts that were created when the North American and European continents split apart in Triassic time (Hickock, 1933; Eugster and Chou, 1979).

Magmatic iron deposits formed by the separation of an immiscible iron oxide melt from silicate magmas. They appear to have originated in magmas of two distinct types, rhyolite and anorthosite. Most deposits are associated with rhyolite, including the huge Kiruna deposit in Sweden, Pea Ridge and Pilot Knob in Missouri, Cerro Mercado in Mexico, and El Laco in Chile (Figures 8-2 and 8-3). Iron is not abundant in rhyolite magmas, and the formation of an iron-rich immiscible magma appears to require highly oxidizing conditions such as might be caused by the contamination of a magma by meteoric water. In addition, large amounts of gas are probably necessary to make the iron oxide magma rise to the surface (Lyons, 1988). Deposits of this type support the Swedish and Mexican steel industries, but are not important on a global scale.

Although there are other types of iron deposits, they are not important to world supplies. For instance, magnetite resists weathering and is heavier than most

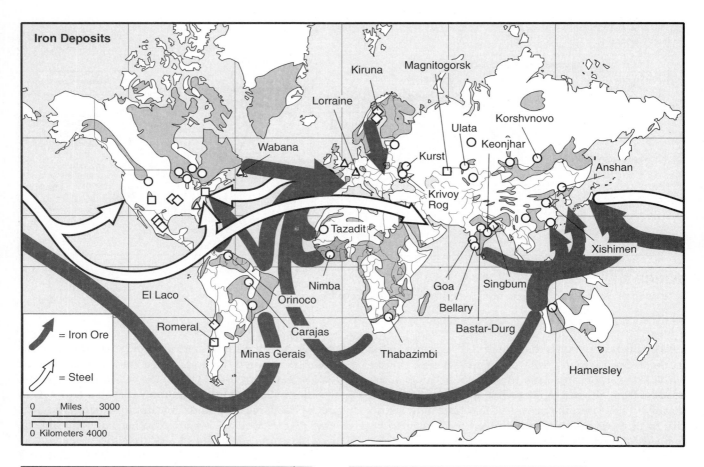

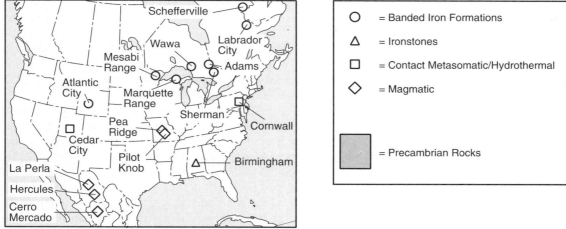

FIGURE 8-3

Location of iron ore deposits and Precambrian rocks that host most banded iron formation, also showing iron ore and steel trade patterns (compiled from data of U.S. Bureau of Mines; Organization for Economic Cooperation and Development).

silicate minerals. It can concentrate in *placer deposits,* where it accounts for most of the black-sand beaches around the world. Deposits of this type are mined in New Zealand. Iron also collects in some *laterite soils,* where it has been mined in northeastern Texas, and it used to be recovered from *bog ores,* which formed where iron collected in swamps and which were a major source of iron during settlement of the United States.

Iron Mining and Processing and Related Environmental Factors

Most iron ore is mined by open-pit methods (Figure 8-4), although Kiruna, Pea Ridge, and a few other deposits are mined underground. Taconite mining is difficult because its high silica content makes it very hard. Weathering removes silica from BIFs to produce *residual ore,* which is cheaper to mine because it is softer and contains as much as 60% iron in contrast to about 25% in fresh taconite. These ores are nearest to the present surface and are being exhausted rapidly. The last residual ore mine in North America closed in 1991, although they remain open in Australia and Brazil.

Iron ores are beneficiated to remove phosphorus and silica minerals, which cause problems in steelmaking (Guider, 1981). Iron mineral concentrates are mixed with clay or organic matter and rolled into small balls known as *pellets* (Plate 7-1B) that are hardened by heating in a kiln. Some coarse-grained ores are baked along with binders to make a related clinker-like product known as *sinter.* Although production of pellets and sinter is costly and energy intensive, these products are preferred as blast furnace feed because they allow increased flow of gas through the furnace and minimize dust. Grinding taconite ores to powders fine enough to produce a clean concentrate of iron minerals requires large amounts of energy and has a strong control on the economics of the operation. In some unmetamorphosed ores of the Mesabi Range, ore mineral grains are only a few micrometers in diameter, and extensive grinding is required. Where the ores have been recrystallized by metamorphism, such as in parts of the Labrador Trough, iron minerals are coarser-grained, permitting lower beneficiation costs that compensate in part for the more remote location.

Mining and beneficiation of iron ores presents environmental challenges largely because it is carried out at such a large scale. The most important recent environmental problem specific to iron mining was the dispute over disposal of tailings at Reserve Mining, one of the largest taconite-producing operations in the Mesabi Range. Under an arrangement approved by state authorities, Reserve was originally constructed with a pipeline to dispose of tailings on the

FIGURE 8-4
Tilden Mine, Marquette County, Michigan, which mines hematite-bearing BIF, has the capacity to produce about 7.2 million tonnes of pellets annually. Note rail loading system with piles of pellets in right foreground (courtesy of The Cleveland-Cliffs Iron Company). See Plate 7-2 for photo of BIF and pellets.

floor of Lake Superior. In response to later public concern about this practice and the possibility that the tailings contained fibrous minerals related to asbestos, the state required Reserve to dispose of tailings on land, a project that required an investment of $370 million (Bartlett, 1980B). These costs and poor market conditions forced Reserve into bankruptcy in 1986, and all assets of the company were sold for $52 million in 1989.

As an element, iron is not a major environmental polluter. In fact, humans have evolved to take advantage of its high crustal abundance by making iron the central component of hemoglobin, as discussed earlier. It is also a constituent of enzymes involved in energy metabolism, and lack of sufficient iron contributes to anemia and reduced resistance to infection. In spite of these generally beneficial attributes, recent studies suggest that excess iron facilitates formation of plaque that clogs arteries, leading to increased risk of heart disease (Altman, 1992).

Iron Ore Reserves, Transportation, and Trade

World iron reserves are estimated to contain 64.6 billion tonnes of iron metal in about 151 billion tonnes of ore, considerably greater than the current annual production of about 900 million tonnes. World resources are estimated to be at least 800 billion tonnes containing more than 230 billion tonnes of iron metal, much of which is low-grade taconite.

Major iron ore deposits in the Ukraine, Russia, Australia, Canada, the United States, and Brazil coincide only partly with the distribution of major steelmaking countries (Figure 8-5). As a result, about 400 million tonnes of iron ore products are shipped annually to global markets, largely by sea (Figure 8-3). Ore vessels operating on the Great Lakes haul ore and pellets from the Lake Superior region to steel mills around Chicago, Detroit, Cleveland, and Pittsburgh. Great Lakes vessels carry up to 65,000 tonnes of ore, whereas, ocean-going ships carrying ore to Japan can take as much as 300,000 tonnes. The movement of iron ore within individual countries is largely by rail, although slurry pipelines are used in Mexico, Brazil, India, and Australia.

Steelmaking

Steelmaking consists of two basic steps (Figure 8-6). The first step converts iron ore pellets containing about 65% iron to *pig iron* containing about 94.3% iron, 4.5% carbon, 0.6% manganese, 0.5% silicon, and trace amounts of other impurities such as sulfur and phosphorus (Lankford et al., 1985; Moore and Marshall, 1991). This is usually done in a *blast furnace* (Figure 4-20A), a large, vertical cylinder that is charged with a mixture of metal ore, coke, and limestone (Peacy and Davenport, 1979). Air is "blasted" in from the bottom to promote combustion of the coke to carbon monoxide, the iron minerals react with carbon monoxide to form pig iron, and the waste minerals react with the

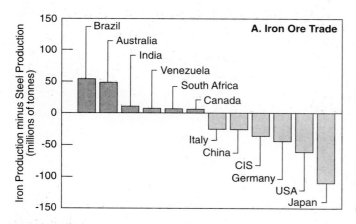
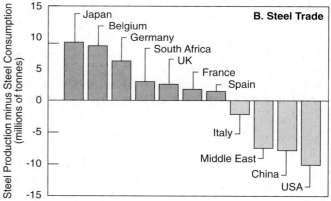

FIGURE 8-5

World trade in **(A)** iron ore and **(B)** steel for 1991, showing the difference between iron ore production and steel production. Countries with excess iron ore production plot above the line, whereas those with excess steel production plot below the line (and use imported ore). This diagram is approximate because iron ore was converted to iron metal by a single factor of 0.75 and steel contains other metals (compiled from data of U.S. Bureau of Mines and Metal Statistics).

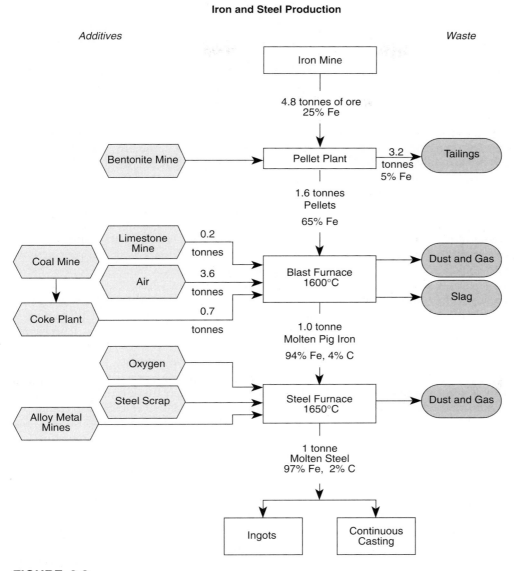

FIGURE 8-6

Schematic illustration of the pig iron and steel production process. Additional materials necessary for the process are shown on the left and waste products are shown on the right (compiled from information of U.S. Bureau of Mines and Moore, 1991).

limestone to form *slag* (Table 4-3). Pig iron liquid, which is much heavier than slag liquid, accumulates at the bottom of the furnace where it is drawn off for further processing. *Direct reduction iron-making (DRI)* is an alternative to the blast furnace. In this method, iron ore is reduced to metal without actually melting, usually by reacting it with carbon monoxide and hydrogen made from natural gas or coal. Its main appeal has been in countries with abundant natural gas and limited coal, such as Mexico, Saudi Arabia, and Venezuela, although about 10% of DRI production

comes from coal-based facilities in South Africa, India, and New Zealand.

In the second step of the steelmaking process, pig iron is converted to steel, which contains no more than 2% carbon. This process is carried out in *steel furnaces,* of which there are many varieties. About 70% of western world steel is made in *basic oxygen furnaces,* in which oxygen is blown through a charge of molten pig iron with about 30% scrap iron, causing the carbon, silicon, manganese, and phosphorus impurities to oxidize. *Electric-arc furnaces* use an electric heat source and can melt charges of up to 100% scrap. Output from steel furnaces is rolled into bars, plates, or rods, cast into ingots, or poured into continuous-casting machines that shape it before it cools.

The most important environmental emissions from iron and steelmaking are slag and gases. Slag, which does not have any of the characteristics of hazardous waste, does not require special handling under the *Resource Recovery and Conservation Act*. Essentially all of the slag from iron and steelmaking in the United States, amounting to about 23 million tonnes annually, is recovered and sold as construction aggregate. Throughout the world, about 80 million tonnes of slag are recycled in this way annually, making a positive contribution to cash flow in most iron and steel operations. Dust and gas emissions, on the other hand, have fewer markets and require more costly cleaning systems similar to those used on coal-burning electric utilities. Largest amounts of dust come from blast furnaces, although steel furnaces also produce significant dust. Whereas dust from blast and basic oxygen furnaces is not considered hazardous, that from electric arc furnaces is sufficiently enriched in cadmium, chromium, lead, zinc, and other metals to pose problems. These metals come largely from scrap that is recycled into the electric arc furnaces in large amounts, and they are of concern because they might be leached from the dusts by surface waters. Most dusts with high metal concentrations must be processed to recover the metals before disposal. The U.S. steel industry is estimated to have spent $400 to $500 million annually since 1972 on pollution control, an amount that totaled over 15% of all capital investment. Although these large expenditures contributed to the poor economic state of the industry, they were not the only factor, as discussed next (Hogan, 1991).

Steelmakers and environmentalists have clashed over several operations in LDCs, especially Brazil, which lacks coal to make the coke needed for steel production. In a compromise intended to use indigenous resources and provide jobs for locals, Brazil attempted to base steel production in the Carajás region on charcoal made from trees in the surrounding forest. As things have turned out, removal of 2 million tonnes of wood each year has damaged forests in the area, causing debate about the relative importance of economic and environmental factors in large industrialization projects and the role of international lending agencies, such as the World Bank, in funding such development (Simons, 1987).

Steel Production, Markets, and Trade

The steel industry has a bad economic reputation (Crandall, 1981; NRC, 1985; Hoerr, 1988). It is often re-

ferred to as a "no-growth" industry in spite of the fact that world steel production has increased about 150% since 1960 (Figure 8-1), a growth rate faster than that of the base metals. Most intense criticism has been reserved for the U.S. industry, where employment decreased from about 650,000 in the 1960s to only about 170,000 in the 1990s. It is less widely known that world primary steel employment decreased from over 2 million to only 1 million during the same period (Figure 8-7A). In Japan and (West) Germany, which are major steel exporters, employment decreased 40% and 45%, respectively, not as large as the 70% decrease seen in the United States, but a big drop nevertheless. This change is due principally to the impressive increase in productivity that has been achieved by world steel producers (Hogan, 1991). Even in the United States, which is often thought of as a laggard, productivity has increased from ten worker-hours per tonne of steel in 1955 to less than five in 1993.

In spite of the growth of world steel consumption, steel production has declined in the United States, whereas it has increased or held steady in all other major steel-producing areas (Figure 8-7B). This occurred because world production capacity expanded faster than markets, thereby putting pressure on more costly producers such as those in the United States. At the same time, domestic markets for U.S. steel came under pressure, particularly in the transportation and construction sectors, which account for almost 60% of domestic consumption. Efforts continue to replace expensive steel in heavy equipment and structures with cheaper materials. Architects would like to build a large high-rise with concrete but no steel frame, and steel has been replaced in automobiles by lighter metals and plastic. If these trends continue demand for steel will decline even if we make the same number of bridges and cars. The only winners will be those who supply steel to manufacturers with an expanding market share. The recent history of steel use in the automotive industry illustrates this effect (Figure 8-7C). Whereas expanding automobile exports created increased steel demand in Germany and Japan during the 1980s, U.S. demand dropped sharply as its market share decreased (Weinberg et al., 1987).

The decline in the state of the U.S. steel industry was aggravated by high wages. In 1982, for instance, average hourly wages in U.S. steel mills were $23 versus $12 in Germany, $10 in Japan, and $8 in the United Kingdom. From 1980 to 1990, the U.S. steel industry lost a total of $4.6 billion on sales of $142.6 billion. According to statistics compiled by the American Iron and Steel Institute and the Japanese Iron and Steel Federation, losses in the U.S. industry between 1984 and 1990 were about 1.58% of cumulative sales, whereas profits in Japan were about 1.82% of cumu-

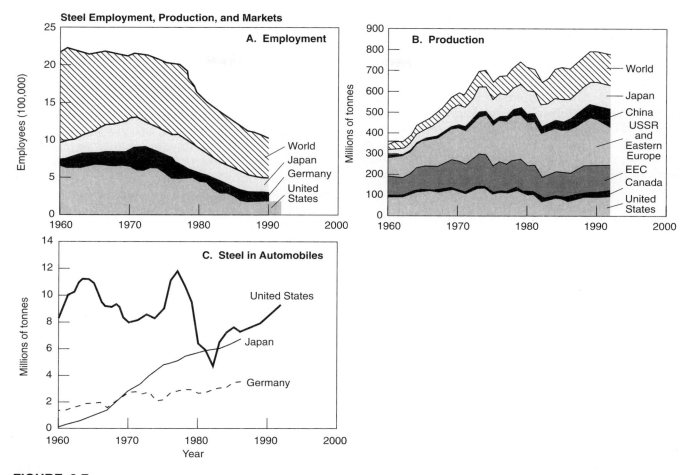

FIGURE 8-7

A. Decline in world steel industry employment showing steeper decline in United States. **B.** World steel production showing gradual contraction of U.S. production, constant production from western Europe (EEC), and increasing production from most other areas. **C.** Changes in market for steel in automobiles, showing weak U.S. steel demand compared to that in Japan and Germany (compiled from data of Weinberg et al., 1987, *U.S. Bureau of Mines Information Circular 9175;* Tilton, 1990; Kramer and Plunkett, 1991; Organization for Economic Cooperation and Development).

lative sales. The poor competitive position of the U.S. industry was also hurt by a generally lower level of government support. Government ownership is common in many steel-producing countries and, even in Japan and Germany where this is not the case, government has provided subsidies, loans, and market controls that have eased financial strain on domestic producers.

Thus, although things have not been good for the world steel industry, a disproportionate share of grief has fallen on producers in the United States. As a result of this, there has been a significant downsizing of the U.S. industry and an increase in imports, leaving the United States almost alone as a major steel importer (Figure 8-5). Only China and Middle Eastern

countries, which use steel for oil and gas production, import similar amounts of steel. The most conspicuous response to this has been a series of maneuvers by the industry and its sponsors in the federal government to curtail imports, as discussed in the chapter on mineral economics.

The situation is even worse in Russia and other CIS countries. The Lenin Steel Works at Magnitogorsk, Russia (Figure 8-3), which has accounted for about one-quarter of CIS production, is a raw material supplier to the Russian industrial belt along the Ural and Volga Rivers (Uchitelle, 1992). It has outmoded, inefficient equipment that produces poor-quality steel, which sells for half of the world price and requires reprocessing in modern facilities in the Far East and

Western Europe. Pollution control facilities at the 60-year old, 41-km^2 plant are either not installed or not working and environmental health problems are ubiquitous among the 450,000 citizens of Magnitogorsk, especially children. Magnitogorsk is one of the last of the large, old steel mills. Similar uneconomic though cleaner plants at Homestead and Aliquippa, Pennsylvania and Lackawanna, New York were shut down by U.S. Steel, LTV, and Bethlehem Steel, respectively, in the 1980s, putting tens of thousands of people out of work in these areas. If Russia and the CIS countries are to compete on the world steel markets, Magnitogorsk may well suffer the same fate.

The final chapter in world steel production has not been written. With plants undergoing modernization and others waiting for attention, the competitive structure of the industry will continue to change. The strong national security interests involved make it certain that the United States and Russia will remain factors in the industry. Whether they are major competitors or simply weak participants depends largely on the degree of modernization and comparative wage structures, and possibly on the level of government involvement.

FERROALLOY METALS

Alloy steels made by combining iron with the ferroalloy metals can be used in many applications for which carbon steel is unsuitable. Alloy metal terminology has evolved along with the materials to provide names for many types of metal combinations. There are *ferrous alloys,* in which steel dominates, and *nonferrous alloys,* which contain little or no steel (but which can involve combinations of the ferroalloy metals with or without aluminum, titanium, magnesium, and other metals). Ferrous alloys are further divided on the basis of their content of alloy metals. *High-strength, low-alloy (HSLA)* steels are based on *microalloying* or the addition of only a few tenths of a percent of the alloy metals. In contrast, *high-performance alloys* contain large amounts of alloy metals. They are further divided into *superalloys,* which retain strength at temperatures above 800°C, *corrosion-resistant alloys,* which contain large amounts of chromium and molybdenum, and *wear-resistant alloys,* which contain tungsten, chromium, and at least 1% tungsten carbide particles. *Stainless steel,* which contains large amounts of chromium and nickel, is the cheapest corrosion-resistant alloy and makes up 1% to 2% of total steel production. The other alloy steels constitute almost 15% of steel production. Protective coatings provide cheaper corrosion resistance, and about 15% of world steel is treated in this way. *Galvanized steel,* which is coated by zinc, is the most important coated steel, although tin-coated and chromium-coated steels are also used.

Manganese

Manganese is brittle and has few uses as the pure metal. It has a large market, however, as an alloying element and as an additive to remove impurities or neutralize them in steel and cast iron. Manganese has been used as an additive in steelmaking ever since the mid-1800s (Jones, 1985). Part of its appeal to steelmakers stems from the fact that it costs even less than steel (though not iron) and can therefore be used in large amounts (Figure 8-1B). The 20 million tonnes or so of world manganese ore production has an annual value of almost $4 billion. If all of it were converted to ferromanganese, the value would be about $10 billion (Figure 8-1A).

Manganese was originally used principally to control the deleterious effects of sulfur and oxygen in steel made by refining molten pig iron. As technology has evolved, however, it has become additionally useful for alloying. About 75% of U.S. manganese consumption goes into carbon steel, with another 20% going into various types of alloy and stainless steels. Smaller amounts of manganese are alloyed with aluminum to provide corrosion resistance, and with copper to provide strength. It is also used in chemicals and other industrial applications, the most visible of which is the use of natural and synthetic manganese-dioxide in dry-cell batteries (Dancoisne, 1991).

Manganese Deposits and Production

Manganese is geochemically similar to iron but has three oxidation states in nature ($+2$, $+3$, and $+4$) rather than just two (Krauskopf, 1967). Under reducing conditions, where it forms Mn^{+2}, manganese remains in solution unless it encounters enough dissolved carbonate or silica to form *rhodochrosite* ($MnCO_3$) or a manganese-bearing silicate mineral. The fact that manganese sulfides are much more soluble than iron sulfides means that manganese remains dissolved in many deep, reduced waters in sedimentary basins. In oxidized solutions, however, Mn^{+3} and Mn^{+4} form oxide minerals including *pyrolusite, psilomelane (romanechite),* and *braunite.*

Manganese forms a wide range of deposits, including the manganese nodules of the modern sea floor, although land-based sedimentary and supergene deposits are the only currently important producers (NRC, 1978; Roy, 1981). *Supergene deposits,* which supply about 25% of world production, consist of manganese oxides and hydroxides that accumulated on deeply weathered manganese-rich rocks. They are, in a way, a special form of laterite soil, similar to bauxite, the aluminum ore. As might be expected, supergene deposits are found largely in humid tropical climates (Figure 8-8), although older, buried deposits are known.

Sedimentary deposits, which supply most of the remaining 75% of world production, consist of layers of manganese carbonates and oxides that were deposited as chemical sediments. A few of these chemical sediments form lenses intercalated with volcanic rocks, and the manganese might have been deposited from warm springs associated with this volcanism, similar to the process that formed Algoma-type banded iron deposits. The largest sedimentary deposits are not associated with volcanic rocks and are found instead in unremarkable limestones and shales that range in age from Precambrian to Cenozoic and were deposited in

shallow arms of the sea that extended onto the continents. Manganese-rich seawater that formed the deposits is thought to have come from deeper parts of the ocean basins where increased organic productivity consumed oxygen, allowing manganese to remain in solution. The manganese could have come from submarine hot springs, reactions between reducing pore waters and sediments, or both. This seawater did not dissolve iron because it contained enough reduced sulfur to precipitate iron sulfides. When this iron-poor, manganese-rich water rose into more oxygen-rich parts of the shallow ocean, dissolved manganese precipitated to form the deposits (Cannon and Force, 1983; Force and Cannon, 1988). These processes are closely related to global cycles, and manganese deposits appear to have formed largely during periods of high temperature and high sea level (Frakes and Bolton, 1992; Dickens and Owen, 1993).

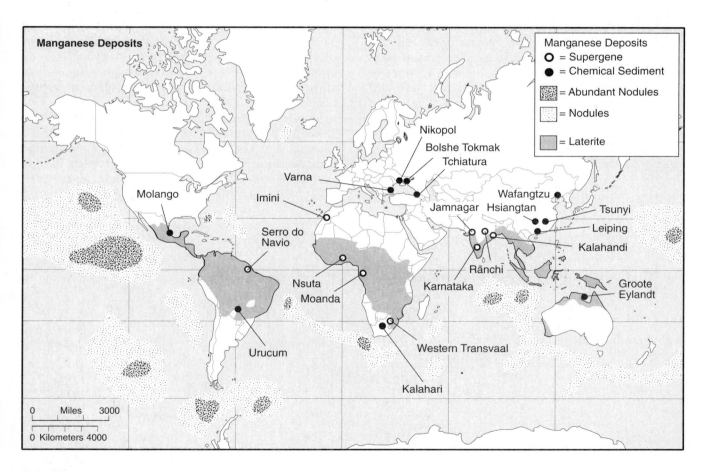

FIGURE 8-8

Location of land-based sedimentary and supergene manganese deposits and manganese nodule concentrations in the oceans. Variation in base and precious metal content of nodules, which is not shown here, will have a strong effect on the location of any future mining. See Figure 5-4 for location of offshore areas claimed as Exclusive Economic Zone by the United States and of manganese nodules claimed by private industry.

Manganese deposits are usually mined by high-volume, open-pit methods similar to strip mines. Ore is beneficiated by conventional methods to make a concentrate that is fed into a blast or electric furnace with iron (usually in the ore), coke, and limestone to make *ferromanganese* with or without silicon. Manganese metal and dioxide are made by electrolytic separation from solutions obtained by dissolving manganese ore in sulfuric acid.

There have been few real environmental concerns related to manganese production. In fact, it is an essential element for life, being used in the formation of tissue and bone, as well as in the development of the inner ear and reproductive functions (WHO, 1981). Estimated average requirements are about 2 to 3 mg/day, which comes largely from food and from inhalation of natural dust. Exposure to excessive amounts of manganese can cause neurological disorders and respiratory problems. Anthropogenic manganese contributions to the atmosphere are small, only about 10% to 20% of natural contributions, and those to waters are even smaller (Nriagu and Pacyna, 1988). Significant problems with manganese exposure have been observed in some mines but more commonly in ferromanganese production facilities, where they are thought to be caused by the inhalation of manganese-rich dust and fumes (Moore, 1991). Organomanganese compounds have been used as additives in unleaded gasoline, although not in the United States. They are toxic, but react to form other compounds during combustion. Emissions from gasoline combustion would probably not increase ambient manganese levels sig-nificantly because of the high average content of manganese in crustal rocks.

Manganese Reserves and Trade

World manganese ore reserves are estimated to contain 800 million tonnes of metal. Unfortunately, most major steelmaking nations lack manganese (DeYoung et al., 1984). North America is in terrible shape, with less than 1% of world reserves even though it has 16% of the planet's land surface. This situation has created an active global trade in manganese ore and manganese alloys. Manganese ore comes largely from Gabon, South Africa, Australia, and Brazil and goes to Japan, France, Norway, and South Korea (Figure 8-9). Ferromanganese and silicomanganese exports come from Norway and the CIS countries (largely Russia) and go to Japan and Germany (Figure 8-9). In spite of their large reserves, Ukraine and Georgia have imported small amounts of high-grade ore, apparently to enrich their low-grade production. Manganese is widely regarded as a *strategic metal* by most importing countries, some of which maintain *stockpiles* for use in emergencies. The U.S. stockpile, for instance, contains almost 3 million tonnes of manganese ore and ferroalloys.

By far the largest manganese reserves are in South Africa and in a zone extending through Bulgaria, Ukraine, and Georgia (Figure 8-8). The South African deposits, mostly in the Kalahari area, are thought to contain about 4.4 billion tonnes of manganese (Nell, 1992). Those in the Varna, Nikopol, and Tchiatura

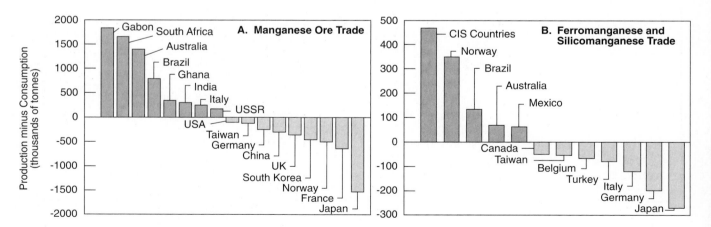

FIGURE 8-9
World manganese trade showing the difference in value of exports and imports for
(A) manganese ore and **(B)** manganese alloys (ferromanganese and silicomanganese) for
1991 (from Manganese Trade Statistics, International Manganese Institute, courtesy of
P.L. Dancoisne).

(Chiatura) deposits are estimated to contain about 500 million tonnes, much of which is either low grade or in the form of carbonate minerals that require additional processing (Jones, 1990; Cannon et al., 1982). Grades at Nikopol and Tchiatura, for instance, range from 13% to 33% Mn, considerably lower than most other manganese deposits of the world, a factor that raises questions about whether they will maintain their historically high level of production. Other large manganese deposits in Australia, Gabon, and Brazil (Figure 8-8) have grades of 44% to 50% Mn, but are at least an order of magnitude smaller than the Kalahari and Nikopol-Tchiatura deposits. The only large deposit in North America, Molango in Mexico, also has low-ore grades. Thus, only a small fraction of global manganese reserves are clearly economic under present conditions and in locations not subject to potential political problems. These facts continue to support interest in deep-sea manganese nodules, which constitute an enormous, untapped resource.

Manganese Nodules

Manganese nodules, which litter Earth's ocean floor as well as the floors of some fresh-water lakes (Figure 8-8), are the most tantalizing of our many ocean resources (Heath, 1978, 1981; Kennett, 1982; Baturin, 1987; Cronan, 1992). They are walnut to grapefruit-sized balls that consist of thin, concentric layers of manganese and iron oxide minerals (Figure 8-10). In addition to manganese and iron, the nodules contain minor but economically important amounts of nickel, cobalt, copper, and other metals. Most nodules are found in areas of the deep-sea floor, in water depths of 5 to 7 kilometers. Other nodules are found on elevated submarine plateaus or in deeper parts of large lakes. The nodules formed where clastic sediments are scarce, and it is thought that they grew by precipitation from the surrounding seawater. The manganese and other metals probably came from mid-ocean ridge hydrothermal vent solutions, alteration of mafic rocks that make up the seafloor, and rivers draining the continents. Related manganese-rich crusts have been found along the sides of seamounts, usually between depths of 1 and 2.6 kilometers (Ballard and Bischoff, 1984). Some of the crusts are rich in cobalt and contain interesting amounts of platinum, cerium, titanium, and lead.

There is a real question about why we see the nodules at all. Isotopic-age measurements on growth layers in the nodules indicate that they grew at rates of 1 to 4 millimeters per million years, which is about 100 to 1,000 times slower than the rates at which clastic sediments are deposited. So, the nodules should be buried by sediment, even in the open ocean where it

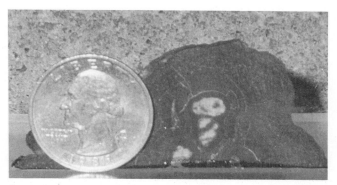

A.

B.

FIGURE 8-10

Cross section (A) and uncut (B) view of a manganese nodule from the North Pacific Ocean (Ocean Drilling Program Site 886) showing concentric growth layers (courtesy of D.K. Rea and H. Snoeckx, University of Michigan).

comes only from windblown dust, or shells of floating organisms. Measurements below the surface show that at least 75% of the nodules are at the sea floor, indicating that some mechanism works to keep them there. Possibilities that have been suggested include ocean currents, which might roll the nodules along the surface or sweep finer sediment away, or the action of burrowing organisms, which would serve to push nodules toward the surface. Whatever their origin, the nodules are definitely not homogeneously distributed. They appear to be most concentrated in the equatorial North Pacific and smaller areas of the mid-Atlantic (Figure 8-8). In areas where they are richest, the nodules actually cover the ocean floor, forming a pavement. The Pacific Ocean alone is estimated to contain about 2.5 billion tonnes of nodules with a grade of

A.   **B.**

FIGURE 8-11
A. Nickel-rich laterite (dark) overlying ultramafic bedrock (light) from which it was
derived by weathering, southwestern Puerto Rico. H.L. Bonham provides scale.
B. Shatter cones on the side of a rock outcrop south of Sudbury, Ontario, showing
radiating pattern pointing upward toward the probable impact level, which is now
eroded (key in lower left provides scale).

about 25% Mn, making them similar in abundance to
low-grade, land-based deposits.

Major efforts have been made to develop a technology for mining the nodules (Bath, 1991). Although one
effort billed as a test of nodule recovery methods
turned out to be an attempt by the United States to
recover a Soviet submarine that sank in the nodule
fields south of Hawaii, other tests have indeed focused
on the nodules. Most equipment employs suction or
scraper systems that remove nodules along with the
upper few centimeters of the sea floor. Large-scale
tests have shown that these methods can produce nodules at sufficiently high rates, although less is known
about the feasibility of processing ores on-board ship
(Bernier, 1984). Nodule mining would cause major disruption of the immediately surrounding sea floor and

comprehensive studies of the potential effects of these
activities are underway (Foell et al., 1992). Commercial production has been stalled, however, by high
costs compared to land-based deposits, uncertainties
of ownership, and concern about the Law of the Sea,
as discussed previously.

Nickel

Present annual sales of raw nickel throughout the
world, amounting to almost 1 million tonnes, are valued at $7 billion. In the United States, slightly more
than half of nickel production is alloyed with steel,
largely to produce *stainless steel* (Kirk, 1990). Another
third is used in copper-nickel alloys and in alloys with
cobalt, aluminum, and other metals, and much of the

rest is used in protective platings for steel and other metals. Nickel metal and alloys are useful largely for their high-temperature strength and corrosion resistance. Thus, the largest uses for nickel products are in turbines and jet engines and in tubing and containers for the chemical industry (Sibley, 1985). Cobalt can substitute for nickel in many of these uses, but it is more expensive.

Geology and Distribution of Nickel Deposits

Nickel is produced from 23 countries and comes largely from laterite and magmatic deposits. Nickel-rich *laterites* form by weathering of ultramafic rocks in tropical climates (Figure 8-11A). Ultramafic rocks form good nickel laterite deposits because they contain nickel substituting for magnesium in *olivine,* their most common silicate mineral. As weathering proceeds, olivine breaks down, releasing nickel, magnesium, and iron into pore waters (Golightly, 1981). If the water contains dissolved oxygen, iron oxides pre-cipitate on the spot and adsorb nickel from solution. If the water remains reducing, nickel remains in solution and percolates to the base of the weathered zone, where it precipitates as *garnierite,* a complex nickel silicate mineral. In both cases, magnesium is simply removed in solution. Nickel-rich laterite can be up to 20 m thick with average grades of 3% Ni and 0.1 to 0.2% Co.

Ultramafic rocks from which nickel laterites form are usually found where slivers of the mantle have reached the surface along obduction zones at collisional tectonic margins. The largest deposits of this type are found in island-arc areas including New Caledonia, Cuba, Dominican Republic, and Indonesia (Figure 8-12). As is the case for bauxites, some older nickel paleolaterites, such as the small operation at Riddle, Oregon, have been preserved from previous periods of tropical weathering and laterite formation (Cumberlidge and Chace, 1967).

Nickel is also produced from *magmatic deposits,* which form in mafic and ultramafic igneous rocks

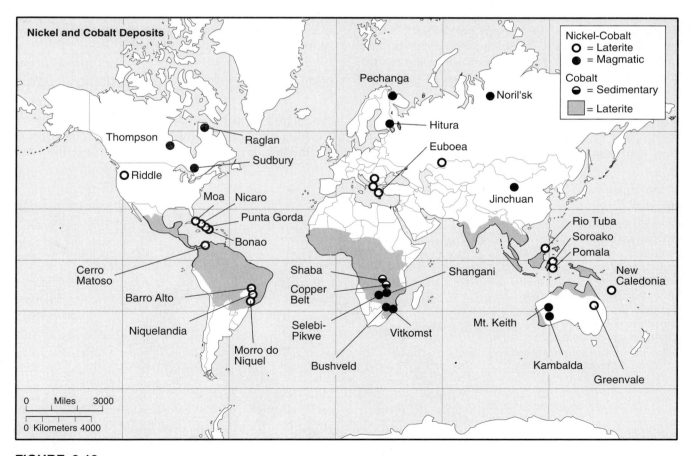

FIGURE 8-12

Location of laterite and magmatic nickel deposits (most of which contain significant cobalt) and of sedimentary cobalt deposits.

(Naldrett, 1981, 1989). Nickel readily combines with sulfur in silicate magmas, and if there is enough sulfur available, an immiscible metal-sulfide magma rich in iron, nickel, cobalt, and other metals will appear as droplets in the cooling ultramafic magma. At lower temperatures, these metal-sulfides might crystallize from the silicate magma as minerals, but the ultramafic magma is so hot that the sulfides remain in liquid form instead. The blebs of metal-sulfide magma are heavier than the surrounding silicate magma and sink to the bottom of the chamber where they coalesce and crystallize to form large bodies of metal-sulfide minerals as the intrusion cools (Figure 2-16). Most nickel in these deposits is found in the iron-nickel sulfides, *pentlandite* and *pyrrhotite*. Where the nickel content of these minerals is high, the deposits are mineable.

Many magmatic nickel deposits form from magnesium-rich ultramafic lava known as *komatiite*, which was formed by the partial melting of the mantle. Komatiite magmas have a higher melting temperature than other common silicate rocks (about 1,600°C) and appear to have formed almost exclusively during Archean time, suggesting that the mantle was hotter during the early part of Earth's history. Komatiite-related nickel-sulfide deposits are found only in older parts of Precambrian cratons in western Australia, southern Africa, and Canada (Figure 8-12).

Magmatic nickel deposits also separated from younger, mafic magmas that were not so rich in magnesium, but only under special circumstances. As it turns out, however, two of these intrusions supply

about 40% of world nickel production. The most unusual of them is Sudbury, Ontario, where a large, funnel-shaped mafic intrusion known as the *Sudbury Irruptive* has magmatic nickel-sulfide deposits along its base (Figure 8-13). Everything about Sudbury is unusual. The irruptive is not ultramafic and it contains a large amount of felsic rock near its top. Its age is only about 1.7 Ga, much younger than komatiite-related deposits, and the rock surrounding it contains extensive breccias and features known as *shatter cones* (Figure 8-11B), which are thought to form only where a strong shock moves through rock. It is now generally accepted that the Sudbury Irruptive was emplaced because an immense *meteorite impact* perturbed the crust and underlying mantle, causing formation of mafic magma that rose and mixed with granitic crust, depositing large amounts of sulfide ore (Pye et al., 1984). The other major exception is the *Noril'sk district* in northern Russia, which is Triassic in age, even younger than Sudbury. Noril'sk formed from basalts that were erupted onto the continent as it attempted to break apart in Triassic time. Although magmas of this type do not usually contain enough sulfur to form magmatic nickel deposits, the Noril'sk intrusions were enriched in sulfur from sulfur-rich evaporites in surrounding wallrocks (Naldrett, 1989).

Nickel Production and Related Environmental Concerns

Production of nickel from its ores involves some interesting trade-offs between laterite and sulfide deposits. Because laterite ores are at the surface, they can

FIGURE 8-13
Geologic map of the Sudbury, Ontario, area showing the location of the Sudbury layered igneous complex, important nickel mines, and the deep Victor deposit. Also shown is the southwestern end of the Temagami (Wanapitei) gravity-magnetic anomaly, which could be a buried feature similar to Sudbury.

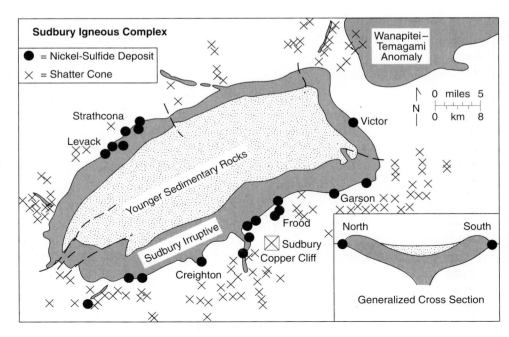

be mined by inexpensive open-pit methods. But it is difficult to beneficiate laterite ores to make nickel mineral concentrates, and it is usually necessary to process the entire volume of ore to recover nickel (Evans et al., 1979). For iron-rich laterites, this is done by hydrometallurgical methods using acid or ammonia to leach nickel from the ore. Magnesium-rich laterites are heated in the presence of sulfur to form nickel sulfides that are then smelted by pyrometallurgy. Sulfide ores, on the other hand, require more expensive underground mining but can be made into low-volume sulfide mineral concentrates that are processed by less expensive, conventional pyrometallurgy. Both processes can produce nickel metal or *ferronickel,* which is used directly in steels.

At present energy prices, nickel-sulfide ores are thought to have the economic advantage because they are easier to treat. Laterite operations that were started when energy costs were lower have had difficulty competing with sulfide ores. Some, including the large Exmibal operation in Guatemala, have been shut down. However, many nickel-sulfide deposits are in MDCs where operating costs are higher. In fact, a recent report estimated that nickel production costs for sulfide ores in Canada in 1992 were actually higher than those for laterite ores in LDCs, almost entirely because of wages and benefits (MMMU, 1992).

Sulfide-nickel ores cause environmental concern because they contain 8 tonnes of sulfur for each tonne of nickel. Nickel smelters at Sudbury are the largest point source of SO_2 in North America, with emissions of almost 840,000 tonnes of SO_2 in 1985. Although these emissions pale beside the total from electric power plants, their concentrated nature has had a profound effect on the Sudbury area, which is surrounded by metal haloes in soil and lakes (Figure 4-22B and Nriagu and Rao, 1987). Sulfur emissions from Sudbury have declined significantly since the 1960s and current regulations require that they be reduced to 365,000 tonnes by 1994. Most of this decrease is being attained by the use of flash smelting to increase SO_2 concentrations in flue gases and by the exclusion of pyrrhotite, which contains much more sulfur per atom of nickel than do other nickel-bearing sulfide minerals. Problems persist for the other major nickel-sulfide producer, Noril'sk, which smelts most of its ore at Pechanga near the Finnish and Norwegian borders in the Kola peninsula. Although the governments of Finland and Norway have offered $100 million toward the estimated $600 million cost of installing flash smelters and other facilities to cut SO_2 emissions at Pechanga, Russia has not gone ahead with the project.

Fossil fuel combustion is also an important source of global nickel emissions. Nickel emissions amounted to about 2,300 tonnes annually from oil-fired boilers in the late 1970s, and 3,000 to 18,000 tonnes annually

from coal combustion as recently as 1986 (Wilson et al., 1986). Where scrubbers have been installed, emissions are lower, but boiler heating probably remains the principal source of environmental nickel. In its inorganic forms, nickel is most readily taken into the human body by way of the lungs (WHO, 1991). It is not an essential element for life, but it also does not appear to be strongly detrimental in small concentrations, as are cadmium and mercury. Thus, the greatest dangers of nickel-related environmental problems are in industrial situations that create nickel-bearing dust and fumes. The most serious such situation appears to be in roasting operations in smelters, where workers exposed to nickel sulfides, oxides, and sulfates have an increased risk of nasal and lung cancer.

Nickel Trade and Reserves

Nickel-sulfide ores at Sudbury and Noril'sk dominate world production (Figure 8-14), both because they are huge deposits and because they are less expensive to process. During the last decade, Russia has been a net importer of ore or raw nickel products from Cuba, but a net exporter of metal and ferronickel to the western world, using nickel sales to earn foreign exchange (Ansah, 1990). Nickel exports go largely to the major steel-producing nations, including Japan, Germany, China, and the United States (Figure 8-14). Norway, which has no significant ore, imports concentrates and uses surplus energy to smelt nickel metal, which is then exported. Cuba is the wild card in the world nickel supply equation, not only because of its large production and reserves, but also because energy for

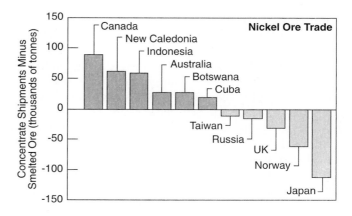

FIGURE 8-14

World nickel ore trade for 1991 showing the difference between nickel contained in concentrate and that in metal produced by each country. Countries that do not produce either commodity are not shown, even if they consume nickel metal. Russia, a major producer, has imported nickel from Cuba (from data of U.S. Bureau of Mines and Metal Statistics).

its nickel production has come from cheap Russian oil, which is no longer available.

Present nickel markets are dominated by stainless steel and the demand for primary nickel is controlled by the amount that is recycled. Recycled nickel comes from manufacturing waste and stainless steel scrap. The relatively low rate of investment in plant and equipment during the last two decades suggests that the supply of scrap will not be sufficient to depress primary nickel markets in the near future. Nickel recycling could be further limited if nickel-bearing scrap metals are classified as hazardous. Proposals of this type could even classify stainless steel as hazardous waste, greatly complicating recycling efforts.

World nickel reserves are estimated to contain 47 million tonnes of metal. Although nickel-sulfide deposits dominate world production, the reserves are dominated by laterite deposits (DeYoung et al., 1985). Energy considerations and the difficulties with Cuban production suggest that these laterite reserves are considerably less dependable than sulfide reserves, which leaves the world with only about 13 billion tonnes of dependable nickel metal in reserve, slightly more than enough for a decade of consumption. Viewed in this way, the outlook for world nickel supplies is not good.

The outlook for additional resources is also challenging. In the United States, the best possibility is a zone at the bottom of the *Duluth Gabbro,* a mafic intrusion just outside Duluth, Minnesota, where sulfur from wallrocks caused the separation of immiscible nickel-copper sulfides. Although this area contains about 6 billion tonnes of nickel-bearing rock, grades are 0.2% Ni, only one tenth those at Sudbury and Noril'sk. The same is true of most other areas of nickel-sulfide ores, including the huge Bushveld Complex in South Africa from which small amounts of nickel and cobalt are recovered during platinum production. Submarine manganese crusts and nodules are a possible reserve, of course, but they are not likely to become economic soon. Thus, attention is forced back to the potential for deeper nickel-sulfide ores at Sudbury and Noril'sk. At Sudbury, exploration at depths reaching 2,700 meters has discovered the Victor deposit, which will probably be the first in a new generation of deep ore bodies (Scott, 1991). Exploration is also taking place on the huge Wanapitei gravity-magnetic anomaly about 50 kilometers to the northeast of Sudbury, which it is hoped will be a second buried irruptive (Pye et al., 1984; Pearce, 1990). If the technical difficulties of deep mining can be solved, deep sulfide-nickel reserves can probably carry us well into the next century.

Chromium

Chromium is the essential alloying element for stainless steel, the least expensive corrosion-resistant metal. Stainless steel is used in household utensils, containers for food and chemicals, and automobile parts. It is by far the most important alloy metal produced in the steel industry, accounting for 1% to 2% of world steel production. With a content of 12% to 36% chromium, stainless steel accounts for about 80% of world chromium consumption. Although other, more expensive metals can be substituted for stainless steel, nothing can do the job at the same price. The remaining 20% of chromium production is used in high-performance alloys and a host of chemicals including green, yellow, and blue pigments, and platings, where a thin film of chromium provides an attractive surface that is highly resistant to corrosion (Papp, 1985, 1990).

Chromite ore production of about 13 million tonnes annually is valued at almost $700 million. *Ferrochromium,* the iron-chromium product used to make steel and the most important form in which chromium is sold, is worth much more, however. Ferrochromium contains chromium valued at about $0.50/pound. At that price, the value of world chromium production is about $5 billion.

Geology and Distribution of Chromium Deposits

Chromium comes from *chromite,* a member of the *spinel* mineral group, which has a composition that can be summarized as AB_2O_4, where A is any combination of Mg^{+2} and Fe^{+2} and B is any combination of

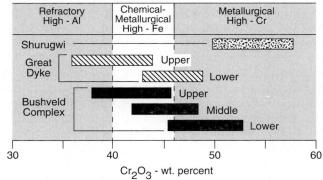

FIGURE 8-15

Range of chromium contents (expressed as Cr_2O_3) of chromites suitable for refractory, chemical, and metallurgical applications, showing compositions of chromites from major deposits in southern Africa (modified from Stowe, 1987).

Cr^{+3}, Al^{+3} and Fe^{+3}. Because of this extensive substitution, the chromium content of chromite varies between about 30% and 60% Cr_2O_3 (Cr_2O_3 contains 68.42% chromium). Obviously, *high-chromium chromites,* which are defined as those with at least 46% Cr_2O_3, are the richest source of chromium metal (Figure 8-15). *High-iron chromites,* which have 40% to 46% Cr_2O_3, can be processed to yield chromium metal and ferrochromium. Chromites with less than 40% Cr_2O_3, which are known as *high-alumina chromites,* melt at higher temperatures and require larger amounts of energy for processing. For this very reason, they are useful as *refractory material* and are frequently used in bricks that line blast furnaces and other hot containers (Mikami, 1983).

Chromite from which chromium metal or ferrochromium can be produced is mined in 22 countries, but about 66% of it comes from South Africa and Kazakhstan. These areas contain the two principal types of chromite deposits, stratiform and podiform (Stowe, 1987). *Stratiform deposits* are found in *layered igneous complexes (LICs),* particularly the Bushveld Complex of South Africa and the Great Dike of Zimbabwe (Figure 8-16). LICs are mafic intrusive rocks consisting in part of layers with very distinct mineral compositions. Some of these layers are made up of essentially one mineral, and the best chromite ore is found in layers or seams, which are called *chromitites* because they consist largely of chromite. The *Bushveld Complex,* by far the largest LIC in the world, contains layered mafic rocks over 9 kilometers thick and a thick sequence of genetically related felsic rocks (Figure 8-17). Even though much of the Bushveld has been eroded, it still covers an area of about 66,000 km^2 (Naldrett, 1989; Sharpe, 1986).

A total of 14 chromite layers is present in the Bushveld Complex, six of which have been mined (Stowe, 1987). These are confined to the Critical Zone, one of several divisions of the Bushveld (Figure 8-17), where they have been divided into three groups, Lower, Middle, and Upper. Current mining focuses on the Steelpoort, Magazine and UG-2 seams

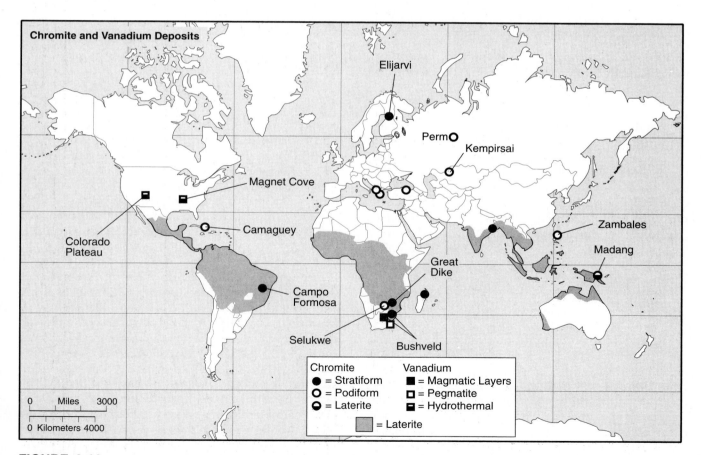

FIGURE 8-16

Locations of world chromite and vanadium deposits (modified from DeYoung et al., 1984 and Stowe, 1987). More detail on southern Africa is shown in Figure 10-7.

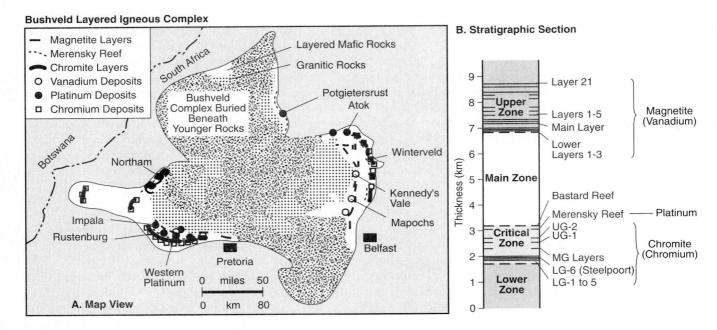

FIGURE 8-17

A. Geologic map of the Bushveld Complex showing the distribution of chromitite, magnetitite, and Merensky Reef layers and producing mines. **B.** Schematic stratigraphic column in eastern Bushveld Complex showing position of these layers. Section omits the Marginal Zone and overlying felsic rocks, which contribute to a total thickness of over 9 kilometers for Bushveld Complex (after Sharpe, 1986; South Africa's Mineral Industry, 1990).

(Figure 8-18). These layers are amazingly continuous; the Steelpoort seam, 1.06 to 1.8 meter thick, can be traced for over 100 kilometers along the surface and probably extends at least 20 kilometers beneath the surface. Although chromite from the UG-2 contains only 42% Cr_2O_3, it also hosts 6 to 7 grams/tonne of platinum-group elements, which is the main objective of the mining. Another layer in the Bushveld Complex, the *Merensky Reef*, also contains chromite but is mined for platinum-group elements, as discussed later.

The *Great Dike* of Zimbabwe, a LIC with a much more elongated form, has large chromite deposits and reserves. It is older (2.5 Ga versus 2.0 Ga for the Bushveld), but its close proximity and similar chromite-rich nature suggest that there was something unusual about the Precambrian mantle beneath southern Africa, in order to produce so much chromium-rich magma. Other LICs are known, including the *Stillwater Complex* in Montana (Page et al., 1985). Although most of them contain chromitite layers, they lack large reserves because the intrusions are smaller and more faulted and eroded. Least studied of these is the Dufek Complex in Antarctica, the second largest LIC in the world (Figure 5-9).

There is considerable debate about just how these curious chromitite seams formed. Why, for instance, do they contain so few silicate minerals such as olivine, pyroxene, or plagioclase, which are abundant in LICs? Most magmas crystallize more than one mineral at a time, and the minerals are often intimately intergrown. Why has that not happened in these LICs? An early solution for this conundrum proposed that the seams were actually immiscible chromite magmas, similar to immiscible iron-nickel sulfide magmas, that separated from the silicate LIC magma and settled to the bottom. But chromite melts at temperatures much higher than those at which LIC magmas form, and the chromite seams do not look like they were ever molten. Another idea held that chromite crystallized from the magma with other silicate minerals but settled to the bottom of the chamber more rapidly because it is heavier, but experiments show this is not an effective process. Most people now feel that the key factor that allowed the formation of the monomineralic chromitite seams was crystallization of chromite when no other minerals were crystallizing from the magma. The periods of "chromite-only" crystallization must have been long enough to permit seams 1 meter thick to form, *and* they must have occurred repeatedly to form the many seams. It turns out that a magma can be made to crystallize chromite alone if it mixes with new magma. This would occur whenever a new batch of magma was injected into the magma chamber, and successive pulses of new magma could account for all of the seams.

Podiform chromite deposits are not found in LICs. They occur, instead, in mafic and ultramafic igneous rocks from the base of the ocean crust and the top of the underlying mantle that have been obducted onto the continent. These are the same rocks that generate

nickel laterite deposits, and podiform chromite deposits also occur at convergent plate margins such as Cuba, Philippines, and Turkey. Chromite in some podiform deposits, such as Selukwe (Shurugwi) in Zimbabwe, is richer in chromium than in stratiform deposits (Figure 8-15). Some podiform chromite bodies are actually faulted and sheared chromitite layers that formed in large mafic magma chambers at mid-ocean ridges. A few podiform deposits are found in circular

FIGURE 8-18
Bushveld chromite deposits, eastern Transvaal, South Africa. **A**. Chromitite layers (black) at Dwars River. **B**. Winterveld chromite mine showing cumulate layers on hillside dipping downward to left. Mine enters at base of the hill to right.

A.

B.

mafic complexes at convergent margins where they could have been conduits for the upward movement of mantle magmas. Although most podiform deposits are smaller than stratiform deposits, the Donskoy mine in Kazakhstan is the largest chromite producer in the world.

Because chromite is resistant to erosion, it can accumulate in *laterites* and *placer deposits.* Some nickel laterite deposits, which form on ultramafic rocks containing disseminated and podiform chromite, have as much as 1% Cr_2O_3, although these have not yet been mined economically (Lipin, 1982).

Chromium Production, Trade, and Reserves

Most chromite is mined by underground methods. Mining is complicated because some seams are only a meter or so thick and their lower surface is locally irregular. Chromite is converted to ferrochromium in electric-arc furnaces, like those used to make steel. Chromium metal can be produced either by pyrometallurgical reduction of chromic oxide or by electrolytic separation of chromium from a solution of ferrochromium and sulfuric acid. Chromium chemicals are also prepared from sulfuric acid solutions that result from kiln roasting of chromite ore. The most important wastes associated with this processing are dust, fumes, and sulfuric acid, which must be caught or neutralized (Papp, 1985).

Environmental problems related to chromium processing are limited because chromium, in the trivalent (Cr^{+3}) oxidation state in which it is found in chromite and other natural minerals, is an essential nutrient. Its principal function is to maintain normal glucose metabolism (WHO, 1981). Chromium deficiencies can lead to problems in insulin circulation as well as possible risk of cardiovascular disease. It is estimated that 90% of U.S. adults have a deficiency of chromium in their diets, an interesting statistic in view of recent experiments suggesting that large amounts of chromium in the diets of animals prolong their lives significantly. There is no evidence that trivalent chromium causes ill effects in workers or the general population in the vicinity of ferrochromium plants, and no longevity studies are available. The hexavalent form of chromium (Cr^{+6}), which is used widely in chemical compounds and has been implicated in skin rashes and lung cancer, occurs in nature only in the Atacama desert and is not produced from trivalent chromium in the human body (Fergusson, 1990). The U.S. EPA considers wastes containing more than 5 mg/l total soluble chromium as hazardous, although the significance of this to chromite-bearing refractory bricks is not clear.

World raw chromium trade involves chromite and ferrochromium production, which must be treated separately (DeYoung et al., 1984; Strishkov and Stebler, 1985). The largest ore and ferrochromium exporter by far is South Africa (Figure 8-19). Other important chromite ore exporters include Turkey, Russia, Kazakhstan, and Albania. Zimbabwe, a major chromite producer, is not an ore exporter but is the world's second largest ferrochromium exporter. The major steelmakers, Japan, Germany, the United States, and United Kingdom, import both commodities, but China imports only chromite ore. Reserves to support world chromite trade amount to about 1.4 billion tonnes of ore. The largest reserves are in South Africa, with Kazakhstan and Zimbabwe distant seconds. In general,

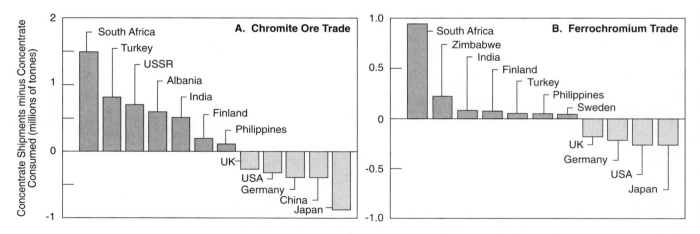

FIGURE 8-19
World raw chromite trade for 1990, showing relations for both (A) chromite ore and (B) ferrochromium. Tonnages are for chromium content of these products (data supplied by John F. Papp, U.S. Bureau of Mines).

South African reserves are lower in chromium than those in Kazakhstan. The reserves in Zimbabwe include both high-chromium chromite from Shurugwi and lower-chromium, high-iron chromite from the Great Dike.

The lack of significant reserves outside southern Africa and Kazakhstan is a persistent worry for European and North American chromium consumers and has been aggravated by the history of chromite exports. The Soviet Union embargoed chromite exports to western countries from 1950 until about 1963. Just as this production became available again, the United States and some European countries embargoed chromite imports from the racially restrictive government in Rhodesia (now Zimbabwe). The Rhodesian embargo was in effect from 1966 to 1972, when it was lifted by Congressional action designating chromite as a strategic metal. It was reinstated in 1977, however, and not lifted again until the Rhodesian government changed in 1980. Throughout this period, chromite production from Rhodesia/Zimbabwe continued, and the ore from these mines was sold to the United States and western Europe through other countries that acted as unofficial clearinghouses. Furthermore, shortfalls resulting from the embargo were made up largely by sales from the Soviet Union, with which the United States and western Europe also had significant ideological differences at the time. As a result of this lesson, chromite was not included in the economic sanctions imposed by the United States on South Africa during the tenure of apartheid.

Long-term chromium reserves will depend on the recovery of lower-grade chromite that is disseminated through LICs or obducted ultramafic rocks that host podiform deposits. Obducted ultramafic rocks are probably best because their chromites generally contain more chromium. Laterites on weathered ultramafic rocks are particularly attractive because they are relatively easy to mine and contain nickel and cobalt. During the strategic metal scare of the late 1970s, unsuccessful efforts were made to develop the Gasquet Mountain laterites in northern California, which were reported to contain 0.85% Ni, 0.01% Co, and 2% Cr. Much larger laterite deposits in the Madang district of Papua-New Guinea have still not been put into production. Thus, the outlook for alternative chromium supplies in the next decade is very poor, and we could well enter the next century completely dependent on southern Africa and Kazakhstan for chromite.

Silicon

Although most of us think of semiconductors when the word *silicon* comes up, we should think of iron and steel. About 70% of world silicon production is used as an additive in steelmaking, where silicon ranks second in volume only to manganese (Murphy and Brown, 1985). Silicon is used as a deoxidant in steel and as an alloy to increase tensile strength and corrosion resistance, largely in the form of *ferrosilicon* alloys containing 50% to 90% silicon. Much of the silicon that is not used in the steel industry goes into alloys with aluminum, copper, and nickel, where it increases fluidity and wear resistance. That leaves a very small amount for a host of special products, including silicones and other semisolids that behave like oil, *silicon carbide*, a widely used abrasive material, and silicon metal, which is used in semiconductors. Annual world silicon production amounts to almost 4 million tonnes worth about $5 billion (Gambogi, 1990).

Silicon production comes largely from *quartz*, which is found in a wide variety of environments. Most common are quartz-rich *clastic sediments* that formed in fluvial and coastal environments, as well as their lithified equivalents, sandstone and quartzite. Others include coarse-grained igneous rocks, principally *pegmatites* that formed during the latest stages of crystallization of granitic magmas, and *hydrothermal veins* consisting almost entirely of quartz. Although these materials are widespread, not all meet the stringent purity specifications demanded by industry, especially the absence of arsenic, sulfur, and phosphorus. Environmental problems associated with quartz mining are discussed in the section on abrasive and refractory minerals.

Ferrosilicon is produced by melting quartz concentrate, coke, and scrap iron in an electric furnace. The process requires large amounts of electric power, which constrains ferrosilicon manufacture to areas of surplus power, usually hydroelectric. In fact, silicon production is similar to that of aluminum, in which the raw material is transported long distances to areas of cheap electrical energy, such as Norway and Canada. Iceland, which lacks quartz deposits because of the dominantly basaltic nature of its crust, even makes ferrosilicon from quartz imported from Norway. Major importers of ferrosilicon include the steelmakers, Japan, Germany, and the United Kingdom. Silicon reserves are simply not a problem. Quartz is one of the most abundant minerals in the crust and useable deposits are widely distributed.

Cobalt

Cobalt's main use, about half of total consumption, is in corrosion and abrasion-resistant alloys with steel, nickel, and other metals in industrial and gas turbine engines and chemical compounds. The remainder is used about equally in magnets, in which cobalt is alloyed with nickel and aluminum (Alnico), as the matrix in which tungsten carbide or diamonds are set to make cutting tools, and as a catalyst. Most of these applications are very important to modern industry and cobalt is widely considered to be a strategic metal. Annual world cobalt production, which amounted to about 25,000 to 35,000 tonnes worth about $1 billion for most of the 1980s, declined in the early 1990s because of problems in Zaire, the main source of cobalt. As can be seen in Figure 8-1, this led to a strong increase in price.

Cobalt Deposits

Strictly speaking, cobalt hardly belongs in the ferroalloy group because so much of it is used in high-performance alloys that contain little or no steel (Kirk, 1985). It usually falls there by default, however, because cobalt is an essential alloying element in steel.

World cobalt production comes from 14 countries (Shedd, 1992) but is dominated by the *sedimentary deposits* of the central African Copper Belt, which extends from Shaba Province in Zaire into Zambia (Fleisher et al., 1976). These deposits are copper-cobalt–bearing sedimentary rocks of late Precambrian age (about 1 Ga) that accumulated around older continents. Copper, largely in chalcopyrite, is actually the dominant metal in these deposits, but cobalt is so highly concentrated that it is a major co-product. Much of the cobalt is in *linnaeite*. Grades of Copper Belt ore in Zaire reach 0.4% Co, making this large sedimentary sequence the most important cobalt reserve on the planet. Exactly how these deposits formed is a subject of active controversy. Early ideas that the deposits were direct chemical precipitates from ocean water have given way to suggestions that the metals were deposited by brines or other solutions moving through the sediments after they were deposited, a concept supported by the occurrence of copper ore in a wide variety of sedimentary-rock types including limestone, shale, sandstone, and conglomerate. Thus, they appear to be similar to other sediment-hosted copper deposits such as White Pine, Michigan, and the Kupferschiefer in Europe (Boyle et al., 1990).

Smaller amounts of cobalt produced by Russia, Canada, Cuba, and Australia, are largely by-products from *sulfide and laterite nickel deposits*. Cobalt is geochemically similar to nickel and will concentrate in similar geologic environments (Mishra et al., 1985). Just as for nickel, large amounts of cobalt substitute for magnesium in silicate minerals, giving mafic and ultramafic rocks cobalt contents as high as 300 ppm. Cobalt also substitutes for nickel and iron in sulfide minerals, particularly pentlandite and pyrrhotite. A special type of magmatic deposit, the platinum-

bearing zones of the Bushveld Complex in South Africa, contain only small amounts of cobalt but yield significant production because of the large amounts of ore that are mined for platinum. Cobalt is also found in *hydrothermal vein deposits* of which the most famous are those in Cobalt, Ontario (Andrews et al., 1986). Although these deposits are important sources of silver, as discussed later, they are not large enough to be an important source of cobalt.

Cobalt Production and Trade

Cobalt mining and processing vary greatly, depending on the type of deposit. In the African Copper Belt, both open-pit and underground mining are used, although underground mines are expected to take over production as near-surface ores are depleted. Mining in this area is plagued by water problems and some mines shut down during the rainy season. Most ore is processed by making a cobalt mineral concentrate, which is roasted to drive off sulfur and leached of its cobalt with acid solutions. The cobalt is then removed from solution by SW-EX, a process that is complicated by copper and zinc in the leach solutions. Although mining and processing operations in the Copper Belt have local environmental problems related in flooding and rock failures, few of them are uniquely associated with cobalt (Van den Steen et al., 1992).

Cobalt is a component of vitamin B_{12}, which has a role in nucleic acid metabolism and the maturation of red blood cells, and has an average abundance of about 0.0015 grams in the human body (Young, 1979; Fergusson, 1990). Excessive cobalt dust, which has been observed in the manufacture of cobalt-cemented tungsten carbide tools, can cause chronic bronchitis, and skin contact with soluble forms of cobalt causes dermatitis. It is classified as "possibly carcinogenic" by the International Agency for Research on Cancer, although specific links to cancer have not been recognized (Moore, 1990; Shedd, 1992).

World cobalt trade is dominated by Zaire, a factor that has created concern about strategic supplies (Blechman, 1985) because this ore is the basis for the production of cobalt metal in a number of industrialized countries that lack domestic ore sources (Figure 8-20). Cobalt reserves containing 4 million tonnes of metal are quite large but are complicated by two factors. First, cobalt is a by-product in almost all of the mines from which it is recovered. Thus, its production rate is controlled by the demand for copper, nickel, or platinum and cannot be increased substantially in response to changes in cobalt demand. Second, the deposits for which this constraint is weakest, those in Zaire, have been crippled by decades of political strife and predatory government management. These de-

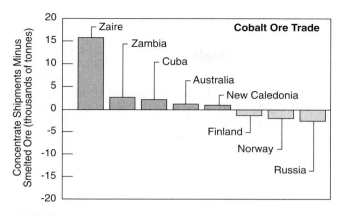

FIGURE 8-20

World trade in cobalt ore and metal (expressed as cobalt content) for 1990. Countries extending above the line are major exporters of ore and those extending below the line import ore for domestic smelters and, in some cases, export metal (data from U.S. Bureau of Mines and Metal Statistics).

posits are almost the only source of foreign exchange in Zaire but have suffered from poor maintenance, limited investment in new equipment, and the departure of most expatriate staff. These factors and guerilla activity in Shaba Province caused cobalt prices to rise to $40/pound in the late 1970s and stimulated a second price rise in 1990 (Figure 8-1). Finally, high mortality rates caused by AIDs have begun to limit the number of healthy workers at operations in Zaire and Zambia and could prove to be the limiting factor in world availability of cobalt.

Efforts have been made to locate more dependable cobalt resources (Crockett et al., 1987). Some success has been achieved in the United States, where major cobalt resources are known in Mississippi Valley-type deposits in the northern part of the Old Lead Belt in southeastern Missouri (Figure 5-11) and in sediments of the Proterozoic-age Belt Supergroup in Idaho. Unfortunately, grades of these deposits are not high enough to permit cobalt mining. A final possibility, cobalt-rich manganese crusts and nodules on the ocean floor, does not appear to have economic potential for the near future (Cronan, 1992). Thus, it is likely that cobalt supplies will remain a major concern to industrialized nations and that the Copper Belt deposits will continue to be the focal point. A ray of hope is offered by the possibility that rapprochement between South Africa and the rest of the continent will allow an influx of investment capital and expertise that might permit these deposits to fill their geological destiny as dominant cobalt suppliers to the world.

Molybdenum

Molybdenum is a quintessential alloy element. About 70% of world consumption goes into steel where it provides hardness and resistance to abrasion, corrosion, and high temperatures. Corrosion-resistant steels with about 5% molybdenum are used in seawater applications, and hard steels with about 10% molybdenum are used in cutting and grinding equipment. The remaining 30% of world molybdenum consumption goes into chemicals such as molybdenum-orange pigment, catalysts for oil refining, and lubricants (Blossom, 1985, 1992). World molybdenum production, amounting to about 110,000 tonnes annually, has a value of less than $1 billion.

Molybdenum is a very scarce element, with a crustal abundance of only about 1 ppm. The most common molybdenum mineral is *molybdenite,* which is recovered largely from *porphyry-type deposits,* either alone or as a by-product of copper production. *Porphyry copper deposits,* which are described in the next chapter, contain as much as 0.05% Mo and are important sources of molybdenum because their copper ore is mined in large volume. They are associated with felsic intrusive rocks with a *porphyritic texture* consisting of large crystals in a matrix of smaller ones. This texture is interpreted to result from a change in the rate of crystallization such as might result from the rapid ascent of the magma from a deeper to a shallower level of the crust. These deposits are found in island arcs and other convergent tectonic margins and they commonly underlie volcanoes. Where the intrusive rock is slightly more granitic, we find *porphyry molybdenum deposits,* such as the Quartz Hill deposit near Ketchikan, Alaska, in which molybdenite is the dominant ore mineral (Figure 9-12).

Most molybdenum, however, comes from *Climax-type molybdenum deposits,* which are associated with granitic rocks relatively far from collisional margins (Figure 9-12). These deposits contain as much as 0.3% Mo, with smaller but important amounts of tin and tungsten, which are recovered as by-products. The large Climax and Henderson mines near Denver, of this type consist of inverted cup-shaped zones of fractured and veined rock that formed over small porphyritic, granitic stocks (Figure 9-11). As in the case of porphyry copper deposits, much of the water that formed these deposits separated from the intrusion as it cooled and crystallized. Porphyry molybdenum deposits are not abundant in older rocks and they cluster along Mesozoic and Cenozoic convergent tectonic boundaries. Although such boundaries extend all the way around the Pacific Ocean, those in Canada, the United States, Peru, and Chile are underlain by larger amounts of continental crust and appear to contain larger molybdenum deposits (White et al., 1981).

Molybdenum is produced in far fewer countries than those producing copper from porphyry copper ore. In fact, many porphyry copper deposits do not recover their by-product molybdenite because it is too scarce to pay for the installation of the necessary facilities. Molybdenite is mined from both open-pit and underground mines. Endako in British Columbia and the renovated Climax mine in Colorado are open pits. Henderson, which is near Climax, is the world's largest-underground molybdenum mine. It began production in 1976 and is an example of highly successful coexistence between mining and other land uses. Although the mine is near Denver and major ski areas to the west, it is essentially invisible to passing motorists. Beneficiation and tailings disposal, which are located in a valley almost 24 kilometers from the mine, are reached by a rail system that includes a 15.4-kilometer tunnel.

Because of their low grades, most molybdenum mining operations produce large volumes of tailings, which are disposed of in impoundment areas that are revegetated when abandoned, as is the case at Henderson. Processing molybdenite concentrate, which usually is not done at mines, involves roasting to convert MoS_2 to MoO_3, the most common form in which it is used in industry. Gases from this process must be recovered to prevent their escape into the atmosphere. Molybdenum itself is an essential element to human life, with a role in the metabolism of copper. It has not been shown to cause toxic effects in mining or metallugrical operations.

Molybdenum trade consists largely of exports from the major producers in Canada, the United States, and Chile to the rest of the world (Palencia, 1985; Blossom, 1992). Important steel producers in Japan and Western Europe are the largest importers. Russian production has been used in the past mostly to satisfy domestic demand and that of eastern European steelmakers. World molybdenum reserves contain about 5.5 million tonnes of metal and the outlook for further discoveries is relatively good. Because of its large reserves, China is likely to become a more important producer and exporter in the future. Additional molybdenum could be recovered from other porphyry copper deposits in which molybdenite recovery circuits have not yet been installed, and from tailings of porphyry copper deposits. The fact that many of these resources are within North America is a source of comfort to strategic planners there and in Europe, and molybdenum is likely to become a metal of choice in important alloys and other applications.

Vanadium

Vanadium provides an example of the strong control that mineralogy exerts on the commercial use of an element. Although it has an average crustal abundance of 160 ppm, more than twice that of copper, it forms few minerals and is used much less widely. Annual world vanadium production of about 25,000 to 35,000 tonnes is valued at a few hundred million dollars, only 0.3% of world copper production (Hilliard, 1990).

Vanadium minerals are so rare because V^{+3}, the form in which it is found in much of the crust, is geochemically similar to Fe^{+3}, an abundant ion that is part of many common minerals. Thus, V^{+3} generally substitutes for Fe^{+3} in minerals rather than forming separate vanadium minerals. The most common vanadium-bearing iron mineral is *magnetite*, which contains as much as 3% V_2O_5 (the chemical form in which vanadium concentrations are quoted, regardless of their form in the material being analyzed). The rarity of vanadium minerals prevented the metal from coming to the attention of early chemists and metallurgists until 1830, and even after that it languished without significant applications for another 60 years because of the difficulty in separating it from most ores (Kuck, 1985).

Vanadium is recovered from magmatic and hydrothermal deposits (Goldberg et al., 1992). By far the most important are *magmatic deposits* consisting of layers or seams of magnetite *(magnetitite)* in the upper part of the Bushveld Complex in South Africa (Figure 8-17). The magnetite layers are largely monomineralic and they are thought to have formed by the same process that formed the chromite layers mentioned previously. They are always found above chromitites, in the upper iron-rich part of LICs. At least 21 magnetitite layers are known, many of which are laterally extensive and have huge reserves. Vanadium is also present in cylindrical pegmatite pipes, such as Kennedy's Vale, that cut across layering in the Bushveld and might be original cumulate magnetite layers that funneled downward from the upper part of the complex (Figure 8-16). The Kachkanar magmatic iron deposits in Siberia are generally similar.

Hydrothermal deposits form when vanadium in rocks and minerals is oxidized during weathering to the pentavalent state (V^{+5}) which is highly soluble. Dissolved pentavalent vanadium will be deposited if the solution is reduced or evaporated. In the Colorado Plateau area of the western United States, vanadium was precipitated when groundwaters encountered reduced zones rich in organic matter. This geochemical process is very similar to that which forms uranium ores, and vanadium is a common constituent of many sandstone-type uranium deposits, often in the mineral *carnotite*. Vanadium is so important in these deposits that one district is even known by the name Uravan. In Western Australia, vanadium was precipitated in the Yeelirrie uranium deposit when groundwater evaporated as it flowed through a near-surface aquifer. Although most hydrothermal vanadium deposits derived their vanadium from weathering, the vein and disseminated ores in the Wilson Springs area around Magnet Cove, Arkansas, surround intrusive rocks and are probably of deeper origin.

Vanadium is also concentrated in many forms of organic material, including coal and oil. In fact, vanadium mining began at the Mina Ragra asphaltite deposit in Peru (Breit, 1992). Although these deposits are not mined at present, vanadium by-products from oil refining and ash from fossil fuel combustion account for over 10% of world production. The largest resource of this type of material is in the tar sands of Alberta. Although the vanadium content of tar sand itself is only 0.02% to 0.05%, fly ash from the processing plants contains up to 5% V_2O_5. Efforts to recover this vanadium have not been economically successful so far, but they represent a large cushion against future demand. Because magnetite is both heavy and resistant to erosion, it can also accumulate in *placer deposits,* although these are not important sources of vanadium.

Vanadium mining is carried out largely by open pit (Figure 8-21), although some Colorado Plateau uranium-vanadium mines are underground. The process used for recovery of vanadium from magnetite, which dominates the supply picture, depends on its vanadium content. High-vanadium magnetite (2% V_2O_5) is usually roasted and leached to release vanadium, which is then precipitated to form vanadium chemicals. These deposits depend entirely on vanadium for revenues. Low-vanadium magnetite is actually smelted in a blast furnace to produce pig iron and vanadium-bearing slag. The slag is usually sold in that form, and it is roasted and leached to release vanadium. Some South African slags contain more than 20% V_2O_5 and are highly desirable, whereas most others are lower grade and are not as much in demand. Most production from China and Russia takes this form, as does some from South Africa. In all of these deposits, vanadium availability is controlled in part by the market for pig iron (Hilliard, 1990).

About 85% of vanadium is used in steel, with about 40% going into HSLA steels in which as little as 0.1% vanadium greatly enhances the metal's strength, toughness, and ductility. Vanadium steels are used in frameworks for high-rise buildings, offshore oil drilling platforms, and pipelines for oil and natural gas. Vanadium is also alloyed directly with titanium and aluminum to improve the strength of these metals. Vanadium chemicals are used in the production of nylon and polyester resins, as well as special glass that will not pass ultraviolet radiation and can be used to protect art objects, fabrics, and furniture.

Environmental problems related to vanadium mining are not significant and those related to processing and fabrication are under control. Problems related to vanadium mining come largely from its close geochemical association with uranium and selenium in the Colorado Plateau uranium-vanadium deposits. Both of these elements are considered toxic, and they

FIGURE 8-21
Mapochs vanadium mine, eastern Transvaal, South Africa. The vanadium-bearing magnetite layers dip steeply toward the viewer and have been mined by stripping off the side of the hill.

are sufficiently enriched in vanadium ores to require special disposal of beneficiation and treatment wastes. The processing of some vanadium ore has caused dust inhalation problems including bronchitis and forms of pneumonia, although these can be largely eliminated by the use of respirators (WHO, 1988). The real problem, if one exists, is in fossil fuel combustion, which supplies almost two orders of magnitude more vanadium to the atmosphere than does metal smelting or all natural sources (Nriagu, 1990b). Although vanadium has no specific role in the human body, it is also not known to be a strong toxin.

Vanadium production comes essentially from the United States, China, Russia, and South Africa. The main exporter is South Africa. Although the United States has domestic production, it imports more ore and slag than it exports. Russian production, which comes from Kachkanar in the central Urals, has been consumed domestically and in eastern Europe. World reserves contain about 10 million tonnes of vanadium, essentially all of which is in South Africa, Russia, and China. In view of the important uses of vanadium and the relatively clouded future for long-term supplies from these countries, exploration is needed to locate vanadium elsewhere in the world.

Tungsten

Tungsten, with annual world production of about 40,000 tonnes worth less than $500 million, is usually classed as a ferroalloy metal, although only about 6% of production is used in steel production. The main use for tungsten is the manufacturing of *tungsten carbide*, one of the hardest synthetic materials used in industry. It is widely used in cutting and wear-resistant materials, particularly those that have to work at high temperatures. One of its most important markets, drill bits for oil wells, takes advantage of this property. Mill products made from pure metal are the second major market, with most being used in the electrical and electronic industries. Tungsten wire forms the filament in many incandescent light bulbs and cathodes for electronic tubes. The metal is used in superalloys with copper or silver and in the chemical industry. It is also used in armor plate and armor-piercing ordnance, although this last use has been replaced in part by depleted uranium.

Tungsten markets are under siege on all fronts. New hard materials such as polycrystalline synthetic diamond, boron nitride, and titanium carbide are being substituted for tungsten carbide. Coatings of these materials are also being put on tungsten carbide to make it last longer, thus cutting down on replacements. As a result, tungsten production is growing more slowly than any of the low-volume ferroalloys, although its unique combination of hardness and high melting temperature will probably hold a core of markets (Yih and Chun, 1978; Stafford, 1985).

The chief sources of tungsten are the minerals *scheelite* and *wolframite,* which are deposited by hydrothermal solutions (Bues, 1986; Kwak, 1987; Elliott, 1992). Most tungsten deposits are closely associated with granitic intrusions from which the metal-bearing hydrothermal solutions were probably derived. There are two main types of hydrothermal tungsten deposits, scheelite-bearing skarns and wolframite-bearing quartz veins. *Scheelite-bearing skarns* are found where granitic intrusions are in contact with limestones. These deposits have a complex metal content, including molybdenum, copper, lead, iron, and even bismuth. The largest skarn deposits are in southeastern China (Shizhuyuan, Figure 8-22), Russia (Tyrnyauz and Vostok-2), and northern Canada (Cantung). Other important producing skarn deposits include Sangdong, South Korea, Pine Creek, California, and King Island, Australia (Figure 9-23). *Wolframite-bearing veins* occur in swarms of veins that cut granite and nearby, noncarbonate sedimentary rocks. The largest deposits of this type, in Jiangxi Province, China, consist of hundreds of cross-cutting, 1-centimeter to 1-meter veins known as a stockwork, which cut a granite intrusion. At the Panasqueira deposit in Portugal, veins surround *cupolas* or protrusions at the top of a granite intrusion. Many tungsten veins are closely associated with tin and molybdenum, which are important co-products at Panasqueira, Portugal and Mount Pleasant, New Brunswick, respectively. Tungsten is an important by-product at the Climax porphyry molybdenum deposit in Colorado and at the Bolivian tin deposits.

Tungsten also forms deposits in the low-temperature environment, although these have not been important metal sources. It has been found in hot-spring deposits and the Searles Lake evaporite brines that contain 70 mg/liter dissolved tungsten, with an estimated resource of 61,000 tonnes of metal.

Tungsten production and reserves are about equally divided between scheelite skarns and wolframite veins. China dominates world production, with Russia a distant second and the rest of the world supplying only a quarter of total output. Most tungsten mining from these deposits involves small-scale underground operations with the attendant safety concerns. Ores are beneficiated by standard methods, although hand-sorting of ore minerals is still employed widely in China. Tungsten concentrates are processed by hydrometallurgical methods to form ammonium paratungstate (APT) [5$(NH_4)_2O \cdot 12WO_3 \cdot 5H_2O$], which can

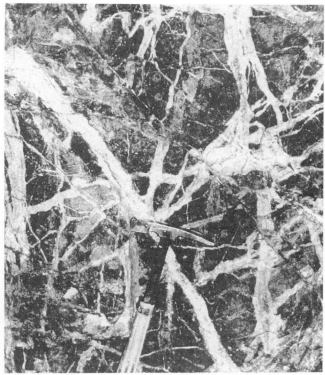

A.

B.

FIGURE 8-22

Shizhuyuan Mine, Hunan Province, China, the largest tungsten deposit in the world. **A**. Skarn (dark) replacing limestone and cut by quartz veins (white) with tungsten, molybdenum, bismuth, and tin minerals. **B**. Beneficiation plant at Shizhuyuan (courtesy of James E. Elliott, U.S. Geological Survey).

be converted into tungsten metal or ferrotungsten for use in steel alloys (Smith, 1990).

Although considerable tungsten is sold in upgraded or intermediate forms, concentrate is still a major trade item and much moves from ore-producing countries to steel-producing countries that lack tungsten ore (Figure 8-23). Even though China is a significant consumer of tungsten, it is the dominant concentrate exporter and its aggressive marketing has caused most other tungsten mines around the world to close or cut production. The main concentrate importers can be recognized by a glance at world steel production, although Japan ranks lower than expected because it has domestic production of tungsten. Total world tungsten reserves contain 2.3 million tonnes of metal, about half of which is in China. North American reserves are sizeable thanks to the large deposits in the Yukon. Large reserves are also found in Russia and Australia,

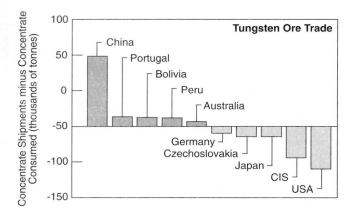

FIGURE 8-23

World tungsten trade for 1990 expressed as tungsten content of ore and metal. Countries plotting above the line are net exporters of concentrate; those below the line are net importers of concentrate and, in most cases, exporters of metal (data from U.S. Bureau of Mines and Metal Statistics).

but European reserves are small (Anstett et al., 1985; Rabachevsky, 1988).

Other Ferroalloy Metals

Columbium (Niobium)

Columbium is an unfamiliar element. Not only is it used in small amounts and in inconspicuous applications, but it also goes by a pseudonym. Outside the United States, the proper name for this element is *niobium,* but columbium remains the industry term in the United States. The main demand for columbium is in HSLA steels, where the addition of as little as 0.1% improves the mechanical properties of carbon steel. Columbium is also used in larger amounts in stainless steel and nickel-based alloys such as *Inconel.* Most columbium is traded in the form of columbium mineral concentrates or as *ferrocolombium* for direct use in steelmaking (Cunningham, 1985, 1990). Annual world columbium production of almost 15,000 tonnes of metal is worth about $100 million.

Columbium occurs in nature largely as the oxide mineral *pyrochlore,* which has a very complex composition best represented as $A_2B_2C_7$, in which A is any combination of sodium, calcium, and cerium, B is columbium and titanium with minor tantalum, and C is oxygen, fluorine, or hydroxyl (OH). Much smaller amounts of columbium are found in *columbite.* Although tantalum is found in both minerals, its concentrations are much lower than those of columbium and it is not usually recovered from the same deposits. The

largest columbium deposits consist of veins and disseminations of pyrochlore in *carbonatites* and quartz-poor, felsic intrusive rocks that were emplaced as shallow intrusions in rifted areas of continents. The only carbonatite complexes from which important columbium production is derived at present are near Chicoutimi, Quebec, and in Zaire. Production in Brazil comes from a carbonatite-syenite complex that has been weathered leaving pyrochlore in the regolith. Much smaller columbium deposits take the form of columbite in veins and pegmatites associated with granites, and small amounts of columbite are produced from weathered rocks of this type in the Jos Plateau of Nigeria (Moller et al., 1989).

World columbium reserves contain about 3.5 million tonnes of metal. Although the reserves are large, their concentration in Brazil is of some concern to European and Japanese markets, but less so to North American users in view of the proximity of the Quebec deposits.

Tellurium

Tellurium is another of the HSLA metals. Addition of only 0.04% tellurium to carbon steel makes it easier to machine. It is also used in alloys with copper, tin, aluminum, and lead (Jensen, 1985). Total world tellurium production has a value of less than $10 million, making it the least important of the ferroalloy metals.

Tellurium forms a large family of minerals known as the tellurides, in which tellurium takes the place of sulfur. The best known of these minerals are the precious metal tellurides, including *calaverite,* which takes its name from Calaveras County, California, the site of Mark Twain's story of the famous jumping frog and an area of gold-telluride–bearing quartz veins. Although calaverite and other telluride minerals are found in many precious metal deposits, particularly the famous Emperor Mine in Fiji, they are rarely a source of tellurium because the ores are not processed in a way that permits its recovery. Instead, almost all of world tellurium production comes from refining copper metal. Tellurium is present in most copper ores both in solid solution in copper sulfides and as minor amounts of copper tellurides. When the ores are beneficiated and smelted, tellurium follows copper and is not removed until the electrolytic refining step (Sindeeva, 1964).

Tellurium is not known to be essential to the human body. In soluble forms it is toxic in moderate to large concentrations, producing first a strong garlic breath, followed by more serious symptoms involving nausea and unconsciousness. Reserves of tellurium depend on world copper ore reserves and the way in which they are smelted and refined. Estimates made by the U.S. Bureau of Mines suggest that these are sufficient for at least 200 years at present production rates.

THE FUTURE OF FERROUS METALS

Our strong global interdependence for essential steel-making minerals is undeniable. Appetites are strongest and domestic supplies weakest in North America, Europe, and Japan. The United States and Canada together are important suppliers of four ferroalloy metals, but depend on outside sources for chromium and manganese. Western Europe and Japan depend on imports for all of the ferroalloy metals, and Japan even supplies its iron ore needs through imports.

The suppliers that have been of continuing concern to strategic planners in Japan, North America, and western Europe are South Africa and the CIS countries, which are major producers of all six of the important ferroalloy metals (not counting silicon, for which reserves are not a problem anywhere). Although changing political alignments in the CIS countries might make some of these reserves more available to western and Japanese markets, transportation costs will still make it difficult for them to compete with South African producers, which are closer to ocean ports. The highly efficient and dependable operation of South African industry has earned it an impressive list of customers over the last few decades.

Partly because of the failure of the Rhodesian (Zimbabwean) chromite embargo, most of these customers did not embargo imports of these strategic metals during the apartheid government in South Africa, although less important imports such as steel were limited. With the advent of new government policies in South Africa come questions about the future of mining, particularly during labor unrest or political instability. It is no comfort to users of these metals that the CIS countries could experience similar difficulties at the same time.

Although steel and ferroalloy metals will probably not become growth industries, their future is a solid one, and all industrial countries must take the necessary steps to assure access to adequate supplies. The main structural change that will affect these steps is *forward integration* in which supplier countries produce ferroalloys rather than simply shipping mineral concentrates to their customers. The response to this in customer countries will probably be to dismantle domestic plants for the production of ferroalloy products from imported ore and concentrates. It should be kept in mind that this step will produce a level of dependence on imports that is considerably more serious than one based on imports of ore concentrates. In times of supply emergency, it is often possible to find small supplies of domestic ore. It will be much more difficult to locate emergency supplies of ferroalloy products.

9

Nonferrous Metals

LIKE IT OR NOT, THE DREARY TERM *NONFERROUS METALS* IS widely used for the many metals that are not closely related to steelmaking. They are divided further into the *light metals,* aluminum, magnesium, titanium, and beryllium, the *base metals,* copper, lead, zinc, and tin, and the *precious metals,* gold, silver, and platinum-group elements. Most of the remaining metals, including antimony, arsenic, bismuth, cadmium, germanium, hafnium, indium, mercury, rare earths, rhenium, selenium, tantalum, thallium, and zirconium are used largely in chemical, electrical, and other industrial applications, and are referred to here as the *chemical-industrial metals.* In this chapter, we will deal with the light metals, base metals, and chemical-industrial metals. Precious metals and gems are discussed in the next chapter.

The total value of annual world light, base, and chemical-industrial metal production, about $70 billion, is over twice that of the ferroalloy metals, but far behind that of steel. The dominant metals are aluminum and copper, followed by zinc and lead in second place, then tin, magnesium, and titanium, and finally the chemical-industrial metals (Figure 9-1A). Prices of the light and base metals have increased cyclically, although titanium and tin prices have not recovered from peaks reached in the early 1980s (Figures 9-1B,C). Primary (mine) production of the light metals is growing faster than steel production, whereas base metal production is growing more slowly, with lead and tin

showing essentially no growth in primary production since 1960 (Figure 9-1D).

LIGHT METALS

The light metals are so named because they have lower densities than most metals (Polmear, 1981). In contrast to densities of 7.87 g/cm^3 for iron and 8.96 for copper, densities of the light metals are only 1.74 for magnesium, 2.7 for aluminum, and 4.51 for titanium. All three light metals are relatively strong, but aluminum and magnesium markets are restricted by their low melting temperatures of 650°C and 660°C, respectively, in comparison to 1,535°C for iron. Titanium, which melts at 1,678°C, competes in high temperature applications. Their combination of light weight and strength has generated strong demand for the light metals, with annual production valued at almost $30 billion.

These metals have high crustal abundances. Aluminum is the third most abundant element in the crust, magnesium is seventh, and titanium is ninth. In that respect, they are similar to iron, the fourth most abundant element in the crust, with which they are sometimes grouped as the *abundant metals.* That the light metals came into widespread use so much later than iron and the scarcer base metals reflects the difficul-

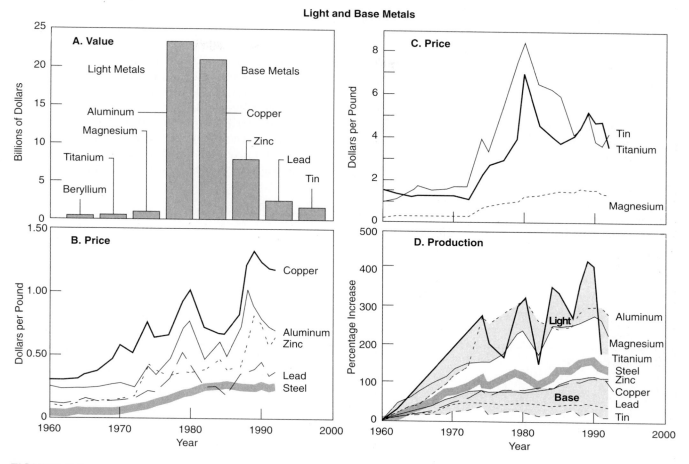

FIGURE 9-1

A. Value of annual world production for major base and light metals.
B and C. Historical price trends for nonferrous metals (note that metals are not divided into base- and light-metal groups). **D.** Growth of world base- and light-metal mine (primary) production (except titanium, which is for United States sponge metal production). Reference to Figure 4-2 shows that none of these commodities has increased in price as rapidly as inflation. Data from U.S. Bureau of Mines with special assistance from Larry D. Cunningham, Deborah A. Kramer, and Langtry E. Lynd.

ties involved in liberating them from their ores, a factor that continues to constrain the growth of light metal markets.

Aluminum

Aluminum was first isolated as a metal in 1825, five millennia after iron began to be used. For several decades after that, it was so difficult to produce that it was essentially a precious metal, used only in jewelry and other ornaments. Although its combination of light weight and high strength was of great interest even then, aluminum did not become available in large amounts until the Hall-Heroult electrolytic process was developed in 1886 (Burkin, 1987). This

process, which involves the use of large amounts of electricity as discussed later, started aluminum on a path of steady growth to the present annual production of 18 to 20 million tonnes worth about $25 billion in 1992.

Geology of Bauxite and Other Aluminum Ore Deposits

Aluminum makes up over 8% of Earth's crust and is a major constituent of many common minerals. Although it can be recovered from common minerals, most production comes from the aluminum oxide minerals *diaspore, boehmite,* and *gibbsite,* which are the major constituents of *bauxite,* the ore of aluminum.

Over 100 million tonnes of bauxite, valued at almost $2 billion, are produced annually. Deposits containing boehmite are common in Europe and northern Asia, whereas those containing gibbsite are common in northern South America. Deposits in Guinea and Jamaica contain both minerals. Bauxite contains 40% to 60% Al_2O_3 (the chemical form in which aluminum abundances are reported), with the rest made up of clay minerals, iron, and titanium oxides, and quartz and other forms of silica.

Bauxite is actually a special aluminum-rich *laterite*. It forms by intense weathering that dissolves away most other elements in the rock, leaving a residuum enriched in aluminum (Patterson et al., 1986; Bardossy and Aleva, 1990). The formation of aluminous laterites and bauxite deposits requires abundant rainfall to dissolve unwanted rock constituents, high temperatures to speed the dissolution reactions, and high relief so that the water will drain away. Because the weathering process is so slow, the formation of large bauxite deposits is favored by extended periods of geologic stability, which allow weathering to affect the same rock for a long time. These conditions are met in parts of the broad equatorial band of tropical and subtropical climates (Figure 9-2). Not all laterites are aluminous, however. Iron, nickel, and cobalt also concentrate in laterites, with the type of metal enrichment depending largely on the type of bedrock that is weathered. Enrichment of aluminum is favored where weathering affects rocks with a high aluminum content, such as felsic igneous rocks and shales (Figure 9-3).

The absence of iron in bauxites is curious. It is present in many rocks that are weathered to produce bauxite and iron-rich laterites are common in many tropical areas. Its absence in bauxites requires special conditions that favor the deposition of aluminum while iron remains dissolved. This appears to take

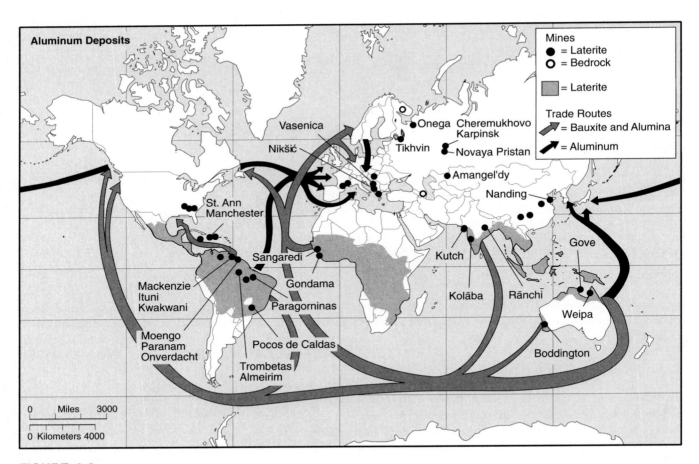

FIGURE 9-2

Distribution of bauxite and bedrock aluminum deposits showing worldwide distribution of lateritic soils and general trade routes for both bauxite and aluminum metal (compiled from U.S. Bureau of Mines; Organization for Economic Cooperation and Development; Patterson et al., 1986; Hosterman et al., 1990).

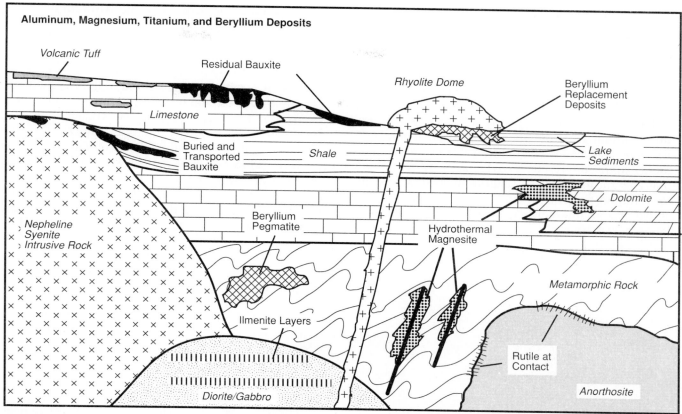

FIGURE 9-3

Schematic illustration of bauxite (black), titanium (vertical lines), magnesium (stipple), and beryllium (cross hatch) deposit types. As shown here, bauxite is *residual* if it is still in the location where it formed by weathering, or *transported* if it has been eroded and deposited as a sediment in another location.

place where the decay of abundant vegetation consumes oxygen in groundwater, thus keeping iron in the more soluble ferrous state discussed in the previous chapter. In areas with less vegetation, ferric iron dominates and iron oxides are deposited in the laterite (Petersen, 1971).

Other less abundant elements also concentrate in bauxite. Much of the world's *gallium,* an element with geochemical properties similar to aluminum, is a by-product of the processing of bauxite to form alumina. A serendipitous example of bedrock contributions to bauxite deposits is Boddington in Western Australia (Figure 9-2), which developed over a hidden *gold* vein deposit in granitic rocks and produces important amounts of gold. Bauxite is also enriched in *titanium,* although it has not yet been recovered economically.

Most bauxite deposits can be found using conventional geologic guidelines. The large deposits around Weipa, Australia, and in Suriname and Guyana (Figure 9-2) formed on sediments rich in kaolinite, an aluminum-rich clay mineral. Other important depos-

its in the Darling Range of Western Australia and Guinea formed by the weathering of felsic igneous and metamorphic rocks. However, some very important deposits do not obey the rules. The belt of bauxite extending through Hispaniola and Jamaica is found on limestone, a rock consisting almost entirely of calcite and dolomite, which do not contain aluminum. The aluminum in these deposits probably came from clay minerals or volcanic tuff that was mixed with the limestones or deposited on top of it (Comer, 1974). Smaller but generally similar deposits are found in Hungary and central Kazakhstan. Although most bauxite deposits are eventually destroyed by erosion, some ancient ones, known as *paleobauxite,* are preserved in older sedimentary rocks. Most of the bauxite in southern Europe, Kazakhstan, and Russia is probably of this type, as are deposits in the Magnet Cove area of Arkansas (Figure 9-2). In some areas, these deposits were eroded, transported, and deposited as bauxite-rich sediment that can also be mined.

As might be expected, bauxite deposits are scarce or absent in cooler climates and in areas where Pleistocene glaciation has scraped the regolith away. Many countries in this situation, including the United States, have considered developing an industry on aluminum-rich minerals or rocks (Bliss, 1976; Patterson, 1977; Hosterman et al., 1990). The only country to have done so in a large way is Russia, which produces aluminum from the aluminum-rich minerals, *nepheline* and *alunite* (Figure 9-3). Nepheline is commonly found in syenite, a quartz-poor intrusive rock that also forms important laterite deposits in Arkansas. Alunite forms in hydrothermal alteration zones above and around some gold deposits, as discussed in the next chapter.

Production of Bauxite and Aluminum

Aluminum ore is produced in 26 countries, almost exclusively from bauxite (Sehnke and Plunkert, 1989). Because most bauxite forms flat layers at or near the surface, it is extracted by strip-mining type methods (Figure 9-4A), although underground methods have been used for deeper bauxite in Arkansas, France, Greece, Hungary, and Kazakhstan. Much of world bauxite production is shipped to a different location for further processing, usually after drying to remove excess water (McCawley and Baumgardner, 1985).

Bauxite is converted to aluminum in a two-step process (Figure 9-5). In the first step, known as the *Bayer process*, a solution of caustic soda (NaOH) is used to

A.

B.

FIGURE 9-4

A. Strip mining of bauxite at MacKenzie, Guyana. Uppermost sand layer is removed by bucket-wheel excavator (upper left) and harder, underlying sandy clays and bauxite are removed by draglines (right and lower left). Mining proceeds from right to left, and waste from new stripping is piled in mined-out area on right (courtesy of Thomas L. Kesler, from Kesler, 1976). **B**. Aluminum refinery at head of deep fjord in Sunndalsöra, Norway, uses hydroelectric power to process imported bauxite.

PLATE 1-1 Hydrothermal alteration (light zones) in wallrock around quartz veins (white) at Mt. Charlotte mesothermal gold mine, Kalgoorlie, Australia.

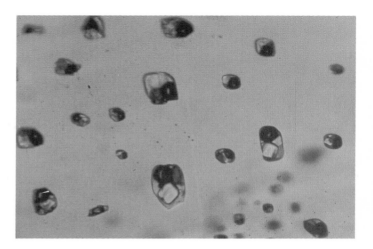

PLATE 1-2 Fluid inclusions inside a crystal of transparent quartz from the Granisle porphyry copper deposit, British Columbia, as seen through a microscope (each inclusion is about 30 micrometers in diameter). The inclusions contain transparent liquid water, a dark vapor bubble, and several daughter minerals (including red hematite). The inclusion was originally trapped as a liquid, but during cooling of the crystal, the liquid shrank to form the vapor bubble and deposit the daughter minerals. If the quartz crystal is reheated under the microscope, the vapor bubble will shrink and the daughter minerals will dissolve, providing information on the temperature and pressure at which the liquid was trapped in the inclusion.

PLATE 1-3 Bitumen coating a cavity in dolomite from the Pine Point Mississippi Valley-type lead-zinc deposit, Northwest Territories, Canada.

PLATE 1-4 Breccia zone in which fragments of limestone and dolomite (dark) have been cemented together by large white dolomite crystals that were deposited from basinal brines (Coy Mississippi Valley-type zinc mine, Tennessee).

PLATE 1-5 (*left*) Felsic volcanic dome at Kirishima Volcano, Japan, emitting water vapor and magmatic gases. Yellow sulfur deposited from these gases coats rocks of the dome.

PLATE 2-1 Soil in humid tropical climate has dark O (hidden by grass) and A zones with organic matter overlying light E zone and darker, orange B zone. This soil is formed by the weathering of copper-rich rock and is enriched in copper.

PLATE 2-2 Copper dissolved in water from a puddle on a weathering copper deposit coats a steel knife blade. This water drains from the deposit to form a trail of copper-rich water and sediment several kilometers long.

PLATE 2-3 Gravels carried into the area by glaciers, such as seen here overlying an iron deposit in Minnesota, cover many parts of northern North America making mineral exploration difficult. Some geophysical methods can be used to "see through" these gravels.

PLATE 2-4 Detailed geophysical (electromagnetic) exploration for metal deposits uses large cable arrays that are laid out across rough terrain.

PLATE 2-5 (left) The final stage of mineral exploration involves drilling to obtain samples of buried rock, either as chips or cylinders called core. These core samples are cut by dark veins of the zinc mineral, sphalerite.

PLATE 3-1 Open-pit mines on deep ore bodies must remove overburden and place it in piles to reach the ore buried beneath. This mine at Boron, California, extracts ore from a borax deposit that is 100 to 200 meters below the surface. Overburden is piled to the right of the pit, the mill for treatment of ore is just behind the pit, and tailings and other mine waste form ponds in the background (*courtesy of C.M. Davis, U.S. Borax and Chemical Corporation*).

PLATE 3-2   Coal strip mine at Falkirk, North Dakota. Mining procedes from right to left. Waste stripped from ore by large dragline is piled on the right, exposing black coal seam. Reclaimed land is outside photo to right (*courtesy of Chuck Meyers, Office of Surface Mining Reclamation and Enforcement*).

PLATE 3-3 This house near Secunda, South Africa, is above a mined-out longwall coal mine from which the entire coal layer was removed. The land surface was lowered by 1.25 meters without significant damage to the structure.

PLATE 3-5 Close-up view of reclaimed wetland on the site of a former phosphate mine in central Florida (*courtesy of L.A. Adams, IMC Fertilizer Corp.*).

PLATE 3-4 Overview of Fort Meade phosphate mine, Florida, showing active mining in back left corner and two settling ponds where land is being reclaimed (lakes in center). Most other land in the photo has already been mined and reclaimed, usually as wetlands, the form in which it began (*courtesy of John Paugh, Cargill Fertilizer, Ft. Meade, FL*).

PLATE 4-1 Contour coal mines.
Above: Contour mines in eastern Tennessee, showing benches that were cut into the hillside to remove coal layers.
Right: Regraded and revegetated contour mine in eastern Kentucky (*courtesy of Chuck Meyers, Office of Surface Mining Reclamation and Enforcement*).

PLATE 4-2 Open-pit mines.
Above: Aerial view of McLaughlin gold mine, northern Napa Valley, California, showing pit in background and overburden piles in foreground, which are being reclaimed as mining progresses (note revegetation of lower two levels of the piles). Older completely revegetated overburden pile is barely distinguishable immediately in front of mine.
Right: Early stage of revegetation program, showing grasses that provide erosion control
(*courtesy of Raymond E. Krauss, Homestake Mining Co.*).

PLATE 5-1   Ice-resistant rigs for Arctic exploration (*courtesy of Jim Rennie, Gulf Canada Resources, Ltd.*).

Kulluk is a conical vessel that is towed into position and held there by anchors. It can drill during spring and fall.

Molikpaq, seen here in the Beaufort Sea in February, is a caisson system that is filled with gravel ballast and lowered onto a gravel berm built on the sea floor. It can drill through the entire year. Note the buildup of pack ice on the upstream side of the rig.

PLATE 5-2   Oil spill and cleanup at Smith Island in Prince William Sound, Alaska (*photos courtesy of Exxon Corporation*).

May 2, 1989 during cleanup.

June 6, 1992 after cleanup. Black bands along rock promontory are natural lichen, not residual oil.

PLATE 5-3 Stockpiles of sulfur recovered from sour gas in Alberta, awaiting shipment from the harbor at Vancouver, British Columbia (*photo courtesy of J. Cadger, Sultran Ltd.*).

PLATE 6-1 Syncrude tar sand operation, Fort McMurray, Alberta, Canada, showing mine in foreground and processing plant in background (*courtesy of Imperial Oil and Syncrude Canada, Ltd.*).

PLATE 6-2 Sandstone uranium ore containing the bright yellow ore mineral, carnotite, which formed when uranium in the sandstone was oxidized by groundwater.

PLATE 6-3 Conglomerate gold ore from the Witwatersrand area, South Africa, showing large pebbles of quartz surrounded by a matrix containing pyrite, gold, and uranium minerals.

PLATE 6-4 Overview of the central part of the Wairakei geothermal field, New Zealand. Pipelines carry geothermal steam to power the plant in the background, with excess steam escaping from the wells and drainage channels.

PLATE 6-5 Geothermal reservoir in the Salton Sea, California, is drilled with a rig similar to that used to explore for oil and gas on land (left). Drilling is carried out from the elevated platform using 60-ft rod segments that are stacked on the right side of the tower. A blow-out preventer (right) below the platform prevents the release of pressurized fluid from the well.

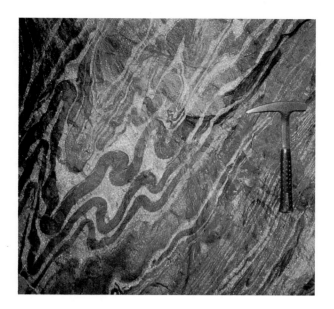

PLATE 7-1 (A) Sedimentary exhalative (sedex) lead-zinc ore, showing folded layers of galena (gray) and sphalerite (brown) in the Sullivan deposit, British Columbia.

PLATE 7-1 (B) Banded iron formation (BIF), showing alternating layers of hematite (gray) and silica (red) from Temagami iron deposit, Ontario. Beneficiation of this ore produces black balls, known as pellets (lower left), which consist of iron oxide grains held together by clay.

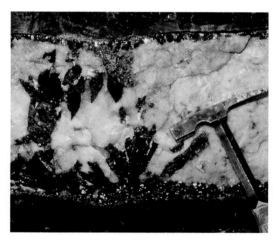

PLATE 7-1 (C) Vein of quartz (white) with crystals of the tungsten ore mineral wolframite (dark) from Dajishan mine, China (*courtesy of James E. Elliot*).

PLATE 7-1 (D) Vein of quartz (white) with gold (yellow) from the Pamour greenstone gold deposit, Porcupine camp, Ontario (*courtesy of E.H.P. Van Hees*).

PLATE 7-1 (E) Veinlets of chalcopyrite that make up the feeder zone beneath the Millenbach volcanogenic (exhalative) massive sulfide deposit, Noranda area, Quebec.

PLATE 7-1 (F) Intersecting veinlets of quartz, pyrite, and chalcopyrite in porphyry copper ore from the Granisle deposit, British Columbia.

PLATE 8-1   Polaris mine on Little Cornwallis Island, Northwest Territories, Canada is the northernmost active mine in the world. It produces concentrates year-round and stockpiles them for a 12-week shipping season that begins in late August. Employees commute from southern Canada and work ten weeks on and two weeks off. Northern native employees can work six weeks on and four weeks off in order to continue traditional pursuits such as hunting and fishing (*courtesy of K.A. Barker, Cominco Ltd.*).

PLATE 8-2   Aluminum potline at Portland, Australia, consisting of 204 cells, each capable of producing about 2 tonnes of metal per day. Exhaust ducts can be seen on right wall (*courtesy of Alcoa*).

PLATE 8-3   Molten copper flowing out of smelter (right) at Okiep, South Africa, and being cast into ingots for later refining (*courtesy of Andrew Lanham, Gold Fields of South Africa*).

PLATE 8-4   Molten gold pouring from a furnace into doré molds at West Driefontein, South Africa.

PLATE 8-5   Doré bars at Pueblo Viejo, Dominican Republic, ready for shipment to a refinery. When composed of pure gold, each bar has a value of more than $250,000 at 1993 prices.

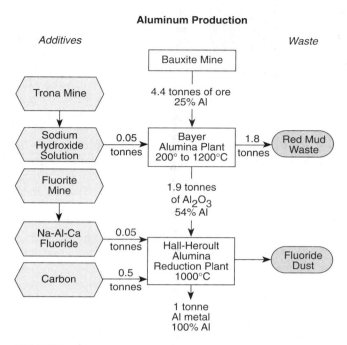

Aluminum Production

FIGURE 9-5

Schematic illustration of typical aluminum production process. Additional materials necessary for the process are shown on the left and waste products are shown on the right (compiled from information of the U.S. Bureau of Mines; Polmear, 1981; Burkin, 1987).

leach aluminum from the bauxite. The dissolved aluminum is then precipitated as $Al_2O_3 \cdot 3H_2O$, which is heated, or *calcined,* to drive off the water, leaving pure *alumina* (Al_2O_3). In the second step, known as the *Hall-Heroult process,* alumina is dissolved at 950°C in a carbon "pot" filled with molten *cryolite* and other fluo-

rides of Ca and Al. Electric current is passed through this molten mixture between an electrode and the walls of the pot, reducing aluminum ions to molten aluminum metal that is cast into ingots or other forms (Plate 8-3).

Aluminum production requires an enormous amount of energy (Table 9-1). Production of alumina uses largely heat and can be carried out in areas with abundant fuel of any sort, but aluminum reduction, which accounts for about 80% of total energy consumption, depends on electricity (Boercker, 1979). In North America, aluminum and steel production use about the same amount of electricity, although almost 30 times more steel is produced.

The need for cheap electricity caused most early aluminum reduction facilities to be located in areas of abundant hydroelectric power, such as the plant near Invercargill, New Zealand, which processes Australian bauxite. In areas such as the Columbia and Tennessee river valleys in the United States, population pressures have increased hydroelectric energy costs driving more recent plants into areas remote from population centers, such as northern British Columbia, Norway, and Iceland (Figure 9-4B). The Middle East also has surplus power in the form of natural gas, and aluminum reduction plants have been constructed in Dubai and Bahrain to use this energy. Further growth in this area can be expected unless liquified natural gas exports consume surplus gas. Some countries with limited energy, such as Japan, have curtailed aluminum production and increased imports (Figure 9-6).

The overriding importance of energy in aluminum production creates considerable debate about the real cost of power, and the nature and degree of alleged

TABLE 9-1

Energy required to produce major metals. These figures will vary greatly for individual operations and are useful largely for comparison between metals (from *U.S. Bureau of Mines Minerals Facts and Problems,* 1985).

Metal	Million Btu/tonne	Comments
Aluminum	125 to 161	bauxite to aluminum metal
Arsenic	3	copper-smelter flue dust to AsO_3
Chromium	13	chromite to ferrochromium
Copper	30 to 40	copper concentrate to metal
Steel	23 to 25	pellets to steel
Lead	30	includes mining
Magnesium	110 to 400	depends on process
Mercury	0.4	includes mining
Silicon	31 to 35	ferrosilicon production
Tin	190	includes mining
Titanium	410 to 460	ore to titanium metal
Zinc	36 to 48	smelting only

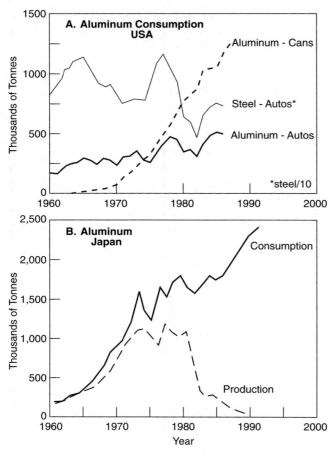

FIGURE 9-6

A. Use of steel and aluminum in automobiles and aluminum in beverage containers in the United States (from data in Tilton, 1990). **B.** Decline in aluminum production in Japan caused by increased energy costs. Note that consumption has continued to rise and is supplied by imports (after Peck, 1988, Figure 4-1, Copyright 1988, Resources for the Future and U.S. Bureau of Mines).

subsidies provided to power production in competing countries (OECD, 1983; Peck, 1988). Estimates made by the U.S. Bureau of Mines in 1989 showed that energy costs varied from lows of 10.9 and 22.8 cents per kilogram of aluminum in Canada and Norway, respectively, to a high of 66.8 cents in India. Recently proposed energy taxes would add several cents to the cost of a pound of aluminum. Energy costs make recycling a particularly important part of global aluminum production because aluminum metal can be remelted and reformed for only about 5% of the energy needed to produce metal from bauxite. Recycling of containers has been most successful, especially where "bottle

bills" require that cans be returned. Recycling currently accounts for about 30% of U.S. and European aluminum consumption.

Aluminum production presents local as well as global environmental concerns. Waste from the conversion of bauxite to alumina includes airborne particulates and tailings containing minerals that did not dissolve. The leach solution is caustic and must be evaporated to dryness before the tailings can be disposed of in piles, a routine practice for alumina producers. These emissions are dwarfed by natural aluminum-bearing sediment and by alumina from the use of *alum* ($Al_2(SO_4)_3 \cdot 18H_2O$) in purification of municipal water supplies. The conversion of alumina to aluminum metal yields particulate and gas emissions. The particulates consist of alumina dust, carbon dust from the electrodes, and particles of fluoride compounds, and the gas emissions consist largely of HF and other fluorine-bearing gases, CO_2, CO, and SO_2. About 60 to 90 kilograms of fluoride gases and carbon and alumina dust are produced for each tonne of aluminum. Most of this is caught by scrubbers and dust collection systems, although as much as 2 tonnes of CO_2 escape for each tonne of aluminum produced (OECD, 1973). Scrubbers also do not stop the fluorocarbons, CF_4 and C_2F_6, which are produced in amounts of 1.5 to 2.5 kilograms per tonne of aluminum by reaction between fluoride and carbon in the cell. Fluorocarbons are among the most potent greenhouse gases being emitted by industry today, about 1,000 times more powerful than CO_2, and could account for as much as 1.7% of global anthropogenic greenhouse effects (Abrahamson, 1992). Little is known about the natural sources of these gases.

Although you might think that aluminum would not be toxic to organisms in view of its high crustal abundance, that does not appear to be the case. High concentrations of dissolved aluminum have been correlated with fish mortality, although the exact form of aluminum that is toxic is not known. Aluminum is also a known neurotoxin in some vertebrates. High levels of aluminum in drinking water and blood have been correlated with Alzheimer's disease in humans, although it has been suggested that the aluminum observed in these tests resulted from sample contamination (Fergusson, 1990; Burros, 1991; Romero, 1991; Kolata, 1992). Aluminum does not cause large-scale problems because it is relatively insoluble in the normal weathering environment (which accounts for the fact that it concentrates in bauxite). It dissolves in larger amounts in acidic water, however, and dissolved aluminum is of concern in acid lakes (Figure 11-5).

Aluminum Markets, Reserves, and Trade

About 90% of world bauxite production is converted to aluminum metal, with the remainder used to produce aluminum oxide abrasives, refractory materials, and chemical compounds. In addition to the low ratio of weight to strength, which makes aluminum so useful in the construction and transport industries, it is also valued for its high electrical and thermal conductivities. In the transportation sector, automobile manufacturing accounts for about 50% of consumption and is growing steadily (Figure 9-6). Airframe manufacture, which accounts for only about 20% of the transportation market, is also a growing market. Aluminum markets are growing largely at the expense of steel, particularly in packaging. Aluminum has also replaced copper in high-voltage electric transmission lines where there is a premium on light weight. Aluminum use patterns differ from country to country, with the greatest difference seen in packaging, which accounts for almost 30% of the U.S. market but less than 10% of the Japanese market (McCawley and Baumgardner, 1985).

World bauxite reserves of about 23 billion tonnes are largely in Guinea, Australia, Brazil, and Jamaica. U.S. resources are small, and only France and Hungary among European countries have significant reserves. The only CIS country with significant reserves is Kazakhstan. Aluminum resources are large and consideration has also been given to recovering aluminum from fly ash produced by coal-fired power plants, which contains enough aluminum to meet present world demand.

About 75% of world bauxite or alumina and about 35% of world aluminum production are traded on world markets, largely by sea. The fact that only 26 countries produce bauxite and alumina, whereas 42 produce aluminum, confirms that it is an important item of world trade. Principal bauxite exporters, some of which also produce alumina and aluminum, are Australia, Guinea, Jamaica, Brazil, and Suriname (Figure 9-2). The largest importers of bauxite and alumina are the United States, Canada, CIS countries, Norway, and Germany, all of which then export aluminum metal (Figure 9-7). Principal importers of aluminum metal are Japan and Taiwan, which receive most of their supply from Australia, and the European countries, which receive most of their supply from Norway and Canada. Since at least 1980, world aluminum capacity has exceeded demand significantly. Appearance of CIS production in western markets has exacerbated this situation and is expected to put near-term pressure on aluminum prices.

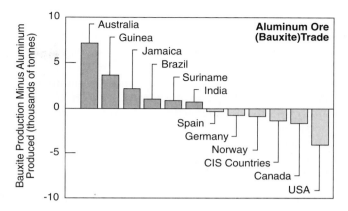

FIGURE 9-7

World trade in bauxite (aluminum content of) and aluminum metal for 1991. Countries plotting above the line produce more bauxite than aluminum and are bauxite exporters; those plotting below the line are net bauxite importers and may or may not export aluminum (data from U.S. Bureau of Mines and Metal Statistics).

Magnesium

Although magnesium is an important metal, its nonmetallic forms account for 70% of total consumption. The principal nonmetallic use is for *refractory* linings in furnaces and molds in the iron and steel industry, which consist of magnesium oxide or silicate. Magnesium carbonates, hydroxides, and chlorides are also used in the manufacture of rubber, textiles, and chemicals. Nonmetallic magnesium is produced and traded in so many different forms that it is difficult to place a simple value on them, although they are worth at least a few billion dollars. The 350,000 tonnes of *magnesium metal* produced annually throughout the world are worth slightly more than $1 billion. Magnesium metal is used as an alloy to harden aluminum and increase its resistance to corrosion, largely in aluminum beverage cans, and as metal components in the automotive industry, where it is valued for its light weight. It is also used as the reducing agent in the production of other metals such as titanium and zirconium from their ores. The real hope for magnesium, however, is as components of motors, seats, and other automobile parts, a growing market (Kramer, 1985; Couturier, 1992).

Magnesium is produced from a wide range of natural sources, which have been exploited in order of declining grade (Wicken and Duncan, 1983). First to be used were deposits of *magnesite*, a magnesium carbonate, and *olivine*, a magnesium silicate, which are found largely in slabs of the mantle that are exposed in obduction zones along convergent tectonic margins. Olivine is a common mineral in ultramafic rocks and most magnesite formed where olivine was altered by CO_2-rich hydrothermal solutions. Magnesite is also found where magnesium-rich hydrothermal solutions replaced limestone or dolomite in sedimentary basins. Olivine deposits are mined in North Carolina, and deposits of magnesite are found at Gabbs, Nevada, Semmerming, Austria, and on the island of Euboea, Greece (Figure 9-8). Attention has also been given to the possibility of producing magnesium from tailings at asbestos deposits, which are found in ultramafic rock. As the demand for magnesium and its compounds grew, attention shifted to *dolomite,* the calcium-magnesium carbonate mineral, even though it contains less magnesium than magnesite. Small amounts of magnesium were also produced from evaporite minerals, which shifted attention to dissolved sources of magnesium. The first such source was *brine* in the central part of the Michigan Basin, which contained about 3% Mg. As technology improved, it became possible to process solutions with lower concentrations of magnesium, such as the

waters of the Great Salt Lake and, ultimately, *seawater,* which contains only 0.13% Mg. At present, seawater and brines account for 60% of U.S. production (Figure 9-8), making it the only metal to be recovered largely from seawater.

Although olivine is used directly as a refractory material, most other forms of magnesium must be processed before they can be used. Magnesite and dolomite are converted to MgO, a refractory compound, by calcining in a kiln to drive off CO_2. Magnesium sulfate (epsom salt) is produced by dissolving MgO in sulfuric acid, and magnesium carbonate is precipitated from this solution by the addition of sodium carbonate. Seawater processing can yield either magnesium chloride or magnesium hydroxide [$Mg(OH)_2$]. Magnesium metal is produced by an electrolytic process in which molten magnesium chloride is put into a 1,292°C bath with carbon electrodes, or by reacting calcined dolomite with silicon. Neither of these processes is as energy intensive as aluminum or titanium production (Table 9-1). Magnesium metal is very easy to shape and machine, although it produces dust that

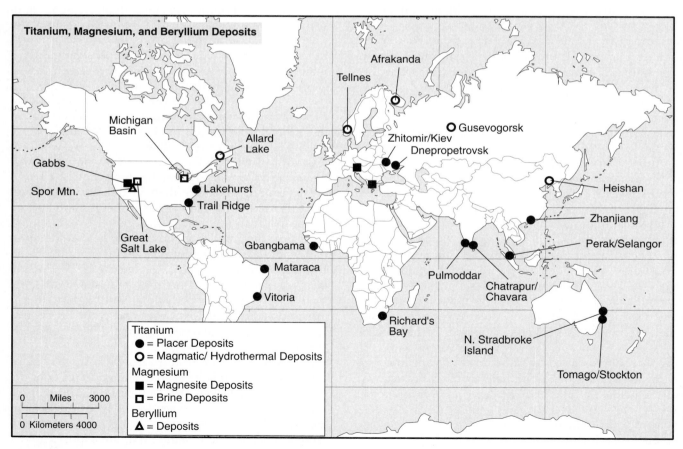

FIGURE 9-8
Location of titanium, magnesium, and beryllium deposits.

will react with water to form potentially explosive hydrogen.

Magnesium production is so small that it has limited environmental impact, although metal-producing plant emissions have been of local concern. The element is essential for life and is used by humans largely to activate enzymes and facilitate protein synthesis reactions (Itokawa and Durlach, 1988). Deficiencies of magnesium have been linked to growth failure and behavioral problems. Magnesium is also important to plant growth as the basic constituent of chlorophyll.

Magnesium reserves and resources are essentially unlimited. World reserves of magnesite are estimated to be several billion tonnes. Backing this up is the entire ocean, which is resupplied with magnesium by weathering and river flow, as well as by alteration of basalt at mid-ocean ridge spreading centers.

Titanium

Commercial use of titanium metal began only in 1906, although it has been used longer in its mineral form (Lynd, 1985). Currently, about 95% of world titanium production is consumed as an oxide (TiO_2), which forms a white pigment. The remaining 5% is used as titanium metal, which has had an unusually rapid market growth. Titanium metal is valued because it has a significantly lower weight than steel, but a similar high melting temperature and strength. Thus, it can be used in high-temperature applications where weight is important, especially in engines and other parts of aircraft and spacecraft. As much as 30% of the weight of modern aircraft consists of titanium. The relatively high cost of the metal has prevented it from making many inroads on other steel or aluminum markets, however, and it has an annual world production of only about 50,000 to 60,000 tonnes worth about $200 million (Figure 9-1A).

Presently active titanium deposits are being exhausted rapidly, mostly because of the large volume of titanium minerals used for pigments, and the outlook for discovering new deposits is not good. Thus, titanium appears to be in the vanguard of metals for which we must find a completely new source in the near future (Force, 1991). Titanium forms two very common minerals, *rutile* and *ilmenite,* as well as *leucoxene,* a form of ilmenite from which iron has been removed by weathering or alteration, thus increasing the titanium content. Rutile and ilmenite are heavy and resistant to weathering and erosion. Not surprisingly, more than half of world titanium mineral production comes from *placer deposits* in which they are combined with other heavy minerals such as monazite, garnet, and zircon (Towner et al., 1988). Most placer deposits containing titanium minerals are beach

dunes and sands that have been uplifted and preserved from erosion (Figures 9-9 and 9-10A). Important deposits of this type include the Trail Ridge area in northern Florida, Richards Bay, along the Natal coast of South Africa, and Stradbroke Island, along the coast of Queensland in Australia (Figure 9-8). Many of these deposits formed as the seas retreated from a major incursion onto the continents in Pliocene time, about 3.5 to 3.0 Ma ago (Carpenter and Carpenter, 1991). A smaller number of titanium placer deposits, including those in Sierra Leone, formed in a stream, or fluvial, environment.

A smaller amount of titanium minerals is produced from *bedrock deposits.* The most common deposits of this type are magmatic in origin and are associated

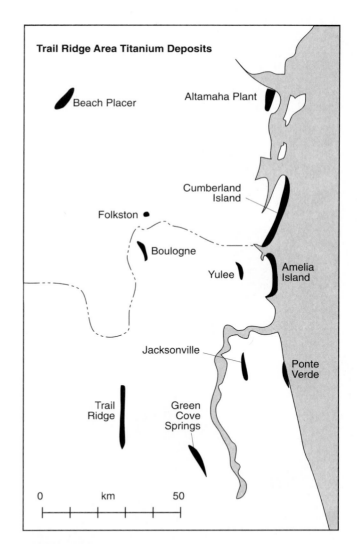

FIGURE 9-9

Map of titanium mineral placer deposits along uplifted beaches in the Jacksonville district, Florida (after Force, 1991).

with mafic intrusive rocks. During cooling of these magmas, ilmenite crystals form layers near the bottom of the magma chamber, probably by the same processes that form chromitite layers. Rutile is also found along the margins of anorthosite, a rock rich in plagioclase feldspar, where it appears to have been concentrated when other elements were melted and removed during intrusion of the anorthosite.

Mining of bedrock titanium deposits requires considerably more energy than placer mining. Bedrock deposits make up for this disadvantage because they have ore grades of as much as 35% TiO_2, much higher than grades commonly found in placer deposits. Mining of placer deposits is done largely by dredges that dig their own lakes, which they fill with tailings from ore processing that is carried out on board (Figure 9-10B). Although dredged land is reclaimed successfully (Watkin et al., 1992), the location of dredging operations in and near beaches in some areas has led to disputes over land use. Controversy has been most severe along the coast of Queensland, Australia, where important rutile placer deposits will probably never be mined.

Processing of titanium ore is usually done by reacting the ore with Cl gas at 800°C to form $TiCl_4$, which is purified by fractional crystallization and then reduced to titanium metal by reaction with magnesium or sodium. This process consumes even more energy than aluminum production (Table 9-1); it is relatively benign environmentally, although excess radioactivity was detected in wastes from pigment production in Japan. The source of this radioactivity is thought to be uranium and thorium in the titanium minerals. An alternative process, in which titanium ore is leached with sulfuric acid, produces a caustic waste solution that must be neutralized. Titanium is not toxic, as shown by its use in medical implant devices (WHO, 1982).

In spite of its high crustal abundance, titanium is mined in only 12 countries, giving it the lowest ratio

A.

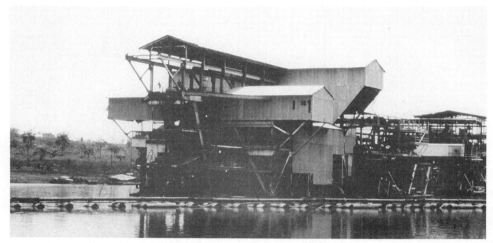

B.

FIGURE 9-10

A. Ilmenite-bearing dune sands from Trail Ridge area, Florida. Inclined layering reflects original dune surface and horizontal layering results from cementing by humic material. Field of view is 2.7 meters across (courtesy of Eric Force, U.S. Geological Survey). **B.** Bucketline dredge (foreground) and scrubbing-screening plant (background), at the Sierra Rutile operation, Sierra Leone. Rutile-bearing sand deposits in the Atlantic coastal plain of Sierra Leone were derived from the weathering of nearby Precambrian rocks and fill a system of valleys that are mined by these dredges (courtesy of Susan Shahan, Nord Resources).

of mines to production volume of any metal. A large proportion of world titanium production comes from Australia, South Africa, Canada, Norway, and Sierra Leone, which are not major consumers. Thus, large amounts of ilmenite, rutile, and other forms of titanium are exported, usually by sea. Among the major importers of titanium minerals are the United States, Japan, Germany, France, and the United Kingdom, which use mineral concentrates to make pigments and, in the case of the United States, Japan, and the United Kingdom, titanium metal. Because they have limited domestic reserves of titanium minerals, the United States and most large European manufacturing nations consider titanium to be a strategic metal. With large reserves located in Canada and Norway, however, supplies are not likely to be in jeopardy.

From a purely numerical standpoint, titanium reserves do not look so bad. In terms of contained TiO_2, rutile reserves amount to about 85 million tonnes and ilmenite reserves are about 200 million tonnes. Unfortunately, high-grade beach placer deposits are being exhausted rapidly and new types of deposits must be found (Force, 1991). Some submerged beach placer deposits on the continental shelf might be mineable, but most of the supply will probably have to come from bedrock deposits. In addition to the magmatic and hydrothermal deposits mentioned previously, production could come from eclogite, mafic metamorphic rock that forms under high-pressure and relatively low-temperature conditions, and which contains locally abundant rutile. By-product rutile will also become important. The two most likely sources are porphyry copper deposits and the tar sands of Alberta. These sources are not particularly rich in titanium, but they are produced in very large volumes and their titanium minerals can be recovered relatively easily.

Beryllium

Almost 400 tonnes of beryllium worth $100 to $200 million are used annually in alloys with copper and other metals that are valued for their extreme light weight (Petkof, 1985). Beryllium is produced largely from the minerals *beryl* and *bertrandite*, which form in pegmatites and hydrothermal veins around felsic intrusions and volcanic rocks (Figure 9-3). Beryl is most common in pegmatites, where particularly good crystals qualify as the precious gems, emerald and aquamarine. Bertrandite is more common in veins and disseminations in shallow rhyolite intrusions and tuffs, such as those at Spor Mountain, Utah (Figure 9-8), the largest active source of beryllium in the world. Reserve estimates are not available for beryllium, but

they are thought to be large in relation to present world consumption.

Production of beryllium is usually carried out by open-pit mining, melting the ore mineral, and then leaching the residue to liberate beryllium. Inhalation of beryllium dust in and around processing facilities causes *berylliosis,* a serious chronic lung disease, although significant problems have been limited by preventive measures taken since it was recognized in the 1940s (WHO, 1990; Rossman et al., 1991).

BASE METALS

The *base metals* were named by medieval alchemists who could not convert them to gold or other precious metals. Since that time, their reputation has been improved through many important applications in modern industry, which support annual world production worth about $35 billion, similar to the value of light-metal production.

It could well be asked why base metals came to the attention of alchemists at all. They have very low crustal abundances and should be rare. The answer is that they commonly form sulfide and oxide minerals rather than substituting in silicate minerals. Even where rocks contain only traces of a base metal, that metal is usually present in shiny sulfide or oxide minerals that would have attracted the attention of early wanderers. Because most of these minerals melt at low temperatures, metallic liquids oozing out of stones around early camp fires probably started society on the metal dependency path that we travel now.

Copper

Copper was among the first metals to be used, largely because it occurs in metallic form at the surface. By about 4000 BC, copper was being smelted from its ores in Israel and other parts of the Middle East. In areas with tin ores, such as Cornwall, England, tools and weapons were made from the harder copper-tin alloy, *bronze,* starting the Bronze Age. In Egypt and other areas where tin was not widely available, the Bronze Age was delayed by centuries. Where zinc was available, especially in the Roman Empire, it was alloyed with copper to make *brass.* For many centuries, copper was the most widely used metal, although the advent of iron and steel limited it to applications for which it is best suited, such as the conduction of heat and electricity, and resistance to corrosion. These markets account for annual world production of about 9 million tonnes worth about $20 billion.

Geology of Copper Deposits

Copper occurs in nature in a wide variety of minerals, of which the sulfide, *chalcopyrite,* is most common. It is found largely in hydrothermal deposits, although magmatic and supergene deposits are significant locally (Bowen and Gunatilaka, 1977). The most important hydrothermal deposits are *porphyry copper deposits,* which formed around intrusions that fed volcanoes (Titley and Beane, 1981). These deposits consist of closely spaced, intersecting veinlets or stockworks containing quartz, chalcopyrite, and other minerals, which surround felsic intrusions (Figure 9-11; Plate 7F). They formed when magmatic water was expelled from the intrusion as it cooled and crystallized, shattering the surrounding rocks and forming a highly permeable zone through which hydrothermal fluids flowed depositing copper. The term *porphyry* refers to a texture in which large crystals that form during early stages of crystallization of a melt are later surrounded by smaller crystals. This change in crystal size can occur when the magma moves upward in the crust, or

FIGURE 9-11

Geologic environments and characteristics of important types of base-metal deposits.

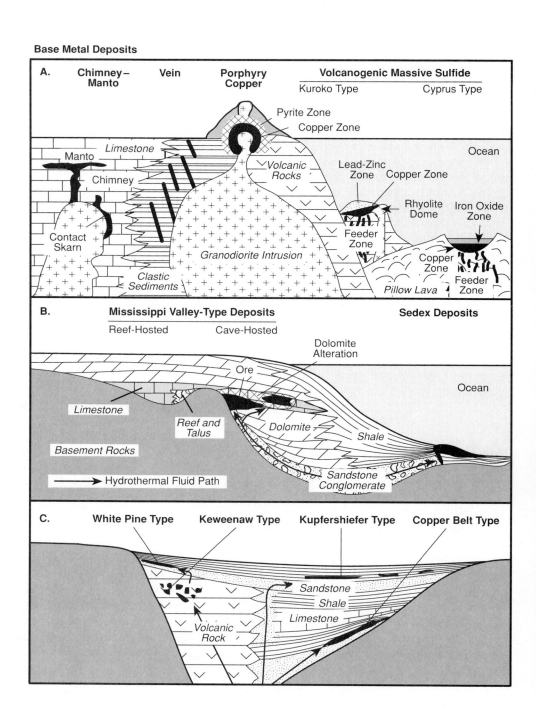

when magmatic water is expelled from it. Fluid inclusions show that quartz in the veinlets was precipitated from very hot (>500°C), highly saline hydrothermal solutions of probable magmatic origin, but that meteoric water invaded the hydrothermal systems as they cooled. Although individual veinlets are small, they are so numerous and closely spaced that the bulk rock averages 0.5% to 2% Cu and can be mined by inexpensive, large-scale methods. In addition to being the principal source of world copper, porphyry copper deposits yield by-product molybdenum and gold. The largest deposits are found at convergent tectonic margins in the United States, Canada, Chile, and Philippines (Figure 9-12).

Other important copper deposits, which formed from seawater hydrothermal systems, are known as *volcanogenic massive sulfides (VMS)*. They consist of massive lenses of iron sulfide minerals, with locally important copper, zinc, and lead sulfides, that were deposited as sediment where hydrothermal systems

vented onto the sea floor at hot springs (Franklin et al., 1981). The massive ore lenses are commonly underlain by a system of veinlets that formed the *feeder zone* through which fluids reached the surface (Plate 7E). Two main types of ancient VMS deposits are named for areas in which they were first recognized. *Cyprus-type* deposits get their name from Cyprus, which was itself named for copper. These deposits formed in the ocean crust at divergent tectonic margins. *Kuroko-type* deposits, which are named for an area in northern Japan, formed at convergent tectonic margins, where island arcs were cut by rift zones. These deposits are underlain by shallow bodies of felsic volcanic rock that supplied heat and possibly even water and metals to the hydrothermal systems. Deposits similar to the Kuroko-type deposits are also common in Precambrian rocks, particularly in eastern Canada (Figure 9-12). The shape of VMS deposits is controlled in part by the relative densities of the venting fluid and surrounding seawater. Where the vent-

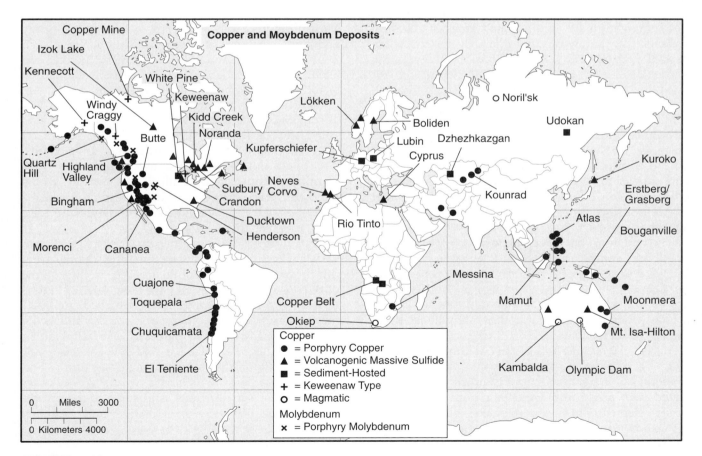

FIGURE 9-12

Location of major copper and molybdenum deposits. Not all deposits shown in this map have been mined. Note the concentration of porphyry copper deposits above subduction zones along the Pacific Rim.

ing fluid is more dense than seawater, it collects in depressions and forms saucer-shaped deposits. Where it is less dense, it rises and forms black smokers (Figure 2-12) that grow together to make a mound.

Basinal hydrothermal systems formed *sediment and volcanic-hosted copper deposits* in sedimentary basins and associated rifts filled with volcanic rock. The largest of these are found in clastic sediments, where they form layers parallel to bedding (Boyle et al., 1990). The copper-rich Kupferschiefer shale underlies an area of 20,000 km^2 extending from northern England to Poland and has been mined for copper in an area of over 140 km^2 in the Harz Mountains of Germany. The White Pine copper deposit in Michigan is hosted by the Nonesuch Shale, which is exposed for over 100 kilometers along the south shore of Lake Superior and extends northward below the lake. The most important area in the world for these deposits is the African Copper Belt, which extends for several hundred kilometers from southern Zaire into northern Zambia and has important co-product cobalt production. Here, copper-rich shales, sandstones, and conglomerates are up to a hundred meters thick. Although early theories suggested that these deposits formed as chemical sediments, it is now thought that they formed when basinal hydrothermal solutions passed through the sediments shortly after they formed, depositing sulfide minerals in pores and replacing the rock with sulfide minerals locally (Figure 9-11). Some basins that contained volcanic rocks formed copper deposits known as *Keweenaw-type* deposits for the famous deposits in the Keweenaw Peninsula of Michigan (Figure 9-12). These ores consist largely of native copper and smaller amounts of silver that fill vesicles and other pores in basalt flows and conglomerates that filled the mid-continent rift zone when North America almost broke apart about 1 billion years ago. The copper is thought to have been deposited by basinal and possibly metamorphic waters that were expelled from the volcanic rocks in the bottom of the rift zones. Similar deposits were mined years ago in the Kennecott area of Alaska and subeconomic deposits are known around Copper Mine Point and on Melville Island in the Northwest Territories.

Magmatic copper deposits consist of immiscible metal sulfide magmas that separated from mafic and ultramafic silicate magmas. Copper is almost always associated with nickel and minor amounts of platinum-group elements in these deposits, particularly at Kambalda, Australia, Sudbury, Ontario, and Noril'sk, Russia, which are discussed in the section on nickel deposits (Naldrett, 1989). Another type of copper deposit that might be related to magmatic processes is Olympic Dam in South Australia (Figure 9-12). This enigmatic deposit consists of iron oxides, copper min-

erals, fluorite, and rare-earth minerals, which were deposited by hydrothermal solutions in a highly fractured granite intrusion (Oreskes and Einaudi, 1991). Although it is similar to porphyry copper deposits, it might also be related to the immiscible iron oxide magmas discussed in the previous chapter. Better understanding of Olympic Dam is important because it is so unusual and so large, with over 11 millions tonnes of copper.

Most copper-sulfide minerals are not stable at Earth's surface and dissolve during weathering. Where weathering acts on hydrothermal or magmatic copper deposits exposed at the surface, it can form two types of deposits. If the dissolved copper precipitates as copper oxides and carbonates, *secondary copper deposits* are formed. If it percolates downward into underlying unweathered rock and precipitates new sulfide minerals (often by replacing preexisting pyrite), *supergene copper deposits* are formed.

Copper Production and Evironmental Concerns

Most porphyry copper deposits, as well as secondary and supergene deposits, are mined in open pits, such as those dotting the landscape around Tucson, Arizona. These operations move hundreds of thousands of tonnes of ore and waste each day and are among the largest earth-moving efforts on the planet. A few porphyry copper deposits, such as San Manuel, Arizona, and El Teniente, Chile, have been mined by large-scale, block-caving underground methods. Higher-grade magmatic, massive sulfide and sedimentary deposits are also mined by open pits where they are near surface, but have been followed to depths of at least 2,000 meters by underground mining (Jolly, 1985; OTA, 1988).

Copper-sulfide ores are beneficiated to produce a sulfide mineral concentrate for smelting. Tailings from this process amount to as much as 98% of the volume of the ore and are disposed of in large piles or valley-fills. Open-pit copper mines are rarely filled after mining, because they are so large and the ore bodies are not lying flat or near the surface as in strip mines. Modern mines are regraded and revegetated when they are abandoned, however, as are waste dumps and tailings piles. Unfortunately, this was not true of older copper mines such as Butte, and some of these are a major source of metal-rich, acid water and sediment, as discussed earlier (Figure 4-22A).

At many mines, low-grade ore and waste are leached by acid waters to dissolve the copper. Much of the acid for this process comes from the natural dissolution of pyrite already present in the rock. Copper is precipitated from these solutions by SX-EW or by reacting it with steel scrap to produce *cement copper,* which is then smelted. In the western United States, cans from municipal waste as far away as the San Francisco Bay area are used to recover copper from deposits. Acid leaching is also used on secondary copper deposits because copper oxides and carbonates are highly soluble.

Copper smelting is carried out largely by *pyrometallurgy* (Gilchrest, 1980). The first step may involve *roasting* to remove sulfur, arsenic, and other easily vaporized elements, but most concentrates are simply dried and melted in a *reverberatory, electric, or flash furnace,* where the concentrate and flux separate into *slag* containing silica and iron, and *matte* containing copper and sulfur (Figure 9-13). The matte is then transferred to a *converter,* where oxygen is blown through to produce sulfur gases and *blister copper.* Blister copper is purified in an *anode furnace,* which removes residual oxygen and the anodes are then *electrolytically refined* to produce pure *copper cathodes* and a residue of gold, silver, and other metals that follow copper through the smelting process. Although *hydrometallurgical* copper smelting methods have less atmospheric emissions, technical problems have limited their application.

Environmental concerns with copper production and use rarely focus on the metal. In fact, copper is an essential element for life. Copper deficiency in humans, which is rare, is manifested in the failure to use iron effectively, causing anemia (Howell and Gawthorne, 1987; Nriagu, 1979). Copper in water tubing is actually beneficial because it discourages the growth of bacteria. Exposure to elevated copper levels can cause symptoms ranging from green hair to death, although even the least severe symptoms are rare, apparently reflecting the ability of the body to cope with an elevated exposure to essential elements, a behavior not observed for nonessential elements such as lead and mercury.

Real environmental concern about copper production has centered on emissions of SO_2 and easily vaporized trace metals such as arsenic, cadmium, and mercury from copper smelters, where impressive reductions have been made. Old pyrometallurgical smelters were major SO_2 emitters because reverberatory furnaces produced gases with SO_2 concentrations that were too low to recover in acid plants or scrubbers, and because lots of SO_2 was lost transferring matte in open ladles from the furnace to the converter. These problems were solved by *continuous smelters,* which put the two steps into a single enclosure, and by *flash smelting* (Figure 4-20), in which an injection of oxygen creates a high-SO_2 offgas that is more easily treated (Table 9-2). The only reverberatory system still operating in the United States, at White Pine, Michigan, actually adds sulfur-bearing minerals to its ore to increase SO_2 contents of its effluent gas enough to make scrubber operation more efficient. The eight major primary smelters in the United States currently recover about 98% of their sulfur emissions, a much better record than that of fossil-fuel power plants (Figure 9-14).

Further improvements are possible, but at greater cost. For instance, the existing smelter system at Bingham, Utah, which has Noranda continuous smelters (Table 9-2, Figure 9-15) that were installed in 1976, recovers 93% of all SO_2 emissions. A new continuous-flash smelting system to be installed by the mid-1990s will recover 99.9% of SO_2 emissions. This system is expected to cost $880 million and it must be paid for by revenues. Ignoring inflation, which would make costs greater, and assuming no change in the price of copper for a decade, this investment amounts to about 15% of total probable revenues, an enormous commit-

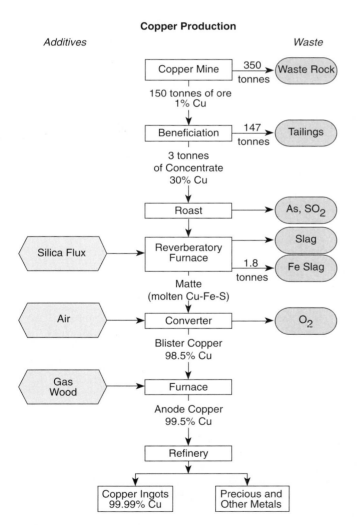

Copper Production

Additives   *Waste*

Copper Mine → 350 tonnes → Waste Rock

150 tonnes of ore 1% Cu

Beneficiation → 147 tonnes → Tailings

3 tonnes of Concentrate 30% Cu

Roast → As, SO_2

Silica Flux → Reverberatory Furnace → Slag

1.8 tonnes → Fe Slag

Matte (molten Cu-Fe-S)

Air → Converter → O_2

Blister Copper 98.5% Cu

Gas Wood → Furnace

Anode Copper 99.5% Cu

Refinery

Copper Ingots 99.99% Cu   Precious and Other Metals

FIGURE 9-13

Schematic illustration of typical copper production process.

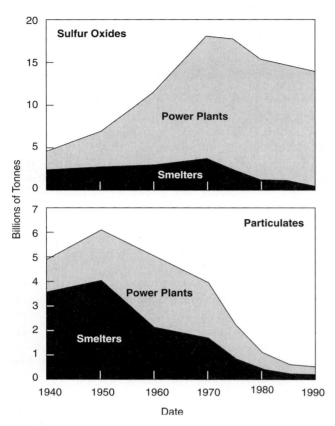

FIGURE 9-14
Comparison of particulate and sulfur oxide emissions from smelters and power plants. Note that both sources have cut particulate emissions, but only smelters have cut SO_2 emissions (from EPA 450/4-91-004).

ment. Environmental costs of this type are not equal in all areas. Sulfur emissions from copper smelters in Canada during the mid-1980s were about 0.7 tonnes per tonne of copper produced, versus 0.11 in the United States, 0.03 in Europe and 0.002 for the new Bingham system. Canadian smelters remain big SO_2 producers because remotely located operations at Thompson and Flin Flon, Manitoba, and elsewhere are far from large urban areas and lack nearby markets for sulfuric acid that would be produced by the recovery systems. Because they are major sources of employment and revenue for northern areas, the government has been slow to act. Recent initiatives are expected to lead to cleaner operations, but they will likely lag progress in the United States, Europe, and Japan. Smelters in many LDCs and CIS countries are just beginning to feel environmental pressures. Russia closed the important Alaverdi and Karabash complexes at the beginning of the 1990s rather than meet the high costs of making them environmentally acceptable.

Copper Markets, Reserves, and Trade

The largest market for copper is in electrical motors, generators, and wiring. Aluminum substitutes for copper in long-distance power transmission lines where lighter cables permit the construction of less expensive towers, and silver substitutes for copper where high electrical conductivity is needed (West, 1982). Most of these battles ended in the 1970s, however, and the new assault on copper comes from glass fibers for telephone transmission lines. Glass fibers have taken over

FIGURE 9-15
Noranda reactor, at Noranda, Quebec, which is shown in a schematic diagram in Figure 4-20B. Oxygen is injected and burned in a reactor chamber (large building at left) producing temperatures of 1,250°C and melting a continuous stream of concentrate and flux. Ducts at ends of the reactor, which are largely for the cleaning of exhaust gases, are more extensive than reactor and make up an important part of construction and operating costs (courtesy of Mireille G. Larouche and Guy Prevost, Métallurgie du Cuivre Noranda).

TABLE 9-2

Sulfur dioxide (SO_2) concentration in effluent gases from different smelting processes (from Office of Technology Assessment, OTA-E-367, 1988). As discussed in the text, effluents with high SO_2 concentrations are easier to clean.

Smelting Process	SO_2 Concentration (%)
Reverberatory furnace	0.5 to 2.5
Converter	15 to 21[a]
Electric furnace	5
Kennecott converter system	8
Flash smelting	10 to 80[b]
Noranda continuous-smelting system	16 to 22
Mitsubishi continuous-smelting system	11

[a]During loading (charging) converter emissions are lower in SO_2.
[b]Higher SO_2 concentrations for pure O_2 air supply.

the trunk lines, although copper remains the cheaper option for local hookups. Copper is also used in construction for roofing, decorative trim, and plumbing, although plastic pipe is already substituted in wastewater lines. Another major copper market, brass for ammunition casings, depends largely on the status of world conflict (Mikesell, 1988). With all of this competition, it should not be a surprise that production of copper is growing slower than that of steel (Figure 9-1D). One of the few bright spots on the horizon is the use of copper in automobiles, which increased by about 40% between 1980 and 1991, as a result of the increased use of motors and electronic equipment.

Paradoxically, the United States is a major exporter of copper concentrates (Figure 9-16A). Although other major copper producers such as Canada and Chile have long been exporters of concentrate, the United States was a net importer of concentrate for decades until the 1980s when efforts to reduce sulfur emissions produced dramatic changes in the industry (Jolly and Edelstein, 1989, 1992; Crozier, 1992). Billions of dollars were spent on emission-cleaning systems and 12 old reverberatory and electric copper smelters were closed. Although some were replaced by modern flash and continuous smelters, overall U.S. copper-smelting capacity declined, leading to an exodus of concentrates in search of smelter capacity. The major importers of copper concentrates are Japan, Germany, CIS countries, South Korea, Spain, and Brazil.

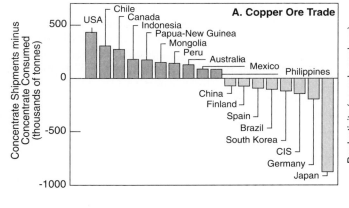
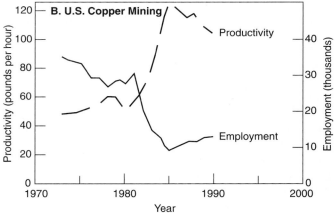

FIGURE 9-16

A. World copper ore trade expressed as copper content of products for 1991. Countries that plot above the line produce more ore than they smelt; those that extend below the line import ore for their smelters. **B.** Relation between productivity and employment in the U.S. copper industry (data from U.S. Bureau of Mines and Metal Statistics).

The change in U.S. copper ore trade status coincided with a gradual increase in the proportion of blister copper exported by many LDC copper producers. The increase reflected their desire to reap the economic benefits of a more refined, higher-margin product rather than the concentrates that they were exporting. Regression of the United States along this economic path is due to its failure to build new smelters to replace obsolete ones, largely caused by environmental concerns. A state-of-the-art copper smelter planned for Texas City, Texas, was canceled in 1992, after a futile, four-year effort to obtain regulatory approval. In response to delays and public concern, many new smelter projects have been located in LDCs, further increasing their capacity to deliver finished copper and decreasing jobs in MDCs. It is particularly ironic that smelters should become so difficult to locate just as the technology and will have become available to make them cleaner.

The balance of future copper production around the world may depend on relative production costs. Costs of this type are important for all commodities, of course, but they are especially pivotal for copper because the competition pits the United States and Canada, with relatively high wages, against South America and Africa, with lower wages (Strishkov, 1982; Porter and Thomas, 1989). As a result of an impressive increase in productivity due to innovative mining and processing methods, the United States and Canada appear to be well positioned for the future. Estimates made in 1990, at the peak of U.S. copper productivity (Figure 9-16B), showed that production costs in the United States and Canada were lower than all other major producers except Chile.

Ore reserves are estimated to contain 310 million tonnes of copper metal, of which Chile and the United States have almost half. Continuing exploration will undoubtedly discover new copper deposits to expand this reserve. For example, Grasberg, which was delineated during the late 1980s in the mountains of Irian Jaya (Figure 9-10), contains about 5.5 million tonnes of copper, more than the known reserves of China (Hickson, 1991). Most copper deposits are much smaller, however, and new ones have to be discovered and brought into production at a regular rate to support world consumption. In North America, important deposits have been found in areas with no history of previous copper production, a clear testimony to the effectiveness of exploration. However, many recent discoveries, such as Crandon, Wisconsin, have not been developed because of environmental concerns. If this trend continues in North America, replacement supplies will have to come from the CIS countries and the Copper Belt of Zambia and Zaire, where operations are inefficient and in need of foreign investment.

In the CIS countries, investment could come from western countries, and South African capital is being courted for the Copper Belt. Future world copper supplies depend on the success of these negotiations.

Lead and Zinc

Lead and zinc are often found together in ore deposits, but have very different uses and biological effects. Zinc is essential for life, whereas lead has no known biological function and is an important environmental hazard. As a result, primary (mine) production of zinc has continued to climb during the last few decades, whereas lead production has actually begun to decline (Figure 9-1D). Present annual world zinc production of 7 to 8 million tonnes, worth about $8 billion, is twice that of lead production of 3 million tonnes worth about $3 billion. In view of its bad environmental reputation, you might ask why lead is produced at all. The answer lies in the important uses to which it can be put, especially in electric storage batteries, which could be our best method of limiting emissions from gasoline engines, as discussed later.

Geology of Lead and Zinc Deposits

Lead is usually obtained from the common ore mineral *galena* and zinc comes from *sphalerite*. They are found almost entirely in hydrothermal deposits, which formed in three major geologic environments (Kesler, 1978; Ohle, 1991). Basinal hydrothermal systems formed the most important lead-zinc deposits, including *Mississippi Valley-type (MVT)* and *sedimentary exhalative (sedex) deposits*. MVT deposits are so named because they were first found in the mid-continent United States in the valley of the Mississippi River (Figure 9-17). Huge amounts of lead and zinc were produced from now-exhausted MVT deposits in the tri-state district near Joplin and the Old Lead Belt near Bonne Terre, both in Missouri (Figure 5-11). Modern production comes from the Viburnum Trend in Missouri, eastern and central Tennessee, the Northwest Territories, Poland, and Spain. These deposits consist of galena, sphalerite, barite, fluorite, and other minerals that fill secondary porosity in limestone and dolomite in much the same way that oil and gas fill other porous sediments (Figure 9-11). The ores were deposited from brines that were expelled from nearby basins, following extensive aquifers, such as the cave system in the southern Appalachians (Figure 4-4), until they reached areas that caused ores to precipitate. Fluid inclusion studies show that the brines had salinities of 15% or more and temperatures of 90°C to 180°C. Generally similar deposits are hosted by porous

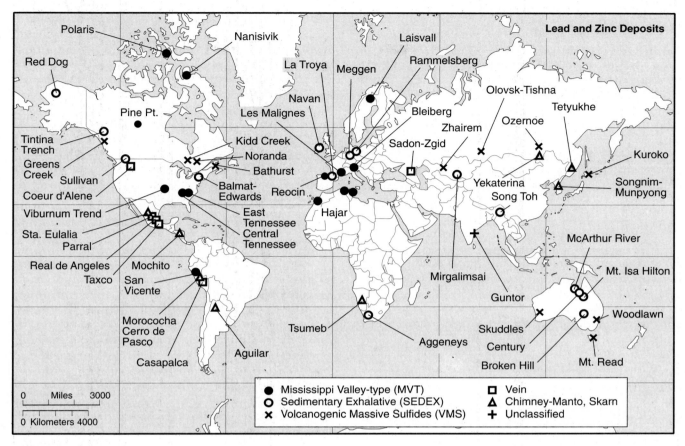

FIGURE 9-17
Location of major lead and zinc deposits.

sandstones in Scandinavia and other areas (Bjørlykke and Sangster, 1981).

Sedimentary exhalative (sedex) deposits consist of layers of lead-zinc-iron sulfides (Plate 7A) that were precipitated by submarine hot springs that flowed into basins filled with fine-grained, clastic sediments (Figure 9-11). In many cases, these solutions did not actually reach the sea floor and instead replaced lenses of sediment just below the water surface, usually shortly after the sediments were deposited. Some of these solutions were hotter and less saline than those that formed MVT deposits, and they probably consisted of a mixture of basinal brines and seawater. Heat for some systems probably came from nearby mafic magmas, although the association between deposits and volcanism is much more tenuous in these deposits than it is in VMS deposits. The largest sedex deposits, including Broken Hill, Mt. Isa, and McArthur River in Australia, and Sullivan in British Columbia (Figure 9-17), are in a system of 1.8 to 1.3 Ga-old rifts that cut the early continents (Young, 1992). Other important deposits are found in

northern Europe, as well as in the Selwyn basin of the Yukon, where their remote location has discouraged mining.

Seawater hydrothermal systems formed *volcanogenic massive sulfide (VMS) deposits,* which were discussed previously as sources of copper. In contrast to copper-bearing VMS deposits, those with lead and zinc are found largely in continents and island arcs and are of the Kuroko type mentioned earlier. The most important deposits of this type are found in the Kuroko district of Japan (Figure 9-17).

Lead and zinc are found in large *vein deposits* at Keno Hill in the Yukon, Coeur d'Alene in Idaho, Hidalgo del Parral in Mexico, and Casapalca in Peru, all of which have important amounts of co-product silver (Beaudoin and Sangster, 1992). In some deposits, fluid inclusions indicate that the ore-forming solutions had temperatures of 350°C or more, were highly saline, and were probably injected into the vein from underlying magmas. In others, the water was less saline and could have been meteoric or even metamorphic water. Some hydrothermal systems were stratified, with a low-salinity fluid above a high-salinity fluid as is seen in some geothermal systems (Simmons et al., 1988; McKibben et al., 1988). *Chimney-manto and skarn deposits* are related to vein deposits and appear to form where the country rock is limestone or dolomite rather than clastic sediment (Figure 9-11). Limestone and dolomite react with hot, hydrothermal fluids to form calcium-rich silicate minerals known as *skarn,* which often contain base-metal sulfide minerals. Some skarn deposits form at the contact between the limestone and igneous rock. Others form *chimneys* (vertical) or *mantos* (horizontal) that extend for hundreds of meters into the limestone away from igneous intrusions (Figure 9-18). Fluid inclusions in some of these deposits indicate that they formed from highly saline, very hot (400°C) fluids that appear to be of magmatic origin. These deposits are best developed in Mexico, Honduras, and Peru (Figure 9-17).

The distribution of lead deposits reflects the average concentration of lead in crustal rocks, which has increased through time due to the decay of uranium. Of the numerous Archean-age VMS deposits, only the huge Kidd Creek deposit produces important amounts of lead. In contrast, Paleozoic-age VMS deposits in Bathhurst, New Brunswick, and Buchans, Newfoundland, are rich in lead, as are Cenozoic-age Kuroko VMS deposits in Japan. Similarly, important sedex deposits first appeared in Middle Proterozoic time, possibly because lead-rich, felsic igneous rocks intruded the continents at that time (Sawkins, 1989). In fact, felsic, continental rocks, also rich in uranium, are considerably richer in lead than rocks of the ocean crust, which rarely host lead deposits.

Lead and Zinc Production and Environmental Concerns

Lead is mined and beneficiated in 47 countries versus 50 countries for zinc, making them among the most widespread metals in terms of primary production (Jolly, 1985; 1990; Woodbury, 1985, 1990). Mining is largely by highly mechanized underground methods (Figure 9-19), although some deposits are mined from open pits. With a few rare exceptions, such as the extremely high-grade ore that was shipped from Pine Point MVT deposits early in their history, lead-zinc ores must be beneficiated to produce concentrates for smelting. Sedex and VMS deposits, which were deposited by the rapid cooling of submarine hot-spring fluids entering cold ocean water, can be too fine-grained for ore and gangue minerals to be separated by conventional beneficiation methods. The McArthur River deposit in Australia contains ore worth about $25 billion but has been idled by these complications since its discovery in the 1960s. Where these deposits have been metamorphosed, recrystallization has produced larger crystals that are easily separated. The outstanding examples of this process are at Broken Hill, Australia, and Aggeneys, South Africa, which contain galena crystals measuring tens of centimeters in width.

Primary smelting of lead and zinc concentrates is carried out in fewer countries than is mining, with about 24 for lead and 34 for zinc. Most primary lead and zinc smelting begins with *roasting,* which drives off sulfur as SO_2 and converts the sulfide minerals to oxides. Conventional pyrometallurgical lead smelting is carried out in a blast furnace, producing *lead bullion,* which must be refined to remove zinc and silver that were also in the concentrate. Zinc oxide from the roasting step is recovered by SX-EW methods using a sulfuric acid solution. When a mixed galena-sphalerite concentrate must be smelted, it is roasted and then put through the *imperial smelting process,* in which zinc is driven off as a vapor (to be condensed elsewhere), leaving less volatile lead in the residue. Secondary

FIGURE 9-18
Ojuela chimney-manto lead-zinc-silver deposit, Mapimi, Durango, Mexico. The chimney at Ojuela reached the surface at several points, including the dark hole at lower right. The suspension bridge leads from the mine town (now abandoned) to the main mine entrance.

less damaging because the lead is relatively immobile and not easily taken up by plants and animals. Lead oxides and sulfates from flue gases are much more soluble, however, and can be moved to deeper levels of the soil by rainwater. They are also dispersed farther from the source because of their small size. The degree to which they pollute areas surrounding smelters depends on the balance between emission rates and the rate at which they are dissolved and moved downward into the soil by rainfall. Although smelter emissions have decreased significantly in the United States in recent years, most have not yet achieved the 1978 EPA limit of 1.5 micrograms of lead per cubic meter. It has been estimated that the cost of full compliance would be about 2.3 cents per kilogram of lead produced, more than can be supported by present profit margins (Isherwood et al., 1988). Similar difficulties beset Canadian smelters; emission controls added $39US million to the cost of a $100US million smelter under construction during 1992 in Flin Flon, Manitoba.

Lead and zinc differ greatly in their effects on humans (Boffess and Wixson, 1978; Nriagu, 1978, 1987; WHO, 1989; Moore, 1991). Zinc is an essential constituent of enzymes, DNA, RNA, and protein synthesis, carbohydrate metabolism, and cell growth. Lead has no biological use and is toxic even in low amounts, causing both physiological and neurological effects in humans. The physiological effects, which are best understood, include nausea, anemia, coordination loss, coma and death, with increasing exposure. The neurological effects of lead range from reduced nerve function to encephalitis. Lead causes many of these effects by competing successfully in important biochemical reactions with other elements that are essential for life. For instance, lead can cause anemia by taking the place of iron in compounds used in the biosynthesis of hemoglobin. It can cause calcium deficiency by accumulating in the bones, the principal reservoir of calcium, and it can interfere with zinc enzymes. In spite of these negative effects, lead is not known to be carcinogenic. It can be removed from the body by use of organic chelating agents such as EDTA, which form stable dissolved lead complexes that are excreted. As discussed later, although the general population has some risk from lead production activities, their risk is greater from lead in widely used products such as gasoline and paint.

Lead and Zinc Markets, Trade, Reserves, and Related Environmental Concerns

Lead is believed to have been used in glazes on pottery by 7000 to 5000 BC and in metal objects and medicines in ancient Egypt and China (Blaskett,

FIGURE 9-19
Remote-controlled, load-haul dump unit removing ore from a stope at Mount Isa mine, Queensland, Australia (courtesy of Mt. Isa Mines).

lead smelters, which process lead metal products, are located in 43 countries, but only 21 countries have secondary zinc smelters, reflecting the large recycling of lead in electric storage batteries.

Lead and zinc mining and processing have been important historical polluters. Older smelters emitted galena particles from the handling of ore concentrates, as well as an aerosol of lead oxides and sulfates that condensed in smelter exhaust gas (Jennett et al., 1977). Smelters built as recently as the 1970s produced measurable emissions from these sources (Figure 9-20). Metal-rich dispersion zones consisting of galena particles, which are usually near the smelter or mine, are

FIGURE 9-20

Distribution of settleable particulate lead and lead in oak leaves and forest litter around lead smelters in Missouri (from Jennett et al. in NSF/RA-770214). Most metal smelting created similar metal haloes (Figure 4-22B) until advanced gas and particulate recovery systems were installed in the 1980s.

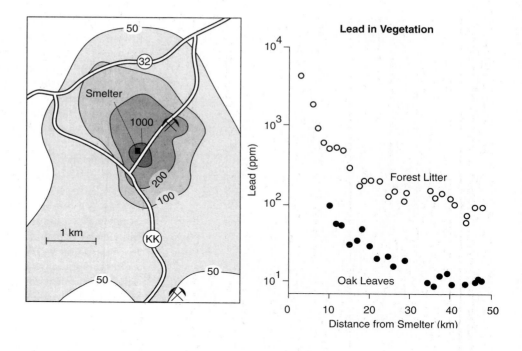

1990). By the time of the Roman Empire, it was used to make pipes, eating and cooking utensils, coins, weaponry, and writing equipment (whence the term *lead* for the graphite in pencils). It was also used to sweeten wine and other foods. Per capita lead consumption during Roman times is estimated to have been equal to that today, but for distinctly different uses. In fact, it has been suggested that lead caused the decline of the Roman Empire, whose later rulers displayed symptoms of lead poisoning. The hypothesis is supported by the elevated concentrations of lead found in the bones of Roman nobility (Gilfillan, 1965; Nriagu, 1983; Patterson, 1987). With a reputation like this it should be no surprise that lead has had to find new uses, a quest with which it has had singular success.

The most important market for lead is in electric storage batteries for fossil-fuel powered cars and trucks, which currently account for 80% of United States lead consumption. Battery-powered electric vehicles and emergency power supplies for large computer systems, hospitals, and similar fail-safe consumers could be the markets of the future. Two percent of the cars sold in California and ten eastern states in 1998 must have zero emissions, which requires that they be powered by electricity from batteries. The proportion is to be 10% by 2003, perhaps as many as 300,000 cars. Whether lead is a part of these batteries depends on the relative merits of lead, nickel-cadmium, and other batteries that are being tested at present. Smaller amounts of lead are

used as sheathing for cables and in plate for construction. It is also the principal component of ammunition.

Three markets in which lead was important a few decades ago are now nearly gone. Foremost of these was the use of *tetraethyllead* [$Pb(C_2H_5)_4$] and related compounds as antiknock additives in gasoline, which has essentially ceased in MDCs, although it continues in some LDCs with domestic refinery capacity, including Mexico and South Africa. The use of lead oxides in interior paint, glass, ceramics, and other chemicals has also declined because of its hazardous characteristics, as discussed previously. Finally, lead is no longer used in plumbing, a factor that is thought to account for the decline in lead content of humans over the last few decades. Lead is still used in solder on copper plumbing, however, and it can cause locally high lead levels in water with which it is in prolonged contact.

Sadly, lead has been so widely used that annual anthropogenic emissions of about 300,000 tonnes have far outweighed natural emissions of less than 20,000 tonnes for years (Nriagu, 1978; Nriagu and Pacyna, 1988; Nriagu, 1990a). The most important historical anthropogenic sources of lead have been the combustion of gasoline and coal, with the primary smelting of sulfide ores a distant third. Lead residues from these emissions are found beside roads and in areas as remote as the oceans and the poles (Figure 9-21). In MDCs, the lead content of gasoline is limited to 0.15 to 0.5 grams/gallon and better recovery systems

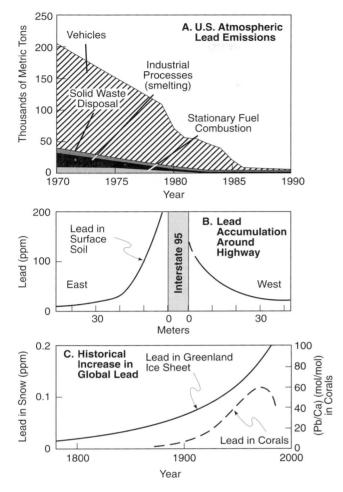

FIGURE 9-21

A. Atmospheric lead emissions in the United States since 1970 showing a dramatic decrease caused by a drop in the consumption of leaded gasoline. **B.** Distribution of lead in the surface soil around Interstate Highway 95 at Beltsville, Maryland, four years after road was opened to traffic. **C.** Change with time in lead content of Greenland ice and corals around Bermuda. Recent decrease of lead in coral is thought to reflect diminished anthropogenic lead emissions (compiled from EPA 450/4-91-004; Murozumi, 1969; Milberg et al., 1980; Shen and Boyle, 1987).

have been installed on electric power plants and smelters, greatly decreasing emissions (Figure 9-21). As a result, lead fallout around roads is down dramatically and upper parts of the ocean contain less lead than underlying levels. Nevertheless, it has been suggested that even low levels of lead exposure can decrease cognitive ability in children. In most cases, the suspected cause is soil that has been enriched in lead from old paint flaking from walls of houses or from automobile exhaust. Eval-

uation of this possible relation has been frustrated by the difficulty of distinguishing effects caused by lead from those caused by other variables affecting the population. Tests around smelters, where the anthropogenic source is most obvious, have been favored for this reason. Of 13 such tests made in four countries since 1970, six have observed differences in cognitive abilities related to body lead content (ICSU, 1987; Moore, 1990), and the director of the U.S. Centers for Disease Control has called lead poisoning ". . . the number one environmental problem facing America's children . . ." (Hilts, 1990). Even if this statement is true, it will have limited effect on mine production because lead is not used in interior paint, gasoline, and other readily dispersed forms in MDCs and these markets are being discontinued in LDCs.

In response to environmental concerns, the U.S. Congress has considered legislation to place a direct tax on lead production. The proposed legislation would impose a tax of $0.75/pound on primary lead production, about 250% of the present price of the metal, effectively stopping domestic lead production within two years (Biviano and Owens, 1992). Although such a tax has some theoretical appeal, it would have unwanted side effects. Because of the close relation between lead and other metals, it would cut domestic production of zinc, cadmium, and bismuth by over 80% and domestic silver production by about 15%.

Zinc was not heavily used until well into the 1800s, although it has penetrated a large number of markets since that time. Zinc's pervasive role in world industry is not widely appreciated by the casual observer because it is used so often in forms other than the pure metal. For instance, about half of zinc consumption goes to make *galvanized steel,* in which thin coatings of zinc are used to prevent corrosion. Galvanized steel is the major component of power transmission towers, culverts, tanks, and nails, where it protects the steel both by insulating it from atmospheric corrosion and through galvanic decomposition, in which two metals form a natural electrolytic cell. Galvanized steel use in automobiles has quadrupled since the early 1980s. *Galfan,* an alloy with 90% zinc, 5% aluminum, and rare earths, outperforms galvanized steel but is more expensive. Sacrificial anodes of zinc are also used to prevent corrosion on ships, oil drilling platforms, and submerged pipelines. Another quarter of zinc production goes into brass, bronze, and other zinc alloys that are used as ammunition shell casings and in motors, refrigeration equipment, and cars, and zinc metal is used in roofing in Europe. The use of zinc in die castings has declined as the size of cars has decreased.

World trade in lead and zinc is similar to that for copper with considerable movement of concentrates from ore-rich countries to those with smelting capacity. Canada, Australia, and Peru dominate the export scene for both metals, and Japan, Germany, Belgium, France, and Italy are the major importers (Figure 9-22). These countries, particularly Belgium, which is a major smelting center, then export metallic zinc and lead. Since about 1990, Poland, Russia, and Kazakhstan have emerged as major zinc exporters, relying partly on their excess smelter capacity and concentrates imported from western mines. Over the same period of time, stocks of lead and zinc in the London Metal Exchange have increased dramatically, suggesting that prices will decline and production will be curtailed. In spite of this short term oversupply, reserves are not particularly good, amounting to about 63 million tonnes of contained lead and 140 million tonnes of contained zinc. The largest reserves are in Canada, Australia, Peru, Mexico, Poland, and Russia. With the exception of the Red Dog sedex deposits in Alaska, reserves in the United States are declining rapidly and could be exhausted within little more than a decade.

Tin

Tin is an element with lots of friends but dwindling markets. Geologists are fascinated by tin because of its highly specialized distribution in Earth, and it was the first commodity to have a global group of producers and consumers to encourage its consumption. In spite of all this, it is the only major metal that has had de-

clining markets since 1960 (Figure 9-1D). Nevertheless, the value of world tin production, amounting to over 200,000 tonnes annually, is a respectable $1.8 billion, even at depressed tin prices (Figure 9-1C).

Geology and Mining of Tin

The most important tin ore mineral is *cassiterite,* which forms ore deposits in only a few parts of the world (Figure 9-23). Three quarters of world production comes from only six countries and three of these, Malaysia, Thailand, and Indonesia, actually represent a single tin-bearing region (Carlin, 1985, 1990).

The reason that tin is found in so few places has been debated for years, with two major competing hypotheses (Lehmann, 1990). One holds that tin deposits are found only in tin-rich parts of the crust, possibly inherited from the early history of Earth. The other holds that tin can be found anywhere that special tin-concentrating processes were active. Regardless of which is correct, we do know that tin deposits are almost always found in veins associated with granite intrusions that probably formed by partial melting of sedimentary rocks (Sainsbury, 1969; Taylor, 1979; Yeap, 1979). The most common deposits, known as *lode deposits,* are in pegmatites, quartz veins, stockworks, and disseminations clustered around protrusions known as *cupolas* at the top of these intrusions. In the Cornwall area of the United Kingdom, tin ore consists of coarse-grained muscovite mica, quartz, and other minerals known as *greisen,* which formed by high-temperature (500°C) alteration of granite. Where the

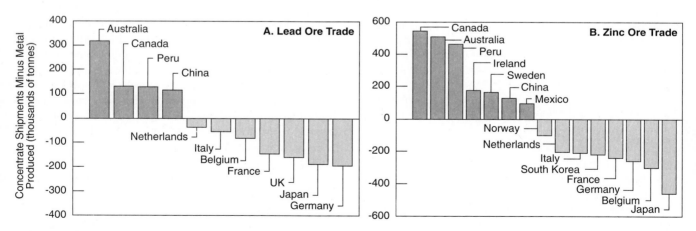

FIGURE 9-22

World **(A)** lead and **(B)** zinc ore trade for 1991 expressed as metal content of products. Countries above the line produce more ore than they smelt; those below the line import concentrate for smelting. The United States is not seen in these diagrams because it is essentially self-sufficient but is not an important exporter (based on data of *U.S. Bureau of Mines and Metal Statistics*).

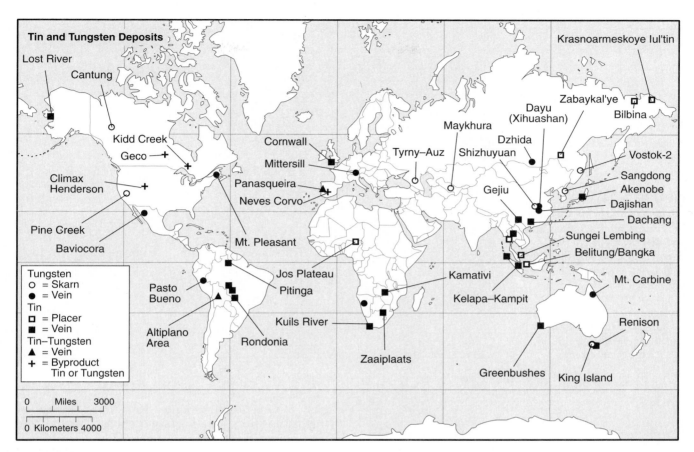

FIGURE 9-23

Location of world tin and tungsten deposits.

wallrocks consist of limestone, such as at Renison, Tasmania and Dachang, China, sulfide-rich, tin-bearing replacement deposits are found (Figure 9-24). Tin is also found in veinlets and stockworks associated with volcanic domes in Mexico and Bolivia, including the huge Cerro Rico tin-silver deposit that is discussed in the silver section. All of these deposits share a common very close association with igneous rocks, supporting the suggestion that the hydrothermal fluids from which they formed are largely of magmatic origin.

Tin is an important by-product from tungsten vein and stockwork deposits and porphyry molybdenum deposits, where it probably has the same magmatic origin. It is also found as a trace constituent in some volcanogenic massive sulfide (VMS) and sedimentary exhalative (sedex) deposits. Interestingly, tin is most common in the largest of these deposits, such as the huge Kidd Creek and Geco VMS deposits and the Sullivan sedex deposit, all in Canada, and the lore of mineral exploration says that you have a big VMS or sedex deposit if it contains tin.

The tin ore mineral, cassiterite, is highly resistant to weathering, forming *placer deposits* during the erosion of tin deposits. Placer tin deposits are, in fact, the dominant source of world tin. Those that have not moved far from their source and consist largely of cassiterite in regolith are called *residual placers*. Those that have moved a short distance downslope are known as *elluvial placers,* and those that are in the drainage system are *alluvial, stream,* or *beach placers.* The most important deposits of this type are in northern Brazil around Pitinga, and in the southeast Asian province, which extends through Thailand, Malaysia, and Indonesia as well as adjacent parts of the sea floor in the Sunda Shelf (Figure 9-25). The Sunda shelf was exposed to erosion during the last global glacial advances because water was trapped in the ice caps, lowering sea level, and streams flowed much farther out onto the shelf than they do now. Thus, the shelf hosts alluvial placer deposits that are submerged beneath only a few meters of water and, in some cases, sediment (Figure 9-25).

A.

FIGURE 9-24
Dachang tin district, Guangxi Province, China, which mines sulfide-rich replacement deposits in Devonian limestone near Cretaceous granite intrusions. **A**. Adit into Changpo mine. **B**. Selective replacement of carbonate layers by sulfide-rich ore (light) (courtesy of James E. Elliott, U.S. Geological Survey).

B.

Tin mining takes different forms, depending on whether it is from lode or placer deposits (Yeap, 1979). Lode mining uses small-scale open-pit and, more commonly, underground methods. Because of the close association of tin and uranium in some granitic rocks, radon can be a hazard in underground mines. Placer mining is done by *gravel pumps* for residual and elluvial deposits, and *dredges* for large alluvial deposits, particularly those on the shelf. Both methods create large volumes of sediment-laden water that must be impounded in the same way that sediment from placer gold operations is treated. Production of tin from cassiterite concentrate is done by heating it to about 1,200°C with carbon in a blast, flash, or electric furnace to drive off the oxygen as CO_2, leaving molten tin metal. Unless the tin concentrates contain sulfide minerals, this process does not cause significant environmental problems.

Tin is not a problematical element from an environmental perspective and it has no essential function in the body. Inhalation or ingestion of relatively high levels are required before negative symptoms are observed. Dust from tin oxide, such as in cassiterite, produces a benign pneumoconiosis known as *stannosis*, which is a potential problem only in tin processing and can be prevented by proper air handling and use of respirators. Organotin compounds are used widely in fungicides and similar chemicals and are toxic, but they do not occur in nature (WHO, 1980).

Southeast Asian Tin Deposits

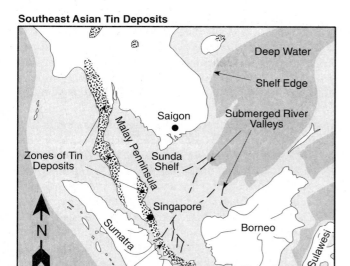

Cross Section Through a Typical Marine Placer Deposit

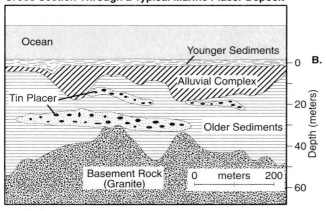

FIGURE 9-25

Tin placer deposits on the Sunda Shelf. **A.** Location of tin-producing areas on and offshore (from Sainsbury, 1969 and Batchelor in Yeap, 1979). **B.** Cross section showing submerged tin placer deposits on the Sunda Shelf (after Batchelor in Yeap, 1979).

Tin Reserves, Markets, and Trade

Tin reserves are estimated to be about 8 million tonnes of metal, most of which is in southeast Asia. In view of declining tin consumption, this is a relatively comfortable margin, especially when compared to other metals in this group.

The main use for tin is in *tinplate*, a protective coating applied to steel and other metals. Tin is best for this application because it will "wet," or cover, most metals easily and it resists corrosion. The largest market for tinplate is on steel cans for food, except spinach, which is too corrosive. Food containers must be sterilized at high temperature and welded shut, processes for which tin-plated steel is particularly well suited. Although glass can also be treated at high temperature, it is not a threat to tin-plated steel because glass containers are heavier and expose food to light, which can change its color. In spite of manufacturing improvements that have decreased the weight of steel cans from 72 grams to 31 grams since 1960, they are still about 60% heavier than aluminum cans, which translates into higher transportation costs. Thus, tin-plated steel has lost markets to plastic and aluminum containers, as well as to chromium-coated steel, particularly in the beverage sector (Figure 9-26A) where tin accounted for only 3.5% of the United States beverage can market in 1992. Tin is also used in solder, a low-melting temperature alloy that is widely used to join pipes, can lids, and electronic connections. Although improvements in soldering techniques have permitted the use of smaller amounts per application, growth in the electronics industry has spurred tin demand. Tin is also an essential component in alloys for bearings to be used at high temperature and there is a growing market in tin chemicals for wood preservatives, antifouling paint, and as an additive to prevent the discoloration and embrittlement of PVC plastics.

Tin concentrates are exported by most of the ore-producing countries except Malaysia, which imports tin concentrates and is the world's most important producer of tin metal (Figure 9-26B). Other important tin-smelting operations are in the United Kingdom and Netherlands (Thoburn, 1981; Baldwin, 1983). Although tin is not critical to defense needs, production of tin is so restricted geographically that it is sometimes regarded as a strategic metal, particularly in North America, which lacks important tin deposits. The relatively small number of important tin-producing countries and the fact that none of them are major consumers led to the formation of a unique cartel, known as the International Tin Council, which included both producers and consumers. As discussed earlier, the council attempted to stabilize the price of tin at a level that would encourage production and consumption but was overwhelmed by market forces (Lohr, 1985).

CHEMICAL AND INDUSTRIAL METALS

Antimony, arsenic, bismuth, cadmium, germanium, hafnium, indium, mercury, rare earths, rhenium, selenium, tantalum, thallium, and zirconium are used

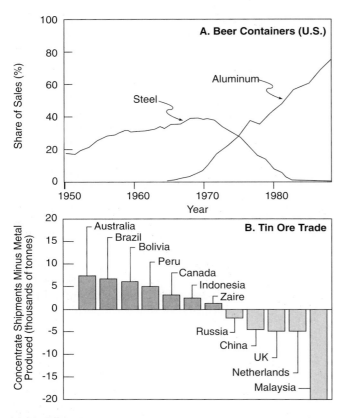

FIGURE 9-26

A. Change in consumption of steel and aluminum in beer containers in the United States. Tin sales have decreased with that of steel cans because tin is used as a coating on them (Tilton, 1990). **B.** World tin ore trade for 1991 based on tin content of ore and metal. Countries that extend above the line produce more tin ore than they smelt. Those that extend below the line import ore for their smelters (from data of U.S. Bureau of Mines and Metals Statistics).

largely in chemical or industrial applications and are not as widely known as the more abundant metals. They do not depend on the same markets and do not trade as a group. Some, such as germanium, have rapidly expanding markets (Figure 9-27). Markets for others, such as mercury and arsenic, are declining. Annual production for all of these metals is valued at only about $1 billion. Part of their importance, however, rests in their environmental effects. Particularly arsenic, cadmium, mercury, and thallium are very toxic and have received more than their share of environmental attention. The elements are discussed next in declining order of annual world sales, as an aid in keeping their overall economic importance in perspective.

Rare Earths

The rare earths (Table 9-3) are a group of closely related metals, from lanthanum to lutetium in the periodic table. Yttrium and scandium are commonly included with the rare earths because of their chemical similarities and tendency to occur in the same deposits (Hedrick, 1985; Hedrick and Templeton, 1990). The largest present markets for rare earths are in the manufacture of catalysts for chemical processes. *Lanthanum, neodymium,* and *praseodymium* are used in petroleum refining and *cerium* and the rare-earth mixture known as *mischmetal* are used in ammonia synthesis. Mischmetal and other rare-earth combinations are also used in special steel alloys. Current annual production of about 50,000 to 60,000 tonnes is worth about $150 million at a price of about $3 per kilogram of mixed rare-earth oxide, an intermediate stage in most rare-earth processing.

The rare earths form two main minerals, *bastnasite,* a fluorocarbonate, and *monazite,* a phosphate, both of which are rich in lanthanum and cerium. *Xenotime,* an yttrium-rich phosphate, is much less common. Bastnasite is usually found in hydrothermal deposits associated with alkaline intrusive rocks or carbonatites. The largest deposits of this type are at Mountain Pass, California, and Bayan Obo, China (Figure 9-28; see pg. 227). Monazite, on the other hand, is a common magmatic trace mineral in some granitic intrusive rocks. Although it does not form many hydrothermal deposits, monazite is resistant to weathering and concentrates in stream and beach gravels, where it is usually a by-product in gold, ilmenite, rutile, cassiterite, or zircon placer mines. Monazite is recovered from titanium placers in Florida and Georgia in the United States and in Australia, Brazil, and India (Figure 9-8), as well as from tin placers in Malaysia. It is also present in the Witwatersrand and Elliot Lake paleoplacer deposits but is not recovered (Moller et al., 1989).

Essentially all rare-earth mining is by open-pit or dredge operations (Shannon, 1983). The elements are recovered from bastnasite or monazite by hydrometallurgy involving acid or base leaching and solvent extraction. Unfortunately, consumption of individual rare earths does not match their relative abundances in ores, making it necessary to stockpile some in order to obtain enough of others. Monazite also contains thorium, which is radioactive and requires special storing and handling. Although mining of rare earths is not associated with significant health hazards, large doses of the pure metals or their compounds cause severe toxic effects that require precautions during production and use (Hedrick, 1985; Viljayan et al., 1989).

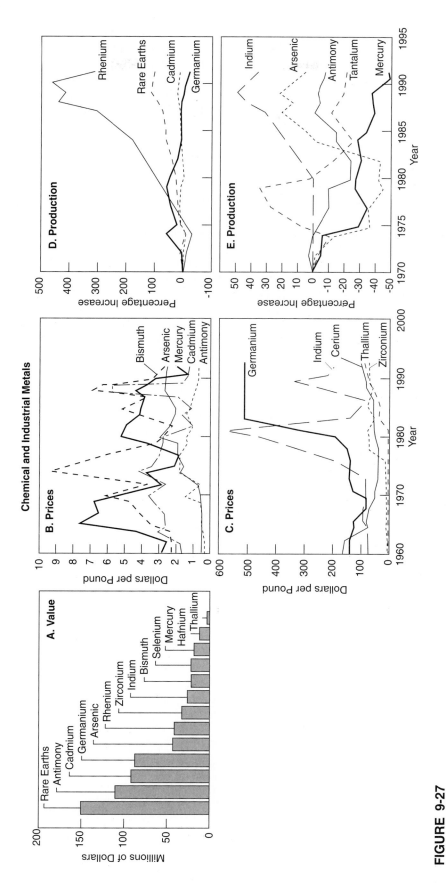

FIGURE 9-27

Chemical and industrial metals. **A.** Value of 1992 annual primary world production. **B** and **C.** Price history. **D** and **E.** History of production (compiled from publications of U.S. Bureau of Mines).

225

The United States and China control about half of world rare-earth production. Because of their very different wage structures, any expansion of Chinese production could squeeze the United States out of the market. Thus, even though world reserves of about 100 million tonnes of rare earths are large, China might dominate future rare-earth supplies.

Cadmium

Cadmium presents us with an environmental challenge. Its main use in rechargeable batteries with nickel, silver, or mercury could power electric cars that will help resolve the fossil-fuel crisis. But, it is highly toxic. Most of cadmium's markets have similar problems. It forms corrosion-resistant coatings for steel and other metals in applications such as washing machines, where cadmium resists the alkaline soap. It is used in yellow, orange, red, and maroon pigments and as a stabilizing agent to prevent PVC plastics from breaking down in ultraviolet light. In all of these applications, the search is underway for substitutes and environmental agencies are considering more stringent limits on cadmium exposure. World production of about 20,000 tonnes is valued at about $100 million, far below even the lowliest of the base metals, tin.

Although cadmium forms the mineral *greenockite*, most cadmium substitutes for zinc in sphalerite, which can have cadmium concentrations as high as 1.3%. Cadmium is recovered from flue dust during the roasting and sintering of sphalerite and in sludge from the electrolytic refining of zinc (Plunkert, 1985B). It is produced in most zinc-smelting countries of the world, a total of about 30. Reserves of cadmium are estimated to be relatively large at 540,000 tonnes.

In an interesting twist of nature, cadmium affects health by interfering with the body's natural uptake and utilization of zinc, the metal to which it is most similar chemically (OECD, 1975; Hutchinson and Meema, 1987). In natural soils and waters, the zinc:cadmium ratio is high enough for zinc to compete with cadmium, but in artificial cadmium-rich environments, cadmium takes over. Cadmium also interferes with calcium, copper, and iron metabolism. The list of health problems associated with exposure to cadmium includes Itai-Itai (ouch-ouch) disease, which involves vitamin D deficiency and resultant softening of the bones. Fortunately for cadmium producers, anthropogenic sources account for only about 15% of total input to global fresh waters and mining activities are a minor part of the anthropogenic component (Nriagu and Pacyna, 1988; Nriagu, 1990). Before the problems of cadmium poisoning were ad-

TABLE 9-3

Rare-earth uses and prices. Alloys and ceramics are important uses for all rare earths and are not listed individually. Prices are given for 99.9% pure form of rare-earth oxide bought in bulk and are useful largely as an indication of the differences in prices of individual elements (information supplied by J.B. Hedrick, U.S. Bureau of Mines).

Rare Earth	Price ($/kg)	Uses
Cerium	20	polishing compounds; radiation shield; glass additive
Dysprosium	100	permanent magnet addition
Erbium	175	fiber-optic amplifier; glass additive
Europium	1400	phosphors in cathode-ray tubes
Gadolinium	118	phosphors, laser crystal
Holmiun	485	dopant in laser crystal
Lanthanum	20	catalyst in petroleum refining; glass additive, nickel-hydride rechargeable batteries
Lutetium	5500	phosphor
Neodymium	18	permanent magnet; glass additive; dopant in laser crystals
Praseodymium	32	ceramic pigment (yellow)
Promethium	NA	fluorescent lighting starter
Samarium	100	permanent magnets
Scandium	3000	metal halide lamps
Terbium	620	phosphors
Thulium	3300	medical isotope (^{170}Tm)
Yttrium	25	phosphor, synthetic gems, superalloys
Ytterbium	220	X-ray source, glass and laser additive

equately understood, however, mine wastes were major local sources of contamination. The worst problems occurred in 1947 downstream from a lead-zinc mine in Toyama Prefecture, Japan, where tailings that had not been properly impounded caused as many as 100 deaths. The installation of a tailings dam stopped the problem (Kobayashi, 1971; Forstner and Wittmann, 1981).

Although current markets for cadmium have not been implicated in important health problems in the general population, there is growing concern about the disposal of cadmium-bearing batteries. A recently passed directive of the European Community, for instance, requires that spent batteries containing more than 0.025% cadmium, 0.4% lead, or 0.025% manganese be collected separately for disposal by 1994. Other products with more than 0.1% cadmium will be banned. The best substitute for cadmium in batteries is lead, which has its own problems, as discussed earlier.

Antimony

The most important use for antimony is in fire-retardant chemicals for addition to plastics and textiles, especially for children's clothing. A decade or so ago, antimony was alloyed extensively with lead in electrical batteries for cars, but this market has dwindled considerably as substitutes have been found. Annual antimony production of about 60,000 tonnes, buoyed largely by demand for fire-retardant chemicals, is worth about $100 million, making it one of the most important of the chemical metals.

Antimony forms the sulfide mineral, *stibnite,* as well as the copper-lead-antimony sulfides, *tetrahedrite* and *jamesonite,* all of which are precipitated from hydrothermal solutions (Plunkert, 1985B). Although stibnite is the dominant mineral in a few deposits, usually with arsenic or mercury, it is more commonly a byproduct of lead-zinc-silver mining. Tetrahedrite is abundant in the Coeur d'Alene veins in Idaho (Figure

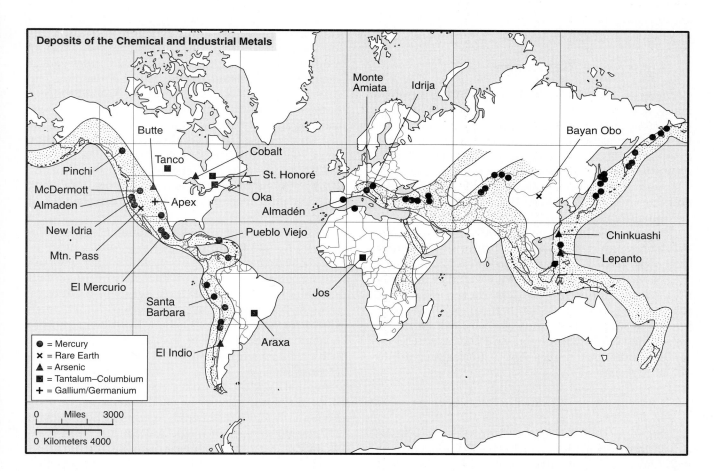

FIGURE 9-28

Location of important deposits of the chemical metals. Also shown is the distribution of deep focus earthquakes along subduction zones, showing the close correlation between mercury deposits and these zones.

9-17) and various antimony minerals are present in smaller amounts in Kuroko-type VMS deposits, chimney-manto lead-zinc-silver deposits, tin-tungsten veins, and tungsten skarn deposits.

Antimony is recovered from these ores by roasting and smelting or hydrometallurgical treatment. It is often the by-product of lead or zinc smelting and is produced in metal ingots or as the trioxide (Sb_2O_3), the major chemical form. Although antimony has health effects similar to but more acute than those of arsenic, problems have not been reported with its use in fire-retardant fabrics. Most global antimony pollution is in the form of atmospheric fallout from fossil-fuel combustion, although antimony contamination has been observed in groundwater surrounding some lead-zinc and precious-metal deposits (Forstner and Wittmann, 1981; Hutchinson and Meema, 1987).

Antimony reserves of about 4.2 million tonnes are very large, with most in China, Bolivia, and South Africa. Efforts by China and Bolivia to control world antimony prices have failed, both because of poor cooperation and because antimony is available from recycled lead-antimony batteries in most MDCs.

Germanium

Germanium is a high-tech element that has survived many market challenges. It was originally used in transistors and other semiconductors, a market that was taken over by silicon in the late 1970s. Since that time, it has found a new use in fiber-optic components, which are replacing copper in long-distance telecommunication lines, as well as in camera lenses and other glasses (Plunkert, 1985C).

Annual world germanium production of about 60 to 80 tonnes is valued at about $100 million, a respectable sum for an element that rarely forms its own mineral. Most germanium is recovered as a by-product of zinc smelting, largely from Mississippi Valley-type ores. It is also found in some copper ores, and one old copper mine, the Apex in southern Utah (Figure 9-28), has been reopened to mine germanium. World reserves are simply not known, and estimated U.S. reserves are only 450 tonnes. Thus, shortages could develop, especially if zinc production declines.

Arsenic

Arsenic is used largely in wood preservatives, herbicides, and insecticides, which take advantage of its well-known toxicity. Although its use as a desiccant in cotton fabrics has been discontinued recently because no acceptable exposure limit could be determined, arsenic will continue to be used for just the properties that make it an environmental concern. An-

nual world arsenic production of about 45,000 tonnes has a value of more than $50 million.

Arsenic deposits are rare, although it forms several important sulfide minerals, including *arsenopyrite, realgar, orpiment, enargite,* and *tennantite* (Edelstein, 1985). It is remarkably common in almost all types of hydrothermal gold deposits, suggesting that it is dissolved and precipitated by the same processes that control gold movement. Arsenic is rarely produced from these deposits, however, because gold can be recovered without treating the arsenic minerals. Enargite and tennantite are common in the upper part of some porphyry copper deposits, where they form large veins that are mined for copper, as at Butte, Montana (Figure 10-4). It has been suggested that vapors escaping from underlying magmas carried arsenic into these hydrothermal systems. Arsenic is also found in veins with cobalt, nickel, and silver in the Cobalt-type silver ores, and in some tin deposits, although it is rarely recovered from them.

Because of its widespread occurrence with other metals, arsenic is recovered largely as a by-product. It vaporizes at only 615°C and therefore is released during the roasting of gold or base-metal ores. Although modern smelters capture most of this vaporized arsenic in scrubbers, older smelters did not and they are surrounded by zones rich in arsenic. Around the Sudbury nickel smelters in Ontario, for instance, arsenic concentrations of lake sediments deposited during the last few decades are several times higher than those from earlier times before smelting began (Nriagu, 1983). Because of its toxic characteristics, there are now stringent limitations on arsenic emissions from smelters, and few smelters will take ore that contains arsenic.

Arsenic is a well-known environmental bad actor. Large amounts (more than 100 mg) induce acute arsenic poisoning resulting in death. Smaller exposure causes symptoms such as white lines on fingernails, and might cause cancer of the lungs and skin. It is also thought that arsenic can be converted to toxic methylated forms by bacterial action. Even before anthropogenic arsenic emissions became a problem, it was known that high natural levels of arsenic were dangerous. Elevated arsenic concentrations in water at Three Forks, Montana, have been attributed to emissions from hot springs in nearby Yellowstone Park. Around the Chinkuashi gold deposits in Taiwan, natural arsenic dispersion in groundwater entered water wells, causing severe arsenic poisoning, including loss of circulation in the hands and feet (Tseng, 1977; WHO, 1981). Early problems with anthropogenic emissions were caused by pesticides used before the 1940s, which contained large concentrations of soluble arsenic. Later concern has centered on tailings and waters from deposits rich in arsenic, such as those associated with the Summitville, Colorado gold deposit. From a global perspective, however, the main source of arsenic is fossil-fuel combustion. Coal commonly contains 1 to 300 ppm arsenic, only some of which is removed with pyrite during coal cleaning, and some coals have as much as 1,500 ppm (NRC, 1979). Reports from China during 1993 indicated that as many as 42 people died and 200,000 were made ill by poisoning attributed to use of arsenic-rich coal.

Because it comes largely from specially equipped smelters, arsenic is only produced in 12 countries, led by France. Reserves are estimated at about 1 million tonnes, a comfortable margin, although actual supplies will be strongly dependent on base-metal and precious-metal production rates.

Rhenium

Rhenium has an illustrious history. Its existence was predicted by Mendeleev in 1860 and it was discovered in 1925 and named for the Rhine River (Blossom, 1985). Discovery of rhenium was no small task. It has a crustal abundance of only about 1 ppb and does not form a common mineral, substituting almost entirely for molybdenum in *molybdenite*. Whereas molybdenite in porphyry molybdenum deposits contains only 10 to 100 ppm rhenium, molybdenite from porphyry copper deposits contains up to 2,000 ppm. Thus, most rhenium production comes from porphyry copper deposits in the United States, Chile, Canada, Russia, and other CIS countries, where it is a by-product of by-product molybdenum production, so to speak.

Annual world rhenium production of about 30 tonnes, which is worth about $40 million, is equally divided between two main markets. One is crude oil refining, where rhenium-platinum catalysts facilitate reforming in production of high-octane, lead-free gasoline. The other is in nickel-rhenium alloys that are used in high-temperature applications, especially jet engines. These markets are increasing and rhenium consumption is expected to continue its impressive growth (Figure 9-27). World reserves are estimated to contain about 2,500 tonnes of metal.

Mercury

Mercury is losing friends and markets. It began as *quicksilver*, prized because it is the only common metal that is a liquid at surface temperatures. Along with the bright red mercury sulfide mineral, *cinnabar*, it was used in medicinal and cosmetic applications by the Egyptians, Greeks, and Romans. Mercury's importance grew tremendously during the 1500s and 1600s, when it was used in the *patio process*, a variety of the

amalgamation method for recovering gold from ore. In the patio process, gold ores were scattered onto a flat patio, covered with mercury and then crushed, often by having mules walk over them. Mercury dissolved the gold from the ore and was then collected and heated to drive away mercury vapor, leaving a gold residue. The patio process became the dominant form of gold ore processing in the New World (Prieto, 1973). Spain, which controlled much of these lands, required that a fifth of any gold that was produced be paid to the Crown. Policing of this requirement was not difficult because most mercury for the process had to come from the Almadén mine in Spain, which was and still is the principal source of mercury in the world (Figure 9-29). Thus, mercury was a key factor in Spain's accumulation of riches from the New World (Flawn, 1966). Unfortunately, mercury has not been able to follow up on this triumph and current annual world production of 5,000 tonnes is worth only about $20 million.

Geology of Mercury Deposits and Mercury Reserves

Mercury often forms deposits on the margins of larger hydrothermal systems, reflecting its relatively high solubility in low-temperature fluids, especially those that are alkaline. It is also vaporized easily and therefore can be carried in vapors that rise above boiling hydrothermal solutions to condense in overlying groundwater (Figure 9-30). Most mercury deposits formed at temperatures of 200°C or less from dilute, meteoric waters and some deposits even contain petroleum residues (White, 1981; Carrico, 1985; Peabody and Einaudi, 1992). On a global scale, they are most common in areas of young volcanism at convergent tectonic margins (Figure 9-28). In northwestern Nevada, for instance, mercury deposits are found in the McDermitt caldera, which formed by explosive rhyolitic volcanism only a few million years ago, and the Monte Amiata district in Italy is on the slopes of a young volcano. Curiously, the world's largest mercury deposits around Almadén, Spain, do not conform to these generalizations. Mercury at Almadén, which hosts more than a third of world reserves, is found largely filling pores in an extensive sandstone layer. There is no sign of a large hydrothermal system in the area and no igneous rocks are known. Although it has been suggested that the deposits formed when mercury-bearing hot springs flowed out on the sea floor (Saupe, 1990), they could be related to a buried hydrothermal system.

Mercury is also recovered as a by-product of gold and some base-metal ores (Rytuba and Heropoulos, 1992). In fact, the large McLaughlin gold mine at the

A.

B.

FIGURE 9-29

Mercury production at Almadén, Spain. **A**. Pouring mercury into a 76-pound flask for shipment. **B**. Ball of iron floating in a tank of liquid mercury.

head of the Napa Valley in California was discovered beneath a mercury mine. During the cyanide-leach process used on most gold ores, mercury follows gold and is recovered just before the gold is poured into ingots. In some old mines, the need to recover mercury was determined when mercury vapor condensed on the ceiling of the gold refinery and rained down on workers. Mercury is also recovered from smelters treating some base-metal ores. In many cases, mercury minerals are not observed in the ores, and much of it may substitute for other metals in sulfide minerals, especially sphalerite.

With the exception of the large mercury deposits at Almadén and Idrija, and smaller deposits in Turkey and Mexico, most mercury production is a by-product of base-metal operations. Large historic producers at Monte Amiata in Italy, Santa Barbara in Peru, as well as those in New Idria and New Almaden in California, are not operating, although they could reopen with higher prices. Known reserves of 130,000 tonnes of contained mercury are relatively large when compared to present annual production.

Mercury Production, Uses, and Related Environmental Concerns

Mercury is produced largely from underground mines, where proper ventilation is a primary concern. After conventional beneficiation, mercury is liberated by heating in retorts or furnaces to form mercury vapor, which is then condensed. Early mercury producers collected the metal in 76-pound *flasks*, and these remain the form in which it is sold today.

Pollution and health effects related to mercury mining and its use in gold production have been impor-

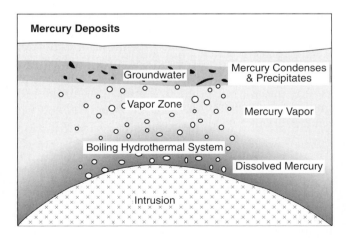

FIGURE 9-30

Schematic illustration of boiling hydrothermal system with mercury transported in the vapor phase and deposited where the vapor condenses in overlying groundwater.

tant problems for centuries (Nriagu, 1979; Fuge et al., 1992; Nriagu et al., 1992). They arise because elemental mercury is more soluble in blood than water, thus allowing it to enter the body easily, where its principal effect is on the nervous system (WHO, 1989, 1991). Workers exposed to high levels of elemental mercury vapor experience symptoms of mercury intoxication, including psychotic behavior, extreme irritability, and tremors. Massive inhalation can lead to death, although this is rare. At Almadén it was a routine procedure to detoxify workers by placing them in a hot room where they would sweat out mercury, and most symptoms disappeared with removal from exposure. Industrial or other exposure to *methylmercury* (CH_3Hg^+) is much more dangerous and prenatal exposure carries particularly high risks of neurological disorders. There is no conclusive evidence that mercury is carcinogenic (WHO, 1990).

The uses of mercury have undergone more changes through history than most metals. The early, mercury-based patio and amalgamation processes of gold recovery were largely displaced by the cyanide-leach process by the early 1900s, leaving large amounts of mercury dispersed through the soil and water around old mining districts. Amalgamation is still used in small operations in Brazil and Indonesia today, where it is an important contaminant, as discussed in the section on gold deposits. Mercury is no longer used in preparing felt, after the true meaning of "mad as a hatter" became apparent. Other recent uses include high-performance electrical batteries, lamps for outside lighting, electrical switches that take advantage of the liquid properties of the metal, and as a catalyst in the manufacture of pulp and paper, and chlorine and caustic soda. Mercury is also used in special exterior paints and coatings to discourage mildew, in thermometers and other measuring devices, and in dental amalgam. Because of its undesirable characteristics, major efforts have been made to limit mercury losses from these processes and, in some cases, to replace mercury altogether. The content of mercury in batteries, for instance, has declined from about 1% in the 1980s to 0.02% now, and it is expected that mercury will be replaced completely by the mid-1990s. More efficient recovery of mercury from chlorine-caustic soda plants or the use of alternative processing methods have combined to cut mercury use in this important market by over 50%. These efforts combined with the ban on the use of mercury in indoor paints have caused world mercury consumption to decline from about 300,000 flasks in 1970 to about 150,000 flasks in 1992.

Estimates of mercury emissions from natural and anthropogenic sources vary greatly, largely because of the difficulty in analyzing the element at the low concentrations in which it occurs (Hutchinson and Meema, 1987). The most important presently known natural sources are volcanic gases, vaporization from soil, and evaporation of water, which are estimated to release 2,700 to 20,000 tonnes of mercury per year. The most important anthropogenic source is combustion of fossil fuels. Although mercury concentrations in coal and oil rarely exceed 1 ppm, all of it is vaporized during combustion and enters the atmosphere unless scrubbing systems are in place. Mercury mining and production, industrial applications such as chlor-alkali manufacturing, metal ore smelting, cement production, and waste incineration are smaller anthropogenic sources of mercury (Fukuzaki et al., 1986; Ferrara et al., 1992). The total output from these sources has been estimated to be 2,000 to 21,000 tonnes of mercury annually (WHO, 1990). The concentration of mercury in Greenland ice has increased by at least 50% since the early 1700s, confirming that anthropogenic sources are important contributors to global mercury reservoirs (Weiss et al., 1971). Recent efforts to recycle mercury-containing consumer items, keeping them out of waste incinerators, are designed to further limit emissions.

Both natural and anthropogenic mercury are released largely as elemental mercury, which is dispersed into the atmosphere in gas form and adsorbed on aerosol particles. Precipitation washes this mercury into the soil and sediment where it is taken up by plants and animals. During bioaccumulation, some of the elemental mercury is converted to methylmercury and other organomercury compounds, a process that might be catalyzed by methylcobalamine, which is synthesized by methane-forming bacteria in the hypolimnion of lakes. Where waters are acid, methylmercury remains in solution and is taken up by organisms. In more alkaline water, it is converted to *dimethyl mercury* (($CH_3)_2Hg$) which is highly volatile and enters the atmosphere (Nriagu, 1979).

The general population's exposure to the increasing global mercury load comes from several sources. The most important one appears to be *biomagnification* of mercury in aquatic organisms, which reflects the fact that fish do not excrete mercury as rapidly as they take it in, partly because some elemental mercury is converted to methylmercury through metabolic processes or bacterial action. Although there is still some uncertainty about preanthropogenic background mercury concentrations in fish, surprisingly high mercury concentrations are widespread in modern fresh-water and marine fish. Some of these mercury enrichments have been traced to industrial activity, including methylmercury disposal at Minimata Bay, Japan, and mercury used in pulp and paper production in Ontario, but most of the mercury appears to be of more distant anthropogenic origin, almost certainly from fossil-fuel combustion (Porcella, 1990). Confirmation of this suspicion has recently been supplied by researchers who showed that mercury contents of lake sediments in northern Minnesota increased by more than four times in the last few decades.

The other major concern about mercury stems from its use in *dental amalgam,* a mixture of about 50% mercury, 35% silver, 12% tin, and smaller amounts of copper and zinc. This alloy is used in dental work because it acts like a liquid for a few minutes but crystallizes into a solid within a few hours. It is well known that amalgam releases mercury during chewing and that some of this is ingested or inhaled and a positive correlation between the amount of dental amalgam and the concentration of mercury in tissue, hair, and bones has also been observed. Although no significant health effect has been correlated with this level of mercury exposure, it has been suggested that mercury encourages the growth of bacteria that are resistant to antibiotics (WHO, 1989, 1990; Kolata, 1993).

Other Chemical and Industrial Metals

The chemical and industrial metals discussed next are used in small amounts and unusual markets, and rarely come to the attention of the general public. Their obscurity is reinforced by the fact that most are by-products from the production of other metals and few of them have important environmental concerns. Most of them have important applications, however, as discussed here.

Tantalum

Although tantalum is geochemically similar to the ferroalloy metal columbium (niobium), it is used largely in capacitors for electronic applications, with a much smaller amount used in ferroalloys (Cunningham, 1985). The value of the annual production of tantalum is difficult to specify because it trades in so many forms and because prices are not widely quoted. In keeping with our effort to compare prices of semifinished metals rather than raw ore, however, world tantalum production of about 400 to 450 tonnes would be worth almost $100 million if it were sold as metal.

Tantalum is found in the minerals *microlite,* which is the tantalum-rich equivalent of the columbium mineral *pyrochlore,* and *tantalite,* the tantalum-rich equivalent of *columbite.* Tantalum deposits are found largely in pegmatites and veins associated with granitic intrusive rocks. Tantalite, in particular, resists weathering and accumulates in placer deposits. Tantalum production comes largely from pegmatites and veins in Brazil, Canada, and central Africa (Moller et al., 1989), al-

though important amounts also come from the slag of tin-producing operations in southeast Asia (Figure 9-28). World reserves contain about 22,000 tonnes of metal.

Zirconium and Hafnium

These two metals are recovered from *zircon,* a heavy mineral that accumulates in placer deposits with the titanium minerals, ilmenite and rutile (Garnar, 1983). Zircon is used in mineral form in refractory products, where it is valued for its high melting temperature of 2,550°C. Some zircon is processed by chemical leaching to yield elemental zirconium, as well as the by-product hafnium, which has similar geochemical characteristics and substitutes for zirconium in zircon (Adams, 1985). The best known use for zirconium and hafnium metals is in nuclear reactors, where zirconium contains the fuel and hafnium forms control rods (Kessler, 1987). The two contrasting applications stem from the fact that neutrons can pass through zirconium easily, whereas hafnium has a high ability to stop neutrons. None of these applications is very large, however, and annual markets for metallic zirconium and hafnium are about $35 million and $12 million, respectively. Reserves of 49 million tonnes of contained ZrO_2 are large compared to present annual production of about 400,000 tonnes.

Indium

Indium is another of the occult metals. It rarely occurs in mineral form, but substitutes for zinc, tin, and tungsten in the crystal lattice of their minerals (Carlin, 1985B). Most indium production comes from sphalerite, which contains 10 to 20 ppm indium. Indium is recovered from the residues left from electrolytic refining of zinc, and it is produced in metallic form. The metal is unusually malleable and even makes a crying sound when bent. It melts at only 156.6°C, a property that creates its largest markets in solders and other alloys with bismuth, lead, tin, and cadmium. It is also used as indium-tin oxide protective coatings on windshield glass, and in semiconductors, breathalyzers, and dental crowns. Although indium is toxic in large concentrations, its production and use are not known to have caused important environmental problems. It is produced by 12 countries, including the major zinc refiners, Japan, Belgium, and the United States. The value of annual world indium production is $28 million and world reserves are estimated to contain 1,670 tonnes of metal, a small amount when compared to the present annual production of 120 tonnes.

Selenium

Selenium's claim to fame comes from its photoelectric properties, which offer hope for solar energy and are used widely in plain paper copying machines. It is also used in glass to prevent the green tint caused by trace amounts of iron, as the active ingredient in dandruff-preventing shampoos, and as a curing agent in rubber (Jensen, 1985). Annual production of 1,800 tonnes is worth about $20 million.

For such a scarce element, selenium is produced in a surprisingly large number of countries because it comes entirely from residues from electrolytic refining of blister copper. Although selenium-rich ore deposits exist, none are mined exclusively for the element (Sindeeva, 1964). Selenium is highly toxic, but its use has not caused significant problems because the toxicity is widely recognized and precautions are applied. The only industrial process in which selenium emissions might be of importance is oil refining, in which some wastewaters are enriched in the element. As discussed earlier, natural selenium deficiencies in rock and soil could be related to the incidence of stroke in humans and excess selenium is related to deformities and disease in animals. World selenium reserves are estimated to contain 75,000 tonnes of metal.

Bismuth

Many of us are familiar with bismuth as the active agent bismuth carbonate in remedies for upset stomach, one of its most important markets. Bismuth is one of the few metals that expands when it crystallizes from a melt, which makes it very useful in casings that require a tight fit at high temperatures. Bismuth oxychloride also provides the luster in many cosmetics.

Most bismuth is a by-product of lead, molybdenum, and tungsten mining and is usually recovered during smelting (Carlin, 1985A). Bismuth grades are too low to have any effect on mining, except in a few deposits in Bolivia and Japan. It is produced by 13 countries, including most of the important lead mining and smelting countries, with world production of about 3,200 tonnes worth about $20 million. Although bismuth has not been implicated in any specific environmental or health problem, it was banned in France in 1978 due to a suspected relation to brain disease and its use in cosmetics was stopped in Austria in 1986. Bismuth reserves are estimated to contain a comfortable 110,000 tonnes of metal.

Thallium

Although thallium forms sulfide and oxide minerals, they are exceedingly rare and most thallium occurs in solid solution or as small inclusions of rare thallium minerals in sphalerite. Thallium has an abundance of only 10 to 40 ppm in most sphalerite and is recovered from smelter flue dust (Plunkert, 1985D). Not all smelters make the effort, however, and thallium production currently comes only from Belgium, France, Germany, Japan, and the United Kingdom. Thallium was originally used as a rodenticide until it was learned that it is toxic to humans, partly because it is retained in the body. The toxicity is thought to result from interference in the metabolism of potassium, an element with which thallium has many chemical similarities (Moore, 1991). Uses for thallium at present are similar to those for indium, solders, and other low-melting temperature alloys, and electronic applications. Thallium-based pharmaceuticals have also been used in the treatment of ringworm, dysentery, and tuberculosis. The value of world thallium production of 16 tonnes is only about $3 million. World reserves are thought to contain 377 tonnes of the metal.

THE FUTURE OF LIGHT, BASE, AND CHEMICAL-INDUSTRIAL METALS

The future of light, base, and chemical-industrial metals is closely linked to the future of transportation. Widespread production of electric cars could greatly increase the demand for copper, as well as lead, nickel, and minor metals. Increased aircraft construction will require aluminum and titanium. Competition for these markets will come from recycled metals and from new, composite materials that will probably require less metal for the same performance. Thus, it is not likely that markets will grow rapidly. This is something of a relief because reserves are not large for arsenic, copper, indium, lead, tin, zinc, and possibly cadmium. Furthermore, many deposits of these metals are small in relation to global demand, and they have to be replaced by new ones at a regular rate. Thus, exploration for these metals is particularly active and a common focus of public concern.

Puerto Rico and British Columbia illustrate the growing problem of nonferrous metal exploration and production in MDCs today. In Puerto Rico, residents have agonized since the early 1960s over whether to permit a mining and smelting operation in the Lares-Utuado area, which contains high-grade porphyry copper deposits. In early newspaper reports, public and church officials claimed that the operations would pollute the area and that profits would flow to foreign owners. Smelter emissions became a major issue, even though studies demonstrated that they would be six times less than emissions from electric power plants in San Juan. News releases detailing the distribution of taxes and profits, as well as plans for tailings and waste disposal, came too late to avert misconceptions among a concerned public and the issue remains unresolved even though unemployment in the area is high and local rivers are filled with trash from picnics and private dumping. Windy Craggy, a huge new copper find in northwestern British Columbia (Figure 9-12), has recently become a similar focus of environmental concern because it is in the Tatshenshini-Alsek area, an area of scenic rivers. In response to public concern, a commission was set up to obtain input from all interested parties and recommend a solution. It was suggested during these hearings that the region be divided into areas with high mineral potential, such as Windy Craggy, and areas with lower potential, where further exploration and development should be stopped. In the end, however, the government of British Columbia closed the entire area to development, even though it is the far northwestern corner of the province far from any population concentration. The history of these two projects is stark testimony to the fact that new, or "greenfields," metal projects are almost impossible to permit and build in many MDCs

today. Since consumption of these metals does not decrease, the long-term consequence of this can only be increased reliance on LDCs.

Although they are less obvious, anthropogenic emissions resulting from our own metal consumption are of much greater concern from a global standpoint. As noted above, lead, mercury, arsenic, and other metals vaporize in coal-burning power plants, smelters, waste incinerators, cement plants, and other furnaces. In the absence of scrubbers, these metals are lost to the atmosphere and analyses of soil (Figure 3-15) and water (Figure 9-12) show that they have been dispersed throughout the globe. Unlike organic materials such as oil and natural gas, metals will not decompose, although some will combine with surrounding elements to form new minerals or become adsorbed into clays and iron oxides. Fortunately, recognition of widespread metal pollution in MDCs has led to remediation methods that have cut down on metal emissions and it appears that soil and sediment with high metal contents are gradually being buried beneath newer, cleaner material. Unfortunately, progress on this has been much slower in LDCs and global attention must shift to these laggards.

10

Precious Metals and Gems

Precious metals and gems are our most valued mineral resources. Whereas you might forget that you own a steel chair, a bag of fertilizer, or even a tank of gasoline, you would not forget that you owned a diamond or a gold ring. These commodities were among the first to be valued by civilization and they have kept their attraction for millennia. It is not hard to see why this has happened. Even in their raw state, these commodities are immediately recognizable as truly special and beautiful. Although other minerals, such as pyrite *(fool's gold)*, might be confused for gold, it is rare to confuse gold for another mineral. Its rich color and luster are simply unique. The same is true for the brilliant play of light and color from diamonds. On top of it all, precious metals and gems are rare. We even use special nomenclature to discuss their abundances. Gold abundances are commonly reported in *Troy ounces,* which contain about 31 grams (Appendix IV). Weights of diamonds and other precious gems are measured in an even smaller unit, the *carat*, which amounts to 0.2 gram. Because these nonmetric units are widely understood, we use them in parts of this discussion.

In spite of their rarity, precious metals and gems are a big business, with world production worth over \$35 billion (Figure 10-1A). Gold is the most important of the group, with diamonds, platinum-group elements (PGE), silver, and the colored gems emerald, ruby, and sapphire following in that order. Whereas PGE production has increased dramatically since 1960, gold and silver production have not kept pace with steel (Figure 10-1C). Gold has lagged because of the declining output from South Africa, which has dominated world gold production for decades but has an uncertain future. The history of gem diamond production is more difficult to trace because publicly available production data were changed in 1983 to include near-gem quality diamonds in the gem category, thus implying a greater increase in gem diamond production than has actually occurred. Even though it is not possible to correct for this change completely, it is clear that diamond consumption has increased dramatically. Confirmation of this can be obtained from data for raw and unset, cut gem diamond imports into the United States, which are reported by the United States Bureau of Mines. These imports, which form the basis for the diamond production curve in Figure 10-1C, have grown at about the same rate as the platinum group metals, faster than almost any mineral commodity.

With such a slow rate of production growth, one could justifiably wonder why gold has been such an important focus of global exploration. The answer lies in its price, which has increased dramatically since 1960 (Figure 10-1B). By the early 1980s gold prices rose to unprecedented levels largely on the basis of speculative buying fueled by increasing global inflation. Although prices have declined since that time, they continue to be supported because so much of the reserves are in South Africa, where events are likely to control world gold, PGE, and diamond markets for many years to come.

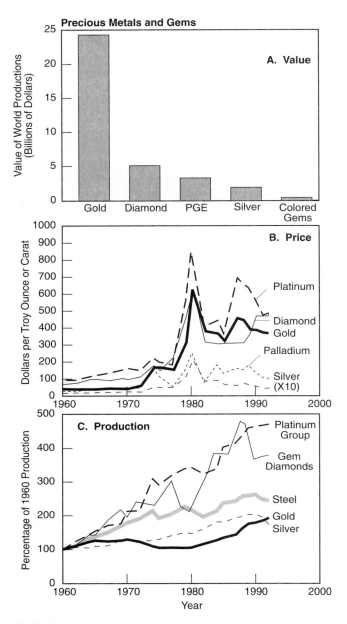

FIGURE 10-1

A. Value of 1991 world production of gold, silver, platinum-group elements, and diamonds. Values shown here are for primary production from mines and do not include recylced or other metals. **B.** Precious-metal and gem prices since 1960. Reference to Figure 4-2 shows that only the prices of gold and gem diamonds have outpaced inflation over this period. **C.** Precious-metal and gem-diamond production since 1960 [compiled from data of U.S. Bureau of Mines and De Beers Centenary reports; Central Sales Organization (CSO) value of annual diamond sales have been increased by 25% to include estimated non-CSO sales (including industrial stones); diamond prices are average price per carat of rough and uncut gems imported into the United States; silver price is multiplied by 10 to fit scale used here; diamond production figure shown here applies to imports of gem (cut and uncut, unset) diamonds into the United States].

GOLD

Unlike most other elements, few people lose track of gold. Only about 18% of the approximately 106,000 tonnes that have been mined throughout history are lost (Grabowski et al., 1991). Of the gold that can be accounted for, 43% is held by central banks (Table 10-1) and the rest is in private stocks of bullion, coins, jewelry, or art objects. As of 1991, more than 83% of world gold consumption went into jewelry and 6% was used in medals and official coins such as the famous Canadian *Maple Leaf* and South African *Krugerrand.* Another 6% was used in electronic equipment, which takes advantage of gold's high electrical conductivity and resistance to tarnish. Dental materials consumed another 2.2%, largely for crowns, which are required to resist corrosion and expand at the same rate as tooth enamel. The remaining 2.8% was consumed in a variety of industrial applications such as reflective coatings on glass and decorative gold foil for everything from plates to capitol domes. Gold is also used in pharmaceuticals for the treatment of arthritis (Goldfinch, 1991). These markets support an annual gold production of about 2,200 tonnes worth almost $25 billion, ranking it along with iron, copper, and aluminum as one of the most economically important metals produced.

Geology of Gold Deposits

Gold is found largely in the native state or in combination with silver in *electrum.* If there is enough tellurium present, it can also form telluride minerals, including *calaverite,* which was discussed as a source of tellurium in the last chapter. It can also substitute in small amounts for other metals in sulfide minerals, particularly in pyrite if arsenic is also present. Because so little gold is needed to form economically interesting concentrations, this type of gold is quite important in some deposits.

The Witwatersrand deposits in South Africa are so much larger than all other types of gold deposits that they completely dominate global production and reserves. This is really strange because they are *paleoplacer deposits,* and none of the numerous hydrothermal gold deposits known on the planet are even half their size, making it hard to understand how placers of this size originated. To show you the problem, we start with a brief description of *hydrothermal gold deposits,* which are divided here into epithermal and mesothermal types (Boyle, 1979).

Hydrothermal Deposits

Epithermal Deposits Epithermal deposits consist of gold-bearing veins, veinlets, and disseminations,

TABLE 10-1
Monetary gold reserves (not geological gold reserves) of major industrialized countries at the end of 1991 (from Monthly Bulletin of Statistics, United Nations International Monetary Fund). During 1992, banks in Austria, Belgium, Canada, England, Netherlands, and other MDCs sold 6 million ounces of gold in what could be a major change in their attitudes toward precious-metal reserves.

Country	Gold Reserves (million of ounces)
United States	262
Germany (West only)	95
Switzerland	83
France	82
Italy	67
Netherlands	44
Belgium	30
Japan	24
Austria	21
United Kingdom	18
Portugal	16
South Africa	16
Canada	14
Russia	10
International Monetary Fund	103
Bank for International Settlements	7
World Total	1141

which formed from hydrothermal solutions cooler than about 250°C that circulated through relatively shallow parts of the crust (Figure 10-2). They can be divided into adularia-sericite and acid-sulfate deposits, which formed from very different types of hydrothermal solutions (Berger and Bethke, 1986).

Adularia-sericite deposits formed from near neutral hydrothermal solutions and are characterized by these two minerals. Classical deposits of this type are associated with felsic and intermediate volcanism in continents or island arcs, where they form veins up to several meters wide and more than a kilometer long. Precious metals in the veins occur in zones called *shoots* that probably formed at levels where the hydrothermal fluid boiled or mixed with shallow groundwater (Figure 2-15). Especially high-grade shoots, which are called *bonanzas,* are among the richest ores known. The veins consist of layers of quartz and other minerals (Figure 10-3) that were deposited by successive pulses of hydrothermal solution made up largely of meteoric water that circulated to depth, was heated by underlying intrusions, and rose through the fracture system. These magmas might also have contributed magmatic waters to the systems intermittently.

These deposits are widespread in western North America. They include Tonopah and Comstock in Nevada, and Cripple Creek in Colorado (Figure 10-4; see pg. 240), which are exhausted now but were important lures for early fortune seekers, as well as the re-cently discovered Sleeper mine in Nevada (Hubbard, 1912; Koschmann and Bergendahl, 1968; Vikre, 1989). Similar deposits are found in the western Pacific volcanic arcs in Japan and New Zealand. Precambrian deposits of this type are rare but include the Mahd Adh Dhahab deposit in Saudi Arabia and, possibly, the large Hemlo district in Ontario. Silver is more abundant than gold in many of these deposits and can be the dominant economic element.

Sediment-hosted micron gold deposits appear to form when epithermal waters or brines pass through carbonaceous limestones. Gold is disseminated throughout altered limestone in these deposits, and it is so fine-grained that it could not be found with optical and electron microscopes, leading to the name *micron gold deposits.* It took the ion microprobe to show that gold was concentrated in thin layers along the outer parts of pyrite grains (Arehart et al., 1993). These deposits are found in Nevada, Utah, and Sonora, and are the most important source of newly mined gold in the United States (Figure 10-4A). The largest deposits, those around Carlin, Nevada, form an amazingly linear zone known as the Carlin Trend and other deposits appear to form similar trends. The deposits are thought to have formed where rising meteoric waters, possibly heated by intrusions, mixed with oxidized groundwater (Hofstra et al., 1991).

Acid-sulfate deposits, which formed from acid hydrothermal solutions, are characterized by minerals such

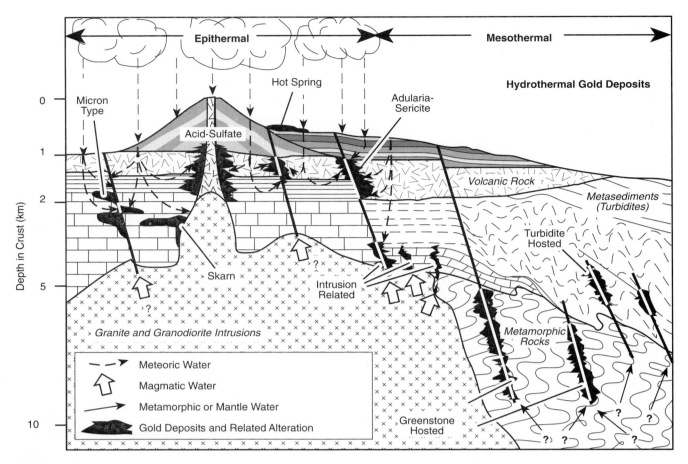

FIGURE 10-2
Schematic illustration of geologic environments in which hydrothermal gold deposits form, with epithermal deposits on the left and deeper mesothermal deposits on the right. Arrows show sources of water thought to have formed these deposits.

as alunite and pyrophyllite. These hydrothermal solutions owe their acidity to relatively large concentrations of gases such as CO_2, SO_2, and HCl, which were probably released from underlying intrusions as they cooled and crystallized. Many acid-sulfate deposits are restricted to small fracture systems that must have channeled these gases upward, but a few large deposits are known, including Goldfield, Nevada, El Indio, Chile, and Pueblo Viejo, Dominican Republic (Walthier et al., 1985; Muntean et al., 1990).

Hot-spring deposits form where adularia-sericite or acid-sulfate fluids reach the surface to form hot springs. Hot springs are usually surrounded by silica-rich deposits known as *sinter* or carbonate-rich deposits known as *travertine*, which are similar to the scale that blocks geothermal wells. Because these deposits form at the surface, they are easily removed by erosion and few of them are known. The largest deposit

of this type is the McLaughlin mine, in the northern part of the Napa Valley in California.

Mesothermal Deposits Mesothermal deposits consist of gold-bearing quartz veins that formed at deeper crustal levels (Figure 10-2). They were deposited by fluids hotter than about 250°C and are usually surrounded by altered rocks containing calcite and other carbonate minerals, indicating that the solutions were rich in CO_2. These deposits are found in three different geologic settings, which form the basis for dividing them into three groups, intrusion related, greenstone hosted, and turbidite hosted, as discussed next.

Intrusion-related mesothermal veins are the simplest of the group. They consist of quartz veins with gold, silver, and base-metal sulfides, which formed around felsic intrusions that were emplaced at depths of 5 kilometers or more in the crust. The fluids that formed these veins are almost certainly a mixture of magmatic

FIGURE 10-3

Gold-bearing quartz vein at Hishikari mine, Japan, the richest gold mine in the world. The vein, which is slightly less than a meter wide here, consists of thin layers of quartz, some of which contain gold. They were deposited by pulses of hydrothermal solution rising through the fracture that became the vein.

water from the intrusion and surrounding deep meteoric water. In most parts of the world, these vein systems are not large enough to be major sources of gold, although they have been mined extensively in Korea and Russia (Shelton et al., 1988).

Greenstone-hosted and *turbidite-hosted mesothermal veins* probably formed at deeper levels, possibly as much as 10 kilometers. Greenstone-hosted veins are so named because they are found in metamorphosed mafic volcanic rocks that contain the green mineral chlorite (Kleppie et al., 1986; Robert et al., 1991). Most greenstone-type deposits are Archean or Proterozoic in age and they are found as quartz veins along offshoots from major crustal fractures that are hundreds of kilometers long and many kilometers deep (Figure 10-2). Younger examples, such as the Mother Lode in California, are along similar fault systems. Related turbidite-hosted deposits are found in thick sedimentary sequences known as *turbidites,* which formed by

erosion of these volcanic rocks. In many of these deposits, gold is concentrated in iron-rich wallrocks rather than in veins, and it deposited when gold-sulfur complex ions in solution reacted with iron to deposit iron-sulfide minerals, removing sulfur from the complex ions. Prevailing opinion favors a metamorphic source for the waters that formed these deposits, probably when thick plates of the upper crust were stacked on top of one another by thrusting, causing a large pulse of fluid to be released. This process apparently took place largely during the later stages of consolidation of the early continents (Groves et al., 1987).

Placer and Paleoplacer Deposits

Placers have been the main source of gold for thousands of years, reflecting the extremely high specific gravity of gold and its resistance to weathering. Most placers are approaching exhaustion, however, with the possible exception of the Siberian deposits, which were sites for prison camps during the Stalin years. Large-scale placer mining last took place in North America during the late 1800s, when gold seekers in California followed them upstream into the veins of the Mother Lode (Figure 10-5). From there, they continued northward through British Columbia into the Klondike goldfields of the Yukon and on into Alaska to Fairbanks and Nome. At the same time, diggers in Australia moved into Victoria to mine the placers at Bendigo and Ballarat. Most of these were typical stream placers, except at Nome, where much gold was in beach placers. Many deposits exhibited the curious feature known as *bedrock placer values* in which gold was concentrated at the bottom of a thick layer of stream gravels. This probably happened when floods disrupted the entire layer, shaking down gold that was originally deposited throughout the sediment (Cheney and Patton, 1967). Another curious feature of placers, *giant nuggets* that are larger than gold in veins from which they were derived, remains controversial to this day. Native placer miners in some areas believe that large nuggets form by the dissolution and reprecipitation of smaller ones and even throw in seeds of fine gold to keep up the harvest. Experiments demonstrate that gold is soluble in water containing organic matter, suggesting that there could be wisdom in this practice (Boyle et al., 1975).

As placer deposits began to be exhausted at the end of the 1800s, the incredible Witwatersrand paleoplacers of South Africa were discovered in 1886. The Witwatersrand is the remains of a late Archean-age clastic sedimentary basin that contains several gold-bearing conglomerate layers, as well as the Carbon Leader, a layer of carbonaceous material, possibly

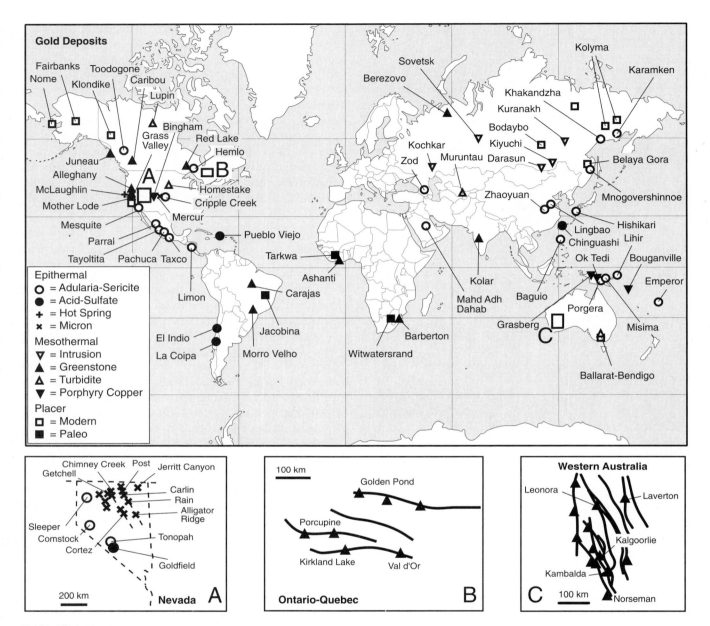

FIGURE 10-4
Location of major gold deposits discussed in the text. Insets show location of epithermal deposits in **(A)** Nevada (including Carlin and Cortez Trends) and mesothermal deposits along major faults in **(B)** Ontario and Quebec and **(C)** Western Australia.

originally an algal mat, that contains gold. Where these layers have been mined along the 300-kilometer northern margin of this basin, they have yielded about 32,000 tonnes of gold and contain an estimated reserve of 20,000 tonnes. These figures are much larger than the 2,000 tonnes of gold recovered from the large California placer deposits or the 5,000 tonnes of production and reserves found in large hydrothermal gold

deposits such as the Carlin Trend, the Homestake mine, and the Porcupine district of Ontario.

This difference has caused doubt in some circles that the Witwatersrand is really a paleoplacer, a skepticism fueled by its scarcity of good gold nuggets and the presence of gold in the Carbon Leader. Most observers, however, feel that the deposits are placers that formed in a stable, long-lived sedimentary basin that

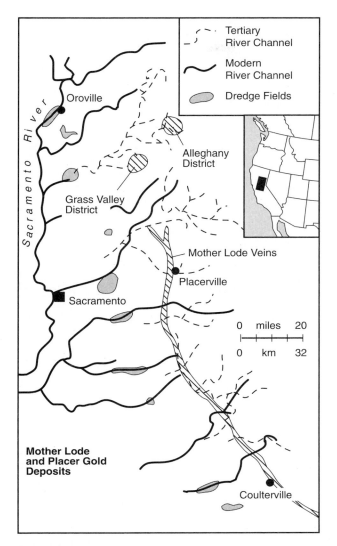

FIGURE 10-5

Distribution of placer gold associated with the Alleghany, Grass Valley and Mother Lode greenstone gold veins. Gold is found in river systems of two different ages, an older Tertiary system that formed in Early Cenozoic time and a modern system that redistributed some older gravels and eroded new gold from some veins. In wider parts of rivers, placer deposits were mined by dredges (compiled from *U.S. Geological Survey Professional Paper 73* and publications of California Division of Mines).

concentrated gold very efficiently. They feel that the lack of nuggets is due to the later migration of basinal and metamorphic water through the sediments, which dissolved and reprecipitated some gold. They argue further that the deposits are so much larger than possible hydrothermal parents because large amounts of gold were eroded from noneconomic gold veins in the drainage basin. This line of interpretation is supported by the presence in the conglomerates of ura-

ninite, another heavy mineral found in paleoplacer deposits at Elliot Lake, Ontario, as discussed in the section on uranium deposits (Phillips et al., 1987; Minter, 1991; Robb and Meyer, 1991; Loen, 1992).

By-Product Gold

Gold is a very important by-product of many base-metal mines, particularly porphyry copper deposits, which are mined in such large volumes that their trace levels of gold (about 0.5 ppm or less) become economically important (Cox and Singer, 1992). The Bingham deposit in Utah produces almost 16 tonnes annually, more than double the production of the nearby Mercur gold mine, and similar amounts come from Ok Tedi and Grasberg in the western Pacific.

Gold Mining and South African Mine Labor

Placer gold mining requires only that gold-bearing stream gravels be diverted over a riffle system to catch the gold. Riffles can be built of wood, rough fabric, or skin. In ancient times, sheepskins were used, whence the name *golden fleece;* now Astroturf is used. Large-scale dredges use more sophisticated gravity-based separation methods.

Although open-pit gold mines are not unusual, underground mines of the Witwatersrand are a different story. These mines must extract millions of tonnes of ore from layers that range in thickness from a few centimeters to only about 2 meters (Figure 10-6). This has meant that the gold-bearing layers must be mined over very large areas. It is difficult and expensive to haul ore great distances along layers to a central shaft. Most early mines used numerous, relatively closely spaced vertical shafts to remove ore, although these are also expensive.

As production continued, Witwatersrand mines have extended to extremely deep levels approaching 4,000 meters, and have encountered very hostile conditions. Even though geothermal gradients in this area are not particularly high, newly broken rock has temperatures of up to 65°C. These temperatures are impossibly hot for workers, particularly in the high humidity created by the water sprayed to prevent silicosis. Pumping chilled air into the workings barely helps because it is compressed by the increasing pressure on its way down, a process that almost doubles the temperature of air from the surface before it reaches miners at 3,000-meter depths. Chilled water undergoes less compression and is widely used, but too much water causes broken rock to flow out of stopes in highly dangerous "mud rushes." Newer cooling efforts include jackets filled with ice or dry-

FIGURE 10-6

Witwatersrand miners loading blasting compound into drill holes in gold-bearing conglomerate prior to blasting the ore. White quartz pebbles can be seen in the conglomerate layer (see Plate 8-2 for a photo of this ore) and its thickness is indicated by the height of the stope. Timber frameworks (behind front miner) are spaced a few meters apart to slow the collapse of the roof (back) of the mine (courtesy of Andrew Lanham, Gold Fields of South Africa).

ice, and plans are being made to manufacture ice underground and pipe it as chips to working faces. In 1985, cooling costs, including training and screening of workers, were about 10% of mining costs and were estimated to rise to 30% by the early 2000s (Collings, 1985).

Deep South African mines also experience *rock bursts* or "bumps," which occur when rock that has been under high confining pressures is exposed in mine workings and fails by explosions, sometimes causing fatalities. Although the depth at which these occur depends on rock characteristics and the type of mining, most deep South African mines have frequent bumps, some of which are large enough to be felt as earthquakes at the surface. Most mines use a seismic system, similar to that used for earthquake detection, to warn workers about rock burst activity.

South African gold mining is highly labor intensive, with over 400,000 unskilled and semiskilled employees drawn from throughout southern Africa (Figure 10-7). About 40% of this labor force comes from Botswana, Lesotho, Mozambique, and Swaziland, and another 35% has come from the homelands of Bophuthatswana, Ciskei, Transkei, and Venda. Fewer come from Soweto and other areas of South Africa near the mines, suggesting that the "push" of homeland poverty is stronger than the "pull" of employment opportunities. These migrant workers are usually employed for a one-year period and live in hostels on the mine properties. Efforts to permit work-

ers to live with their families were impeded by apartheid laws and the sheer size of the work force, although some progress has been made recently (Crush et al., 1991; COM, 1992).

South African mines have been widely criticized for paying too low a wage to their unskilled and semiskilled laborers, a charge that deserves critical examination (Wilson, 1985). Wages that can be paid to any gold miner will be limited by the amount of gold produced. As can be seen in Table 10-2, which compares major gold mines in North America and South Africa, average annual gold production per employee is 484 ounces in North America versus only 44 ounces in South Africa (Table 10-2). This enormous difference reflects the highly mechanized nature of North American gold mining and is not a comment on the skill of the South African miner; no one could do better under the same geologic conditions. But it does control wages. Thus, if North American wages average $20,000 to $40,000 annually, equivalent South African wages should be $1,818 to $3,636, assuming that the companies have similar profit margins. According to the South African Chamber of Mines, the average wage for unskilled and semiskilled workers in the mines in 1991 was about $3,800 (11,400 Rand), proportionally more than the North American miner. This higher wage probably accounts in part for the lower operating profits in the Witwatersrand mines (Table 10-2), and it suggests that there is little opportunity to increase South African mine wages without decreasing em-

FIGURE 10-7

Major mineral deposits in southern Africa, showing the locations of placer and bedrock (kimberlite pipe) diamond deposits, Bushveld Complex and Great Dike layered igneous complexes (LIC) and Witwatersrand sedimentary basin. Detail on Bushveld Complex is shown in Figure 8-17; Witwatersrand mines are along the northern margin of the basin. Countries and homelands supplying labor to the mines (percent of total supply given for each country) from South African Chamber of Mines Report (1991).

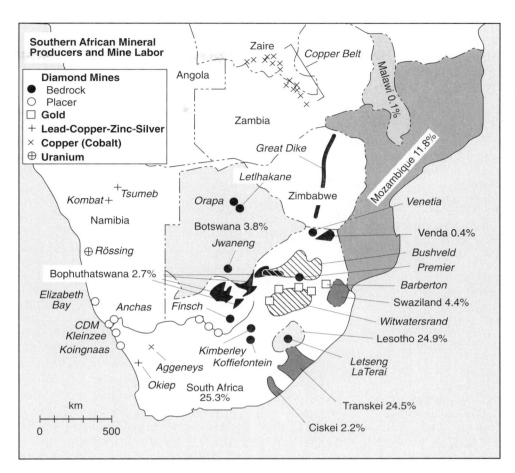

Gold Production and Related Environmental Concerns

Placer gold mines were early environmental offenders. The greatest concern was expressed in California in the late 1800s over the use of *hydraulic mining,* a method that applied high-powered blasts of water from hoses to break up gravels and move them into gold-concentrating systems. Sediment stirred up by these operations in the Sierra Nevada reached as far as San Francisco Bay. In 1883, Congress passed the *California Debris Commission Act* requiring that wastewaters be impounded until sediment had settled out, and many placer mines were closed because they could not support the added costs (Koshmann and Bergendahl, 1968).

Gold is recovered from ores by two main methods, both of which involve environmental concerns (Salter, 1987; Stanley, 1988). Early recovery used *amalgamation,* in which ore is mixed with mercury that selectively dissolves the gold and is then removed by

ployment or making the South African mines significantly less profitable than their global counterparts.

evaporation. This approach, known as the *patio process,* was widely used during Spanish settlement of the New World, as discussed in the section on mercury. Mercury from these operations was never recovered and remains as a pollutant in many old mining areas. Mercury concentrations of up to 3.7 mg/kg have been observed in stream sediments around gold deposits in northern Wales, where mining ceased in 1916 (Holden, 1990; Fuge et al., 1992). The situation is worse where amalgamation is still being used, largely in Latin America and the Pacific Rim. Particular concern has focused on the Madeira river basin in Brazil, from which native miners known as *garimperos* have produced more than 3 million ounces of gold by this method. Nriagu et al. (1992) suggest that 90 to 120 tonnes of mercury are discharged into this drainage basin annually. Mercury contents of some fish, which are a major part of the local diet, exceed the 0.5 microgram/gram limit set by the World Health Organization, and at least 84 cases of mercury poisoning have been diagnosed (Koren, 1992). Efforts to stop this have been thwarted by popular resistance because gold mining is one of the few means of local employment. Adoption of simple mercury retorts, in which vapor-

TABLE 10-2

Gold production of major South African and North American producers in relation to employees and profitability. Production is given in ounces (1 kg = 32.105 oz) and gross profit margin refers to operating margin before taxes (data for 1990 from annual reports).

Company	Number of Employees	Gold Production	Ounces per Employee	Gross Profit Margin
South African Producers				
Kloof	17,516	1,020,000	58	40.5
Western Deep Levels	29,374	1,350,000	46	30.0
Driefontein	32,500	1,600,000	49	46.7
Free State Consolidated	117,842	4,000,000	34	12.5
Blyvooruitzicht	10,711	330,000	31	6.8
Total	207,943	8,003,000		
		Average	44	27.3
North American Producers				
American Barrick	1,730	790,000	457	45.7
Battle Mountain	1,500	960,000	640	16.0
Echo Bay[a]	1,900	740,000	389	29.5
Hemlo	450	470,000	1,044	50.5
Homestake	2,225	1,045,000	470	15.3
Newmont Gold	2,550	1,420,000	557	45.6
International Corona	570	720,000	1,263	41.0
Lac Minerals[b]	3,600	880,000	244	28.2
Total	14,525	7,025,000		
		Average	484	34.0

[a]Significant production comes from high-cost, fly-in mine at Lupin, NWT, Canada.
[b]Significant production comes from El Indio mine, Chile.

ized mercury is condensed and recycled, would be a major improvement.

The *cyanide process,* which replaced amalgamation in the early 1900s, is based on the fact that precious metals form soluble complex ions with cyanide (CN⁻) anion (Constantine, 1985; Popkin, 1985). Because cyanide will not dissolve quartz, iron oxides, and other common gangue minerals, it yields a relatively simple gold-bearing solution, which is known as a pregnant solution. In some gold mines, gold is dissolved from the ore by crushing and grinding it and then mixing it with cyanide solution in large vats. In arid areas, ore is put in piles onto which cyanide solutions are sprayed and pregnant solution is collected as it leaks out the bottom of the pile (Figure 10-8). This process, known as *heap leaching,* requires less investment than conventional processing using crushing and vats, and has made it possible to mine low-grade gold ores that would otherwise be uneconomic (Wargo, 1989). Gold is sometimes recovered from pregnant solutions by adding zinc to form soluble zinc cyanide and a pre-

cipitate of gold and silver. The pregnant solution can also be passed over activated carbon, which adsorbs dissolved gold. Gold from either process is cast into bars of *bullion* (Plates 8-4 and 8-5), known as *doré* when it contains silver, which must be further refined to remove impurity metals such as Hg, As, and Cu.

Some ores cannot be treated by the cyanide process because their gold is in small inclusions or even in solid solution in minerals such as pyrite. This gold is generally recovered by *roasting,* which converts pyrite into porous iron oxides containing small grains of gold that can be dissolved by cyanide. Roasting is considerably more expensive because it requires a special oven and releases SO_2 and As gases that must be recovered by scrubbers. It has been used successfully at several mines, however, and will be required to mine much of the deeper, sulfide-rich ore reserves in the United States. Biological decomposition of ores might take the place of roasting at some point in the future.

Cyanide is a highly toxic compound. Although it is found in common plants such as almond and cassava,

A.

FIGURE 10-8
The Gold Quarry sediment-hosted micron gold mine **(A)** in Nevada, produces over 1 million ounces of gold annually from heap-leach pads **(B,** middle background) and a more complex beneficiation plant for ore that will not leach effectively (foreground) (photographed by Manley-Prim, Tucson; courtesy of Newmont Gold Company).

B.

TABLE 10-3
Solvents that might be used in place of cyanide in processing of gold ores, showing their toxicities. Toxicity figure given here is LD_{50} in mg/kg measured on rats by ingestion of compound, as discussed in Chapter 3. Smaller numbers indicate greater toxicity (from Hiskey and DeVries, 1992).

Compound	Composition	Toxicity
Thiosulfate	$S_2O_3^{-2}$ ion	7500
Sodium iodide	NaI	4340
Sodium bromide	NaBr	>3500
Bromochlorodimethylhydantoin "BCDMH"	Complex	600
Thiourea	$CS(NH_2)_2$	125
Malononitrile	$CH_2(CN)_2$	61
Sodium cyanide	NaCN	6.4

concentrations in ore solutions are higher and require special handling. During ore treatment, the pH of cyanide solutions must be kept above about 11 to prevent the cyanide from reacting with the hydrogen ion to produce HCN, a deadly gas. Even when dissolved, only about 50 to 250 mg of cyanide ion can cause death in humans (Moore, 1991). Most solutions used to treat gold ores contain about 250 mg/liter of dissolved sodium cyanide. Nevertheless, accidental deaths resulting from the use of cyanide in gold recovery plants are extremely rare, considerably more rare than deaths due to traffic accidents during commuting to work, for instance. In spite of this generally good record, cyanide is coming under increasing environmental scrutiny and initiatives to restrict its use have been made in some states and provinces. Although less toxic substitutes for cyanide are known (Table 10-3), it is not yet clear that they will be cost effective or that they are not also environmentally damaging.

Cyanide oxidizes to produce H_2O, CO_2, and nitrogen compounds, which are not toxic. Originally, cyanide-bearing wastes were oxidized in holding ponds and then discharged into local streams. To speed up the process and remove other metals from solution, modern plants agitate waste solutions in the presence of SO_2 or microbes (Devuyst et al., 1991). This is particularly helpful in arid areas, where migrating birds use ponds as resting spots, in spite of aggressive prevention efforts including screens, plastic flags, remote-control chase boats, and noisy cannons. Although this problem has been highly publicized, the number of birds killed by cyanide is less than 0.1% of that killed by hunters each year (Kennedy, 1990; Conger, 1992). Runoff or groundwater recharge from abandoned heap-leach pads could be an environmental hazard, particularly in wet areas where oxidation is slow. Recent concern has centered on the Summitville mine in Colorado, where snow melt creates a heavy spring runoff over abandoned leach pads, some of which were not positioned properly with respect to existing drainages. Because the original operator went bankrupt in 1992, supervision of the property reverted to the Environmental Protection Agency, which lacks funds and personnel for this type of work. The original problem, which resulted from inadequate state supervision and careless operation, is a lesson that major mistakes occur, even in an era of environmental awareness (Posey et al., 1993).

Gold Production and Reserves

Although gold is produced in 67 countries, South Africa dominates the picture, accounting for 30% of production. As can be seen in Figure 10-9, however, production is declining. This is partly due to declining grade, which went from 0.4 ounces per ton (13 grams /tonne) in 1970 to about 0.2 ounces per ton (5 grams /tonne) in 1990. Although ore mined annually increased by 50%, the declining grade caused productivity to decrease significantly to less than 30 ounces per employee. Whether this is the prelude to the end of South African dominance of world gold mining will be a complex function of geologic factors, mining costs, politics, and gold reserves (Pretorius, 1981; Thomas, 1984). There is no question about this last factor. In spite of nearly a century of production, almost half of world gold reserves of 44,000 tonnes are in the Witwatersrand. Even though U.S. gold reserves doubled between 1974 and 1990, and gold reserves elsewhere also increased, combined U.S., Canadian, Australian, and Brazilian reserves are still less than half of those in South Africa. And much of the U.S. reserve is in deeply buried ores that will require expensive underground mines and high-cost pyrometallurgical extraction methods.

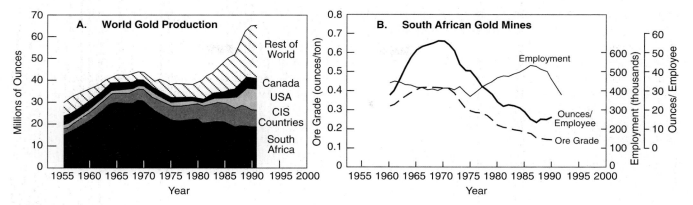

FIGURE 10-9

A. World gold production. Note the decline in production from South Africa and the increase from the United States. **B.** Change in grade of ore mined and employment in South African mines (compiled from data of U.S. Bureau of Mines and South African Department of Mineral and Energy Affairs).

The future of mining in South Africa is clouded by political changes including proclamations by the African National Congress that the gold industry should be nationalized. This seems particularly unwise in view of the dismal record of national mining companies in Africa, but it could well be the pivotal factor in future world gold supplies. The wild card in all of this remains CIS gold production and reserves. Production from these countries has been static for almost a decade and most of the operations are antiquated and poorly run. Recent agreements between CIS governments and western mining groups for cooperative exploration of major gold deposits such as the Newmont-Muruntau effort in Uzbekistan could increase CIS production, partly offsetting any decline in South African production.

All in all, the gold reserve picture is not comforting. World reserves are adequate for only about 20 years of production at present rates and much of this is in South Africa. In spite of intense exploration for a century, no province similar in size to the Witwatersrand has been found. Other discoveries have all been much smaller. For instance, the largest discoveries of recent years, the Post-Betze micron gold deposit in the Carlin Trend and the Lihir epithermal vein deposit in Papua New Guinea, contain less than 1,000 tonnes each, less than 5% of remaining Witwatersrand reserves. Thus, unless demand for gold changes dramatically in the near future, the price must rise significantly to make lower-grade reserves economic. This, in turn, will keep South Africa in the world gold limelight for the foreseeable future.

Gold Trade and Related Monetary Matters

Gold differs from all other elements in its historical relation to money. It was among the first metals used in coins, where it gradually became dominant because of its rarity, beauty, resistance to corrosion, and extreme *malleability*, the property that permits it to be shaped easily. As international trade increased, gold became the medium for settling debts and by 1816 the abundance of newly mined gold led the United Kingdom to establish the *gold standard*. In its purest form, the gold standard set the price for gold in the currency of the country and allowed free import and export of the metal. Thus, the amount of money that a country could put into circulation was controlled by the amount of gold in its central bank, a powerful deterrent to inflation. Other countries, including the United States, followed the United Kingdom into the gold standard, which endured until the economic crises following World War I. On January 31, 1934, the U.S. government increased the official price of gold from the previous $20.67/ounce ($643/kg) to $35/ounce ($1,090/kg). In 1968, a two-tier international market was established, an official market, and a private market in which the price was determined by supply and demand. Responding to various international economic stimuli, the price of gold began an upward spiral that peaked at about $800/ounce ($24,918/kg) in 1980 (Figure 10-1). Although there is no formal link between gold and money at present, government gold stocks (Table 10-1) are still widely regarded as an indication of the health of their economies and the value of their currencies (Hawtrey, 1980; Drummond, 1987; Crabbe, 1989). The U.S. gold stock is valued officially at $42.22/ounce ($1,315/kg).

Gold has been regarded historically as a secure haven for wealth in troubled times. Although the allure of gold hoarding has diminished in MDCs, it remains important in some areas. In 1990, for instance, an estimated 265 tonnes of gold were hoarded in the Far East, largely Japan, Thailand, Taiwan, and Vietnam, with another 44 tonnes in the Middle East, dominantly Saudi Arabia (Murray et al., 1991). Because hoarded gold does not earn interest, it must compete with other investments solely on the basis of price appreciation. Viewed in this way, gold has been only a mediocre investment during this century, increasing in value by about 1,900% between 1925 and 1990, in comparison to 750% for the Consumer Price Index and 2,700% for the Dow Jones Industrial Average. However, when inflation became a problem between 1960 and 1990, the return on gold was 1,100%, versus only about 400% for the CPI and DJI. Thus, investment demand for gold will probably continue, particularly in countries with unstable currencies. At present, however, jewelry makers in Italy and India dominate world gold trade (Figure 10-10). Although most jewelry enters world markets, some is hoarded in India. The combination of gold markets and investment demand appears to be strong. So unless a major new discovery is made or the price rises enough to render deep Witwatersrand ore economic, gold may return to its status as a scarce metal.

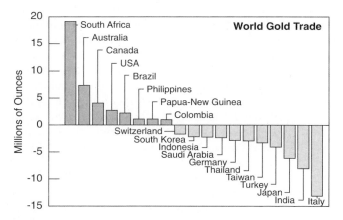

FIGURE 10-10

World gold trade, showing the difference between mine production and gold used in fabrication for 1991. Gold used in investments, hoarding, and coinage are not included but are minor in comparison to the amount used in fabrication. Gold obtained from scrap and from industry stocks is included in the fabricated gold total (compiled from Murray et al., 1991).

SILVER

Silver is evolving from a precious metal to an industrial metal, a path also being followed by platinum. Its usage differs significantly between the MDCs and the LDCs. In LDCs, where consumption rates are low, silver is still a precious metal and is used largely in jewelry and tableware. In MDCs, it is also used in industrial applications and coins, and consumption rates are higher. The main industrial market for silver is photographic film, in which silver halide compounds embedded in film are reduced to native silver, with the amount of silver reduced being proportional to the amount of light exposure. Silver also finds a wide variety of applications in the electrical and electronics industries, depending on its very high electrical and thermal conductivity. It is also used in pharmaceuticals, in highly reflective mirrors, in batteries with zinc, and as a major component in dental amalgam with mercury. These markets support annual world production of almost 15,000 tonnes worth almost $2 billion (Figure 10-1).

Geology of Silver Deposits

Silver occurs in *electrum,* the silver sulfide *argentite,* and several complex sulfide minerals containing lead, copper, antimony, and arsenic, such as *tennantite-tetrahedrite* (Boyle, 1968). It is found in a wide variety of hydrothermal deposits as well as in a few placer deposits containing electrum. It is the most important metal in three deposit types but is more commonly a by-product.

The most familiar silver-dominant deposits are the *epithermal vein deposits* discussed previously as sources of gold (Figure 10-11). Many of these veins in the United States and Mexico have silver:gold ratios greater than 100, causing their economics to be dominated by silver (Simon, 1991). The other two types of silver-dominant deposits are less familiar. The *Cerro Rico type* (Figure 10-12), which includes the largest silver deposit in the world, is best developed in the Bolivian tin-silver belt (Turneaure, 1971). Cerro Rico, at Potosi, Bolivia, consists of quartz-silver-tin veins that cut a dome of silica-rich volcanic rock (Figure 10-13; see pg. 251). The veins began to deposit ore minerals at temperatures of 500°C from very saline fluids, and continued to precipitate ore down to temperatures of less than 100°C as the solutions became less saline (Sillitoe et al., 1975; Suttill, 1988). This pattern suggests that the hydrothermal system began with hot magmatic waters and that cooler meteoric waters invaded and finally dominated the hydrothermal system as time went on.

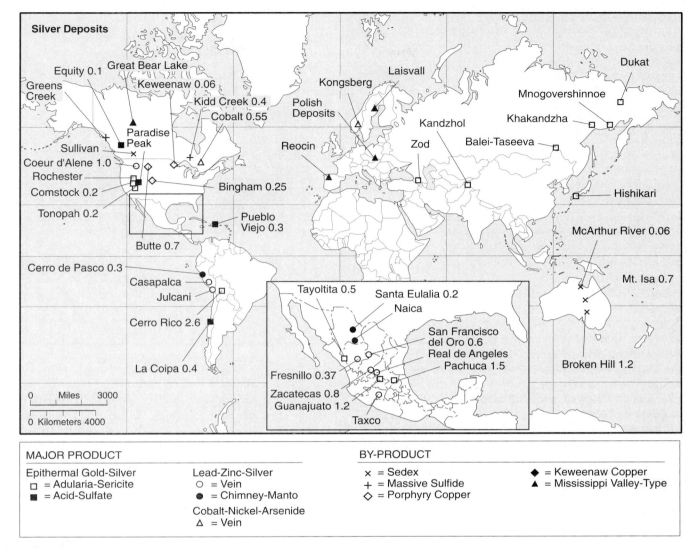

FIGURE 10-11

Location of important silver deposits showing historic production plus reserves in billions of ounces (compiled from U.S. Bureau of Mines data; Boyle, 1979; and reports on individual districts).

Cobalt-nickel-arsenide veins are also important sources of silver. The largest such deposit, at Cobalt, Ontario (Andrews et al., 1986), consists of veins with native silver and cobalt, nickel, and iron arsenides in calcite and quartz. The veins are found in sediments above and below a large, tabular gabbro intrusion and they appear to have formed from basinal brines that were heated by the intrusion (Figure 10-12).

By-product silver comes largely from gold and base-metal deposits. In most of these deposits silver minerals form small inclusions in base metal sulfide minerals and they are difficult to separate. The huge Proterozoic-age sedex deposits at Broken Hill and Mt.

Isa in Australia are major silver producers (Both and Stumpfl, 1987), as are the Bingham porphyry copper deposit in Utah and the Kidd Creek massive sulfide deposit in Ontario. The Greens Creek massive sulfide deposit in Alaska was the biggest silver producer in the United States at the end of the 1980s (Reese, 1990). Silver is also important in chimney-manto deposits, such as Santa Eulalia and Naica in Mexico and Cerro de Pasco in Peru, where it can dominate economics (Megaw et al., 1988), and it is found in almost all types of lead-zinc vein deposits, including the large Coeur d'Alene district in Idaho. Mississippi Valley-type ores are commonly low in silver, although the Old Lead

FIGURE 10-12

Schematic illustration of silver-dominant deposits including Cerro Rico-type epithermal tin-silver veins around a rhyolite volcanic dome and Cobalt-type silver-cobalt-nickel-arsenide veins around flat, sill-like mafic intrusions.

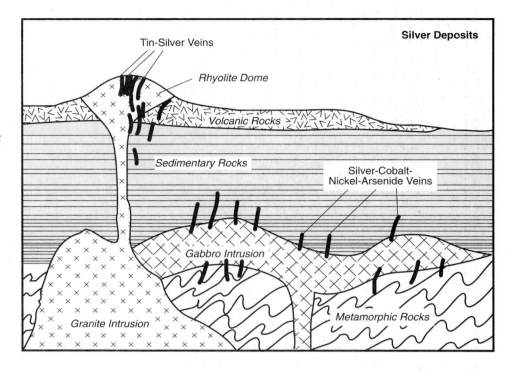

Belt and Viburnum Trend in Missouri have produced significant silver.

Sediment-hosted copper deposits at White Pine in Michigan and the Kupferschiefer in Germany and Poland have by-product silver, as did the Keweenaw native copper deposits in Michigan (Olson, 1986). Because silver has an even higher electrical and thermal conductivity than copper, copper metal with minor amounts of silver from these deposits was particularly desirable. In fact, the term *Lake-brand copper,* named for the Keweenaw area on Lake Superior, refers to a premium brand of copper that contains up to 400 grams of silver per tonne. Silver is not found in all copper deposits of this type; it is not an important constituent of the Copper Belt ores of Zaire and Zambia, for instance.

Silver Mining, Production, Markets, Trade, and Reserves

Silver is produced from underground and open-pit mines that have no unique operating or environmental problems. Production of silver from its ores depends largely on whether it is present with gold or with base-metal sulfides. Where silver is in electrum and other gold-silver minerals, usually in epithermal deposits, it is recovered by cyanide leaching, as discussed in the section on gold. Where it is present in base-metal sulfides, as in lead-zinc and most copper deposits, it must be recovered by a special step in the smelting process in which a separate, silver-rich mineral concentrate is formed. Silver recoveries in all of these processes are relatively low and it is very rare for all silver in a deposit to actually make it to market.

Silver is produced in 56 countries, with most of it coming from the United States, Canada, Mexico, and Peru (Mohide, 1992). The United States is the only major silver producer that is a net importer of the metal (Figure 10-14A). The main competition for industrial silver markets are silver-free photographic film, video tape, and xerography. So far, silver-free film is available only in black and white, and other image-preservation methods lack the resolution and color quality of photographic film. Combined industrial, jewelry, and silverware markets constitute about 95% of world silver consumption, with the remainder being used in coins.

The use of gold and silver in coinage creates a temptation for governments that maintain precious-metal stockpiles, such as that held by the United States in the National Defense Stockpile. By using this material in coins, governments can show a "profit" on the coins that would not be possible if they purchased the metal on the market in the same year. The allure of this approach is strongest in years of high budget deficits. For instance, all of the 187 tonnes of silver used in the U.S. American Eagle coin have come from the stockpile, which was reduced by 50% by minting between 1986 and 1992.

FIGURE 10-13
Cerro Rico, seen from the city of Potosi, Bolivia, is a dome of volcanic rock cut by quartz veins containing silver and tin. Top of the dome (dark) was altered by the addition of silica. During the sixteenth and seventeenth centuries, this was the second largest city in the New World, contributing greatly to the wealth of Spain (courtesy of C.G. Cunningham, U.S. Geological Survey).

Silver production has fallen short of consumption, frequently by a wide margin (Figure 10-14B). Most of the shortfall is made up by recycled silver from old scrap and fabrication waste, which accounts for up to 50% of consumption in the United States. During periods of peak demand, however, silver comes from other sources. The largest of these are silver jewelry in India and bullion, coins, and silverware in MDCs, particularly the United States, which were estimated to total 118,360 tonnes at the end of 1991. Silver holds a special allure in India, where it is used in everything from jewelry to edible silver foil, 31 tonnes of which are consumed annually as an aphrodisiac and health aid (Keating, 1992). Figure 10-14B shows that silver consumption by nonindustrial sources, largely investors and hoarders, was particularly high during the

1960s. Much of this silver was purchased as a hedge against inflation, but some of it was accumulated as a gamble that the shortfall in silver mine production would lead to a dramatic price increase. The culmination of this speculation, which took place when the price of silver ran to $48/ounce ($1,495/kg), was the Hunt brothers' silver debacle, which was discussed earlier.

World silver reserves of 280,000 tonnes are adequate for only about 18 years of production at present rates and are highly dependent on by-product silver from lead-zinc and copper deposits. Most silver-dominant deposits form at shallow levels of the crust and are relatively easy to discover compared to more deeply buried deposits. Thus, it is likely that a larger proportion of them have already been found. This means that an increasing proportion of future silver production is likely to come from by-products of lead-zinc and copper mining. Here, too, the outlook is poor. Important by-product deposits, such as the silver-rich sedex deposits at Mt. Isa and Broken Hill, Australia, are being exhausted, and new deposits such as McArthur River contain much less silver. It is unlikely that by-product silver from porphyry copper and massive sulfide deposits will be able to supply world demand. Thus, only a major change in silver markets will prevent prices from rising, along with gold, from their decade-long decline (Figure 10-1).

PLATINUM-GROUP ELEMENTS

The platinum-group elements *(PGE)* include *platinum, palladium, rhodium, ruthenium, iridium,* and *osmium,* which have similar chemical properties and occur together in nature. Platinum and palladium, with crustal concentrations of about 5 ppb each, are about as scarce as gold and the other PGEs are even scarcer. Compared to gold and silver, the PGEs are late-comers. Platinum was recognized as an element only in 1750, with osmium the last to be discovered in 1844. PGEs were first observed in some placer deposits, where they formed curious steel-gray nuggets that looked like silver but behaved like gold. In fact, the name *platinum* comes from *platino,* the diminutive term in Spanish for silver, which was used to describe nuggets found in the San Juan and Atrata rivers in Colombia (Loebenstein, 1985).

The main market for PGEs, that of a *catalyst,* is definitely industrial. Catalytic converters in automobile exhaust systems account for 40% of global platinum and palladium demand and 85% of rhodium demand. These converters speed oxidation reactions that convert hydrocarbons, nitrous oxides, and carbon monoxide in exhaust gases to less harmful carbon dioxide,

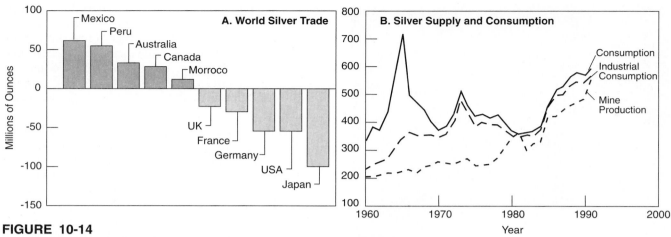

FIGURE 10-14

Silver trade. **A.** Balance of exports and imports for major silver producers and consumers for 1992. **B.** Historic trends in silver mine production and consumption showing long-term excess of consumption (compiled from data published by *Engineering and Mining Journal* and *Metal Statistics*).

nitrogen, and water. Another 7% of U.S. palladium-based catalysts are used increasingly in diesel-powered vehicles, although they are more readily contaminated by lead than those consisting of platinum and rhodium. Manganese, chromium, and copper have also been tried, but are not robust enough for the long-term service required in cars and trucks. PGE consumption is used as a catalyst in oil refining, in the production of nitric acid, and in fuel cells.

The other big market for PGEs in the United States, constituting about 30%, is the electrical and electronic industry. Here PGEs are used in electrical contacts, in high-resistance wires, in memory devices, and in special solders. A growing use in this sector is the application of PGEs in sensors that measure the oxygen content of automotive fuel and adjust the mixture for more efficient combustion. Dental and medical applications consume about 10% of the PGEs, largely to strengthen crowns and to treat cancer and arthritis. Other applications for the PGEs, constituting about 15% of total consumption, include special manufacturing hardware such as nozzles through which glass and ceramic fibers are extruded. World PGE production of almost 300 tonnes is worth about $3 to $4 billion, somewhat more than silver.

Geology of Platinum Deposits

Although minor PGE production still comes from placer deposits, most comes from magmatic deposits associated with mafic igneous rocks. These deposits can be divided into three groups. Dominant among them, in terms of both production and reserves, are the large *layered igneous complexes (LICs)* that were discussed as our main source of chromium and vanadium. As with these metals, the Bushveld Complex contains the most important PGE deposits (Figure 10-15). Present PGE mining in the Bushveld focuses on the Merensky Reef, which is found between the chromitite and vanadium-bearing magnetite layers (Figure 8-17), as well as the UG-2 chromitite, and the Platreef in the Potgietersrus area. The *Merensky Reef* is a complex layer consisting of coarse-grained, mafic silicate minerals, and one or more chromitites a few centimeters thick, which contains PGE-bearing sulfide minerals and alloys (Figure 10-16). Although the Merensky Reef forms a well-defined layer, 10% to 20% of it consists of depressions known as *potholes* where the reef did not accumulate or was remelted (Ballhaus, 1988). The entire reef usually contains 3 to 20 grams of PGEs per tonne, similar to the grade of gold in the Witwatersrand ores. These PGE concentrations are generally thought to be the result of magmatic immiscibility similar to that which formed magmatic nickel deposits (Naldrett, 1989). However, the extremely high concentrations of PGEs, in comparison to background abundances in the Bushveld rocks, has led to suggestions that hydrothermal solutions scavenged PGEs from deeper parts of the Bushveld and deposited them in the Merensky Reef (Boudreau et al., 1986).

The *UG-2 chromitite*, which is also mined widely for PGEs, differs from the Merensky in that it consists largely of chromite. It also contains over three times as much rhodium per tonne as the Merensky Reef. Be-

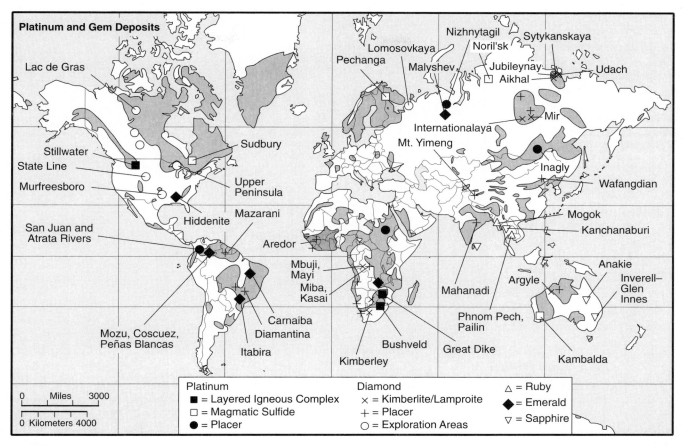

FIGURE 10-15

Locations of important deposits of platinum-group elements and gems. Shaded areas
show distribution of Precambrian cratons that host most diamond deposits.

cause rhodium sells for about eight times the price of
platinum, there is a great economic incentive to mine
the UG-2 Reef. The *Platreef* is found at the base of the
Bushveld on its northern side (Figure 8-17) and it
consists of blebs and veinlets of PGE-bearing sulfide
minerals in ultramafic rocks that contain large, partly
digested fragments of wallrocks. This ore is lower
grade, but larger tonnage than that in the UG-2 or
Merensky Reef, and was the last to be mined by
modern efforts.

PGEs are also found in other LICs, but in smaller
amounts. The best known of them is the Stillwater
Complex in Montana, where platinum production began only a few years ago from a layer known as the
J-M Reef. Although the J-M and Merensky Reefs have
many important geological similarities and appear to
have formed in basically the same way, the J-M averages about 21 grams of PGE per tonne, as much as
three times higher than the Merensky Reef. But plati-
num:palladium ratios in the J-M are 0.33:1 versus
about 3:1 in the Merensky, and its platinum:rhodium

ratios are about 32:1 compared to only 15:1 in the Merensky. The larger demand for platinum and rhodium
and their higher prices give the Merensky Reef an economic edge. Furthermore, reserves in the J-M Reef are
considerably smaller than in the Bushveld reefs because the Stillwater Complex has been extensively
folded, faulted, and eroded, and its limits are not as
well defined, with some PGE values extending above
and below the main layer (Naldrett, 1989).

The least important source for PGEs is *nickel-copper
sulfide ore* that formed by the separation of immiscible
sulfide magmas, as discussed in the section on nickel.
PGE concentrations in the sulfide minerals that make
up these ores are considerably lower than in the LICs.
As a result, PGE production from these deposits is feasible only where there is large nickel production, as
at Sudbury and Noril'sk. Sulfides in these deposits are
ten times higher in PGEs than sulfides in other nickel
deposits such as the Kambalda district in Australia,
but they are still 50 times lower than concentrations
seen in the Merensky Reef (Naldrett, 1989).

FIGURE 10-16

Close-up view of the Merensky Reef in the Rustenburg Platinum Mine showing coarse-grained nature with abundant chromite (dark grains). Reef increases in thickness abruptly toward the left as it enters a pothole.

Production of the Platinum-Group Elements

PGEs are produced from the Bushveld Complex almost exclusively by underground mining. These mines have the same problem that faces the Witwatersrand gold mines, that of extracting enough ore from thin layers to meet production requirements, and the consequent need to mine over large areas. The effort is further complicated in the Merensky Reef by the potholes, which require expensive diversions of working levels, causing lower overall ore recoveries than would be attained from a truly planar layer. To meet production quotas, mines have been opened throughout the Bushveld Complex wherever PGE ore layers are of adequate thickness and grade (Figure 8-17). Most of these mines are working at depths of 500 to 1,000 meters, and one project, Northam, has been developed to extract ore from the Merensky Reef below 1,200 meters. Although these mines are only about half as deep as the Witwatersrand gold mines, the geothermal gradient in the Bushveld Complex is twice as great as it is in the Witwatersrand, making operations at depths of only 1,500 meters roughly equivalent to the most hostile conditions seen in the gold mines.

Production of PGE metals differs from operation to operation but always involves the separation of a concentrate containing sulfide minerals and PGEs. In some cases, a special PGE-rich concentrate is formed before the other concentrate and is sent directly to the smelter. Once a PGE-rich matte is formed, it is dissolved and different metals are separated by ion exchange. Because there are so few PGE-producing facilities, various intermediate PGE products are sent all over the world for processing, with most travel among South Africa, Canada, Norway, and the United Kingdom (Loebenstein, 1985).

PGE smelters and refineries are subject to the same environmental problems as those affecting base-metal smelters. Although some forms of the PGEs are toxic (for instance, only a few ppm of palladium sulfate or nitrate can cause blood to coagulate), the elements are so rare that difficulties are observed only in refinery operations or chemical plants (Gowland, 1992). The main problem observed there is platinosis, which involves respiratory and dermatological symptoms from exposure to PGEs (IPCS, 1991).

Platinum-Group Element Markets, Trade, and Reserves

PGEs are produced in only ten countries, one of the smallest numbers for any mineral commodity. Most platinum production comes from South Africa, while most palladium production is from Russia. Smaller contributions come from LICs in the United States and Zimbabwe, magmatic nickel sulfide deposits in Australia, Canada, and Finland, and placer deposits in Colombia and Ethiopia. Production also comes from smelters in Japan that are supplied by imported nickel-copper ore. Most of this production goes to the large car manufacturers in Germany, Japan, and the United States.

PGE production has increased dramatically since 1974, when the automotive catalyst market opened. From 1983 when the United States required catalytic converters to 1992, the world PGE market almost doubled and was one of the fastest growing of world metals (Figure 10-1). Further increases can be expected because converters became mandatory in the European Community only in 1993 and some LDCs, including Brazil, Mexico, and South Africa, do not yet use them. Unfortunately for producers, supply increased even faster, reaching an amazing 10% annual rate by 1991, as more mines came on stream to supply projected future demand. The resulting oversupply caused PGE prices to decline and could put significant short-term stress on PGE producers. Platinum and palladium prices have been somewhat weaker than rhodium, which is restricted in abundance because it was strictly a by-product at all mines and could not be produced in large quantities until 1991.

Many people feel that the real future for PGEs lies in the jewelry and investment markets. Investment demand was very important in the United States during the late 1980s, but has since lessened. The jewelry market has never been strong in the United States but is a major market in Japan and Europe. In 1986, jewelry and investments accounted for 30% and 16% of world PGE demand (Feichtinger et al., 1988).

About 90% of world PGE reserves of 56,000 tonnes of contained metal is in South Africa and Russia, where both geological and political uncertainty cloud the supply picture (Sutphin and Page, 1986). In South Africa, much of the reserve is in deeper levels of the Merensky Reef or in the UG-2 or Platreef, where mining will not be as simple and new processing methods will be needed. Although the UG-2 contains significant chromium, it is currently discarded, and PGE production has to bear the complete cost of mining. In addition, much of the South African PGE production is in Bophuthatswana, one of the independent homelands that has not been widely recognized outside South Africa. Continued good relations between the PGE companies and Bophuthatswana will depend in part on the future of negotiations for a new government in South Africa. The supply situation in Russia is similarly cloudy, with major uncertainties surrounding the future of the outdated Noril'sk operation. Although producers would welcome some shortages to boost PGE prices, extended shortages would encourage substitution and lead to the loss of the pivotal catalytic converter market, which is already under pressure from electric cars.

GEMS

Although diamonds are the best known of the world's gem stones, over 150 natural compounds have been used as gems (Jahns, 1983; Pressler, 1985; Austin, 1991). Of these, diamond, emerald, and ruby consistently sell for the highest prices, with alexandrite and sapphire only slightly behind (Table 10-4). Amber, aquamarine, jade, opal, pink topaz, spinel, and tourmaline have intermediate values. Many other minerals, from agate and amethyst to zircon and zoisite, make up a third tier of gems that are usually inexpensive, although special examples can be quite expensive, a fact that has discouraged use of the term *semiprecious stone.*

As can be seen in Table 10-4, gems do not have distinctive chemical compositions. Instead, it is their appearance and rarity that confer value on them (Hurlbut and Kammerling, 1991). The crystal structure of most gems imparts clarity, color, and brilliance, including "fire," which is produced by the refraction of white light into different wavelengths and colors. Other gems, such as amber and pearls, are valued for their luster. The highest-priced gems are very hard, measuring 8 to 10 on the 10-point Mohs scale for mineral hardness (Table 10-4). In fact, diamond is the hardest natural or synthetic substance known. Because they are so hard, precious stones resist scratching and preserve their polish, brilliance, and value.

Most of the common gems of the world are retailed in cut and polished forms that differ from the natural shape of the gem crystal. For instance, the most common natural crystal form for diamond is an octahedron, but it is commonly cut into other shapes that enhance light refraction and color, and which change with fashion (Figure 10-17). Thus, the world gem industry consists of two parts, mining and cutting-polishing, which are usually followed by jewelry manufacture (Green, 1981). The most important cutting-polishing centers are Israel and Belgium, which import and export up to $15 billion worth of diamonds annually. Diamond cutting is also carried out in New York, and India is increasing its production of small cut stones. Important colored stone preparation is centered in India, Hong Kong, Thailand, and Brazil.

Gems form by a wide variety of processes, many of which have not been adequately investigated because of the rarity and small size of their deposits. Even diamonds, which have been mined on a large scale in several locations, are still only partly understood. The following summary focuses on the most important of the gems but, for lack of space, cannot cover many other interesting ones.

TABLE 10-4
Major gem stones and their mineralogical classification. Many other stones, including agate, apatite, coral, epidote, feldspar, garnet, hematite, kyanite, lazurite, malachite, olivine, pectolite, prehnite, sphene, and zircon are used widely in jewelry and especially striking specimens can have high values (compiled from Jahns, 1983; Sauer, 1988; Sinkankas, 1989; Austin, 1991).

Gem Stone	Mineral	Hardness	Major Elements	Important Sources
Highly Valuable				
Alexandrite	Chrysoberyl	8.5	Be, Al, O	Russia, Brazil
Diamond	Diamond	10	C	South Africa, Namibia
Emerald	Beryl	7.5-8	Be, Al, Si, O	Colombia, Brazil, Russia
Ruby	Corundum	9	Al, O	Cambodia, Myanmar (Burma)
Sapphire	Corundum	9	Al, O	Australia, Thailand, Sri Lanka, Brazil
Less Valuable				
Amber	Not a mineral	Very low	Hydrocarbon	Russia, Austria, Dominican Republic
Amethyst	Quartz	7	SiO_2	Many locations
Aquamarine	Beryl	7.5-8	Be, Al, Si, O	Brazil
Jade	Jadeite	6.5-7	Na, Al, Si, O	Myanmar, China
	Nephrite	6.5-7	Ca, Mg, Fe, Si, O, H	Russia, China
Opal	Opal	5.5-6.5	Si, O, H	Australia, Brazil, India
Pearl	Calcite	2.5-4	$CaCO_3$	Many locations
Topaz	Topaz	8	Al, Si, O, H, F	Brazil, Russia, India
Tourmaline	Tourmaline	7-7.5	Na, Mg, Fe, Al, B, Si, O, H	Namibia, Brazil, United States
Turquoise	Turquoise	5-6	Cu, Al, P, O, H	United States, Egypt, Australia

Diamonds

Gems are the ultimate in value among mineral commodities. Even though a carat weighs only 0.2 grams, good-quality one-carat diamonds wholesale for many thousands of dollars. In fact, all of annual world production of about 50 million carats worth about $5 billion (Figure 10-1) would occupy a cube only about 2.86 meters on a side. In contrast, the same fortune in gold would occupy a volume about 10 times larger. Diamonds are clearly the best natural form in which to concentrate wealth, if you can find them.

Geology of Diamond Deposits

Diamonds are metastable at Earth's surface. As can be seen in Figure 10-18, the range of temperatures and pressures in which diamond is stable does not include surface conditions (Kirkley et al., 1992; Boyd and Meyer, 1979). As the diagram shows, carbon takes the crystal form of *graphite* at surface conditions and transforms to the crystal form of diamond only at very high temperatures and pressures (Meyer, 1985). Fortunately for anyone with a diamond ring, the rate at which this transition takes place is vanishingly small at normal surface temperatures, where diamond persists as a metastable compound.

Diamonds are found as xenocrysts, or foreign fragments, in an unusual potassium-rich, ultramafic rock known as *kimberlite* (Glover and Harris, 1985; Mitchell, 1986). Diamond-bearing kimberlite is found almost always in thick continental crust or craton of Archean age, where it forms cylindrical intrusions known as *kimberlite pipes* that appear to extend upward from deeper tabular, dike-like bodies (Figure 10-19). Kimberlite pipes include fragments of many rocks and minerals, including diamonds, that are held together by kimberlite magma. Many of the rock fragments are pieces of ultramafic rock from the mantle, known as *mantle nodules.* Combined with the diamond stability relations shown in Figure 10-18, this suggests that kimberlite magmas originated in the mantle and that they brought up pieces of it, including diamonds, when they were intruded. These magmas probably came from a depth of about 150 km or more, the depth at which the geothermal gradient under cratons reaches the diamond stability field in Figure 10-18. In the Kaapvaal craton of South Africa, where most diamond-bearing kimberlites have been found, the pipes formed over more than 2 billion years of Earth history, suggesting that the mantle and crust beneath this area provided special, long-lasting favorable conditions for kimberlite generation (Groves et al., 1987).

The fact that diamonds are present in kimberlites

FIGURE 10-17

Natural and cut diamonds. Natural diamonds (light) have an octahedral form, consisting of pyramids attached at the base. Cut diamonds (dark), shown here in brilliant cut, have 58 facets, 33 on top and 25 below, and reflect much more light (courtesy of De Beers Centenary).

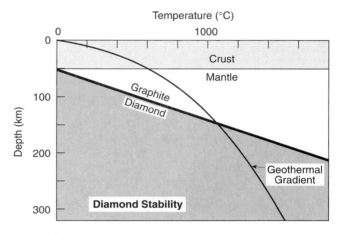

FIGURE 10-18

Schematic cross section through continental crust and upper mantle showing the downward increase in temperature (geothermal gradient) and stability fields for graphite and diamond, both crystalline forms of carbon (C). Note that graphite is stable at the surface of the Earth, and that the geothermal gradient reaches the diamond stability field in the mantle at a depth of about 150 km. Thus, diamond is metastable at Earth's surface.

FIGURE 10-19

Kimberlite pipes before and after mining. On left is unmined pipe surrounded by a ring of extruded breccia at the surface. The top of the pipe is filled with lake sediments, which are underlain by kimberlite breccia and then massive kimberlite, which grades into a dike at depth. On right, kimberlite is being mined by block caving (upper part) and sublevel caving (lower part). Block-caved ore is drawn out at the main haulage level below. Sublevel ore is drawn out on each level, dropped through the ore pass to the crusher, and hauled to the ore bin. Ore from both parts of the mine is lifted up the shaft (based on diagram of De Beers Corporation).

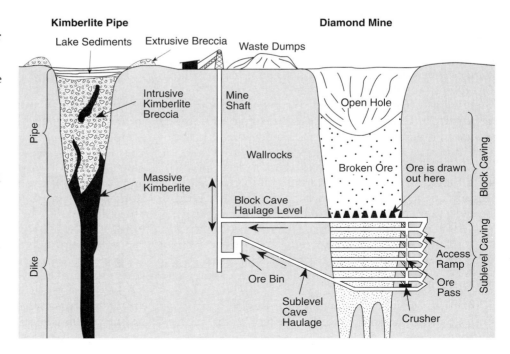

that reached high levels in the crust requires that they rose rapidly to this level from their source in the mantle 150 km below. If they had risen slowly, the diamonds would have reverted to graphite. Such a rapid magma ascent is thought to have been aided by high H_2O and CO_2 contents in the kimberlite magma, which lowered its viscosity and provided explosive gas pressure to help drill through the crust. That all diamonds did not rise so rapidly is shown by the recent discovery of graphite with the external crystal shape of diamond in mantle rocks that are exposed at the surface in Spain and Morocco (Pearson et al., 1989). In what might be regarded as the reverse of this process, it has been suggested that some aggregates of small black diamonds known as *carbonado* were actually formed by meteorite impact that converted organic material at Earth's surface to diamonds (Smith and Dawson, 1985; Kirkley et al., 1991).

Diamonds are not found in all kimberlite intrusions. Of the 5,000 or so kimberlite bodies that are known, about 30% are estimated to contain diamonds and only about 1% of them contain diamonds in mineable concentrations. Until the 1950s, diamonds had been mined successfully from kimberlite pipes only in the Kaapvaal craton of southern Africa (Figure 10-15). Since then, however, they have been found in kimberlites in other parts of the world and even in other rock types. One surprising discovery was the Argyle deposit in northern Australia (Figure 10-15), where diamonds are hosted by *lamproite,* a rock that was not known previously to contain them (Lang, 1986; Mitchell, 1991). Lamproites are similar to kimberlites but contain more potassium and flourine and less CO_2. The fact that they host mineable concentrations of diamonds suggests that future exploration should not be confined to kimberlites. Recent exploration spurred in part by this discovery has found apparently rich diamond-bearing kimberlites and lamproites in the Northwest Territories of Canada (Figure 10-15). Diamonds have even been found in metamorphic rocks in Australia, China, and Kazakhstan, suggesting that diamond mining will become more widespread in the near future (Xu et al., 1992; Levinson et al., 1992).

Diamond placer deposits have a wider geographical distribution than diamondiferous kimberlite pipes and are not necessarily confined to thick Archean crust because they can be transported long distances by rivers. Some diamond placers have been traced to pipes that are too low grade to mine, such as those near Murfreesboro, Arkansas (Bolivar, 1986), and the sources of others are not known. Erosion of known diamond pipes has been surprisingly deep. Around Kimberley, where it all began as discussed earlier, about half of the original vertical extent of the pipes has been eroded. Because the grade and quality of diamonds are greatest in the upper part of these pipes, a sizeable fraction of their diamonds is now in placers, along the Orange River and the Atlantic coast (Figure 10-7). Elsewhere in Africa, placer deposits may be derived from deeply eroded kimberlites in Angola, Ghana, Guinea, Sierra Leone, Tanzania, and Zaire (Ellis, 1987).

Diamond Mining and Production

Diamonds are mined by both open-pit and underground methods (Reckling et al., 1983). Along the western coast of southern Africa, beach placers are mined by building large dikes to hold out the surf (Figure 10-20). Mining is also carried out in near-offshore areas by divers using suction devices mounted on shore or small boats. More deeply submerged beach zones from Pleistocene glaciation are mined from offshore vessels (Gurney et al., 1991). Diamond placer mining is extremely thorough. Even though large equipment is used to remove the sand and gravel, every last grain has to be swept from irregularities in the bedrock surface to find the diamonds.

Kimberlite mining usually begins with an open pit but changes to an underground operation as it gets deeper (Figure 10-19). Most underground diamond mining is carried out by block caving or sublevel caving, which is needed to supply the large volumes of rock required to meet diamond production goals. Diamond grades reported in 1991 by De Beers Mines for kimberlite mines in southern Africa varied from 1.57 carats per tonne at Jwaneng to 0.08 carats per tonne at Koffiefontein, a surprisingly large range for any commodity. The volume of diamonds in a tonne of ore with these grades would be about 0.01 cm^3, a truly small volume to recover.

Diamonds were recovered from ore in the early days by concentration in rotary pans followed by hand-sorting, an arduous task that produced security problems, limited the scale of the operation, and yielded relatively poor recoveries (Wagner, 1971). The discovery in 1896 that diamonds adhere to grease opened the way for large-scale mining and processing. Modern kimberlite diamond mines begin by crushing ore to fragments small enough to liberate most diamonds. The ore is then passed through pans, similar to those used to pan gold, and cyclones, similar to those used in gas scrubbing systems (Figure 3-18), which concentrate heavy minerals. Fine-grained heavy mineral concentrates from these processes then go to grease tables and coarse-grained material goes to X-ray sorters, which scan each grain and use air jets to remove any diamonds that are detected. It is this stage of the process that concentrates the diamonds

most effectively. Large stones can show up at any point in this process. The largest diamond of all time, the 11 cm by 6 cm, 3,026-carat Cullinan diamond, was spotted by a mine worker in 1906 on the side of the Premier mine pit, but the 426.5-carat Ice Queen was found during the final stages of concentration.

Diamond Classification and Gem Diamond Trade and Reserves

Natural diamonds are hand-sorted and graded according to *size, quality, color,* and *shape* at central locations such as Oppenheimer House in Kimberley (Figure 6-1). The processes are guided by a set of representative diamonds kept at major grading centers and which is itself compared against a master set of diamonds kept in London. Diamonds are first divided into 11 groups on the basis of size, and then sorted according to type or shape, quality, and color. The main divisions of the classification are between *gems* and *industrial diamonds,* which are widely used in abrasives and cutting tools. An intermediate class of near-gem diamonds depends on markets for the other classes. In South Africa and most other areas, diamonds that have been graded are examined by government inspectors to assure that they meet specifications (Optima, 1985).

All stones, including the occasional enormous gem (Table 10-5), are cut into a wide variety of shapes to enhance their appearance. The size and number of gems that can be cut from a single stone are controlled by its shape and the distribution of imperfections. For instance, when the second largest stone in the world, the 995-carat Excelsior, was cut into 21 gems of which the largest was only 69 carats, about 63% of the stone was wastage caused largely by black spots. The great Cullinan diamond was cut into 96 small brilliants, nine polished fragments, and nine major stones, including the 317-carat Lesser Star of Africa and the 530-carat Greater Star of Africa, which belong to the British Crown and are the largest cut diamonds in the world.

Diamonds are produced in 21 countries, with Australia, Botswana, and Zaire producing a larger proportion of industrial diamonds than South Africa and Namibia. In fact, Australian production is said to contain only 5% to 10% gem-quality diamonds. Production of gem diamonds from placers is small in other African countries, as well as Brazil, China, and India, so that South Africa, Botswana, and possibly Russia are the main sources of world gem-diamond production. This distinction is important because gem diamonds control the profitability of mines through their much greater value and because their price is more

TABLE 10-5
Important large diamonds of the world (after Joyce and Scannell, 1988).

Name	Size (Carats)	Source
Cullinan	3,026	Premier, South Africa
Excelsior	995	Jagersfontein, South Africa
Star of Sierra Leone	969	Sierra Leone
Great Mogul	793	India
Vargas	728	Brazil
Woyie River	770	Sierra Leone
Jonker	726	Elandsfontein, South Africa
Jubilee	634	Jagersfontein, South Africa
Lesotho	601	Lesotho
Goyaz	600	Brazil
Centenary	599	Premier, South Africa

strongly controlled, as discussed next. The ratio of gem to industrial diamonds varies widely from deposit to deposit and has been reported to be as high as 50% at Koffiefontein and 80% at Letsang LaTerai in Lesotho (Anders, 1993).

Most of the world's diamonds are sold by the *Central Selling Organization (CSO)*, which is operated by De Beers Centenary (van Zyl, 1988). This company perpetuates the early philosophy of Cecil Rhodes that a stable diamond market benefits everyone from producers to consumers. Before Rhodes imposed his vision on the industry, it was plagued by price fluctuations spurred by miners who competed for market share by producing more diamonds, often far exceeding demand, which fluctuated considerably with economic conditions. Although several groups have competed with the CSO over the years, most have eventually joined it. The CSO even handled sales for the old Soviet Union and, since then, has contracted to represent the Russian province of Yakutia (Sakha), the major source of CIS diamonds. Production from Zaire, Australia, and other locations is also represented by the CSO, which is estimated to control about 80% of world diamond sales.

The CSO is widely regarded as the most effective cartel in the mineral industry. Between 1949 and 1990, it increased the asking price for rough diamonds sold to cutters and polishers by 1,800% (Austin, 1991). In contrast, the price of industrial diamonds has actually fallen during this period because of competition from synthetic stones. The gem-diamond market is rigidly controlled, with the CSO selling packets or groups of uncut diamonds for its clients at "sights" in London, Lucerne, and Kimberley that are attended only by approved bidders. It has also expanded the market for its diamonds through very effective advertising campaigns and maintained a $2 to $5 billion stockpile of surplus gem diamonds. Even with this support, the price of diamonds is barely on the high side of the consumption-price correlation followed by most mineral commodities, including gold and the colored gems (Figure 10-21). Furthermore, retail gem and near-gem diamond prices fell significantly during the mid-1980s. Thus, the main function of the CSO appears to be to stabilize and maximize wholesale diamond prices, smoothing irregularities in supply and partly insulating its clients from the cyclic swings typical of most mineral commodities. Whether it has enough financial strength to continue this effort in the

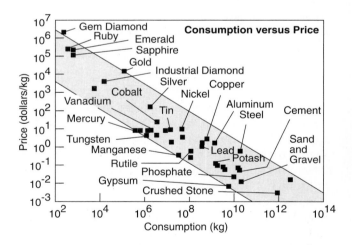

FIGURE 10-21
U.S. consumption versus price for most mineral commodities, as reported for 1991 by the U.S. Bureau of Mines. Note that all commodities fall along a trend of increasing price with decreasing consumption (Lund and Goldstone, 1969). The fact that gem diamonds do not fall significantly above this trend suggests that their prices are generally consistent with current demand and availability.

face of increasing production from Angola and Yaku-tia remains to be seen.

World gem and near-gem diamond reserves and resources are estimated to be about 300 million carats, only about six times higher than annual world production. This figure has not changed significantly in official reports for almost a decade, raising some doubts about its reliability, although there is general agreement that reserves are small. As discussed further in the section on industrial diamonds, it is not likely that synthetic gem diamonds will help make up this impending shortage. This relation has not gone unnoticed by the world mining community and exploration is underway to capitalize on it.

The Beryl Group—Emeralds and Aquamarines

Beryl, a relatively common beryllium-aluminum silicate (Figure 3-2), forms several important gem stones where it develops crystals with good color and few imperfections, including green *emeralds,* pale blue or bluish-green *aquamarines,* yellow *heliodor,* and pink *morganite.* The different colors of these beryls are commonly ascribed to differing contents of trace elements, which are known as *chromophores,* substituting in their crystal structures. The most common chromophores are chromium and vanadium for emerald, iron for aquamarine, manganese and iron for morganite, and manganese, iron, and titanium for heliodor (Sinkankas, 1989). Prices of these stones vary greatly but averaged $2,750 per carat for high-quality emeralds in 1990 (Austin, 1991).

Gem beryls are found in beryllium deposits, which have already been discussed. The best gems come from pegmatites and hydrothermal veins in which slow cooling probably permitted the growth of the near-perfect crystals typical of gems. The famous emeralds of Colombia are found in narrow calcite veins that cut carbonaceous shales. In some areas, availability of the chromophore appears to be an important factor. In Brazil, for instance, emeralds are found where veins and pegmatites around granitic intrusions cut ultramafic rocks that are relatively rich in chromium (Figure 10-22). In other deposits, such as Santa Terezinha de Goias in Brazil, the beryls are not found in well-developed veins and appear instead to have been deposited by hydrothermal solutions that pervaded the entire rock (Guiliani et al., 1990). For reasons that are not entirely clear, emeralds tend to be smaller and to contain more inclusions than aquamarines, which have been found in very large stones.

Emeralds are mined largely from Brazil, Colombia, Russia, and Zambia (Figure 10-15) and aquamarines

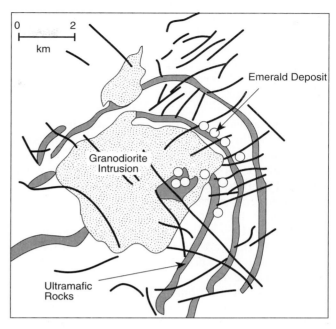

FIGURE 10-22

Schematic geologic map of the Carnaiba emerald district in Brazil showing the concentration of deposits in a chromium-rich ultramafic rock that surrounds part of a granodiorite intrusion (from Guiliani et al., 1990).

come largely from Brazil. Most production is from bedrock deposits or related regolith, although some aquamarines come from placers. In all of these deposits, the gems are distributed very irregularly, making it essentially impossible to determine the size and grade of an ore body before mining. In most cases, the veins and pegmatites are too small for use of mechanized mining and most work is done by hand, an arrangement that encourages theft of stones and requires constant supervision by management personnel. As a result, most emerald mines tend to be small, and to be family owned and operated. Emerald mining in Colombia has been carried out on a slightly larger scale, with unencouraging results (Mowatt, 1992). Most of the mines in Colombia are owned by the government, which also handles emerald sales, but pilferage is a major problem, even forcing intermittent closing of mines.

Emerald reserves are simply not known, even in the deposits that are being mined. In the absence of the discovery of a large, new deposit that can be mined by mechanized methods, emerald and aquamarine production will remain a small business. Most small operations lack the capital to deal with the complexities of deeper mining, suggesting that production will decrease as near-surface deposits are exhausted. Thus, the outlook is not good for future emerald supplies.

The Corundum Group—Rubies and Sapphires

Corundum, the oxide of aluminum, forms gems when it occurs in well-developed transparent crystals with good color (VanLandingham, 1985). Red corundum is known as *ruby,* and blue corundum is known as *sapphire. Star sapphire* and the less common *star ruby* are varieties in which a six-rayed cross or star in the mineral is caused by the pattern of light refraction. Chromophores are thought to be chromium for ruby, and iron and titanium for sapphire. Many other colors of corundum are known, some of which form gems. Average prices for fine-quality, one-carat stones were $4,200 for rubies and $1,600 for sapphires in 1990 (Austin, 1991). In addition to its use as a gem, corundum is widely used in industry because of its extreme hardness (Table 10-4).

Under most conditions in Earth's crust, corundum (Al_2O_3) is not stable in the presence of quartz (SiO_2), with which it will react to form the minerals *kyanite, sillimanite,* or *andalusite,* all of which have the same composition (Al_2SiO_5). If quartz and corundum react in the presence of water, they can form *kaolinite.* Because quartz is so common in crustal rocks, these reactions limit the geologic environments in which corundum can occur to very special, quartz-free rocks where aluminum is abundant. The most common such rock is bauxite, the ore of aluminum, although the surface temperatures and pressures at which bauxite forms are not high enough to form corundum. Corundum can form if bauxite is metamorphosed, but this is a truly rare occurrence. More commonly, corundum forms in low-silica mafic rocks such as peridotite and in limestones that have been altered by hydrothermal solutions near the contact of silica-poor igneous intrusions, such as syenite.

Ruby and sapphire production are largely confined to Australia, Cambodia, Myanmar, Nigeria, Sri Lanka, and Thailand (Figure 10-15). In most of these areas, production is from placer deposits, which take advantage of the relatively high specific gravity (about 4) of these gems. With the exception of Australia, production is dominated by poorly capitalized, small operations and is highly dependent on local politics. During the fighting for control of Cambodia in the 1980s the Khmer Rouge, which controlled the western part of the country, opened part of the Kanchanaburi ruby fields to Thai miners in an effort to finance their ac-

tivities. Similar uncertainties afflict most other ruby-producing areas, making its supply the least secure of the precious gems. To make matters worse, reserves are not known (Shigley et al., 1990).

THE FUTURE OF PRECIOUS METALS AND GEMS

The real question about precious metals and gems is whether we have reached a major crossroads in their history. For millenia, they have been valued for their beauty and as a safe haven for monetary Cassandras. During this time, we have grown to accept the slightly revised golden rule, which states that "She who has the gold, makes the rules." At the same time, industrial applications for these commodities have increased until they dominate markets for the PGEs and silver, and impact those for gold and diamonds. There is even talk that the world has outgrown the need for mineral commodities as investment vehicles, and possibly even as ornaments and art objects. The availability of so many other financial instruments is supposed to siphon money from potential gold investors, and synthetic gems are supposed to satisfy the world's gem buyers.

Predictions of this type have been made periodically for centuries, however, and society still covets precious metals and gems. In fact, if we do turn our backs on these commodities, sociologists will probably look back on the event as one of the most fundamental changes in history. As egocentric as all generations are about the significance of their own, it is hard to believe that we have arrived at that point. So we will probably continue consuming these commodities as we have before, with increasing industrial uses adding pressure to the demand from the arts and investments.

To satisfy this demand, however, exploration is badly needed. No group of mineral commodities has a dimmer reserve outlook than the precious metals and gems, particularly in the face of uncertainty about the future of deep mining in South Africa. Making matters worse is the fact that relatively few unconventional resources are known for these mineral commodities. Thus, future precious-metal supplies may well depend on improved recoveries as by-products from other types of metal mining.

11

Fertilizer and Chemical Industrial Minerals

FERTILIZER AND CHEMICAL MINERALS ARE THE MOST IMPORTANT of the numerous *industrial minerals,* a term that includes almost any commercial Earth material that is not a metal or fuel (Harben and Bates, 1984). In spite of their critical role in modern society, most of the industrial minerals are not well known and many, such as cement, come to us in forms that are not ordinarily thought of as minerals. Some are used in a large variety of products, such as limestone, an important source of cement, crushed stone, and lime. For the purposes of this book, industrial minerals are divided between two broad fields of use: the fertilizer and chemical minerals of this chapter, and the construction and manufacturing minerals of the next chapter. Although some minerals, such as gypsum, have uses related to both groups, almost all of them have dominant markets on which this grouping is based. The value of annual world production for the industrial minerals as a whole totals almost $150 billion and ranges from a high of about $65 billion for cement to a few million dollars for some of the manufacturing minerals.

The $34 billion world fertilizer and chemical industry is based on the fossil fuels and on nitrogen, limestone, phosphate, potash, sulfur, and salt, which are reviewed in this section. The value of annual world production of these six industrial minerals is dominated by nitrogen, at almost $13 billion, with the rest worth $3 to $4 billion each (Figure 11-1A). Other mineral commodities used in smaller amounts in the fertilizer and chemical industries include fluorite, iodine, sodium carbonate, and bromine.

Production of these commodities since 1960 can be divided between nitrogen, which has grown very rapidly, and everything else, which has not (Figure 11-1C). Within the laggard group, lime and salt have not even kept pace with steel. Sodium sulfate production, which forms the lower limit of the laggards, has declined in the United States, although world production continues to grow. Prices of the commodities also fall into two groups. Phosphate and salt prices are relatively low and have not varied much, whereas nitrogen, potash, and sulfur prices have varied tremendously from year to year.

With the exception of nitrogen compounds, deposits of the major fertilizer and chemical minerals are formed during the normal evolution of a *marine sedimentary basin,* that is, a sedimentary basin filled with seawater. Limestone and phosphate deposits are largely *organic sediments,* whereas salt and potash are *chemical sediments,* or *evaporites,* that formed by the evaporation of water. Most sulfur deposits formed by bacterial modification of sulfate-bearing evaporite minerals. The distribution of these mineral resources reflects the location of ancient oceans, particularly shallow areas that extended onto the continents (Reading, 1986). Some of the less important minerals in this group formed by the evaporation of nonmarine lakes.

Fertilizer and Chemical Minerals

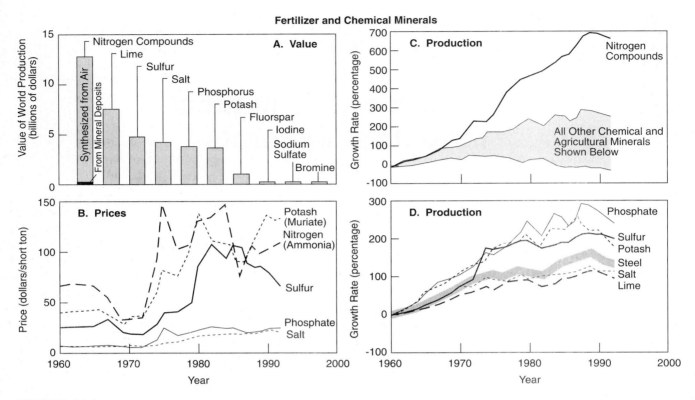

FIGURE 11-1

Fertilizer and chemical minerals. **A.** Value of 1992 world primary production (including that used outside fertilizer and chemical markets). **B.** Change in prices since 1960. None of these mineral commodities has increased in price as rapidly as inflation over this period, as shown in Figure 4-2. **C and D.** Growth in production since 1960. Note that production of nitrogen compounds has grown more rapidly than other commodities (from data of U.S. Bureau of Mines).

The fertilizer and chemical minerals are associated with an unusually large number of environmental concerns. Most reflect the fact that the minerals are highly soluble and are consumed in large amounts. As can be seen in Figure 11-2, consumption of fertilizers and deicing salt for roads, both highly beneficial but important environmental culprits, has increased tremendously, even though the area of farms and length of roads has not changed significantly. The resulting dispersion of these constituents into soil, surface water, and groundwater is a growing problem, forcing us into a more careful comparison of their advantages and disadvantages.

cal applications, the most important of which is *lime* (CaO) (Pressler and Pelham, 1985). About 5% of U.S. crushed stone goes into the production of lime. World figures are not available but would amount to almost $1 billion in stone, if world consumption has the same relation to GDP that it does in the United States. World lime production, which amounts to about 150 million tonnes, is valued at about $7.5 billion. This high cost, relative to other industrial minerals, reflects the energy involved in heating limestone to make lime. Additional markets for limestone include agricultural additives to control soil acidity, and the manufacturing of glass and paper (Carr and Rooney, 1983).

LIMESTONE, DOLOMITE, AND LIME

The common sedimentary rocks limestone and dolomite account for about 70% of U.S. crushed stone production, as noted in the next chapter. They are also our main source of calcium and magnesium for chemi-

Limestone and Dolomite Deposits and Production

Limestone is a rock made up of *calcite* and *aragonite*, which have the same composition (CaCO$_3$) but slightly different crystal structures, and *dolomite*

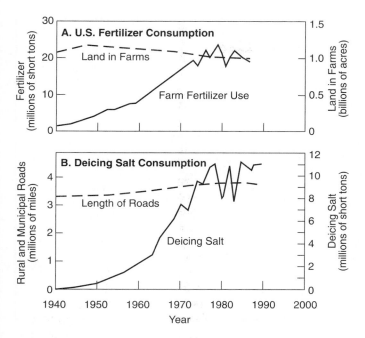

FIGURE 11-2

A. Fertilizer consumption versus farmland in the United States. **B.** Deicing salt use versus length of highways and rural roads (compiled from CEQ, 1989).

[CaMg(CO$_3$)$_2$], with lesser amounts of chert, apatite, pyrite, hematite, and clastic sand, silt, or clay. Pure limestone is known as *high-calcium limestone,* and rock with a high proportion of dolomite is known as *high-magnesium dolomite.* If clastic silicate impurities dominate, the rock is *marl* (Carr and Rooney, 1983; Reading, 1986).

Most limestones are the direct or indirect product of organic activity (Figure 11-3), although small amounts of calcite will precipitate from seawater as it evaporates (Table 11-1). Since most forms of life require light, this means that limestones usually form in relatively shallow water. Some limestones consist of the skeletal remains of coral, molluscs, and algae, which form *reefs* and similar structures. Others are

TABLE 11-1

Minerals precipitated from seawater (as percentage of weight of total precipitates).

Mineral or Group	Percentage
Calcite	0.3
Gypsum	4.5
Halite	77.2
Sylvite	1.4
Other Salts	16.5

composed dominantly of fine-grained material known as *micrite,* which consists of fecal matter, shells of small organisms such as algae, and calcite that was precipitated from seawater, possibly by organic activity. In general, limestones that form in areas of wave action, known as *high-energy environments,* contain smaller amounts of clastic silicate sediment. They sometimes contain *oolites,* spheres made up of concentric zones of calcite that grew over a small nucleus. Limestones that form in quiet *lagoons* or the open continental shelf, which are *low-energy environments,* often contain significant proportions of windblown silicate impurities and cannot be used for more specialized markets. Limestones that form in deep water consist of the shells of small floating organisms that sink to the bottom to form a *carbonate mud.* In water depths below about 4,300 meters in the present ocean, calcite and aragonite dissolve, preventing the accumulation of carbonate mud in extremely deep seawater (Blatt et al., 1980; Reading, 1986).

Dolomite is not usually deposited as a primary sediment. Instead, it is an alteration product of limestone (Figure 11-3), which undergoes a wide range of chemical changes, collectively known as *diagenesis,* after it is deposited. The most important of these is *dolomitization,* the process by which Mg-bearing water invades limestone and replaces it to form dolomite. Solutions that can form dolomite include evaporated seawater, such as might form in a saline lagoon, and fresh water that reacts with seawater-saturated limestones, such as might form where meteoric water collects on an island (Figure 11-3). Dolomite can also form by hydrothermal alteration where Mg-rich hydrothermal brines invade limestones at depth. Other reactions that can affect limestones include the deposition of pyrite and silica (usually as chert), both of which are undesirable impurities from a commercial standpoint (Blatt et al., 1980).

Even with these quality restrictions, there is plenty of suitable material available, and limestone and dolomite mining for chemical purposes is widespread. In North America, the best deposits were formed when shallow seas flooded the continents during Paleozoic time and when reefs formed along continental shelf margins in late Mesozoic time. Most limestone is mined from open pits, but a growing number of mines, mostly in Illinois, Missouri, and Kentucky, have moved underground to minimize environmental disturbance.

Lime is made by *calcining,* the process in which a gas is driven off from a solid by heating. In this case, limestone is heated to temperatures of 700° to 1,100°C to drive off CO$_2$ (Boynton, 1980; Boynton et al., 1983). Depending on how much magnesium was present in the original limestone, this leaves either *quicklime*

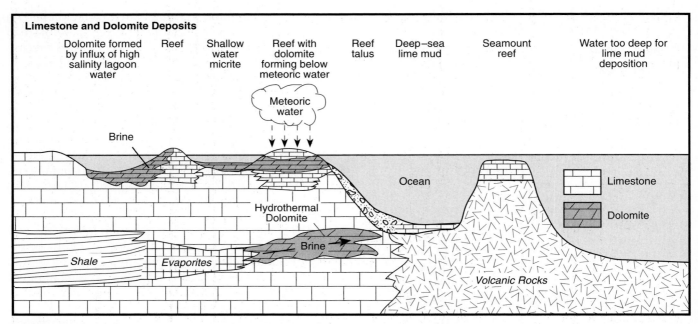

FIGURE 11-3

Schematic illustration of geologic environments in which limestone and dolomite form.

(CaO) or *dolomitic quicklime* (CaMgO$_2$). Water can be added to these to produce *hydrated lime* [Ca(OH$_2$)] or its dolomitic equivalent. The manufacture of lime produces particulates, SO$_2$, and CO$_2$, but emissions are considerably less than those from smelting, cement production, and refining, not to mention electric power generation. The main concerns of the industry are energy consumption in lime kilns (Hsieh, 1983) and the gradual loss of mineable land to population growth, as discussed in the next chapter in connection with aggregate mining.

Limestone and Lime Markets and Trade

Lime's largest market is in steelmaking, where it is used as the flux that lowers melting temperatures and scavenges impurity elements such as phosphorus, silicon, and aluminum into the slag. As shown by Figure 11-4 declining capacity in the U.S. steel industry has cut this market by about 50% during the last 15 years. Lime is also widely used to control the pH of natural and industrial processes because it consumes H$^+$ (Table 3-4). It has been used for this purpose in processes ranging from the beneficiation of copper ores to the purification of drinking water. Lime is also essential in paper and aluminum manufacturing and is the main source of calcium for the manufacture of calcium hypochlorite bleaches and other chemicals. It has recently been used to make precipitated calcium carbonate, a very fine-grained precipitate of calcite. Al-

though it might seem self-defeating to mine limestone, calcine it, and then convert it back to calcite, the shape and size of the newly precipitated grains can be controlled so well that they make an excellent filler and extender, a market discussed in the next chapter (Pressler and Pelham, 1985). The only lime markets with significant growth potential are flue-gas desulfurization (FGD) of power plant and smelter gases and water purification. Current consumption of lime for

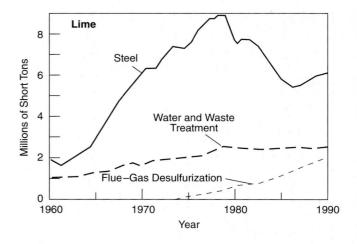

FIGURE 11-4

Lime use in the United States since 1960 (from *U.S. Bureau of Mines Minerals Yearbooks*).

FGD, almost 2 million tonnes (Figure 11-4), is expected to triple in the next decade. Water purification, though a smaller market, will be boosted by corrosion control provisions of the *Safe Drinking Water Act,* which are most easily met by adjusting the pH of the water with lime (Miller, 1993).

A small but growing amount of limestone is used for *liming,* the process by which the pH of acid lakes is brought near neutral. Although limestone is favored because of its low cost, lime, quicklime, soda ash, and sodium carbonate have also been used for this purpose. Material is usually added to a lake or its tributaries in slurry form, with dosages ranging up to 50 tonnes per hectare of lake surface. Liming results in an abrupt increase in pH and dissolved calcium. The pH change causes a decrease in dissolved aluminum, one of the major agents of fish mortality in acid lakes (Figure 11-5). The duration of this effect depends on a wide range of factors, including residence time of water in the lake, acid storage in the lake (usually as H^+ adsorbed on fine-grained sediment), and the acidity of precipitation *(acid loading).* A rule of thumb is that liming lasts for about three to five times the residence time, which is less than a year for most shallow northern lakes. Despite the cost and inconvenience of liming, it has been carried out on over 5,000 lakes in Canada, Norway, Sweden, and the United States with results that are widely regarded as favorable (Hasselrot and Hultberg, 1984; Lessmark and Thornelof, 1986; Wentzler and Aplan, 1992).

In spite of its expanding market, the future is not all rosy for lime. For one thing, it must compete with its own raw material, limestone, which is used for many of the same applications even though it reacts more slowly and weighs more. Increasing energy costs favor limestone because of the heat needed to produce lime. It must also compete with fly ash, dust recovered from cement manufacturing, and other industrial wastes that have many of the same properties and that are themselves searching for a market now that they cannot be spewed into the environment. As a result, lime manufacture and marketing will remain highly competitive and will be especially constrained by energy costs.

PHOSPHATE

The importance of the element phosphorous to life is clearly shown by its role as a component of the genetic material DNA (Emsley, 1980). Its role is so important, in fact, that Isaac Asimov once wrote "We may be able to substitute nuclear power for coal power, and plastic for wood, and yeast for meat, and friendship for isolation—but for phosphorus there is

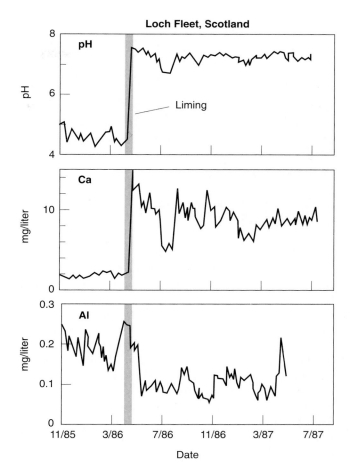

FIGURE 11-5

Changes in pH and dissolved calcium and aluminum in 1 km² Loch Fleet, Scotland, after the addition of 300 tonnes of limestone to streams and bogs draining into the lake. Note abrupt decrease in aluminum and increase in calcium and pH (Longhurst, 1989).

neither substitute nor replacement." As much as 95% of world phosphate production of about 150 to 160 million tonnes is used as a fertilizer. The remainder is used in phosphorus chemicals, which go into detergents, fire retardants, toothpaste, and other products (Stowasser, 1985). At current prices, world phosphate production is worth almost $4 billion and the value of fertilizers prepared from them is many times that.

Geology of Phosphate Deposits

Phosphorus is found largely in the calcium phosphate mineral *apatite,* which is a common constituent of most rocks, as well as skeletal material in many organisms, including humans. Phosphate deposits, which are concentrations of various forms of apatite, are found in three very different geological settings.

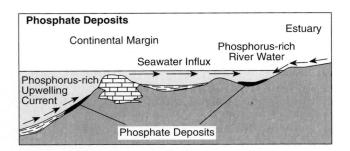

FIGURE 11-6
Schematic illustration of processes that form phosphate deposits in the marine environment.

About 80% of world phosphate production comes from *sedimentary deposits,* which consist of accumulations of apatite that formed by biologic activity (Emigh, 1983; Slansky, 1986; Notholt et al., 1989). These deposits form on the continental shelf or slope where organic productivity is high and there is a limited influx of clastic sediment (Figure 11-6). In some cases, productivity is stimulated by upwelling currents carrying phosphorus-rich cold waters from deeper ocean levels. Ancient deposits of this type are found where deep-water shales grade into shallow-water limestones along the margin of an old continental shelf. Stimulated productivity and phosphate deposition can also take place in estuaries and isolated arms of the sea that are fed by phosphorus-bearing river water. These deposits are not usually as rich in phosphate and cannot be mined until the phosphate has been concentrated by wave action or some other sedimentary process that removes other sediment.

Most sedimentary phosphate deposits consist of organic debris including fish bones, shark teeth, and plankton shells that are interbedded with other chemical deposits, such as chert and diatom-rich silica sediment, as well as fine-grained black shale. As the phosphorus-rich organic material is buried, some phosphorus is dissolved and reprecipitated to form *pellets* of fine-grained apatite. Wave action winnows out nonphosphate sediment and moves the pellets into layers of mineable grade and thickness. Where deposits are exposed to later weathering, phosphate will be further concentrated because it is less soluble than the limestone in which it is usually enclosed.

The formation of sedimentary phosphate deposits is highly dependent on ocean currents, which are controlled by the distribution of continents and land. Plate tectonic processes have had a strong effect on the location of ancient phosphate deposits. Many ancient deposits occur in long zones that were once along continental margins (Figure 11-7). Deposits in Idaho, Montana, and adjacent states are in the Permian-age Phosphoria Formation, a 330,000-km^2 area that formed the shelf along the western margin of North America at that time. In Florida, Georgia, and the Carolinas, phosphate deposits are found in younger, Miocene-age sediments that accumulated in small basins along the margin of the continent (Cathcart et al., 1984). Along the northern and western margins of Africa, a major phosphate province occupies the shelf that formed as North and South America separated and as Africa separated from Europe. The largest of these deposits is in Western Sahara, a former Spanish colony administered by neighboring Morocco, which was attracted to the area by the potential revenues from phosphate production (Figure 11-7).

Additional phosphate production comes from *igneous deposits* that are rich in apatite (Krauss et al., 1984; Stowasser, 1985). Apatite is present in trace amounts in most igneous rocks, but it is abundant only in rocks that are poor in silica, such as syenite and the carbonate-rich igneous rocks known as *carbonatite.* These rocks usually form along rift zones that cut older continental crust and they are mined at Jacupiranga and Araxa in Brazil, in the Kola peninsula of Russia, and at Phalaborwa in South Africa (Figure 11-7).

The final type of phosphate deposit, *guano,* used to be the dominant producer but has largely been exhausted. Guano is bird and bat excrement. Before European settlement of the Americas and Australasia, it covered the floors of caves and the entire surface of many small ocean islands in tropical areas. These deposits were mined widely during the 1800s and formed the main source of fertilizer to the sugar cane growers of northern Australia and the West Indies. The largest deposits were on Christmas Island in the Indian Ocean, where production ceased in 1991, and on Ocean and Nauru islands in the Pacific Ocean, where phosphate production is the main source of local income. In 1990, per capita income to the 9,000 citizens of Nauru amounted to about $20,000, largely from phosphate mining.

Phosphate Production, Use, and Related Environmental Concerns

Phosphate is extracted largely by strip-mining methods with extensive and highly successful reclamation (Plates 3-3 and 3-4; Figure 11-8). Limited underground mining has been used in Russia, Tunisia, and Morocco. Phosphate ore is processed by standard beneficiation techniques to form concentrates. Because plants can-

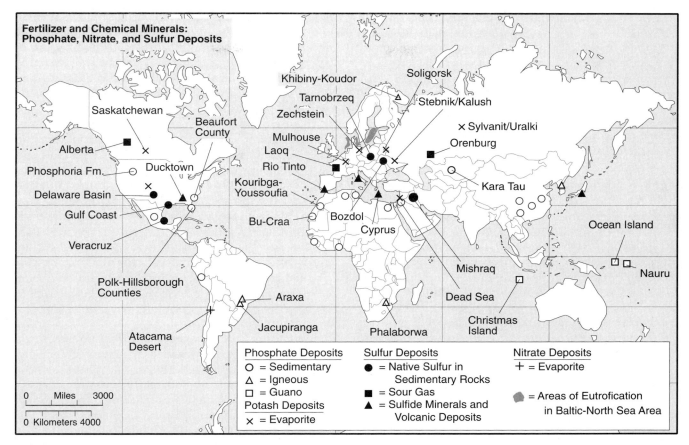

**Fertilizer and Chemical Minerals:
Phosphate, Nitrate, and Sulfur Deposits**

Khibiny-Koudor
Soligorsk
Tarnobrzeq
Stebnik/Kalush
Zechstein
Sylvanit/Uralki
Saskatchewan
Beaufort County
Orenburg
Alberta
Mulhouse
Ducktown
Laoq
Phosphoria Fm.
Rio Tinto
Kara Tau
Kouribga-Youssoufia
Delaware Basin
Gulf Coast
Bu-Craa
Bozdol
Veracruz
Cyprus
Ocean Island
Nauru
Polk-Hillsborough Counties
Mishraq
Araxa
Dead Sea
Atacama Desert
Jacupiranga
Phalaborwa
Christmas Island

Phosphate Deposits	Sulfur Deposits	Nitrate Deposits
O = Sedimentary	● = Native Sulfur in Sedimentary Rocks	+ = Evaporite
△ = Igneous		
□ = Guano	■ = Sour Gas	⬟ = Areas of Eutrofication in Baltic-North Sea Area
Potash Deposits	▲ = Sulfide Minerals and Volcanic Deposits	
× = Evaporite		

0 Miles 3000

0 Kilometers 4000

FIGURE 11-7

Distribution of phosphate, sulfur, potash, and nitrate deposits showing the location of areas of eutrophication in the North Sea and Baltic Sea.

FIGURE 11-8

Dragline moving overburden from phosphate ore in central Florida (courtesy of Lisa Adams, IMC Fertilizer, Northbrook, IL).

not take up raw phosphate easily, apatite concentrates destined for fertilizer markets are dissolved in sulfuric acid to make phosphoric acid and other soluble forms of phosphate known as *superphosphate* (16-21% P_2O_5) and *triple superphosphate* (43-48% P_2O_5). Concentrates intended for chemical markets are heated in enclosed electric furnaces in which phosphorus vaporizes and is then condensed to make phosphoric acid (Emigh, 1983; Stowasser, 1985).

Phosphate production in Florida, one of the major sources of world phosphate, takes place in rather densely settled land and deals with a wide range of environmental challenges (McFarlin, 1992). Tailings from ore processing have only about 5% solids and must remain in settling basins for about a year to reach a level of 20% solids that will support grazing and other low-level agricultural uses. This difficulty reflects the very fine-grained nature of the phosphate ore, which contains clay minerals that form a colloidal suspension in the wastewaters. Efforts to speed the dewatering are complicated by the fact that any reagent added to the tailings to flocculate the colloids could also become a contaminant. Mining and processing wastes also contain uranium and fluorine. Some fluorine is recovered during processing as H_2SiF_6, which is used largely in municipal water treatment. Uranium is associated with its entire suite of radioactive decay products, including radon gas. Even if by-product uranium is recovered, as it was during the 1970s and 1980s when uranium prices were high, daughter products remain in the wastes. Radon accumulations were observed in homes built on older reclaimed land, but this problem has been alleviated by the installation of adequate ventilation. Reclamation in modern mines largely eliminates the problem because most of the natural radioactivity is concentrated in a stratum found just above the ore level, and this radioactive rock is put in the bottom of the reclaimed pit, buried at a level even deeper than it is in unmined ground.

The final concern of phosphate producers is *phosphogypsum,* a waste material formed during the production of superphosphate and other fertilizer materials. Phosphogypsum is similar to gypsum, except that it contains about 1% P_2O_5, 1% F, and 10 to 30 times more radon, none of which is desirable. These properties and the radon scare of the 1980s resulted in a 1989 EPA ruling that phosphogypsum is unsuitable for sale as common gypsum. Production of each tonne of P_2O_5 yields about 5 tonnes of phosphogypsum. About 500 million tonnes of phosphogypsum have already accumulated in Florida and a similar amount is expected by 2000. Phosphogypsum waste is stored in large piles, known as gyp stacks, that must be under-

lain by an impermeable clay layer and a plastic liner, and they have to be covered with a plastic membrane to retard the escape of radon and leaching by groundwater. Water used to transport gypsum slurries to the gyp stacks is acidic, with high dissolved solids, and must be recirculated to the plant or treated with lime before being released.

The use of phosphate has also come under scrutiny. Most attention has been given to its role in stimulating the growth of algae and other organisms in surface water, a process known as *eutrophication* (Bowker, 1990; Manahan, 1990). Eutrophication can be a deleterious process if it causes blooms of algae, which consume dissolved oxygen when they die. The process is widespread in lakes near populated areas and is even observed in shallow, isolated arms of the ocean (Figure 11-6). Although all nutrients contribute to organic productivity and eutrophication, phosphorus is usually the limiting one because its natural concentration is the lowest, making additions from pollution that much more important.

As it turns out, phosphate fertilizers are probably not the only or even the major cause of phosphate-induced eutrophication. Fertilizer phosphate does not leach readily from the soil, partly because it is adsorbed onto clay minerals or combines with calcium in soil waters to form apatite (Bockman et al., 1990). In fact, after 100 years of fertilization, phosphate in some soils has migrated downward only about 20 cm. Instead, a major culprit appears to be municipal wastewater, which contains 25 mg/liter of dissolved phosphates coming largely from detergent, in contrast to permissible concentrations of less than 0.5 mg/liter for surface water. One of the best ways to remove this phosphate is through the addition of lime, which causes precipitation of apatite, although the procedure is relatively costly and has not been widely applied. Instead, the use of phosphate in detergents has been discouraged.

Phosphate Trade and Reserves

Phosphate production comes from about 38 countries, although about two thirds of the total comes from only three areas, the United States, CIS countries, and Morocco-Western Sahara (Figure 11-9). U.S. and Russian consumption are just about equal to production, which makes Morocco the main source of world phosphate exports, most of which goes to Europe. Although world reserves of about 12 billion tonnes of mineable phosphate rock are large, Morocco-Western Sahara holds almost 50% of these reserves, and local sovereignty disputes make is unclear whether they

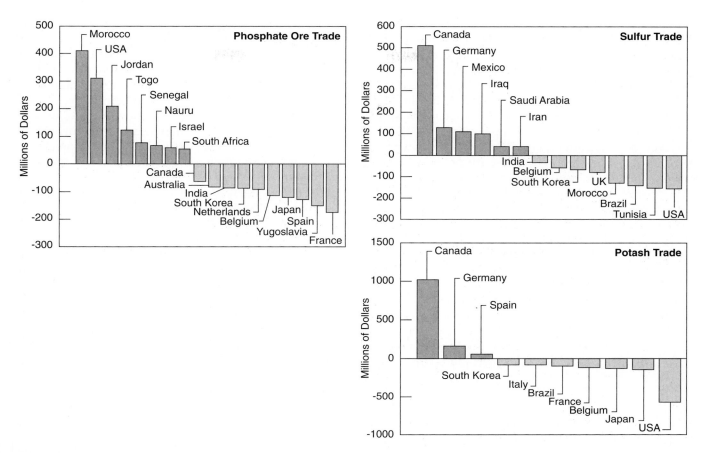

FIGURE 11-9

World trade in fertilizer and chemical minerals showing the difference in value of exports and imports in 1990 for phosphate, potash, and sulfur. These diagrams lack data for CIS countries and are complicated by the fact that these commodities are traded in so many different forms (compiled from *United Nations Trade Statistics Yearbook*, 1992).

will remain easily available to world commerce (Brown, 1988).

SALT

Salt protects your car at the same time that it destroys it, with about a quarter of U.S. production used to remove ice from roads in winter. A larger use for salt is in the preparation of sodium and chlorine chemical products, largely chlorine gas and caustic soda (NaOH), which consume about half of U.S. production. The remaining quarter of salt production is used as a food additive, in agricultural products, for water purification, and other applications (Multhauf, 1978; Morse, 1985A). Over 200 million tonnes of salt are pro-

duced in the world each year, with a value of about $3.9 billion (Figure 11-1).

Salt Deposits and Production

Salt is extremely abundant. Dissolved sodium and chlorine make up about 3.5% of the ocean and the mineral *halite* (NaCl) is found in many natural deposits. Not surprisingly, most halite is in *evaporite deposits* that are found in two main forms, bedded salt and salt domes (Lefond and Jacoby, 1983; Warren, 1989). *Bedded salt* consists of layers of salt that preserve the original depositional form of the sediment. Bedded-salt deposits can be hundreds of meters thick and their genesis presents a real geologic problem. For instance, to produce a layer of salt only 150 meters thick it is necessary to evaporate 10,000 meters of seawater, which would be deeper than any part of the modern ocean. To overcome this limit, it has been proposed that evaporites form in *silled basins,* which are arms of the sea that are isolated from the open ocean by barriers known as sills, which restrict the influx of seawater. Basins of this type in arid regions would lose more water through evaporation than they received from incoming seawater or fresh water, thus becoming saline enough to deposit evaporite minerals. Silled basins might form thick evaporite deposits in several ways. One possibility is that water depths in the basin remained shallow throughout its life with continuous deposition of new evaporites. Another possibility is that water depths in the basin were deep and that evaporites deposited from dense brines that formed by continuous evaporation at the surface and sank to the bottom of the basin. Finally, the deep basin might have evaporated to dryness to make a deep, salt-covered plain fed by a cascade of seawater flowing over the sill from the open ocean, as has been proposed to account for salt deposits in the Mediterranean Sea (Borchert and Muir, 1964; Schmalz, 1991).

Several different geologic environments appear to have created evaporite-depositing silled basins (Figure 11-10). In Paleozoic time, the ocean flooded much of the continents and arms of these shallow seas were isolated by reefs in the Williston basin of Saskatchewan and North Dakota, the Michigan basin, the Permian basin of west Texas and New Mexico, and the Zechstein basin of northern Europe (Figure 11-11). During Mesozoic time, evaporite basins formed where the continents rifted apart, forming valleys that were flooded slowly by the sea. Basins of this type include the Danakil depression, which is the southern continuation of the Red Sea rift between Egypt and the Arabian peninsula, and the Rhine graben, which now contains the valley of the Rhine River.

Evaporite deposits have also formed from freshwater lakes where the influx of water fell short of the amount removed by evaporation. The most common environment in which these lakes formed were rifts in the continents that never reached the ocean (Figure 11-10). In desert areas, lakes that formed in these rifts evaporated to form salt accumulations known as *playa evaporites* (Figure 11-12). The composition of playa evaporites is influenced by the chemistry of

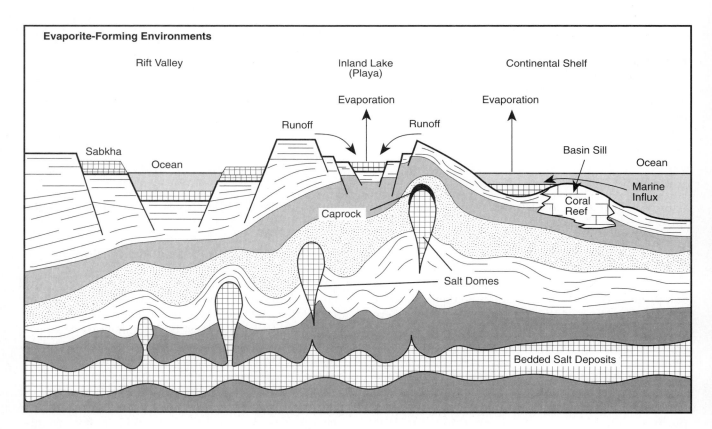

FIGURE 11-10
Schematic illustration of processes that form evaporite deposits.

FIGURE 11-11

Distribution of evaporite basins in North America showing the location of major salt and potash deposits.

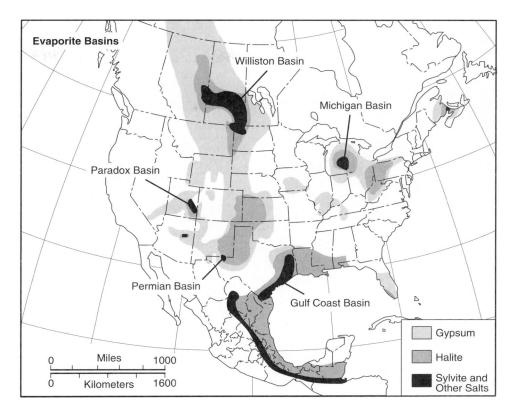

precipitation and rocks in their drainage basins. In areas far from the ocean, where airborne salt water aerosol is absent, rock chemistry dominates and the playas can be rich in boron, lithium, and sodium carbonate.

Salt domes, the other major type of salt deposit, are tabular, tubular, and teardrop-shaped bodies of salt that have intruded upward from buried bedded salt into overlying sediments. Salt domes form because the specific gravity of halite (2.16) is more than that of un-

FIGURE 11-12

Playa salts (white) covering the bottom of Clayton Valley, Nevada. Lithium is produced from brines pumped from beneath the playa (from Kesler, 1976).

A.

FIGURE 11-13
A. Aerial view of Avery Island salt dome rising above the swamps of southern Louisiana; Gulf of Mexico in left background. In addition to a salt mine, oil is produced from the flanks of the dome at depths of 1.5 to 5 kilometers and the surface hosts an entertainment park as well as farms of chili peppers for tabasco sauce. **B.** Large stopes in Avery Island salt mine (from Kesler, 1976; Molly M. Mangan, Akzo Salt Inc.).

B.

consolidated sediment but less than that of most consolidated sediment. As salt layers and newly deposited clastic sediment are buried, compaction causes the clastic sediment to become more dense than the salt, which can then begin to rise. Salt domes are most common where the original bedded salt has been relatively deeply buried, as around the North American Gulf Coast and the Persian Gulf (Figure 11-13A). In addition to these rock deposits, salt can be obtained directly from the ocean, saline lakes, basinal brines in wells, and salt springs that form where meteoric water passes through salt layers on its way to the surface.

Salt is extracted from conventional open-pit and from underground mines with unusually large stopes (Figure 11-13B). It also comes from solution mines, in which it is dissolved in situ by water circulating through a system of wells (Morse, 1985A). Sea salt and some spring and playa salt are also harvested from evaporation ponds. Abandoned salt mines have become important storage areas. They began as storage for local processing and industrial wastes but are now used to store oil and natural gas in areas of high consumption, allowing more efficient use of pipelines and other transportation facilities during off-peak periods. The U.S. Strategic Petroleum Reserve is located in salt domes in Texas and Louisiana. Salt deposits have also received attention as disposal sites for nuclear wastes, as discussed earlier.

Salt processing begins with crushing, grinding, and sizing. Impurities such as gypsum, anhydrite, and shale are removed by scanning a falling curtain of small grains and ejecting nontranslucent grains with an air jet, or by heating all of the grains, causing darker waste grains to become hotter and stick to a resin-coated belt (Bleimeister and Brison, 1960). Salt for human consumption is dissolved and recrystallized in a multistep vacuum process that uses steam from earlier steps to heat later ones. Wastewater from processing creates important disposal problems and is sometimes injected back into unused parts of salt mines.

Salt Markets and Trade

The advantages and problems of salt center on its use in removing ice from roads (Terry, 1974; Carlson, 1986), an application that has grown by almost 8,000% in the United States since 1940 (Figure 11-2). The use of deicing salt is based on its ability to depress the freezing temperature of water (Figure 11-14). Grains of salt put on roadways react with small amounts of water, gradually melting surrounding ice.

Deicing salt has made an immense contribution to the safety and convenience of winter driving. A recent study concluded that road salt decreased the frequency of injuries in winter traffic accidents by 88% and the average cost of an accident by 10% (Bertram, 1993). Unfortunately, this has come at a price. Present bare-pavement policies require applications of up to 0.3 tonnes of salt per kilometer of highway for an average snowfall. This results in saline runoff that corrodes cars, bridges, and other metal structures, kills plants along the roadside, and flows into streams and groundwater where it has become a major contaminant (Schell, 1985; Rowe, 1988). Chloride contents of the lower Great Lakes, for instance, have more than doubled since the beginning of the century. The concentration of chloride ion in Irondequoit Bay, which borders Rochester, New York, on the east and is crossed by an interstate highway has increased from 30 mg/liter in 1950 to 120 mg/liter in 1970, paralleling deicing salt consumption (Bubeck et al., 1971). Salt increased the density of water in the bay so much that it stabilized the hypolimnion, impeding convective overturn. Difficulties have also been observed around open-air salt storage areas where saline run-off recharged fresh water aquifers. Current EPA regulations in the United States require that salt stockpiles be protected from precipitation except when adding or removing salt (Bertram, 1993). The NaCl concentration in U.S. drinking water is currently limited to 250 mg/liter, although some organizations urge limits as low as 25 mg/liter for people on low-sodium diets. These effects have spurred the search for a salt substitute.

Several substitutes for deicing salt are available, but they also have disadvantages. Calcium chloride is more effective because it melts ice at lower temperatures (Figure 11-14) and produces heat during the reaction. It suffers from the same corrosion and contamination problems, however, and is more expensive than salt because it is manufactured from limestone and salt. Of greater interest is calcium-magnesium acetate (CMA), another manufactured compound, which has the added advantage of being biodegradable into CO_2, water, and calcium-magnesium oxides such as lime (Boice, 1986). Unfortunately, CMA costs about 30 times as much as salt and is less effective on a ton per ton basis. Proponents of CMA argue that this price difference is balanced by the cost of salt damage to cars and highway structures, although this argument has little effect on municipal governments, which can postpone repair expenditures but must remove snow and ice each winter.

Salt is produced by 97 countries, with major production coming from MDCs, China, and India. U.S. imports come largely from Canada, Mexico, and the Bahamas. Australia exports to Nigeria and Brazil, and Europe obtains additional supplies from China and India. Salt reserves are essentially limitless, and produc-

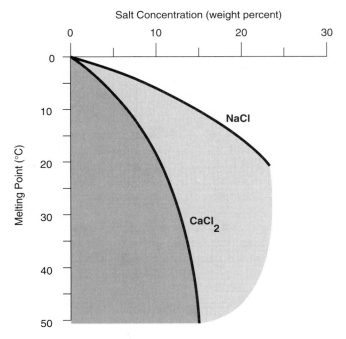

FIGURE 11-14
Depression of freezing point of water by NaCl and $CaCl_2$.

tion is available from solar evaporation even if energy costs become prohibitive.

POTASH

Potash is an industry term that refers to a group of water-soluble salts containing the element potassium, as well as to ores containing these salts. This ambiguity developed because potassium salts were originally recovered from large pots in which plant ashes were leached with water that was then evaporated, a logical source for an element that is essential for plants. The most common mineral in potash is *sylvite*, potassium chloride, which chemists call *muriate of potash*. Potash abundances are quoted in terms of yet another compound, K_2O.

The vast bulk of potash consumption, about 95% in the United States, is used in fertilizers, with the rest going into potassium compounds (Searls, 1985). World potash production, which comes from only 13 countries and amounts to 25 to 30 million tonnes, has a value of almost $4 billion, about the same as the other major agricultural and chemical minerals.

Potash Deposits, Production, and Environmental Concerns

Modern potash production comes from *evaporite deposits* known as *sylvinite*, which consists of a mixture of halite, sylvite, carnallite, and other potassium, magnesium, and bromine minerals known as *bittern salts* because of their bitter taste in comparison to halite (Adams and Hite, 1983). Sylvite, carnallite, and the bittern salts are among the last minerals to crystallize from evaporating seawater and they make up a relatively small fraction of total seawater evaporite (Table 11-1). Thus, potash-bearing evaporites are not as common as halite or gypsum, and they usually occupy a more central position in evaporite basins (Figure 11-11).

Sylvinite deposits are so soluble that they are not found at the surface and must be mined from deeply buried deposits that were protected from groundwater. Unfortunately, sylvinite is not strong enough to support mine openings at great depth (Coolbaugh, 1967; Haryett, 1982). Rock stability has been a particular problem in the deposits 1,000 meters deep in the Williston basin in Saskatchewan, where openings created by mining close by flow and rock failure within a month or so. Most mines in the area use continuous miners but recover only 35% to 45% of the ore, leaving the rest in place to prevent the collapse and destruction of the mine. Additional problems are presented by groundwater, which can dissolve sylvinite

and destroy the mine. The water comes largely from formations directly above and below the potash levels, and it enters the mine through faults and collapsed areas created by mining. To make matters worse, shafts that reach the sylvinite must pass through an overlying unit, the Blairmore Formation, which contains water under high pressure. Shafts can be excavated through the Blairmore only by freezing the rock and lining it with steel shields, which must remain sealed to prevent the influx of water (Prugger and Prugger, 1991). As a result of these problems, five of the 17 mines in the area have experienced major water inflow and one has been abandoned.

These problems have given lower quality ore in mines at shallower depths some room to compete. For instance, mines at depths of 200 to 450 meters in the Permian basin of New Mexico can extract up to 90% of their thinner ore layer without collapse and water problems. Solution mining has been used on sylvinite layers in Saskatchewan and Utah, where water leakage, collapse, or tilting of the layers prevent conventional mining, and it will be the main method for mining deep reserves (Adams and Hite, 1983).

Potash can also be recovered from *brines* that fill pores in marine evaporites, as well as nonmarine evaporite such as Searles Lake, California, and the Bonneville Salt Flats west of Great Salt Lake, Utah, which are discussed in the sections on boron and soda ash. In a few areas, including the Great Salt Lake and the Dead Sea, saline lake waters are processed for potash and other elements.

Potash processing usually involves crushing and grinding, followed by flotation to separate waste sediment and halite from sylvite. More complex ores are dissolved and sylvite or other potassium salts are precipitated from the resulting solutions. Potassium-rich fertilizers are used for wheat, corn, soy beans, rice, and cotton. Potassium sulfate is used on plants such as grapes, citrus, tobacco, and vegetables that cannot tolerate the chloride released when sylvite dissolves.

Potassium lost from fertilizer applications is not as big an environmental contaminant as phosphorus or nitrates. Excess potassium is usually adsorbed onto the surfaces of clay minerals in fine-grained soils, although sandy soils have less ability to retain the element. Increases in dissolved potassium have been observed in surface and groundwaters in agricultural areas, sometimes above permissible potassium contents of about 10 mg/liter for drinking water. However, this limit was set because sewage and other municipal wastes are relatively high in potassium. The element itself is not a significant problem (even milk contains about 1,400 mg/liter of potassium) and it is not usually a major cause of eutrophication (Bockman et al., 1990; Manahan, 1990).

Potash Trade and Reserves

The most important potash producers are Canada, Germany, and the CIS countries, Russia (53% of CIS total), Byelorussia (42%), and Ukraine (5%) (Figure 11-9). German production, which is equally divided between the east and west, comes from the Zechstein basin around Göttingen, Hannover, and Stassfurt (Figure 11-7). World trade in potash is highly dependent on transportation costs because so many of the important sources are not near port facilities. Canadian deposits in Saskatchewan have an unfortunate location in this respect but compensate by being high grade and close to the North American prairie agricultural belt. Uralkali and Sylvanit in Russia are worse, with low-ore grades and long distances to ports, especially Vanino and Vostochnyy on the Pacific Rim (Figure 11-7). Because of the difference in cheap sea transport and more expensive rail freight, potash from Europe competes with U.S. potash as much as 250 kilometers inland from ports on the eastern United States (Searls, 1985).

Potash reserves of 9.4 billion tonnes of contained K_2O are enormous in relation to present annual production. Resources are even larger and include extensive deposits below 2,000-meter depths, where information is limited and comes largely from oil and gas drilling. The deeper parts of basins judged to lack hydrocarbon potential remain poorly explored. In terms of immediately mineable potash, Canada rules the roost, with almost 50% of world reserves. The large potash reserve position has led to a severe excess of global potash capacity, forcing most producers to scale back production. For instance, Potash Corporation of Saskatchewan, a provincial company that owns about half of Canadian production, operated at about 40% capacity during the early 1990s. The overcapacity situation has been further aggravated by a 15% decrease in potash consumption in the MDCs between 1979 and 1990 (Searls, 1993).

SULFUR

About 90% of U.S. sulfur consumption is used to make sulfuric acid, the most important industrial acid and the basis for the sulfur chemical industry. Most sulfuric acid is used to make superphosphate fertilizers, with about 70% of total U.S. sulfur consumption going for this purpose. A distant second market for sulfur is in catalysts for production of high-octane gasoline during petroleum refining. Smaller amounts are used in acid solutions for base-metal refineries and in sulfur chemicals (Morse, 1985B). World sulfur production amounts to about 50 to 60 million tonnes worth about $5 billion.

Sulfur is produced from a wide array of sources, including *mined sulfur*, which comes from conventional, solid mineral deposits, and *recovered sulfur*, which is a by-product of other mineral products. Three sources of mined sulfur have been used over the last century. The most important are *sulfur deposits in sedimentary rocks*, which contain *native sulfur*, or sulfur in the elemental state (Figure 11-15). These deposits are commonly found in areas rich in the calcium sulfate mineral, *gypsum*, and they are thought to form by the action of *sulfate-reducing bacteria*, which remove oxygen from sulfate, forming H_2S that reacts with dissolved oxygen to precipitate native sulfur (Barker, 1983; Wessel and Wimberly, 1992). Native sulfur deposits are found wherever gypsum is abundant. Most sulfur production comes from *caprocks* of salt domes, the porous rock consisting of relatively insoluble gypsum and limestone that accumulates on the top of salt domes as the salt is dissolved by fresh groundwater near the surface. Sulfur-bearing salt domes are not common. Less than 1% of the domes in the large Gulf of Mexico province (Figure 11-7), the site of the only salt-dome sulfur mines, contain commercial sulfur deposits (Ackerman, 1992). Generally similar sulfur deposits are found around bedded gypsum-rich evaporite deposits in the Delaware basin of west Texas, and in Poland, Ukraine, and Iraq.

Native sulfur in volcanic rocks is considerably less common but has been mined in Japan, New Zealand, and a few other areas of recent volcanism (White, 1968). This sulfur also forms by oxidation of H_2S, although the gas comes from underlying magmas rather than from the bacterial reduction of sulfate. As a "last resort" sulfur can be obtained from *metal sulfide minerals*, usually *pyrite*. Deposits of this type have been mined at Ducktown, Tennessee, Rio Tinto, Spain, and in Russia, Japan, and Cyprus. Pyrite concentrates from the cleaning of coal are used as a source of sulfur in some parts of Europe.

Over the last few decades, mined sulfur has had to compete with *recovered sulfur* from industrial processes. The most important such source is *sour gas*, which was discovered in large amounts in Alberta during the 1950s. As discussed earlier, even small amounts of H_2S must be removed before gas can be transported and sold. Because sulfur from sour gas is a by-product, it can be sold at lower prices than sulfur from mined sources. The advent of sulfur production from the Alberta gas fields had a major impact on world sulfur prices, which was mitigated only partly by the long rail transport needed to get the sulfur to shipping points in Vancouver (Plate 8-4). As environmental regulations grew, sulfur was recovered in

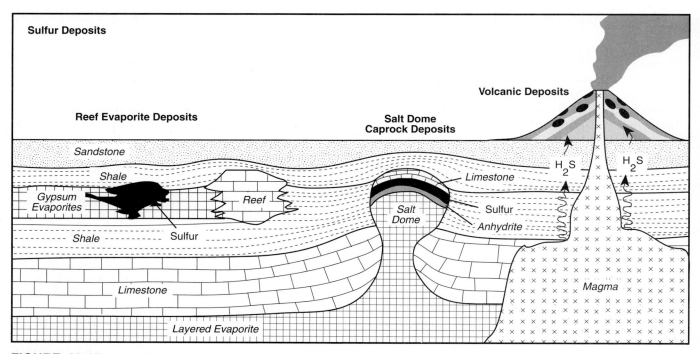

FIGURE 11-15
Schematic illustration of geologic processes that form sulfur deposits.

a marketable form from other oil and gas processing wastes, gases emitted during the heating of coal to make coke, and smelter gases. By the early 1980s sediment-hosted native sulfur deposits had lost their lead, probably permanently, to these recovered sources (Figure 11-16).

Although some sulfur is mined by conventional open-pit and underground methods, most native sulfur in sedimentary rocks is mined by a solution mining technique known as the *Frasch method*. This process depends on the fact that native sulfur melts at 112°C and becomes a less viscous liquid as temperature rises to about 160°C (Lee et al., 1960). Thus, superheated water or steam pumped into porous rock containing native sulfur will melt it to form a liquid that can be lifted to the surface by pumping compressed air into the deposit, making a froth of liquid sulfur. The Frasch method works best on ore that is surrounded by impermeable rock, conditions met by most salt-dome and other sediment-hosted sulfur deposits. The method has made it possible to mine offshore deposits from drilling platforms in the United States and Mexican Gulf Coast. Native sulfur obtained from these mines, as well as pyrite and H_2S, are heated in air to make SO_2, which is then reacted with water to make sulfuric acid.

The future of world sulfur mining depends on recovered sulfur supplies, which have grown greatly in recent years in response to increased environmental regulation. In fact, sulfur producers now distinguish between SAF (sulfur in all forms) and SOF (sulfur in other forms), which refers to sulfur in acid, rather than elemental sulfur or *brimstone,* as it is known in the trade. At the same time, the sulfur market has not grown as rapidly and marginal sulfur miners have been squeezed out. This cycle has largely been completed in the MDCs. The only important sulfur emissions that are not recovered in these countries come from electric power plants, which seem unlikely to contribute to the market soon. Most presently used FGD systems at power plants recover sulfur as some form of sulfate mineral that is not usually used as a source of elemental sulfur, or concentrated sulfuric acid. Further pressures could come from LDCs, however, if they enforce environmental regulations similar to those in the MDCs.

Current sulfur trade consists largely of exports by Canada, Germany, Mexico, Russia, Poland, and the Middle Eastern natural gas producers to the United States, Tunisia, Brazil, and Morocco, which make sulfuric acid for the treatment of phosphate ore (Figure 11-9). With the increase in by-product sulfur from the processing of oil and gas, it is becoming more difficult for sulfur production to be adjusted to demand. The Gulf war, for example, stopped about 1.7 million tonnes of annual sulfur production from Iraq, creat-

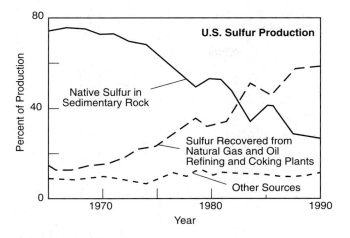

FIGURE 11-16

Sources of sulfur since 1965 showing the decline in sulfur production from native sulfur in sedimentary rocks and the corresponding increase in sulfur recovered from sour gas, oil refining, and coke plants (from *U.S. Bureau of Mines Minerals Yearbooks*).

ing a short-term squeeze on the market because other producers could not increase production sufficiently. Thus, sulfur mines will continue to be of strategic interest, as a buffer against disruptions of this type, which is good news to native sulfur producers in the Gulf of Mexico. Further good news comes from a surprisingly short reserve position. In spite of its many possible sources, sulfur reserves are small, particularly in comparison to the large reserves common for most of the fertilizer and chemical minerals. World sulfur reserves of 1.4 billion tonnes amount to only about 20 years' supply at present consumption levels. Thus, unless major sulfur-recovery efforts are undertaken at oil and gas refineries and metal smelters in LDCs, shortages could result. Such efforts are most likely to occur in CIS countries where pollution abatement is badly needed, although the lack of local funds will almost certainly require injection of foreign capital, as in the case of the proposed Noril'sk improvements discussed in the nickel section.

NITROGEN COMPOUNDS AND NITRATE

Nitrogen is the most important and fastest growing fertilizer and chemical commodity (Figure 11-1), with more than 85% of world production used in fertilizers. Although nitrogen is very abundant, making up 75.5% of the atmosphere (Table 3-1), it must be converted to *ammonia (NH₃)* or *nitrate (NO₃⁻)* before plants can use

it efficiently. In nature, this process known as *nitrogen-fixing* is carried out by bacteria. In modern agriculture, the process is assisted by the addition of nitrogen fertilizers, usually *ammonium nitrate (NH₄NO₃)* or *urea [CO(NH₂)₂]*, to the soil (Foth, 1984; Brady, 1990). The 15% of world nitrogen production that is not used for fertilizers goes into nitric acid, an important industrial acid, ammonium sulfate [(NH₄)₂SO₄], and other compounds that are used in explosives, plastics, resins, and synthetic fibers. One of the newer uses for nitrogen is sodium azide (NaN₃), the propellant in automotive airbags (Davis, 1985). World nitrogen production, which comes from about 66 countries, amounts to about 100 million tonnes annually worth an estimated $13 billion, making it by far the most valuable of the chemical and industrial minerals. However, very little of this comes from mineral deposits.

Nitrogen Deposits and Production

Almost all of world nitrogen production comes from the *Haber-Busch process* in which air reacts with hydrogen from natural gas to make ammonia. Before this process became commercially feasible in the 1920s, nitrogen was obtained from natural deposits, including *guano*, the bird and bat excrement that was a major source of phosphate, and *nonmarine evaporite deposits*. Nitrate evaporite deposits are found only in the Atacama desert of northern Chile (Figure 11-7), one of the driest parts of Earth, with an average annual rainfall of less than 0.1 cm. Extreme dryness is essential for the formation of nitrate minerals, which are highly soluble.

The Chilean nitrate deposits occupy a belt 15 to 80 kilometers wide extending along the eastern side of the coastal mountain ranges for about 700 kilometers. Unlike boron, iodine, and lithium, which form in playas, nitrate deposits form crusts and cements up to 2 meters thick in soil and rock debris along the sides of valleys (Figure 11-17). The crusts consist of about 25% nitrates, with halite, sodium sulfate, and other evaporite minerals containing calcium, magnesium, bromine, iodine, and chromium. They are thought to have formed as a special form of *caliche*, a term usually used for calcitic cements that are precipitated by the evaporation of groundwater from pore spaces in regolith in arid areas. Some of these evaporites were dissolved and redeposited by sea fog and episodic rain showers. The source of these elements is not well known but probably involved sea spray to supply the halite. Other elements probably came from volcanic emanations, weathering of local bedrock, and decomposition of organic material, a possible source of the nitrogen. The fact that the present ultradry climate has persisted since Miocene time is thought to have been an impor-

FIGURE 11-17
Aerial view of nitrate mines
(darker areas) on lower slopes
of hills in Tarapacá Province,
Chile. Note the abandoned
processing plant on the hill in
background (courtesy of George
E. Erickson, U.S. Geological
Survey).

tant factor in allowing this process to continue long
enough to make commercially important deposits.

Production from the Chilean nitrate deposits involves direct open-pit mining followed by a complex
solution-reprecipitation process that concentrates nitrates. Reserves are enormous in relation to production, but the future of the deposits depends on their
ability to withstand competition from synthetic
sources. Bromine, iodine, and sodium sulfate are all
by-products of these deposits, which are also the only
natural occurrence of abundant chromate (Ericksen,
1981).

Nitrogen Fertilizers, Trade, and Related Environmental Concerns

The use of fertilizers has changed the global nitrogen
budget. Estimates of the magnitude of natural nitrogen fixation are similar to anthropogenic fixation of
about 100 million tonnes annually (Bockman et al.,
1990). Fertilizer applications are estimated to provide
up to 80% of total fixed nitrogen to agricultural soils,
with much of the rest derived from the atmospheric
aerosol and natural biological fixation. Most of the
nitrogen fertilizer that is added to the soil is oxidized
to form soluble nitrate, the main form in which
nitrogen nourishes plants. Unfortunately, soils retain

little of this nitrogen and any that is not taken up by
plants is usually released into waters. Even nitrogen
taken up by plants and subsequently eaten by animals is excreted as ammonia and related compounds
and adds to the load of nitrates in surface waters
(Figure 11-18).

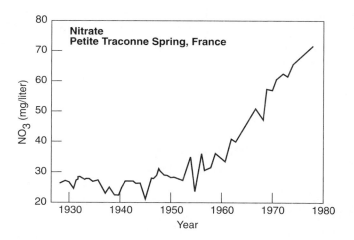

FIGURE 11-18
The increase in nitrate content of Petite Traconne spring,
Brie region, France, is related to the use of nitrogen
fertilizers (from OECD, 1986).

In some areas of heavy fertilizer use, nitrate contents of drinking waters are approaching the acceptable limits of about 50 mg/liter (Lee et al., 1992). This is of concern because nitrate is reduced by bacteria in saliva and the stomach to nitrite (NO_2^-), which has been linked to *methaemoglobinaemia,* a potentially fatal condition that limits the oxygen-carrying capacity of hemoglobin in the blood. The problem, also known as *blue baby syndrome,* is most common in babies fed with bottled milk made from nitrate-rich water. A link between lymphoma and groundwater with high nitrate contents from fertilizer use has been suggested in eastern Nebraska (Weisenberger, 1987). Carcinogenic compounds formed by reactions between nitrite and food have also been suggested as a cause of gastric cancer, although a recent World Health Organization study found no such link and noted that gastric cancers have decreased during the last few decades, the same period that nitrate concentrations in drinking water have increased (Bockman et al., 1990).

LDCs now host about 75% of world fixed nitrogen capacity. Along with this change and increased demand, the United States has gone from exporter to importer, with a trade deficit in nitrogen compounds of about 2.8 million tonnes of contained nitrogen. The main exporters to the U.S. market are Canada, Trinidad (relying on their large oil and gas production), and the CIS countries.

OTHER AGRICULTURAL AND CHEMICAL MINERALS

Fluorite

Fluorite has gone from one environmental storm to another. Just as concern about fluorine in drinking water has begun to wane, it has been implicated in the disappearance of Earth's ozone layer. *Fluorite,* or *fluorspar* as it is called in the industry, is the mineral from which we obtain the element fluorine. Over 60% of U.S. fluorspar consumption goes into hydrofluoric acid, which is the basis for the world fluorine chemical industry. Among the most important products prepared from hydrofluoric acid are elemental fluorine that goes into UF_6 for the enrichment of uranium, synthetic cryolite for use in aluminum production, and chlorofluorocarbons (CFCs) for refrigeration and propellant fluids. Fluorspar itself is used as a flux in steelmaking and in glass and ceramics (Pelham, 1985). World fluorspar production to supply these markets amounts to about 3 to 5 million tonnes annually. Prices vary depending on the purity of the product, with *acid-grade* material containing 97% CaF_2 as the most expensive and *metallurgical-grade* material with at least 60% CaF_2 as the least expensive. Averages for all grades are in the range of $200/tonne, giving annual world production a value of almost $1 billion (Figure 11-1).

Most fluorspar deposits consist of fluorite that was deposited by low- to medium-temperature hydrothermal solutions (Figure 11-19). The source of these solutions and the nature of the deposits vary greatly. Deposits in the Cave-in-Rock area of southern Illinois (Figure 11-20), the most important area of fluorite production in the United States, appear to be Mississippi Valley-type deposits that were contaminated with fluorine, possibly from a nearby intrusive rock (Grogan and Bradbury, 1968). Veins in the Burin peninsula of Newfoundland, from which important Canadian production has come, were deposited from magmatic hydrothermal solutions from a granitic intrusion, which mixed with meteoric solutions (Strong et al., 1984). In both of these deposits, as well as the Pennine district of the United Kingdom, fluorite is found with barite, and lead and zinc sulfides. In contrast, large deposits in the San Luis Potosi area of Mexico are at the contact between limestone and shallow rhyolite intrusions and appear to have formed when meteoric water circulated through the hot rhyolite shortly after it was emplaced, leaching out fluorine and depositing it as fluorite where the solutions encountered calcium-rich limestone (Ruiz et al., 1985). The most unusual fluorite deposits in the world are found at Vergenoeg in the upper part of the Bushveld complex in South Africa, where immiscible magmas of iron oxide and fluorite were formed as the magma cooled (Crocker, 1985).

Most fluorite production comes from China, Mongolia, Mexico, and South Africa, with very little coming from MDCs. The United States, for instance, is almost 90% dependent on imported supplies. Fluorite mining is done by both open-pit and underground methods and processing involves simple crushing, grinding, and beneficiation.

Fluorine is toxic in high concentrations but beneficial in smaller ones. At concentrations of 0.7 to 1.2 mg/liter, the amount used in most fluoridated drinking water in the United States, it substitutes for OH^- in *apatite,* the phosphate mineral in teeth and bone, making it less soluble and decreasing the incidence of dental caries. In 1985 about 120 million people in the United States used water to which fluoride had been added (Easley, 1990) and 9 million had enough natural fluoride in their water. Although fluoride has been under attack ever since it began to be used in water in 1945, the only significant health problem with which it has been linked is *fluorosis,* a disease that involves dental defects and bone lesions (Osterman, 1990; Mahoney et al., 1991; Hamilton, 1992). This problem is caused by concentrations of fluoride that are

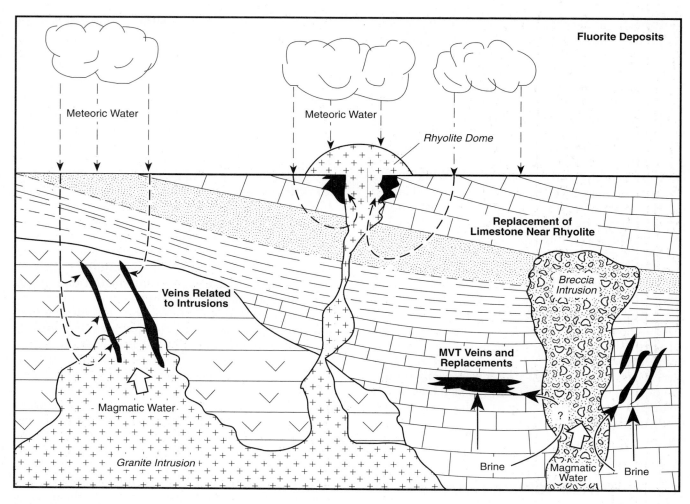

FIGURE 11-19
Schematic illustration of geologic processes that form fluorite deposits.

much higher than those in municipal water supplies. It has been observed most frequently in grazing animals around older industrial facilities that use fluorite or in people and livestock that use farm wells with high natural fluorine concentrations, usually from phosphate-rich rock or fluorine-rich volcanic glass (Felsenfeld, 1991). Fluorite particle emissions can come from mining operations, brick works, and aluminum processing facilities, and gaseous HF or SiF_4 emissions can be produced by phosphate processing plants and other industrial operations (Thompson et al., 1979; Polamski et al., 1982; World Health Organization, 1984).

Fluorine is at the center of the controversy about CFCs, which are thought to destroy the atmospheric *ozone* layer that protects us from ultraviolet radiation, a major cause of skin cancer (Glas, 1988). CFC emissions climbed rapidly in the 1970s and 1980s, shortly before the "ozone hole" was observed at the South Pole (Figure 11-21). Even though the United States banned the use of CFCs in most aerosol propellants in 1978, other countries did not and global restrictions on CFC emissions took force only in the 1980s with several international agreements, including the 1988 Montreal Protocol. The CFC problem is not simple, for fluorite or the environment. Hydrofluorocarbon (HFC) and hydrochlorofluorocarbon (HCFC) compounds, which have been developed as alternatives to CFCs, require more hydrofluoric acid than CFCs and are expected to boost fluorite consumption in the short term. However, they also appear to be greenhouse gases and might also be phased out. The same problem afflicts butane and other volatile organic compounds that are used in place of CFCs as propellants in aerosol containers. Even if CFC emissions are stopped, present gases will take up to ten years to

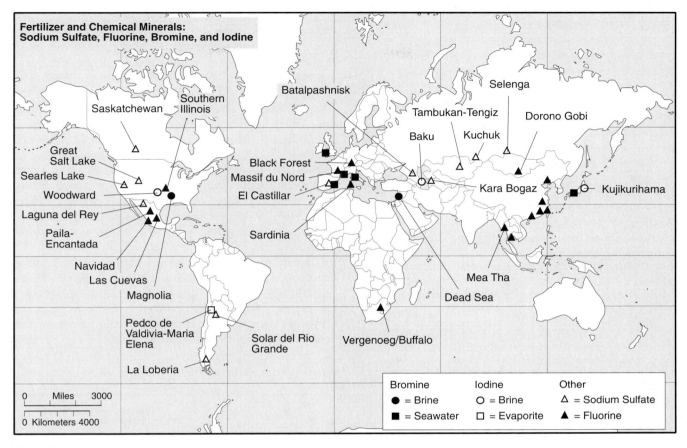

FIGURE 11-20
Distribution of fluorite, sodium sulfate, iodine, and bromine deposits.

reach the upper atmosphere, where they could persist for a century. In view of these constraints it is likely that fluorine consumption will not rise greatly and that presently known reserves of 210 million tonnes are adequate for many decades.

Iodine

The world iodine industry was boosted with the discovery in the 1820s that iodine prevented goiter (Warren, 1989). Food additives directed at this problem for both humans and livestock remain a major market. Other markets include silver iodide, the light-sensitive compound on X-ray and graphics film, and various compounds for use as germicides, wood stabilizers, catalysts, and pharmaceuticals (Lyday, 1985). World iodine production, amounting to about 16,000 tonnes, has a value of about $160 million (Figure 11-1).

Iodine, like bromine, occurs in *brines and evaporites*, sometimes as a by-product of bromine. The world's largest iodine supplier is Japan, which obtains most of its production from brines associated with the

Kanto natural gas fields in the Chiba peninsula (Figure 11-20). These brines, which contain up to 160 mg/liter iodine, actually produce twice as much bromine as iodine. In the United States, the only important iodine production comes from brines of the Pennsylvanian-age Morrowan Formation around Woodward, Oklahoma, which contain 150 to 1,200 mg/liter iodine. Unlike most other iodine sources, the Woodward brines do not contain enough bromine to support economic extraction. Iodine is also a by-product of the Chilean nitrate evaporites, which were discussed above (Ericksen, 1981, 1983; Jan and Roe, 1983).

Iodine is the perfect example of an element that is essential for life but toxic in high concentrations. In spite of its therapeutic effects on goiter, elemental iodine is even more toxic than bromine. Iodine vapor in working environments is limited by the U.S. EPA to concentrations of 0.1 ppm, although its major applications have not been implicated in important environmental problems and its markets remain relatively undisturbed. The only environmental problem of sig-

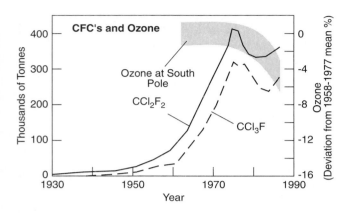

FIGURE 11-21

U.S. fluorocarbon emissions since 1930 and the change in ozone concentration in the atmosphere at the South Pole since 1960 (compiled from Environmental Trends, 1989, Interagency Advisory Committee on Environmental Trends, Washington, D.C. and GEMS Monitoring and Assessment Service, 1989).

nificance associated with iodine production is subsidence around brine wells in Japan. Iodine reserves are large and are augmented by both the ocean, with 0.05 ppm iodine, and some forms of seaweed, which burn to produce ash with 1.4% to 1.8% iodine that was the main source of iodine prior to the discovery of brine and evaporite sources.

Sodium Sulfate

Sodium sulfate is a common chemical compound that is recovered in about equal amounts from mines (natural) and as a by-product or waste product from the manufacture of rayon, cellulose, lithium carbonate, boric acid, and paper (synthetic). Natural sodium sulfate is used largely in the production of brown paper (45% of the U.S. market), soaps and detergents (36%), and glass (10%). World production of natural sodium sulfate amounts to about 2.5 million tonnes valued at about $300 million (Figure 11-1).

Natural sodium sulfate comes largely from *nonmarine brines* and evaporites that are found in playas (Weisman and McIlveen, 1983; Kostick, 1985). Many inland lakes are far enough away from the ocean to avoid major contributions from sea spray aerosol, allowing their composition to be dominated by elements weathered from rocks in the surrounding watershed. Eugster and Hardie (1978) divided inland lakes into four main types on the basis of their water compositions. Lakes containing Na-CO_3-Cl and/or Na-SO_4-Cl waters are most common and form soda ash and sodium sulfate evaporite deposits, respectively. Less

common lakes contain Na-Mg-Ca-Cl or Na-Mg-SO_4-Cl brines and evaporites.

Playa evaporites almost always form in enclosed valleys with no easy outlet for incoming water. Most such valleys form by faulting, as did the Basin and Range province in Nevada and adjacent states. The floors of many of these valleys, including the Imperial Valley and Death Valley in southern California, actually reach below sea level. These valleys contained large lakes during Pleistocene time when continental glaciation created humid climates over much of North America (Figure 11-22). Then the area became arid and the lakes evaporated to form playas. Enclosed valleys also form in northern areas where sediment freed by the melting of glaciers forms a hummocky surface with an irregular drainage pattern.

The deposition of sodium sulfate in playa lakes is controlled by its main mineral *mirabilite*, which varies in solubility from low to very high as the temperature increases. Thus, mirabilite precipitates from brines as they are cooled and redissolves as they are warmed, either in a daily or seasonal cycle. It can also precipi-

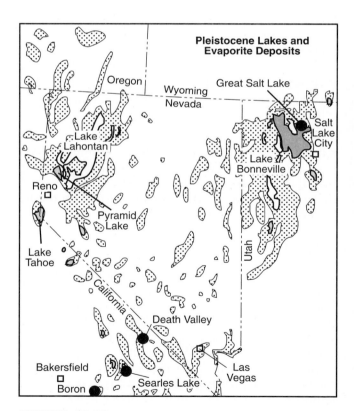

FIGURE 11-22

Distribution of Pleistocene lakes formed by increased rainfall during continental glaciation in the western United States, showing the location of playa evaporite and brine deposits (distribution of lakes from Hamblin, 1989).

tate from water that cools as it flows into the ground below the playa, forming layers of crystals that fill pores in underlying sediment. Resulting sodium sulfate deposits, some of which come and go with the seasons, can be tens of meters thick and cover large areas. The main area of sodium sulfate mining in North America is in Saskatchewan, which is part of a belt of glacial lakes that extends from Alberta to Montana and North Dakota (Figure 11-20). Sulfate in these lakes could be from springs and groundwaters draining gypsum-rich parts of the underlying Williston basin evaporite deposits that were discussed earlier. In the western United States, sodium sulfate is produced from brines of the Great Salt Lake and Searles Lake, both of which are the remains of much larger Pleistocene glacial lakes. The concentration of sodium sulfate in brines at these lakes, 12% and 35% respectively, makes it a major product from both operations.

Sodium sulfate mining is split among open-pit mining of near-surface layers, solution mining of layers that are too deeply buried to merit open-pit mining, and recovery from brines pumped from lakes or wells drilled below playa evaporites. The only problems specific to U.S. sodium sulfate production are the need for large areas of flat land in solar evaporation ponds around the Great Salt Lake and the difficulty of disposing of waste solutions in that area.

The major sodium sulfate producers are Mexico, Spain, Russia, Kazakhstan, Canada, and the United States (Figure 11-20). The world sodium sulfate industry is not expanding rapidly and U.S. and Canadian production are actually decreasing. Recycling and substitutes have decreased the paper market from about 60% of U.S. consumption to only 45%. Some of the lost market has been recovered by an increase in the soap and detergent market, where sodium sulfate acts as a filler and makes up 25% to 40% of the volume of the product. New "compact" detergents contain little or no sodium sulfate, however (Kostick, 1992). As a result, the outlook for expanded sodium sulfate production is not good, and reported world reserves of 3.3 billion tonnes, which are adequate for more than 1,000 years at present production rates, could last even longer.

Bromine

Bromine is beset by environmental complications that are having a strong impact on its markets. Its first major market was in methylene bromide (CH_3Br) and ethylene dibromide ($C_2H_4Br_2$) that were added to leaded gasoline to prevent the formation of lead coatings on cylinders and spark plugs. An average of 1 gram of bromine was required for every 3.86 grams of lead in these fuels, creating a substantial demand.

This market has essentially disappeared with the ban on leaded gasoline in most MDCs and bromine has found markets in fire-retardant compounds in plastics and epoxy in electrical and electronic equipment, soil fumigants and other agricultural chemicals, oil-well drilling mud, and compounds for the treatment of water and sewage, all of which are under environmental attack to various degrees (Lyday, 1985A). The value of world bromine production is difficult to determine because much of it is sold under long-term contracts or within vertically integrated chemical companies. At quoted prices for elemental bromine, world production of about 400,000 tonnes would be worth about $530 million, although the actual value is probably closer to $200 million (Figure 11-1).

In the United Kingdom, France, Spain, and Italy, where other sources are not available, bromine is recovered from seawater, which has a concentration of about 60 mg/liter. In the United States, bromine has been produced largely from basinal brines of the Devonian Detroit River Group in the Michigan basin and the Jurassic Smackover Formation in the northern part of the Gulf Coast basin in Arkansas, which contain up to 5,000 mg/liter bromine (Jensen et al., 1983). The bromine:chlorine ratio of these brines is also higher than that in seawater. The bromine:chlorine ratio of seawater does not change during evaporation until halite begins to precipitate, taking out chlorine. From then on, the brine is enriched in bromine. Thus, brines with bromine:chlorine ratios greater than seawater could be residual solutions from the late stages of evaporation of seawater. Saline lakes such as the Dead Sea, which are also enriched in bromine, probably formed in much the same way.

The production of bromine from brine wells is complicated by the relatively high concentrations of H_2S found in many of these wells. As in the case of the production of sour gas, this H_2S must be recovered. Most bromine extraction facilities reinject spent brine into the formation from which it was pumped originally. Reinjection of this type is strictly monitored and can be curtailed if the brine quality or formation fluid movement is not judged to be acceptable. The separation of elemental bromine from brines involves the use of stone or cement towers through which the brine falls as it reacts with chlorine or sulfuric acid. As discussed earlier, iodine is a common by-product of most bromine brine operations (Lyday, 1985A).

Elemental bromine and its compounds are toxic and must be treated with great care. Concentrations of bromine-bearing vapors above 1 ppm are considered a health hazard and concentrations of 500 to 1,000 ppm can produce death. The highest risk from these compounds is confined to bromine-producing and manufacturing facilities, although there is concern about the possible effects of bromine products on the general population. The U.S. EPA suspended the use of ethylene dibromide in soil fumigants in 1983, and studies are underway to evaluate the toxicity of smoke from materials treated with bromine-bearing fire retardants. The only major application in which bromine lacks satisfactory competitors is in additives to so-called "clear drilling muds," which dissolve high-molecular-weight, bromine-metal compounds in brines to achieve a high specific gravity without the use of barite or other conventional solid materials. These fluids are used in completion operations on oil and gas wells. Bromine has been dispersed into the atmosphere from the combustion of leaded gasoline and is a common constituent of the upper part of soils, especially in urban areas (Sturges and Harrison, 1986A and B).

U.S. bromine production has declined during the last few years because well fields in the Michigan basin were shut down in response to decreased demand. World bromine production comes largely from the United States and Israel, with smaller amounts from the CIS countries, France, and Japan. Reserves for future production are essentially limitless, with the entire ocean as a back-up to the many bromine-rich brines that have yet to be tapped.

THE FUTURE OF FERTILIZER AND CHEMICAL MINERALS

Continuing demand for fertilizer and chemical minerals is assured. We simply cannot do without them. Reserves vary from alarmingly small for sulfur to enormous for potash, indicating that exploration for many of them must go on. Although substitutes for most of these commodities are not widely used at present, it is likely that some of the fertilizer market will be taken over by *sewage sludge*. This material, which is the purified solid residue from municipal sewage treatment, is an unusually large-volume waste product and disposal of it is becoming more and more difficult, particularly since ocean dumping is no longer allowed in some areas. In an innovative solution, New York and a few other municipalities are giving the material away as a fertilizer. In the future, it may be possible to pelletize the material and actually sell it in competition with fertilizers.

Aggravating the reserve situation for most of the chemical and fertilizer minerals is the fact that few of these minerals can be recycled in a meaningful way. Many of them are used in highly soluble forms that dissolve and disperse in ionic form into surrounding soils, rocks, and waters. If we do not find a way to limit this dispersion, we may well have to limit their use. This problem is even more serious because of the small amounts of metals and other trace elements that are also released during applications of salt and fertilizers. Because of their greater capacity to be adsorbed by clays and other soil minerals, many of these elements are retained in the soil. In fact, soil metal levels appear to be climbing gradually and could become sufficiently enriched in the future to create problems for plants growing in them. In view of the extremely beneficial nature of these minerals, it is unlikely that this will cause their use to decline greatly. Instead, we must find ways to apply them that minimize their dispersion beyond the desired area. This will require the use of purified materials and possibly the incorporation of compounds that adsorb excess amounts of the elements as they dissolve.

12

Construction and Manufacturing Industrial Minerals

CONSTRUCTION AND MANUFACTURING MINERALS ARE THE LEAST known of the world's minerals, even though they are essential for housing and all types of public works (Harben and Bates, 1984). The average U.S. home contains 60 tonnes of concrete products, 25 tonnes of sand, gravel, and stone, 7 tonnes of gypsum products, and 0.1 tonne of glass. In spite of this, these mineral products rarely come to mind. Furthermore, their unit values are relatively low and transportation to market can increase costs by as much as 100% (Juszli, 1989). For this reason, mines and processing facilities for most of these commodities are usually as close to market as possible, making them much more widespread than gold mines and oil wells. The good news about this is that production of industrial minerals is high in most MDCs and is an important source of employment. The bad news is that deposits of these commodities are a common focus of conflict over land and environmental concerns.

World production data for construction and industrial minerals such as crushed stone, sand and gravel, and common clay are not tabulated and have been estimated here by assuming that data for U.S. consumption, which are available, have the same relation to world consumption that U.S. GDP has to world GDP (Figure 1-5B).

CONSTRUCTION MINERALS

The most common mineral materials used in construction are cement, gypsum, dimension stone, common clay, and *aggregate*, which includes sand and gravel, crushed stone, and related materials (Guthrie, 1992). Cement, a processed material, bears the same relation to its rock source as steel does to iron ore. The value of annual production for these commodities, which totals about $100 billion, is the highest of the industrial mineral groups (Figure 12-1). Rates of growth in production for these minerals since 1960 have diverged significantly, with only cement outpacing the growth of steel. Data on the world production of aggregate and common clay are not available, but U.S. production of aggregate has not kept up with steel, and that of dimension stone and clay has actually fallen since 1960. Prices of all of the commodities increased during the late 1970s and early 1980s in response to increasing energy prices but have leveled off since then.

Cement

Cement is a world-class, high-tech business camouflaged as a stodgy old material. The 1.14 billion tonnes

287

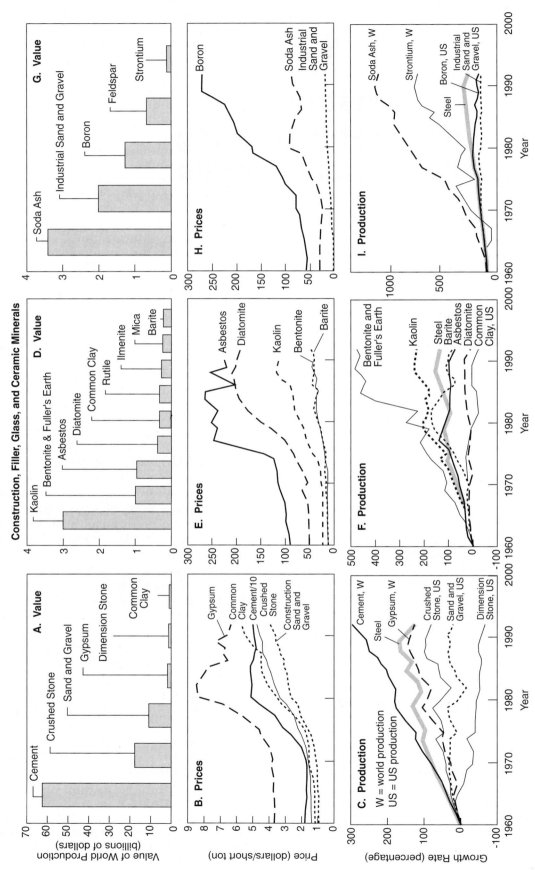

FIGURE 12-1

Construction, filler, glass, and ceramic minerals, showing the value of 1992 production (**A, D, G**) and changes in price (**B, E, H**) and production (**C, F, I**) since 1960. Compiled from data of U.S. Bureau of Mines. Value of world production for crushed stone, sand/gravel, and dimension stone is estimated from U.S. data as discussed in text. Bentonite-fuller's earth and kaolin production is for the United States before 1980 and world thereafter. U.S. and world kaolin production has grown at a similar rate since that time, but world production of bentonite+fuller's earth has outpaced that in the United States.

288

of cement produced annually have a value of about $65 billion, highest of the industrial minerals and among the highest of all mineral commodities (Figure 12-1). Cement is produced by essentially every country in the world, a reflection of the fact that its raw materials are very widespread.

Cement Raw Materials and Production

Cement is produced from *cement rock,* a silty limestone consisting of calcite, clay minerals, and small amounts of iron oxide. When cement rock is heated, CO_2 is driven off, leaving the desired mixture of calcium, aluminum, iron, and silica. The preferred cement raw material is *cement rock limestone,* which has the right combination of calcite, clay, and iron oxide to form cement directly. It usually forms in areas where clastic sedimentation and carbonate-producing organic activity took place together, such as where carbonate reefs were breached by silty rivers (Ames and Cutcliffe, 1983). Where cement rock is not available, limestone, shale, sand, and iron oxide are mined separately and mixed in the proper proportions.

Certain relatively common elements and minerals can render potential cement raw materials useless. Most problematical is magnesium, usually in the form of dolomite, which is not permitted in concentrations above about 5% MgO. Iron can also be a problem if it is present as pyrite, which will create SO_2 during heating. Where limestone is not available, other sources have been used. Shells dredged from offshore areas are the main source of calcium in some metropolitan areas. Silica and alumina can be obtained from fly ash, red mud waste from alumina production, and slag. In rare cases, gypsum and calcium feldspar have been used as the primary source of calcium (Johnson, 1985).

Cement production requires crushing and grinding the raw material to make a fine-grained powder followed by "burning" or heating the mixture in a rotating cylinder known as a *rotary kiln* (Figure 12-2). Temperatures in the kiln range from about 1,000°C, at which calcite *calcines,* or loses CO_2 to become CaO, to about 1,500°C at which the CaO reacts with silica to form calcium silicates. Calcium silicate comes out of the kiln in chunks known as *clinker* that must be ground into a powder to make cement. Gypsum is usually added to the clinker to control the rate at which the cement sets when it is mixed with water (Neville, 1987).

Cement production is a major energy consumer but has made impressive progress in decreasing its energy dependence. In 1972, production of a tonne of cement in the United States required an average of about 7.3 million Btu (8.43 Kj). By 1982, that figure had dropped by 24% and by 1990, it was down another 8%. Some

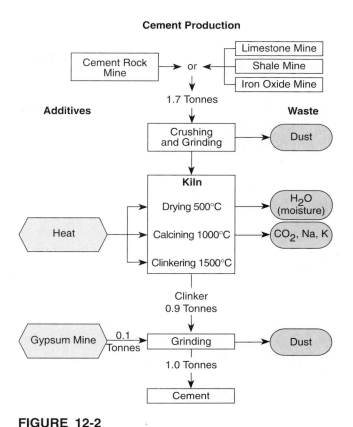

Cement Production

FIGURE 12-2

Schematic illustration of cement-producing process (compiled from Ames and Cutliffe, 1983 and Johnson, 1985).

of the most modern cement plants in Japan, Korea, and Europe use only 3.3 million Btu (3.8 Kj) per tonne (Ullman, 1991). About 80% of energy consumption goes into heating the kiln, with the remaining 20% going largely into grinding the clinker. Coal is the main fuel used for cement production, although efforts are being made to burn motor oil and other wastes (Johnson, 1985; Capone and Elsinga, 1987). Some cement kilns have even been used to dispose of toxic organic compounds by burning at high temperature although emissions from these units has been a focus of recent concern.

Cement production is the only major anthropogenic source of CO_2 other than fossil fuel combustion, and its SO_2 emissions rank behind only petroleum refining and metal smelting. Particulate emissions from cement plants in MDCs have declined considerably but remain high in many LDCs (Figure 12-3). SO_2 emissions have not changed, even in the United States, because concentrations are too low for efficient recovery (Figure 12-4A). Cement plants also emit trace elements that are vaporized by high temperatures in the kiln. Although most cement raw materials are not suffi-

A.

B.

FIGURE 12-3
Operating cement plants in mid-1980s in Texas, USA **(A)** and nearby Nuevo Leon, Mexico **(B)** showing large particulate emissions from Mexican plant compared to U.S. plant (photo of Texas plant courtesy of Portland Cement Association, Skokie, IL).

ciently enriched in trace elements for this to be an important problem, high mercury emissions have been observed (Haynes and Kramer, 1983; Fukuzaki et al., 1986; Arslan and Boybay, 1990).

Cement Markets and Trade

Most cement is known as *hydraulic cement* because it hardens under water. The most common type of hydraulic cement, *portland cement,* makes up about 95% of U.S. consumption and is named after a building stone mined at Portland on the south coast of England. Although cement can be used alone, it is used more commonly in *concrete,* which is a mixture of sand, gravel, or other mineral materials held together by cement. Concrete is essentially a synthetic rock that will take any desired form, and it is used widely in all

types of building projects. About three quarters of U.S. consumption goes into ready-mix concrete, with another 12% going into concrete block, pipe, and other products and the remainder being used in larger buildings, highway construction, and similar projects. Concrete use varies greatly from place to place, depending on competition from other building and paving materials (Johnson, 1985). Although 93% of U.S. highways are paved with asphalt, concrete covers about 40% of the 67,000-kilometer Interstate System. Concrete is not used widely in single-family homes, but it is a major component of apartment buildings, and new high-strength versions compete with steel framing for office buildings. Plans have even been made for a 125-story concrete building in Chicago without the conventional steel framework (Sedgwick, 1991).

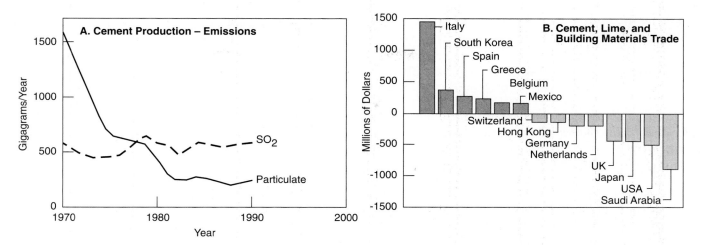

FIGURE 12-4
A. Change in particulate and SO_2 emissions from cement plants in the United States since 1970 (from EPA 450/4-91-004). **B.** World trade in cement, lime, and building products for 1990 showing the difference in value of exports and imports (from *United Nations Trade Statistics Yearbook*, 1992).

The leading cement producers include China, CIS countries, Japan, and the United States. Because limestone and clay are expensive to transport in relation to their value, most cement plants are near sources of these materials. The large amounts of energy required for clinker production give cement producers in areas of cheap energy an edge. At a price of about $55/tonne for cement, a further advantage goes to producers located near consumers or with access to cheap barge or ship transport. World cement trade statistics are complicated by the inclusion of lime and other building materials in the same category (Figure 12-4B) but show that Italy is a leading exporter, taking advantage of good access to ocean transport. Energy costs may become a big factor in future cement trade. Recent antidumping decisions against cement exports from Japan and Mexico have been based in part on the possibility that their energy costs are subsidized (Ullman, 1991).

Environmental Damage to Concrete

Concrete's reputation has been damaged by the alarming decay of bridges, highways, and other structures around the world. Many of these, such as the Interstate System in the United States, were built with the expectation that they would have a long life. Instead, the concrete is crumbling and governments are faced with large repair costs. This has made many people question whether concrete is the miracle rock that it was originally thought to be. The answer lies in a bet-

ter understanding of what concrete actually is and how it reacts to its environment.

Portland cement hardens when water is added because elongated crystals of hydrated calcium silicate grow into an interlocking matrix that has very high strength (Erling and Stark, 1990). When these crystals form, they leave small pores that were originally occupied by water, and it is in these pores that the damage begins. Water seeping into them freezes and expands, breaking the concrete. Minerals such as gypsum, which precipitate as water evaporates, have the same effect. Deicing salt on highways reacts with steel-reinforcing bars in the concrete to form rust that breaks up the concrete even more. Acidic water also dissolves the concrete, particularly in concrete sewer pipe where bacteria make the water very acidic. Interestingly, this acidification has become more problematical in recent years, since metals that poisoned the bacteria were banned from sewers (Sedgwick, 1991; Mays, 1992).

New approaches are being tested to deal with these environmental challenges to concrete. The corrosion of reinforcing steel rods is being limited by epoxy coatings and the use of low-voltage DC currents to reverse oxidation reactions, and fiberglass rods are being tested. Organic and silicon compounds are being added to fill pores left during the consolidation of concrete. Although many efforts are still experimental, it is widely felt that concrete will soon see improvements similar to those that began in 1970 for metals and ceramics. With these developments and essentially inexhaustible raw material reserves, cement should have a good future.

Construction Aggregate—Crushed Stone and Sand and Gravel

Construction aggregate is fragmental rock and mineral material that is used alone as fill and in combined form with concrete, asphalt, and plaster. Natural aggregate is very inexpensive but is used in such huge volumes that it has the highest value of the unprocessed industrial minerals (Figure 12-1). U.S. production of construction aggregate depends on economic activity but averages about 1.1 billion tonnes annually from about 8,900 operations. It includes almost all of the sand and gravel and about half of the crushed stone produced (the other half of crushed stone is largely limestone used to make cement, lime, and other chemicals). The value of this construction aggregate, including only the half of crushed stone that is used in construction, is about $5.5 billion or about $5/tonne (Figure 12-1). Although statistics on world construction aggregate production are not available, it would amount to 3 to 4 billion tonnes worth about $20 billion if production in other countries has the same relation to GDP as it does in the United States. The extremely low unit value of aggregate means that it cannot be transported more than a few kilometers from its source without becoming prohibitively expensive. As discussed later, this factor is of critical importance to future land-use decisions in MDCs.

Geology and Production of Aggregate

The favored type of aggregate is *sand and gravel* (Davis and Tepordei, 1985), most of which comes from stream, beach, and glacial deposits (Figure 12-5). Deposits of this type are common along rivers and streams, where they tend to collect in *point bars* that deposit on the inside part of bends. Where rivers reach the ocean, sand and gravel are deposited in *deltas* and in *beaches,* and ocean currents can move beach sand to form *sand spits.* During Pleistocene time, sand and gravel were deposited in piles of material, known as *moraines,* at the ends of widespread glaciers. Sand and gravel also accumulate in *alluvial fans* where rivers cross faults into valleys. Where sand and gravel are not available, *crushed stone* is produced from massive rock (Schenck and Torries, 1985). The best rock for this purpose is limestone, which makes up about 70% of production largely because it is relatively soft and easily mined. Harder granite and basalt supply another 20% and sandstone, quartzite, and other materials

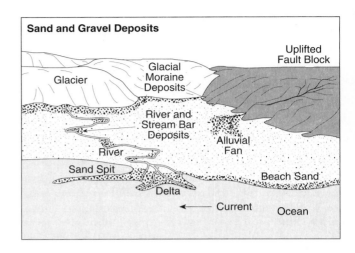

FIGURE 12-5

Schematic illustration of geologic environments in which sand and gravel deposits form.

make up the rest. Even though crushed stone requires more processing and is about 20% more expensive, it supplies slightly more than half of U.S. aggregate production because sand and gravel deposits are scarce in so many areas (Langer, 1988).

Not all sand and gravel or massive rock deposits meet construction aggregate specifications (Tepordei, 1985). Some of the more important specifications include resistance to abrasion, chemical attack, and splitting due to freezing water. Even small amounts of pyrite, a common trace constituent of many rocks, can render the material completely unacceptable because it oxidizes to produce a rusty color. Thinly layered shales are also useless because they are easily broken apart by the freeze-thaw cycle. As a result, many large areas of rock or gravel cannot be mined.

Although 95% of U.S. aggregate comes from open-pit mines, interest in underground aggregate mines is rising as a measure to minimize environmental impact. This trend actually began long ago in Europe where cities found it easier to mine downward than to transport aggregate from distant locations outside town. Much of central Paris was undermined by operations of this type, which collapse occasionally. Some modern underground aggregate mines are used as commercial storage facilities. The processing of sand and gravel consists essentially of washing and screening the material to produce different size fractions. In an interesting twist, sand and gravel operations in areas downstream from gold mining areas, such as those around the Mother Lode in California, recover small amounts of placer gold as a by-product.

FIGURE 12-6

Aggregate quarries on the outskirts of Monterrey, the third largest city in Mexico. Although the quarries were located perfectly from an economic standpoint, they created too much particulate pollution (note plumes of dust coming from the quarry at left) and most were closed during the late 1980s.

Environmental Impact of Aggregate Production

Aggregate deposits are being squeezed out by other land uses in MDCs, presenting a critical problem to regional planners and construction projects. No one wants to have aggregate mines near them, but everyone wants cheap aggregate. With each 5- to 10-kilometer distance adding about $1 per tonne in transportation costs, efforts to replace nearby deposits with more distant ones will increase construction costs. At the same time, the mines are often unattractive and can have particulate emissions (Figure 12-6).

The situation is particularly critical around Los Angeles, where aggregate supplies have come largely from alluvial fans along the south side of the San Gabriel Mountains (Figure 12-7A). In areas such as Irwindale, which is on a fan 13 kilometers long that extends into the Los Angeles valley, mining became increas-

ingly difficult as population pressures caused land prices and reclamation requirements to increase (Figure 12-7B). Mining also interfered with groundwater in the alluvial fan, which is an important local aquifer. Alternative aggregate deposits are farther away from Los Angeles and add considerably to transportation costs (Henderson and Katzman, 1978; Goldman and Reining, 1983). Cities in the northeastern United States are also experiencing shortages; the price of aggregate constitutes as much as half of the cost of cement. In some parts of the United States, Denmark, the Netherlands, Japan, and the United Kingdom, adequate land sources are no longer available, and aggregate is dredged from the continental shelf, which contains a huge resource for future use (Williams, 1992). Although aggregate is not a major factor in international trade, Japan, the Netherlands, and Switzerland have significant imports, and the United States imports small amounts of aggregate from Bahamas, Canada, and Mexico (Tepordei, 1993) (Figure 12-8). Long-distance, ocean transport is still limited, although crushed stone was exported from Scotland to Houston, Texas, in 1985, largely because of the lack of material in the area and the availability of cargo for the return trip (Zdunczyk, 1991).

Before concluding that society will simply have to make the change to more distant aggregate sources, give some thought to its implications. In a recent study of this matter, it was concluded that a 5-mile increase in the distance between aggregate mine and consumer in the United States would add $450 million to total transportation costs, require 2 million additional barrels of oil for fuel, and greatly increase truck traffic and highway accidents (Goldman and Reining, 1983). An estimate of transportation costs in the Los Angeles area during the early 1980s indicated that the cost of aggregate delivered to the harbor area could vary by as much as 100% depending on the source (Figure 12-7A). In view of this, most governments include aggregate mining in land-use plans. Nevertheless, suburbs have already covered important aggregate deposits in many areas and conveniently located reserves are becoming scarce. There seems to be no way of avoiding an increase in aggregate costs in the near future, although the size of the increase depends on regional land-use decisions.

Dimension Stone

Dimension stone, which includes decorative slabs, large blocks, monuments, and tombstones, is the quintessential shrinking business, at least in the United States. In the early part of the 1900s, dimension stone accounted for about half of the stone pro-

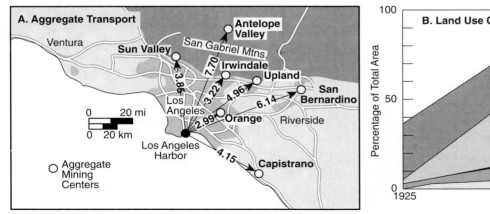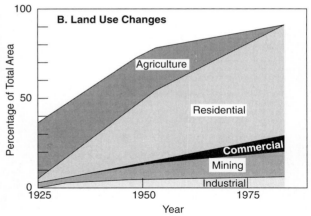

FIGURE 12-7
A. Distribution of aggregate mining deposits in the Los Angeles area showing cost in 1981 of transporting a short ton of aggregate from each mine to the Los Angeles harbor area (after Goldman and Reining, 1983). **B.** Change of land use in the San Gabriel alluvial fan area between 1925 and 1978 (from Henderson and Katzman, 1978).

duced in the United States. Since that time, crushed stone, concrete, and other materials have taken much of the dimension stone market and the 2 million or so tons of dimension stone consumed annually in the United States amounts to only about 0.2% of crushed stone consumption. Over the last few decades, dimension stone production in the United States has declined slightly to a value of about $200 million (Figure 12-1). Using the GDP relations discussed previously, world production might amount to about 6 million tonnes worth about $700 million.

The most common types of rock used for dimension stone are *granite* and related intrusive igneous rocks, *limestone* and its metamorphosed equivalent *marble*,

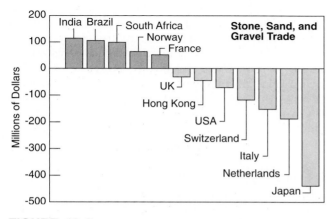

FIGURE 12-8
World trade in stone, sand, and gravel for 1990 (from *United Nations Trade Statistics Yearbook*, 1992).

sandstone, and *slate,* a form of metamorphosed shale that splits into large, flat slabs. Their most common use is in *cut stone* that forms a decorative veneer on buildings but is rarely part of the supportive framework of the building. Smaller blocks are used in window sills, door jambs, and lintels. Larger blocks of *rough stone* are used as foundations for bridges and other structures, and walls are built of rectangular pieces known as *ashlar. Slabs* of slate and other stone are also used for paving (Power, 1983; Taylor, 1985).

Dimension stone must resist abrasion as well as corrosion during weathering. Slate, granite, brick, and clay products are highly resistant to corrosion by acid, whereas steel, limestone, and marble have poor acid resistance. Copper, aluminum, and painted wood fall between these extremes. Urban areas in humid environments have particular problems with corrosion caused by acid rain and discolored gypsum coatings caused by dry deposition from the atmospheric aerosol. Even granite is damaged under these conditions, as shown by the incredible deterioration of Egyptian obelisks that were moved to London and New York by early explorers. Although SO_2 emissions have declined since that time in most cities (Figure 12-9), problems are still being encountered, particularly with limestone and marble, which are more easily corroded. For instance, marble that covered the Amoco building in Chicago weakened due to corrosion and had to be removed in the early 1990s. It was replaced by more resistant granite from Mt. Airy, North Carolina, at a cost of $60 to $80 million (Taylor, 1991). Special coatings for stone that have been developed recently are likely to solve many corrosion problems.

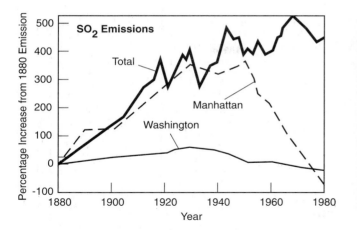

FIGURE 12-9

SO_2 emissions in cities versus total emissions since 1880, showing that total emissions have continued to climb even though urban emissions have declined. This change reflects the wider use of clean natural gas in home heating but widespread emissions of SO_2 from coal-fired electric generating plants (modified from *U.S. National Acid Precipitation Assessment Report, 1990 Integrated Report*).

FIGURE 12-10

Yule marble mine near Aspen, Colorado, showing step-like walls resulting from the removal of large blocks such as seen on the floor of the mine. People provide scale (courtesy of James Rudolph, U.S. Bureau of Mines and J. Groeneboer).

The mining and processing of dimension stone require special deposits and techniques. Most mines are open pit, although underground quarries are common. With the exception of slate, dimension stone deposits must lack closely spaced joints or fractures, which are actually very common in most rocks. The rock must be removable in large blocks that can be cut and shaped without breaking apart (Figure 12-10). These blocks were originally broken out by drilling a line of holes and driving in wedges to split the rock, or by the use of wire saws and chipping machines. More recently, granitic rock has been cut by jet channeling units, which use heat from a fuel oil-oxygen torch to cut a groove by spalling off chips of rock. This method will not work on limestone and marble, which calcines under the intense heat, and it has come under environmental attack because it is loud and creates dust. High-speed water jets might take over the rock-cutting process. Once the stone has been removed, it must be cut and polished into slabs and other shapes, which is usually done with the abrasive materials discussed later.

In spite of its relatively low unit value, *Carrara marble* from northern Italy is exported throughout the world. This marble, which has an amazingly uniform texture, has been used in some of the most famous sculptures in the world, including Michelangelo's Pieta. Although bathrooms lined with "Italian marble" are often assumed to be a sign of luxury, Carrara marble is almost always the cheapest form of cut stone available in most MDCs and it even competes in cost with tile. Another widely traveled stone is the black, iridescent *larvikite,* a type of mafic rock named for the small town in Norway from which it is quarried. In the United States, limestone from Indiana is the most widely used stone and quarries that dot the landscape around Bloomington supplied much of the stone used for the Federal Triangle in Washington, D.C.

The outlook for dimension stone depends more on public taste than other factors. By its very nature, dimension stone varies from piece to piece. Architects are understandably nervous about this, not wanting to construct a building with abrupt differences in color or texture. This imposes high quality control on producers, causing increased waste and higher prices that make way for competitors such as synthetic polished stone made of rock fragments in special cements. Regardless of what happens on the demand side, dimension stone reserves should be adequate.

Lightweight Aggregate and Slag

Lightweight aggregate and slag are produced in smaller volumes than natural aggregate and have special markets. *Slag* comes largely from blast furnaces and steel mills. About 23 million tonnes of it are sold annually in the United States at an average price of about $5 to $6/tonne. *Lightweight aggregate* is any aggregate that has a relatively low weight but retains the strength of natural rock. Aggregate of this type is especially useful in construction applications where weight is an important factor. It is also used in insulation and as an additive to soils and other materials where increased porosity is desired (McCarl, 1985).

Volcanic rocks make the best *natural lightweight aggregate*. The lightest such aggregate consists of pumice, in which abundant vesicles lower the bulk density of the rock so much that it will float in water. Vesicular andesite and basalt, as well as scoria, which also can be used, rarely float but have greater strength. Some forms of *volcanic tuff* with tiny angular glass fragments also have high enough porosity to be considered lightweight aggregate. *Manufactured lightweight aggregate*, which is more important than natural material, is produced from minerals and rocks that contain water that expands on heating to create pore space. These include *clay-rich shales*, the mica mineral, *vermiculite*, and volcanic glass known as *perlite*. Water is held in the crystal structure of vermiculite and is in the structure of clay minerals in shale (Figure 3-3). In perlite, it is trapped as molecules that did not have time to coalesce into vesicles and escape from the cooling magma. Whereas shale is relatively widespread, pumice and perlite are found only in areas of rhyolitic volcanic rocks and vermiculite forms deposits only where magnesium-rich igneous rocks have been altered by hydrothermal solutions. Some slag and other industrial wastes also qualify as *synthetic lightweight aggregates*.

The lightweight aggregate market is small but will probably grow rapidly because of the advantages of transporting and using strong, light construction materials. Presently, only 3 to 4 million tonnes per year of lightweight aggregate are produced in the United States, with slightly less than half of that consisting of expanded shale. Annual world production of pumice, perlite, and vermiculite, not all of which goes into lightweight aggregate, amounts to about 11, 1.7, and 0.5 million tonnes annually, with values of $275, $55, and $50 million, respectively.

Gypsum

Gypsum is the main raw material from which we manufacture *plaster,* an essential construction material.

Smaller amounts are made from *anhydrite*, the anhydrous equivalent of gypsum. About 100 million tonnes of gypsum and anhydrite are produced annually from at least 80 countries. The value of this production, based on U.S. prices, would be about $700 million.

Gypsum deposits are almost entirely *marine evaporites* that formed by one of two processes. Thick deposits of gypsum associated with halite and potash are thought to have formed during the early stages of classical evaporation of seawater, as discussed in the previous chapter. Other thick gypsum deposits, which are not associated with salt, are thought to have formed in salt flats, or *sabkhas,* such as those along the margin of the Persian Gulf. Sabkhas are tidal flats. During high tide, seawater flows onto them and is trapped in pools. During low tide, water in the pools is concentrated by evaporation and sinks into the ground where it deposits gypsum. Some salt is also deposited, but it is usually dissolved during the next tidal influx. Through time, this process can deposit a large thickness of gypsum-rich evaporite. As discussed in the section on sulfur, gypsum also accumulates at the top of salt domes where they rise into and are dissolved by fresh groundwater.

Gypsum deposits are widespread but are rare in older Archean rocks because the lack of oxygen in the Earth limited the sulfate content of its oceans and because old evaporites are too soluble to be preserved (Reading, 1986; Warren, 1989). Gypsum also loses water to become anhydrite when it reaches temperatures of about 50°C. Thus, as gypsum is buried beneath younger sediments, where it becomes hotter, it converts to anhydrite. As it is exposed again during erosion, it can convert back to gypsum.

Gypsum is usually mined by open-pit methods, although a few underground mines are in operation. Some of the caverns under Paris resulted from gypsum mining, from which the term *plaster of Paris* was derived. About 75% of mined gypsum is calcined at temperatures of 100°C or so to produce the *hemihydrate,* $CaSO_4 \cdot \frac{1}{2}H_2O$, which is the true form of plaster of Paris. When mixed with water, this material forms a network of gypsum crystals that makes a strong, soft solid. Plaster was originally used as a coating to finish other building surfaces but is now used largely in wallboard, a sandwich of gypsum between two sheets of heavy paper (Appleyard, 1983).

In spite of its low cost of about $7/tonne for raw gypsum (Figure 12-1) or $17/tonne for calcined material, gypsum is relatively widely traveled. This is largely because domestic production in the United States, Japan, and many European countries is not adequate to satisfy demand. Metropolitan areas in the northeastern United States are supplied in part by gypsum brought in on ships from Mexico, Canada,

Dominican Republic, Jamaica, and Spain, which is cheaper than gypsum moved by train from deposits in the Great Lakes area. Many cities in northern Europe are supplied in a similar way by Spain, and Japan's extra needs are supplied largely by Thailand. Although well-delineated gypsum reserves of about 2.6 billion tonnes amount to only about 20 years' present production, unexplored reserves are certain to be much larger and more than adequate for the next century, particularly with underground mining.

FILLERS, EXTENDERS, PIGMENTS, AND FILTERS

Fillers and extenders are used to provide special characteristics or to cut the cost of materials (Severinghaus, 1983). The list of mineral *fillers* is very long and their list of applications is even longer, with many curious examples (Table 12-1). Paint containing the heavy mineral barite has been sprayed inside car doors to make them close with the proper sound, and small amounts of finely ground mica provide the sheen in lipstick. Materials that also provide color are known as *pigments*. Others are used largely as *filters* through which liquids and gases can be strained. The distinction between these uses can be a fine one, and some commodities discussed elsewhere in this book are also used as fillers, extenders, pigments, or filters. Lime, for instance, competes with kaolin as a filler in paper. Finally, rutile and ilmenite, which are our sources of titanium metal, are also important white pigments.

Annual world production of these commodities is valued at about $8 billion and is dominated by the clays (Figure 12-1). Production has grown rapidly for

bentonite and fuller's earth clays, but has been slower for most other commodities. Asbestos and barite consumption have been impacted by special problems as discussed later.

Clays

Clay is a term with many meanings and the potential confusion is greatest when talking about clay as an industrial mineral (Ampian, 1985). To the lexicographer, clay is any fine-grained material that becomes plastic when mixed with water. To the geologist and environmentalist, it is a layered silicate mineral (Figure 3-3) and important constituent of soils, as well as the finest grain size for clastic sediments, regardless of their composition. To the miner, it is any saleable clay mineral material that can be used in the filler, extender, and related markets, currently valued at $5.5 billion annually (Figure 12-1).

Clay minerals form where water leaches cations from feldspars and other minerals, leaving aluminum, silicon, and oxygen. This occurs in two main environments, the weathering zone and hydrothermal systems (Patterson and Murray, 1983). Clay formation in the *weathering zone* is part of the soil-forming process and clay-rich soils are particularly common in humid, subtropical, and tropical climates where acid waters leach most of the cations from the soil (Figure 2-5). They were formed in large amounts during the last major global-warming period from Cretaceous to Eocene time. Soils themselves rarely form good clay deposits, but they can be eroded to form fine-grained, clay-rich sediments known as *shale,* which is mined widely for common clay, the material from which we make brick and tile, as discussed later. Even richer deposits form where special sedimentary processes oc-

TABLE 12-1

Mineral raw materials used as fillers and extenders in the United States, listed in order of declining value (all data in short tons; compiled from *U.S. Bureau of Mines Minerals Yearbook*).

Raw Material	Volume Used in Filler/Extenders	Value ($ millions)	Percentage of Total Production
Kaolin	4,000,000	400	>45
Diatomite	500,000	136	100
Fuller's earth	1,800,000	132	>80
Barite	1,400,000	52	100
Bentonite	1,400,000	49	>40
Talc/Pyrophyllite	1,100,000	22	66
Asbestos	20,000	5	52[a]
Mica	100,000	5	100

[a] Proportion of asbestos used as a filler will be larger in many developing countries.

cur, such as the kaolinite deposits of the coastal plain of Georgia (Figure 12-11). These deposits are lenses of pure kaolinite that were deposited in meanders of rivers that drained the Appalachians (Figure 12-12A). Some of the kaolinite came directly from soils of the Appalachians, but most of it appears to have formed by the weathering of feldspar grains in the coastal delta, the clay being washed into the meanders later. *Bentonite* deposits, on the other hand, are found in thin, very extensive layers that are thought to be the weathered remains of fine-grained volcanic glass that was deposited as ash (Kesler, 1956; Spencer et al., 1990).

Hydrothermal alteration forms clays where acid, hydrothermal solutions cooler than about 200°C flow through feldspar-rich rock-leaching Na^+, K^+, Ca^{++}, and other cations and replacing them with H^+. Most clay deposits of this type are associated with meteoric hydrothermal systems that derived heat from intrusive or volcanic rocks (Figure 12-12B). Large kaolinite deposits in Cornwall in southern England are in the outer parts of hydrothermal systems associated with granitic batholiths and formed at depths of several kilometers (Bray and Spooner, 1983). Other kaolinite deposits in young volcanic rocks along convergent margins are like those that host epithermal gold-silver veins, and probably formed at shallower depths (Kesler, 1970).

Almost all clay minerals are produced from open-pit mines, most of which are strip mines (Figure 12-12). Reclamation of mined land is a major factor in clay mining, with costs often exceeding the original cost of the land. The processing of clays for market ranges from simple drying and washing for common brick and tile clays to complex beneficiation involving flotation for kaolinite clays.

The clay trade can be broken down into three groups. In order of decreasing value but increasing volume, these are the kaolinite group, the smectite group, and the common clays. The first two of these groups reflect the two main groups of clay minerals, kaolinite and smectite (or montmorillonite) (Figure 3-3), which have different properties and different markets. Kaolinite clays do not change composition

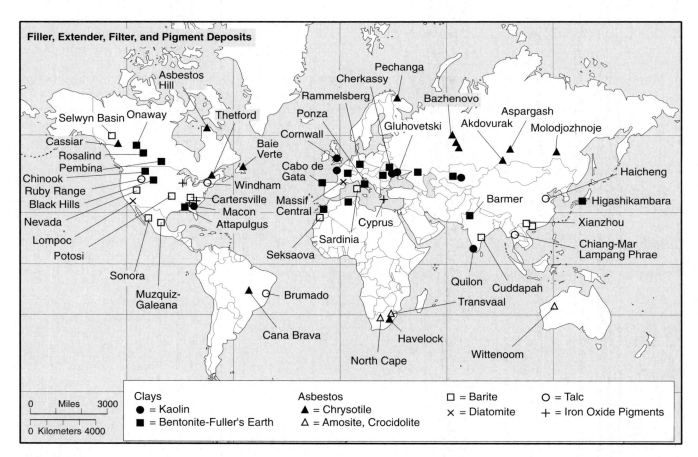

FIGURE 12-11

Location of important filler, extender, filter, and pigment deposits.

A.

B.

FIGURE 12-12
A. Kaolin deposit in the Georgia coastal plain. Four layers can be seen in this photo. The upper layer, which is behind and to the left of the trees at the top of the mine, is a pile of overburden. Immediately below the trees is the darker, Tertiary-age Twiggs clay, which overlies lighter-colored Cretaceous sands. The whitest layer at the bottom of the wall and making up the floor of the pit is kaolin (from Kesler, 1956, Figure 5). **B.** Kaolinite-rich volcanic rocks formed by hydrothermal alteration, Michoacan, Mexico. Note steam from active hot springs (courtesy of Thomas L. Kesler).

The main *kaolinite-group* products are *kaolinite* and *halloysite,* a generally similar mineral with a tubular structure in its dehydrated state, as well as less pure forms known as *refractory clay* and *ball clay. Kaolin,* the industry term for clay material consisting largely of kaolinite, is used as a filler and coater in paper, rubber, and paint, as the main component in refractories and fine china, and as a catalyst in petroleum refining and other manufacturing processes. Factors that affect kaolinite quality and value are its crystal size and brightness, with iron staining particularly undesirable for many applications. *Coating clays,* which are used to make paper smooth and glossy, are mostly less than 2 micrometers in diameter, whereas *filler clays* are coarser. World kaolin production of about 27 million tonnes comes largely from Georgia and southern England. Although the total value of this production is not tabulated, it would be about $3 billion based on an average value of U.S. kaolinite production.

The main *smectite-group* products, *bentonite* and *fuller's earth,* are used largely in drilling mud and as a binder in foundry sand and iron ore pellets. Bentonite makes the drilling mud act like a solid when it is not agitated so that it will support rock chips cut from the hole if pumping stops. This property of a suspension, known as *thixotropy,* is very important. Without it, the drill hole would have to be abandoned if pumping of the mud was stopped because rock chips would sink to the bottom trapping the drill bit and pipe. Fuller's earth is a similar swelling clay that is valued for its ability to absorb other liquids. It is used as a cleaning compound and as a carrier for liquid pesticides and other liquids that must be distributed in solid form.

when immersed in water, but smectite clays undergo cation exchange. They also adsorb water because it is a polar molecule, which causes them to swell. The commercial smectite clays, bentonite and fuller's earth, are valued for these properties.

Attapulgite, a special form of these minerals, comes from deposits around Attapulgus, Georgia. World bentonite and fuller's earth production is about 12 million tonnes, 80% of which is bentonite. At average 1992 U.S. prices for these products, this would have a maximum value of $1 billion.

Common clay accounts for most of the volume of world clay production, but not the value. World production of common clay is not tabulated but would be about 100 million tonnes annually, if production in other countries is proportional to their GDP. The value of this production, at U.S. 1992 average prices would be about $500 million. Common clay is used largely for brick, drain and roof tile, sewer pipe, and other construction materials. Much of it is also used in lightweight aggregate, as discussed earlier. The most common clay deposits are shales that are low in iron and calcium.

Although world clay trade statistics are complicated by the inclusion of all commodities in a single class, they are dominated by kaolinite because it is so much more expensive. Major exports come from the United States, United Kingdom, and Brazil, and imports go to Japan, Italy, Finland, and Canada (Figure 12-13).

Asbestos

Asbestos is a general term for any fibrous mineral with a thread-like or *acicular* shape, usually with a length greater than three times its width. Commercial asbestos must also be flexible, resistant to acid, and have a very high tensile strength (Schreier, 1989; Skinner et al., 1988). Long fibers of commercial asbestos are woven into fabrics that resist heat and corrosion, and short fibers are used as a filler to impart strength to cement pipe, asphalt shingles, and automotive brake

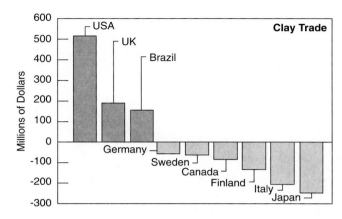

FIGURE 12-13
World clay trade for 1990. Although the data include all clay products, the high value of kaolinite dominates the diagram (compiled from *United Nations Trade Statistics Yearbook,* 1992).

parts. The importance of many of these applications and the lack of ready substitutes made asbestos a rapidly growing mineral commodity until the 1970s (Figure 12-1).

Then the asbestos scare hit, and it has been downhill ever since. Recent world asbestos production, which amounts to almost 4 million tonnes, has a value of almost $1 billion and is projected to continue its downward trend. Whether this actually takes place depends greatly on the outcome of efforts to sort out the health effects of the many different forms of commercial asbestos, as discussed later.

Geology and Production of Asbestos

Commercial asbestos comprises six distinct mineral groups, including the *serpentine asbestos* mineral, *chry-*

TABLE 12-2
Classes of asbestos minerals showing cumulative world production (1870 to 1980) of each type along with mortality rates from all lung diseases (from Ross, 1987; Schreier, 1989).

Mineral Name	Chemical Composition	Production (tonnes)	Mortality Rates (all diseases)
Serpentine Asbestos			
Chrysotile	$Mg_6[SiO_5(OH)_4]$	90,000,000	low to none?
Amphibole Asbestos			
Crocidolite	$Na_2Fe_5[Si_4O_{11}(OH)]_2$	2,700,000	very high
Amosite	$Fe_7Si_8O_{22}(OH)_2$	2,200,000	very high
Anthophyllite	$(Mg,Fe)_7[Si_4O_{11}(OH)]_2$	350,000	very high
Actinolite	$Ca_2Fe_5[Si_4O_{11}(OH)]_2$		small
Tremolite	$Ca_2Mg_5[Si_4O_{11}(OH)]_2$		small

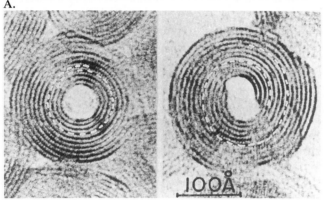

FIGURE 12-14

A. Photograph of crysotile asbestos fibers filling vein from Thetford, Quebec. **B.** High-resolution transmission electron microscope cross-sectional view of chrysotile asbestos fibers from Transvaal Province, South Africa, showing spirals formed by sheets of silicate tetrahedra and magnesium octahedra (from Yada, 1971; courtesy of K. Yada and F.J. Wicks).

sotile (or white asbestos), which makes up about 95% of world production, and five *amphibole asbestos* minerals of which only *crocidolite* (blue asbestos) and *amosite* (brown asbestos) are of economic significance (Table 12-2). All of these are essentially hydrous silicates of magnesium and/or iron, and their fibrous form is a direct reflection of their crystal structure (Figure 12-14). Chrysotile consists of separate layers of silicate tetrahedra and brucite [$Mg(OH)_2$] that are rolled into hollow tubes. Amphibole asbestos consists of two ribbon-like chains of silicate tetrahedra linked by Mg^{+2}, Ca^{+2}, Fe^{+2}, Na^+, and OH^- ions. Exactly why these minerals form elongated fibers is not clear, although it is thought that aluminum substitution increases this tendency (Campbell et al., 1977).

Almost all asbestos deposits form by hydrothermal alteration of rock that is rich in magnesium and/or iron (Riordan, 1981). *Chrysotile asbestos deposits* are usually found in altered ultramafic igneous rock that came from the lower part of the ocean crust or upper mantle. These rocks are found along ancient convergent tectonic margins, including a suture formed by the collision of North America and Europe in Paleozoic time, which hosts the largest asbestos deposits in North America near Thetford Mines, Quebec (Figure 12-11). Asbestos occurs in these deposits as veins or as mats of disoriented short fibers that replaced the massive rock during hydrothermal alteration. Asbestos-forming hydrothermal solutions probably came from seawater that circulated deep into the ocean crust or magmatic water from granites that intruded the deposits. A few chrysotile deposits are found in other hydrothermal alteration environments. The large Asbestos Hill deposit in the Ungava peninsula of Quebec is in a sequence of magnesium-rich, komatiite volcanic rocks that hosts magmatic nickel deposits. Deposits at Msauli, South Africa, and adjacent Havelock, Swaziland, are in similar rocks. Asbestos can also form in dolomite, the magnesium-rich carbonate rock, where it is altered by solutions containing silica, although few deposits of this type are large enough to mine. The only important *amphibole asbestos deposits*, which are in northern Cape Province, South Africa, are in magnesium-rich layers of banded iron formations, where they overlie dolomite. The hydrothermal solutions that formed many of these deposits were basinal brines and meteoric waters heated by the Bushveld layered igneous complex, the intrusion that formed the platinum and chromium deposits discussed earlier.

Asbestos is mined largely from open pits, although some South African amosite mines are underground (Clifton, 1985). Asbestos fibers are usually classified according to length, which can vary from rare fibers more than about 2 centimeters long to those only a few millimeters in length (Mann, 1983). Fibers are usually concentrated from the ore by dry methods in which ore is shattered by a rapid, sharp impact and the liberated fibers are then blown or sucked away. Large amounts of air are used in this process and all of it must be thoroughly filtered before release. Wet processing would make it easier to suppress fiber release, but it has been used only in deposits consisting largely of short, matted fibers. Asbestos is shipped in closed bags, and new products consisting of asbestos embedded in resins or plastics are being developed to minimize dust during transport.

Asbestos Markets, Trade, and Environmental Concerns

In the late 1970s, U.S. consumption of asbestos went into asbestos-cement pipe (35%), flooring products

(20%), friction products such as brake linings (15%), roofing shingle (10%), and insulating materials, with total consumption of 550,000 tonnes. By the late 1980s consumption had decreased to about 300,000 tonnes with declines in all categories. Continuation of this steep decline was almost assured on July 12, 1989, when the United States banned the manufacture, importation, processing, and distribution of most asbestos-containing products. The ban was to have taken place in three stages. The first stage, set for August 27, 1990, applied to asbestos-containing floor and roof felt, asbestos-cement sheets, vinyl asbestos floor tile, and asbestos clothing. The second stage, scheduled for August 25, 1993, banned asbestos-containing gaskets, transmission components, and original-equipment drum and disk brake components. The third stage, to begin August 26, 1996, applied to asbestos-cement pipe, shingle, and after-market brake components. In what is likely to be a continuing saga of litigation, this ruling was overturned in the U.S. Court of Appeals in 1992 because consideration was not given to the cost of the ban, health risks posed by asbestos substitutes, and related factors.

Concern about asbestos centers on its role as a cause of lung diseases. Research into this problem has been hampered by the long period between asbestos exposure and symptoms, known as the *latency period,* which can exceed 30 years. Further complications come from the difficulty in diagnosing the diseases and fragmentary medical records. Results that have been obtained so far suggest that chrysotile, the most widely used asbestos, is much less dangerous than amphibole asbestos minerals. *Asbestosis* is a chronic affliction resulting from the inhalation of asbestos fibers and is commonly seen in workers who have been associated with high levels of asbestos dust. It involves an increase in the amount of fibrous protein in the lung, which ultimately decreases its flexibility and oxygen-absorbing capacity, sometimes leading to death. The incidence of severe asbestosis has been declining for years and should continue to do so because of dust control in the workplace. *Lung cancer,* which is associated with the increased inhalation of fibrous minerals of all types, is not necessarily associated with asbestosis. Although the incidence of lung cancer among nonsmoking asbestos workers is normal to slightly high, it is much higher among miners who are smokers as well as those in some textile plants. This difference could reflect the capacity of fibrous minerals, including asbestos, to facilitate the movement of hydrocarbons from tobacco smoke or plant fibers into the lungs. *Mesothelioma,* which can be either benign or malignant, is a tumor in tissue that encases the lung. Malignant mesothelioma has been shown to be a sta-

tistically significant cause of death in only two mining districts in the world, northern Cape Province, South Africa, and Wittenoom, Australia (Figure 12-11), both of which produced crocidolite, the amphibole asbestos known as blue asbestos. Mortality rates were high among these workers, their families and even people with short exposure periods to crocidolite. Although data for amosite (the other commercially important amphibole asbestos) miners are limited, mortality rates among manufacturing workers are also high. No significant correlation between mesothelioma mortality and chrysotile miners has been demonstrated (Skinner et al., 1988; Ross, 1987).

The asbestos problem is almost certainly a wider issue involving all fibrous materials, both natural and man-made. Strong evidence for this came from observations that residents of two small towns in Turkey exhibited high mesothelioma mortality with no significant asbestos exposure. This mortality is thought to be caused by *erionite,* a fibrous zeolite mineral discussed later, which was present in the air in concentrations of 0.01 fiber/cm^3. By contrast, Quebec asbestos miners were originally exposed to chrysotile concentrations of up to 21 fibers/cm^3 and the present U.S. government limit on asbestos in the workplace is 0.2 fibers/cm^3 (Ross, 1987). Although it is too early to be certain, fiber size and shape appear to be important characteristics that determine the danger of fibrous substances. Fibers shorter than about 5 micrometers are easily suspended in air and inhaled. Short, straight fibers make their way deep into the lung system where they cause more damage, whereas curly fibers or clusters are stopped in the upper part of the bronchial tract. From a mineralogical standpoint, this suggests that chrysotile should not be a major offender because of its curly fiber shape. It also suggests that other mineral and synthetic fibers could be of concern, a possibility that is supported by observations that mortality rates among workers in synthetic fiber manufacturing facilities are higher than they are among asbestos miners in Quebec (Mossman et al., 1989).

It should be clear that the decision to ban asbestos is not simple. Whereas crocidolite and amosite appear to be strongly linked to disease, chrysotile is not. One problem in applying this finding to commercial applications of asbestos is in the purity of the product. Even small amounts of a bad type of fiber can make an innocuous fiber product very dangerous. Amphibole asbestos minerals such as *tremolite,* which are present in small amounts in chrysotile deposits, have been suggested as the cause of malignancies observed in groups exposed to chrysotile. Compounding the confusion is the fact that many early experimental studies on the medical effects of asbestos were undertaken with a blend of different types of asbestos from different areas, including mixtures of serpentine and amphibole asbestos, making it difficult to determine what caused many of the effects that were reported.

Just as the scientific community has begun to grasp the wide diversity of fibrous materials and their effects, the public has come to think of all fibrous minerals as asbestos and as highly dangerous. This has resulted in a clamor to remove asbestos from buildings where it is found in insulation around pipes and in ceiling and floor tiles, and a mini-depression in the sale of large buildings known to contain asbestos. However, asbestos fiber concentrations in air in buildings, even those with damaged asbestos-containing materials, are usually much lower than currently permitted asbestos levels in the workplace and barely distinguishable from outside air. Some European countries have responded to these findings by banning amphibole asbestos but permitting other types to be used, whereas the United States considers all asbestos to be the same and advocates a more extensive ban. At present, not enough is known about fibers that would substitute for asbestos or about dangers that would be introduced by possibly inferior products such as brake pads manufactured without asbestos. As discussed in the section on talc, this problem could affect all of the natural fibers (Magon, 1990; Cohen, 1991).

These uncertainties have hurt asbestos producers, causing a big decline in employment in the asbestos mining and manufacturing industries. More importantly, producers at all points in the chain have been faced with a barrage of litigation focused on present and anticipated asbestos-related problems (Sand, 1992). The frenzy culminated in 1982 when about 300 companies were sued in U.S. courts, including some that only contracted to install a product manufactured by someone else (Joseph, 1982). The prospect of billions of dollars in claims for medical and asbestos removal costs prompted many producers, including Manville Corp. the dominant U.S. supplier, to file for Chapter 11 bankruptcy, setting up trust funds to settle suits made up largely of company assets with several hundred million in payments from insurers of the companies (ME, 1990). Even though overlapping cases were consolidated to cut down on court time and fees, the legal system was completely overwhelmed (Labaton, 1991). Worse yet, it failed to help those most affected by asbestos, workers who were exposed 20 to 40 years ago before asbestos problems became widely recognized. In 1992, the Keene Corporation, a defendant in some asbestos cases, estimated that only about 35% of the $9 billion of asbestos litigation funds had

reached plaintiffs, with the rest going to legal costs. Whether asbestos will survive as an industry remains to be seen. At the present pace, it could end just at the time that data become adequate to judge the complex relations between fibrous minerals and lung disease.

Diatomite

Diatomite is a rock consisting of tiny siliceous fossils known as *diatoms,* which are single-celled aquatic plants similar to algae (Kadey, 1983). Only 1.6 million tonnes of diatomite are produced each year throughout the world, but its relatively high unit value makes it worth almost $400 million (Figure 12-1). Most of this is used as a filtering agent, taking advantage of the unique cellular form of diatomite, with its numerous very small holes (Figure 12-14). Almost every liquid that you can imagine, including beer, medicine, fruit juice, organic chemicals, swimming pool water, and varnish, is filtered through diatomite to remove suspended solids. Smaller amounts of diatomite are used as fillers and extenders in paints, paper, rubber, and plastic and as the abrasive ingredient in some polishing compounds.

Diatomite deposits form as sediments in all types of water, from marine to fresh. The best environment is a relatively shallow body of water with a good supply of dissolved silica to make diatoms, and limited contamination from clastic sediments. The most common source of silica appears to be glassy volcanic rock that is easily dissolved during weathering. The presence of other elements such as boron in trace amounts in diatoms suggests that they, too, are needed for growth.

Diatoms have become an important source of information in the war against acid rain. It turns out that diatoms and another species of algae known as chrysophytes have very specific tolerances to water acidity (Figure 12-15). Because these organisms form in most lakes and their shells are well preserved after they die, they can be used to determine the pH of lake and stream water prior to the time when direct measurements began (Meriläinen, 1986; Dixit et al., 1992). This sort of historical information is critical to assessments of the importance of anthropogenic sources of acidity. As seen in Figure 12-16, diatom studies suggest that lake acidification is a relatively recent phenomenon that coincides with increased industrial activity.

Deposits of diatomite are mined largely in the United States, France, Romania, and the CIS countries (Figure 12-11). Mining is by open-pit methods in the

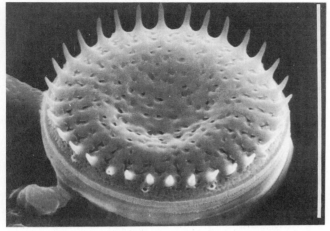

A.

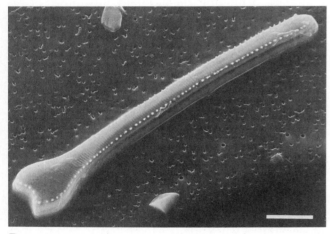

B.

FIGURE 12-15

Examples of diatom morphologies. **A.** *Stephanodiscus sp.* is a centric diatom that grows in the plankton of large, basic, hardwater lakes. **B.** *Actinella punctata* is a pennate diatom that grows attached to underwater surfaces in very softwater, acid lakes and bogs (bar is 10 micrometers in length in both photos; courtesy of Eugene Stoermer and Mark Edlund, University of Michigan).

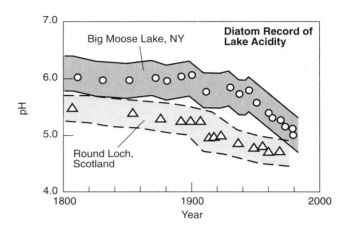

FIGURE 12-16

Historic record of pH changes in Round Loch, Scotland and Big Moose Lake, New York, as recorded by diatoms in lake sediment. Band surrounding data points shows measurement uncertainty (modified from *U.S. National Acid Precipitation Assessment Report, 1990 Integrated Report* and Longhurst, 1989).

United States, although underground methods are used in Spain and dredging is used in Iceland. Processing consists largely of heating to dry the material, followed by grinding and sizing. Additional heating can be used to fuse smaller particles together, changing filtration characteristics, if necessary. Because of the risk of silicosis, which is discussed further later, processing areas must be kept free of dust. Diatomite reserves have not been determined in many areas, but are estimated to be at least 800 million tonnes. About 250 million tonnes of that are in California and adjacent parts of the western United States (Meisinger, 1985). These reserves are more than adequate, al-

though they are not located close to many of their markets.

Talc and Pyrophyllite

Talc may have been the first mineral that most of us came into close contact with. It is the main ingredient in talcum powder, which is widely used to prevent diaper rash. Most of talc's markets result from its sheet-like crystal form, which is similar to that of clay or mica. All can be cleaved into thin laminae, those of mica firm and resistant and those of talc soft and easily pulverized. Talc powders are used as fillers in paints, paper, ceramics, cosmetics, plastics, roofing materials, rubber, and textiles. Rocks with large amounts of talc, known as *soapstone,* are used for sculptures. *Pyrophyllite,* an aluminum silicate that is physically similar to talc, has many of the same uses, with a greater proportion of sales to the ceramics industry (Clifton, 1985). World production of these commodities amounts to about 9 million tonnes worth about $250 million.

Talc is a hydrous magnesium silicate that forms by hydrothermal alteration of magnesium-rich rock, an origin similar to that of asbestos (Chidester, 1964; Anderson et al., 1970). Talc deposits are most common where fault zones cut mafic and ultramafic igneous and metamorphic rocks or where silica-bearing hydrothermal solutions flowed through magnesium-rich dolomite. Some talc deposits contain the noneconomic asbestos minerals *tremolite* and *actinolite,* which has led to concern about the environmental effects of talc use, although no major problems have been identified (ME, 1990).

Pyrophyllite deposits are found in similar hydrothermal settings where the original rock was rich in aluminum, or where hydrothermal solutions leached all cations from the rock. As noted earlier, pyrophyllite is a common alteration mineral around many acid-sulfate epithermal gold-silver deposits. These deposits do not normally contain asbestos-like minerals. Large amounts of talc and pyrophyllite are produced in the United States, Japan, and China, and reserves are very large in relation to present annual production.

Barite

Barite is almost a one-use commodity. More than 90% of U.S. consumption is used as the main filler in *drilling mud,* the water-rich slurry of finely ground barite, bentonite, and other compounds that is used in oil and mineral drilling (Ampian, 1985). Barite's high specific gravity (about 4.5) makes it an ideal material for drilling mud because it significantly increases the bulk density of the mud, causing rock chips to float. World barite production of over 5 million tonnes comes largely from China, India, Turkey, and the United States and is valued at about $200 million.

Barite deposits are of three very different types, all of which are hydrothermal in origin. Many smaller deposits were precipitated from hydrothermal solutions that circulated through veins and other open spaces. They are often found in Mississippi Valley-type lead-zinc-fluorine deposits or in veins around intrusions. The really large deposits, however, consist of lenses and layers of barite in shale, with or without associated lead and zinc. They are similar to the previously discussed lead-zinc *sedex deposits* and take that name. Sedex barite deposits are associated with lead and zinc in Paleozoic-age shales around Meggen and Rammelsberg, Germany, and in the Selwyn Basin in the Canadian Yukon (Figure 12-11), where they appear to have formed when barium-rich hydrothermal solutions vented into sulfate-rich seawater. Other sedex barite deposits consist entirely of barite, with no lead, zinc, or other metals, such as those in the Devonian-age Slaven chert in Nevada and in the Mississippian-age Stanley shale in Arkansas (Papke, 1984; Mitchell, 1984). These deposits probably formed when cool, reducing pore waters, possibly created by high organic productivity, dissolved barium from the sediment and then vented into seawater containing sulfate. They are similar in origin to the extensive sedimentary manganese deposits discussed previously (Brobst, 1983; Ampian, 1985; Orris, 1992). Maynard and Okita (1991) have suggested that deposits consisting largely of barite formed in continental margin settings, whereas those with lead and zinc formed where continental rifting took place. Because barite is almost inert during normal weathering and soil formation, weathering of barite-bearing rock forms *residual deposits* in which the regolith is enriched in barite. Deposits of this type, which developed on Mississippi Valley-type deposits that cropped out at the surface, are common in the Potosi area of Missouri and near Cartersville, Georgia (Kesler, 1950).

Barite is mined largely by open-pit methods, although some sedex and vein deposits are mined by underground methods. Processing to make pure barite involves gravity-separation methods and grinding to make a fine powder. Barite is so inert that it is not an important environmental problem, although the lead mineral, galena, which is associated with some barite, can be of concern. As discussed earlier, most drilling mud is recycled and there is no evidence that it is an important environmental contaminant. Barite reserves of 170 million tonnes, which are adequate for many decades, are concentrated in the United States, India, and China (Figure 12-11).

Mica

Mica was valued by aboriginal people because of its sheen, which comes from light reflected off its hexagonal, platy crystals. The fact that large crystals could be cleaved into very thin sheets that were nearly transparent led to its early use in small windows in dwellings. Its main use now is as fine-grained fillers and extenders in paint, rubber, wallpaper, cement, and drilling mud, where it provides body and sheen. Smaller amounts are used as sequins and even as the gloss in lipstick. Larger crystals of mica are used as insulators and windows in special applications that require electrical and thermal resistance. World production of all types amounts to about 200,000 tonnes annually, worth less than $100 million even in ground and pulverized form.

Mica is produced largely in the United States (Chapman, 1983; Davis, 1985). Most commercial mica is muscovite, the colorless, potassium-rich mica that is named for Moscow. Fine-grained muscovite mica comes from aluminum-rich metamorphic rocks that were derived from shales. Coarser-grained mica comes largely from granitic pegmatites. Significant fine-grained mica is produced as a by-product of kaolin and feldspar mining. Most mining focuses on deposits where the surrounding rock has been partly decomposed by weathering, leaving mica grains that are easier to free and concentrate by washing the rock. Mica reserves have not been measured in many areas but are large.

Zeolites

Zeolites are a family of hydrous silicate minerals with peculiar crystal structures that allow them to adsorb or trap other atoms or molecules. Industrial and environmental applications for zeolites are growing rapidly and world sales amount to about 1 million tonnes of zeolites annually, mostly from the CIS countries, Cuba, Germany, and the United States. The value of this production has not been tabulated because products and prices are so varied, but it is probably more than $100 million (Jacobs and van Santen, 1989).

Although the crystal forms of zeolites vary greatly, they are characterized by two important features. First, they are made of silicate tetrahedra, many of which contain aluminum rather than silicon. Substitution of Al^{+3} for Si^{+4} in the tetrahedron produces a charge imbalance that is satisfied by other cations that are held loosely to the structure. Second, the structure itself is very open with passages that allow atoms and ions to enter. These properties are applied in two main ways. First, the structures are used in *adsorption* of ions and gases. One of the first such applications involved water softeners, where sodium ions on the zeolite crystals exchanged places with calcium ions in the hard water, producing a sodium-bearing soft water. When all the exchangeable sodium ion on the zeolite had been removed, salty water was passed over it to replace calcium ions with sodium ions and it could be used again. More recently, the same principle has been used to remove lead and other undesirable elements from surface waters and process waters in chemical plants and smelters (Groffman et al., 1992).

The second major use of zeolites is in *molecular sieves* in which the size of holes in a specific crystal are used to allow certain ions or gases to pass through while holding others behind. Mol sieves, as they are called, are widely used in the separation of gases from one another and in petroleum refining (Bekkum et al., 1991). Some *natural zeolites* are used in less sophisticated applications, ranging from kitty litter to fillers in cement. As it turns out, *synthetic zeolites* can be made with a much wider range of crystal structures than are seen in natural zeolites (ACS, 1989). Many of these have larger holes in their structure that are especially useful for separating large organic molecules, and they are used widely in crude oil refining and petrochemical manufacturing.

Natural zeolite minerals form in low-temperature, hydrothermal, and sedimentary environments (Mumpton, 1977; Olson, 1983). The main ions present in them other than silicon and aluminum are sodium, calcium, and potassium. Although there are many natural zeolites, only analcime, chabazite, clinoptilite, and mordenite form large enough deposits to warrant mining. Zeolites are among the first new minerals to form as igneous and sedimentary rocks are buried and begin to undergo diagenesis and metamorphism, but they form mineable deposits only where the original rock was very rich in feldspar or volcanic glass. Most deposits are found in volcanic tuff layers, especially where they were deposited in old lake beds. The deposits contain more than one type of zeolite and are not as pure as synthetic zeolites.

Concern about the widespread use of zeolites was raised with recognition of the high incidence of mesothelioma associated with erionite, as discussed in the section on asbestos. Mordenite has also been implicated in increased lung fibrosis, but not lung cancer (Guthrie, 1992). Erionite differs from other zeolites in having a fibrous form, suggesting that crystal morphology will be the important factor in determining the acceptability of zeolite products.

Zeolite reserves are not known and are difficult to quantify because different applications have very different requirements. Even without new markets, demand is likely to increase substantially. If ways can be found to separate mixed zeolites in natural deposits, demand will probably grow even more.

Mineral Pigments

Mineral pigments in the form of aboriginal paintings and ornamental coatings were among the first uses for minerals. Although many minerals are used as pigments locally, the only natural pigments that are present in large enough amounts to be produced commercially are the iron oxides (Severinghaus, 1983). Present annual world production of natural iron oxides is about 300,000 tonnes. Based on the average price of U.S. production, world production would be valued at about $80 million.

Natural iron oxides have a wide range of colors that reflect the main mineral present. *Limonite* has a yellow color, *hematite* has a red color, and *magnetite* has a brown to black color. These iron oxides are rarely found in the pure state in nature and the minerals with which they are mixed can have a strong influence on their color. Of particular importance are manganese oxides and organic material, both of which are black. The most common trade names used for iron oxide pigments are *ocher*, the yellow pigment rich in limonite, *sienna*, a reddish pigment containing hematite, and *umber*, a purplish pigment that also contains manganese oxide.

Iron oxide deposits form largely by the weathering of iron sulfides and possibly by direct sedimentation of iron oxides from submarine hydrothermal vents. Major deposits of umber and ocher associated with the VMS copper deposits of Cyprus (Figure 12-11) appear to have formed as fine-grained sediment during the waning stages of hydrothermal vent activity after the deposition of metal sulfides. The ocher deposits actually overlie the sulfides and grade downward into them and the umber deposits occupy depressions in the host submarine lavas. Ocher and umber are also mined in the Cartersville district, Georgia, where locally manganiferrous iron formation has been extensively weathered. The deposits are almost always mined by open pits, and the material is prepared for sale by washing and separation into different grain sizes.

The largest market for natural iron oxide pigments is in construction material, where it is used to color brick, tile, and cement products. Paints and other coatings are the second largest market. The disadvantage of natural pigments, and it is a big one from the standpoint of modern manufacturing, is their variability. Just a glance at the side of a hill will show how the color of weathered rock varies from place to place. In large deposits, blending can be used to maintain a specific color, but most deposits will not support this type of production for long. The suitability of natural pigment is also affected by its grain size, which affects color. As a result, there is a relatively large market for *synthetic iron oxide pigments*. In the United States, synthetic pigment pro-duction is almost twice as large as natural pigment production and it is valued at almost ten times as much because of its better, more consistent quality. In spite of the technical advantages of synthetic pigments, and their increasing availability as a result of environmentally induced iron oxide recovery in steel-making and other processes, there will always be a place for the cheaper natural pigments. Reserves for future production are not particularly large, however.

GLASS RAW MATERIALS

Glass is an amorphous solid without a well-defined crystal structure (Zarzycki, 1991). Most glass is made by melting quartz and other minerals and rocks (Mills, 1983), and then cooling the melt in a way (usually rapidly) that prevents it from crystallizing. Glass is almost exactly analogous to steel and cement in that it consists largely of a processed mineral raw material, industrial sand, with some important mineral additives that decrease the temperature necessary to melt the mixture and impart desirable characteristics (Table 12-3). The total value of annual production for all of these commodities, based in part on estimates from U.S. production, is about $7 billion (Figure 12-1). Production of soda ash and strontium have grown enormously since 1960, whereas boron and industrial sand and gravel production, at least in the United States, have been almost static.

TABLE 12-3

Raw materials used in glass and ceramics manufacture in the United States listed in order of decreasing value (compiled from U.S. Bureau of Mines Mineral Commodity Summaries, 1992 and Mills, 1983).

Commodity	Amount Used in Glass (tonnes)	Value ($ millions)	Percentage of U.S. Production
Soda ash	3,027,000	$326	49
Boron	318,000	255	58
Industrial sand	9,545,000	170	41
Feldspar	296,000	24	100
Lime	259,000	14	1.7
Limestone	1,727,000	8	0.25
Sodium sulfate[a]	32,000	4	10
Fluorspar	4,200	1	1
Arsenic	700	1	4
Selenium	140	1	30
Nepheline syenite[b]	250,000	NA	NA
Lithium[a]	1,800	NA	50
Strontium	NA	NA	70

[a]Approximate.
[b]Imported from Canada.

Quartz Sand

Quartz, which comes from *industrial sand and gravel,* is the basic constituent of glass and many other commodities. About 40% of U.S. industrial sand and gravel production goes into glass, with most of the remainder going into foundry sand, and smaller amounts sold as abrasive sand and hydraulic fracturing sand that is used to enhance porosity in oil reservoirs. The total value of world industrial sand and gravel production, extrapolated from the U.S. value of $415 million, would be about $1.5 billion.

Quartz sand deposits suitable for glass manufacture are much scarcer than common sand and gravel deposits used for construction (Heinrich, 1980, 1981; Zdunczyk, 1991). Elements such as iron that would impart color to glass must be avoided, as must refractory minerals such as cassiterite, corundum, kyanite, chromite, or zircon, which would remain unmelted, creating imperfections in the glass. The best North American glass-sands are found in *blanket sand deposits* of Paleozoic age that cover parts of the eastern United States. These sands were deposited in beaches and dunes along marine coasts and are especially pure because they were derived from the erosion of sandstones that were deposited during Precambrian and earlier Paleozoic time. Thus, they have been purified by two and sometimes even three cycles of erosion and sand concentration. The largest of these are the Ordovician St. Peter, Silurian Clinch, and Devonian Oriskany and Sylvania sandstones, which are mined from Oklahoma to Pennsylvania. Smaller glass-sand deposits are found in beaches and dunes that formed along the Atlantic, Pacific, and Great Lakes coasts during Pleistocene glaciation. Quartz suitable for some glassmaking is also found as large crystals in veins and pegmatites.

The largest glass-sand producers are the United States, Netherlands, and Argentina. Most U.S. and Argentine production comes from land-based, open-pit mines, whereas Netherlands production is dredged from the North Sea. Land-based mining of dunes is in direct competition with recreational land uses and has been excluded from many geologically favorable areas, particularly in the Great Lakes region. Processing involves some crushing and grinding and, where necessary, gravity separation and flotation to remove impurity minerals. The main environmental concerns during all of this are the escape of silica dust, which can cause the serious lung disease silicosis, as discussed in the next section. Conservation of energy used in melting the material is also of great concern (De Lucia and Manfrida, 1990). Reserves of sand for future glassmaking are enormous, as long as land to mine them is not closed by zoning regulations.

Soda Ash

Soda ash is the industry term for sodium carbonate (Garrett, 1992). In addition to glassmaking, which consumes about half of soda-ash production, it is also used in chemicals, soaps and detergents, paper manufacturing, flue-gas desulfurization systems, and water treatment. World production amounts to about 33 million tonnes annually, only about 28% of which comes from natural sources, largely in the United States. The rest is synthesized from salt and limestone in chemical plants. The total value of world soda-ash production, including both synthetic and natural, is about $3.3 billion, although natural deposits account for only about $1 billion of this.

Natural soda ash comes from extensive deposits of complex sodium carbonate minerals such as *trona,* which are found in lacustrine evaporite deposits where ocean water and even sea spray had little effect on water compositions. As discussed in the section on sodium sulfate, inland lakes can be divided into four compositional groups, of which the sodium-carbonate type is one of the most important. The largest trona deposits in North America are in the Eocene-age Green River Formation of Wyoming, which hosts the oil shale deposits discussed earlier (Figure 7-25). In the Green River basin in Sweetwater County, Wyoming, these sediments contain at least 25 sodium carbonate beds more than 1 meter thick, only five of which have been mined. In the Piceance Creek basin in Utah, these sediments contain thick layers of nahcolite and other sodium carbonate minerals that have not yet been mined. Sodium carbonate is also produced from playa brines at Searles Lake, California (Figure 12-17) and from playas and springs along the east African rift system and in Botswana. U.S. soda ash reserves in the Green River deposits make up 96% of the world reserve of about 24 billion tonnes (Mannion, 1983; Kostick, 1985).

Soda-ash mining in Wyoming is carried out by underground methods very similar to those used on coal layers. Most mining involves continuous-mining machines that are reinforced to cut the abrasive trona and shale. Mining must be carried out in a way that limits the collapse of overlying rock in order to keep methane in nearby oil shale layers from entering the mine and causing an explosion. During treatment, ore is heated to drive off CO_2, and then dissolved and reprecipitated in a purer form. The main environmental problems associated with trona mining and processing involve tailings settling ponds containing highly alkaline (pH = 10.5) water, which leaches oil from the feathers of migratory birds that land on the ponds. Water birds have also been covered by salt that crystallized during cold evenings, preventing them from

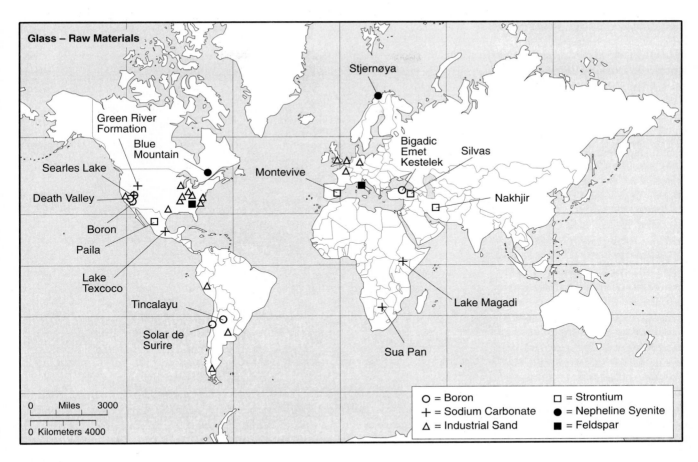

FIGURE 12-17
Location of important deposits of minerals used in the glass and ceramic industries.

flying or even floating. Although extensive efforts are made to prevent birds from using the ponds, bird rescue and rehabilitation is a common practice.

World soda-ash trade is basically a competition between natural material from the United States and synthetic material largely from Europe and Mexico, where long transport distances from the mines in Wyoming leave room to compete. The main problem for the synthetic soda-ash plants is the disposal of large volumes of calcium chloride waste. Substitution of waste glass, known as *cullet,* is a real threat to both types of soda ash as recycling programs gain momentum in MDCs. Nevertheless, soda ash is so widely used and world reserves are so large that production is not likely to decline.

Boron

Almost 60% of U.S. boron production is used as a flux in the manufacture of glass, glass fibers, and insulation. The huge volume of glass manufactured in the United States makes this the dominant world market,

although large amounts of boron are used in Europe as a filler in soaps and detergents. Present world boron production amounts to about 2.5 million tonnes of boric oxide equivalent. Using the value of U.S. boron production, this would have a value of about $750 million.

Boron comes almost entirely from lacustrine evaporites. The largest such deposits, which are in southern California, consist of hydrous sodium borate minerals. *Kernite,* which contains only four water molecules in its structure, is stable at higher temperatures than *borax,* which has ten, and apparently forms as the evaporites are buried beneath younger sediment. The boron in these deposits is thought to have come from hot springs that flowed into the lakes during evaporation; some borate deposits in South America are depositing from hot-spring waters of this type at present. Boron is also enriched in the Searles Lake brines in California, where it is thought to have come from hot springs (Kistler and Smith, 1983).

Boron production was made famous by the 20-mule team wagons that were used to haul borates out of

FIGURE 12-18
Borax ore processing plant at Boron, California, showing the large wet scrubbers that recover dust from pulverized ore. These and other pollution-abatement equipment account for 50% of plant electric load (courtesy of C.M. Davis, U.S. Borax and Chemical Corp.).

Death Valley during the early part of the 1900s. Present operations in the area are much more modern, and have included underground mining carried out to limit surface disturbance at the edge of the Death Valley National Monument. Elsewhere in southern California, mining is by conventional surface methods. Mined material is crushed and ground to make a powder that is dissolved to recover borax and boric acid (Figure 12-18). Boron production from brines simply avoids the crushing, grinding, and dissolving stages, but has to deal with a more complex, dilute solution.

Natural borates are not very toxic to animals, but can be toxic to plants even though low levels of boron are essential for plant life. Boron-hydrogen compounds known as boranes, which do not occur in nature, are highly toxic and have caused problems in some industrial applications (Lyday, 1985). The most important suppliers of boron are the United States and Turkey, and world reserves amount to about 160 million tonnes of ore of varying grades.

Feldspar, Nepheline Syenite, and Aplite

Feldspar, one of the most common silicate minerals, and the feldspar-rich rocks nepheline syenite and aplite, are used almost exclusively in glass and ceramics, where they act as a flux and source of aluminum. Annual world feldspar production is just above 5 million tonnes worth about $210 million. World production for nepheline syenite and aplite are smaller but

probably increase the annual world value to about $350 million.

Although feldspars are found in almost all types of igneous rocks, they are mined largely from special rock types that are depleted in mafic and other minerals, which would not melt easily during glassmaking or which might add iron or other undesirable elements to the mix. The most common feldspar-rich rocks of this type are *aplite, alaskite, and pegmatite,* all of which form during the latest stages of crystallization of granitic magmas, after most of the mafic minerals have already formed. Deposits of these rocks are mined at Spruce Pine, North Carolina, and the Black Hills of South Dakota. Feldspar-rich beach sand deposits have also been mined in Spain. Nepheline syenite, a feldspar-rich intrusive rock similar to granite but with very little quartz, is also used in glassmaking and as a source of feldspar. The largest deposits of nepheline syenite are at Blue Mountain, Ontario, and on the island of Stjernøya in Norway (Figure 12-16).

Lithium

Lithium is also a glass additive, with smaller amounts used in lubricants and greases, storage batteries, and in lithium brines in refrigeration systems (Ferrell, 1985). Lithium additives produce ceramic products that do not expand or contract as temperature changes, a valuable characteristic for everything from spacecraft to cookware. World production of about

5,600 tonnes of contained lithium is probably valued at only $50 million even in intermediate forms such as lithium carbonate.

Lithium comes from two very different types of deposits (Anstett et al., 1990). *Pegmatites* in North Carolina and Zimbabwe formed from residual magmas during the crystallization of large granitic intrusions. The main lithium mineral recovered from these deposits, *spodumene,* can be used directly in glass and other ceramics, but it must be processed into another form such as lithium carbonate for other markets. *Lacustrine brines* and *playa evaporites* in Nevada and Chile also contain lithium. In Chile, these deposits are in the same area as the nitrate evaporites discussed earlier and it is likely that they were enriched in lithium by the weathering of nearby rocks (Kunasz, 1983). Because the lithium is already in solution in brine deposits, they are less expensive to process and are likely to capture an increasing share of lithium markets in the future. The effect of this will be to make the United States and Chile even more important as sources of lithium to the rest of the world. World reserves are about 2.2 million tonnes of contained lithium.

Strontium

Strontium has probably prevented an entire generation of viewers from being damaged by radiation from color television sets. The discovery in the late 1950s that glass containing strontium blocked radiation from color television tubes without damaging the quality of the image produced a new market for strontium. Presently, about 80% of world strontium consumption is used in glass, with most going into faceplates of color television tubes (Ferrell, 1985; Ober, 1992). Strontium also provides the familiar red color in fireworks, is mixed with iron oxide to produce strontium ferrite for ceramic magnets, and is used in electrolytic zinc smelting. World strontium production, which has a value of about $20 million, comes largely from Mexico, Turkey, Spain, and Iran (Figure 12-17).

Deposits of strontium consist of layers and disseminations of the strontium-sulfate mineral, *celestite,* almost always in limestone. In the best deposits, celestite forms extensive, nearly pure layers, known as *mantos,* that have apparently replaced layers of limestone. The replacement appears to have taken place where groundwaters containing dissolved strontium came into contact with sulfate-rich waters. The groundwaters were enriched in strontium when they reacted with *aragonite,* the crystal form of calcium carbonate that was originally deposited in the limestone, converting it to calcite, the crystal form in which the calcium carbonate is found today. Aragonite accommodates larger amounts of strontium substituting for calcium than does calcite and releases it when it changes to calcite. The sulfate-rich waters with which the groundwater reacted to precipitate celestite probably came from salt-flats or sabkhas. Most deposits of this type are found in and around Cretaceous-age reefs (Bearden, 1988).

Celestite is mined by both open-pit and underground methods. Where inexpensive labor is available, the more selective underground mining is widely employed (Figure 4-14A). The main forms in which strontium is used are strontium carbonate and strontium nitrate, which are produced by heating celestite in kilns with coke. Future demand for celestite depends in part on just how soon the color television industry begins using flat screens, which do not require shielding from radiation. Flat screens are desirable because they require less space and weigh less, thus saving transportation costs and opening new opportunities for design. When flat screens are perfected, strontium demand will decrease unless other markets are developed. Whatever the case, world celestite reserves of at least 6.8 million tonnes are very large.

ABRASIVE AND REFRACTORY MINERALS

We all know from experience that pushing a rock across on a smooth wooden table is not a good idea; abrasives are simply the commercial application of this concept. You are familiar with the use of abrasives in sandpapers, whetstones, grinding wheels, and in powders for blasting paint and corrosion from metals and discoloration from brick and stone surfaces. But did you know that they are also used to provide the cleaning ability of soaps, detergents, toothpastes, cleansers, car polishes, and silver polishes (Hight, 1983; Crookston and Fitzpatrick, 1983)?

The most valuable natural abrasive material is industrial diamond, the hardest substance known, with silica sand a distant second and other products such as silica stone, garnet, tripoli, emery, feldspar, and diatomite bringing up the rear. The abrasive market is considerably larger, however, because so many synthetic abrasives are available. The most important of these, synthetic industrial diamonds, are produced in South Africa, China, the United States, Russia, Sweden, and Japan and have an annual world value of about $250 million. The other major synthetic abrasives are *fused aluminum oxide,* which is prepared from bauxite, and *silicon carbide,* which is prepared from quartz and coke. Although world consumption statistics are not tabulated, they are probably in the range of $500 million to $1 billion, somewhat larger than the natural abrasives market.

Refractory materials are a similar motley lot. They provide heat-resistant bricks, blocks, and other forms that are used in a wide range of industrial applications. Many of the materials used in glass are also used in refractories, including industrial sand itself, about 20% of which is used to make foundry sand. In addition to the new materials that are discussed here, other important refractory raw materials that have already been discussed include magnesite, magnesium-rich chromite, zircon, and bauxite.

Industrial and Synthetic Diamond

Industrial diamonds are those that cannot be used as gems. They range from imperfect stones of several carats to very fine-grained diamond powder (Reckling et al., 1983). Although many industrial diamonds come from natural deposits, at least four to five times as many are *synthetic diamonds*. The annual value of total world natural industrial-diamond production is not tabulated directly, although U.S. consumption for natural and synthetic diamonds is given by the U.S. Bureau of Mines for all classes except "miners diamond." If this category, which accounts for about 20% of U.S. consumption, is supplied equally from both sources, then the value of U.S. natural industrial-diamond consumption (as reported to U.S. customs) would be about $70 million. If the rest of the world consumes natural industrial diamonds in the same ratio to GDP as does the United States, then world consumption would be worth about $250 million.

The main industrial markets for diamond depend on its hardness, which is greater than that of any other substance. Large stones are used in tools and bits for cutting rocks and small stones, known as dust, grit, and powder, as well as in grinding wheels, saws, and other devices for cutting and polishing rocks, minerals, metals, ceramics, and other hard substances. Diamond-tipped knives, for instance, are used in everything from surgical procedures to pasta cutting. In addition to its uses as an abrasive, diamond's hardness makes it an excellent form or *die* through which metal can be drawn to make special wires. Although 98% of natural diamonds and almost all synthetic diamonds contain small amounts of nitrogen substituting for carbon atoms, the small fraction that is free of nitrogen has very special thermal, light, and electrical properties. One type of nitrogen-free diamond is used as a heat sink in electronic devices, as windows and lenses in analytical equipment, and as semiconductors (Pressler, 1985).

Diamond synthesis, which began in 1955, is carried out by placing carbon in a heated press that recreates the extreme pressures and temperatures of Earth's mantle where diamond is stable (Figure 10-17), or by subjecting carbon to an explosion that generates instantaneous high pressures (Mackay and Terrones, 1991). The first of these methods is most common and makes the largest stones. Presently, single diamond crystals as large as 0.01 carats can be synthesized economically and the market for this smaller-size material is dominated by synthetics. Larger synthetic crystals, up to about 2 carats in size, are available for specific applications but are not sold as gems, and synthetic polycrystalline aggregates can also be manufactured (Austin, 1992). It has been shown recently that thin films of diamond can be deposited from a vapor onto surfaces of other materials at considerably lower temperatures and pressures than those used for traditional diamond synthesis. These diamond coatings could make glass and other materials resistant to abrasion, but their potential for making surfaces smooth is even more attractive. Thicker diamond films might also be used as semiconductors or sound conductors. Unfortunately for the natural-diamond market, these diamond coatings will be synthesized from other forms of carbon.

Natural industrial diamonds come from the mines that produce gem diamonds, with the largest production from the placer deposits of Zaire and the Argyle lamproite pipe in Australia (Figure 10-14). Of the natural diamond producers, only South Africa and Russia also produce synthetic stones, accounting for about 40% of world synthetic diamond production. Currently, natural industrial stones larger than 20 carats constitute less than 10% of the U.S. market, with the rest dominated by synthetics. In fact, natural and synthetic diamonds are in strong competition and prices for natural diamonds have declined steadily since 1980, even though world consumption has increased (Figure 12-19). Although prices for synthetic diamonds also declined during this period, natural diamond prices declined more, putting them at an historical discount.

Natural industrial diamonds are likely to remain in heavy demand because of the scarcity of competitively priced synthetics in the large-size ranges. World reserves of industrial diamonds, which are estimated to be about 980 million carats, are concentrated in Australia, Zaire, Botswana, Russia, and South Africa. The increasing rate of production in recent years suggests that they will be exhausted relatively rapidly, putting further pressure on the search for new diamond reserves, as discussed in the previous chapter.

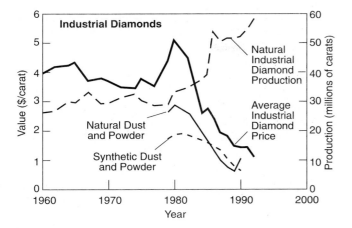

FIGURE 12-19

Change in price of natural and synthetic diamonds and world production of natural industrial diamonds since 1960. Note convergence of prices for natural diamond dust/powder and synthetic dust/powder by the late 1980s. Diamond prices are calculated from U.S. Bureau of Mines import data (into the U.S.) for these categories.

Other Natural Abrasives

Silica sand, the second most important abrasive material, comes largely from *industrial sand deposits,* which were discussed earlier. Because of the corrosion of metals by salt, sand used for abrasive purposes is mined only from deposits that formed in fresh water. *Silica stone* and *tripoli* are forms of *chert, flint,* and other amorphous to microcrystalline forms of silica that are found largely in sedimentary rocks. Some of these deposits formed as accumulations of siliceous organisms such as diatoms in deep marine environments that were not contaminated by clastic sediments. Others formed by the redistribution of silica shortly after the deposition of the sedimentary rock. *Tripoli,* an industrial term derived from a deposit of this type near Tripoli, Libya, is a porous rock that forms by the removal during weathering of carbonate and other minerals that were intergrown with the silica. The most famous of the many natural silica abrasive rocks is *Arkansas novaculite,* a Mississippian-age sediment that makes excellent whetstones.

Garnet is a common metamorphic mineral that becomes abundant enough to mine in a few rocks. In some locations, such as Gore Mountain, New York, it forms crystals as much as a foot across that must have formed by hydrothermal recrystallization of original metamorphic minerals. *Emery* is an impure form of corundum, the aluminum oxide that forms ruby and sapphire, which is also found in metamorphic rocks.

The high aluminum content of these rocks makes them relatively rare; they are likely to have been derived from metamorphism of a soil, laterite, or shale. *Wollastonite* is a calcium silicate mineral that forms during metamorphic reactions between quartz and calcite. It is used in special refractory ceramics such as the white tops of spark plugs.

The main environmental concern associated with the use of silica-rich abrasive products is the respiratory disease *silicosis* (Table 4-3). Silicosis is caused by the inhalation of crystalline silica or quartz. It causes nodular areas in lung tissue and is commonly associated with tuberculosis, which it accelerates. Although it has been a serious problem in many mines that contain silica, including chert, its greatest effects have been seen in manufacturing settings that use silica in powder form, usually as an abrasive or ceramic ingredient. Valiante and Rosenmans (1989) have noted that silicosis is relatively difficult to diagnose, often being confused with tuberculosis, and they have suggested that it is more widespread than commonly thought. According to Logue (1991), silica sand remains an important contaminant in the workplace, partly because fine-grained sand remains suspended in air long after its use, where it can be inhaled by persons without respiratory protection. In spite of these dangers, it is the first choice for most abrasive and polishing applications because of its price and because the dangers of substitute materials are not adequately known.

Graphite

Graphite is the crystal form of carbon in which the atoms form hexagonal plates (Figure 3-2). It is used as the "lead" in pencils because of its color, softness, and crystal form, which causes it to cleave or smear off into lines of black platelets when drawn across paper. Other uses for graphite include facings for metal foundries, batteries that supply low-current levels, bearings in high-heat environments, brake linings, and lubricants. These uses depend in part on graphite's high temperature resistance and its excellent cleavage (Taylor, 1985). World graphite production of about 600,000 tonnes is probably worth about $300 million.

Most graphite production comes from Korea, India, and Mexico, with some high-quality material also coming from Sri Lanka and Madagascar. Graphite deposits form layers, disseminations, or veins in metamorphic rock or contact-localized skarn deposits near granitic intrusions. Some of this graphite is the greatly decomposed remains of original organic sediment such as coal, and layers found in ancient Precambrian rocks could be accumulations of algae and other early life forms. Other graphite deposits probably formed by direct deposition from hydrothermal solutions containing dissolved carbon. The crystallinity of material sold as graphite can range from excellent to almost nonexistent, or amorphous. High-quality deposits in Madagascar consist of well-formed crystalline flakes and clusters of graphite in gneiss and schist. Deposits in Korea consist of almost amorphous material in layers that were originally coal. Graphite mining is largely by open pit, with simple processing to purify the material (Graffin, 1983; Krauss et al., 1988).

Kyanite and Related Minerals

Kyanite, sillimanite, and *andalusite* are different crystal structures of the aluminum silicate, Al_2SiO_5. These minerals are valued largely for their refractory characteristics and over 90% of U.S. consumption goes into furnaces for steel, other metals, and glass. A smaller amount is used in high-temperature glasses and ceramics. The value of world production, which amounts to about 340,000 tonnes, is not tabulated but would probably be only about $40 million, based on current prices for U.S. production.

These minerals form by the metamorphism of aluminum-rich rocks that are poor in other cations such as sodium, calcium, or iron. The three minerals form under different temperature and pressure conditions, prevailing in different parts of the crust. Andalusite is found in low-pressure environments such as intrusive rock contacts and regional metamorphic rocks that have not been deeply buried, whereas kyanite is found in more deeply buried, regionally metamorphosed rocks. Most metamorphic rocks that contain these minerals started as shale or some other aluminum-rich rock. Because they are resistant to erosion, these minerals can also be mined from weathered overburden or from beach sands derived from the erosion of the overburden.

Kyanite is the most widely used of the minerals. Most of it is mined by open-pit methods and processing includes crushing and grinding. Kyanite expands during heating and all three of the aluminum silicates react to form *mullite* and a silica-rich glass when heated to about 1,300°C. Kyanite production comes largely from the United States and South Africa. Reserves in these areas and elsewhere have not been quantified but are large.

THE FUTURE OF CONSTRUCTION AND MANUFACTURING MINERALS

Construction and manufacturing minerals will remain very important for the foreseeable future. As composite materials are developed for specific applications, we will use less and less of the raw materials directly, but they will remain the basis for the more complex forms of cement, plaster, and glass. With the exception of glass, none of these materials are amenable to recycling. Thus, like the chemical and fertilizer minerals, we will have to rely on new deposits for these minerals for the foreseeable future. Fortunately, reserves for most of them, except diamonds, are relatively large, as long as we can resolve land-use disputes related to them.

13

Global Mineral Reserves and Resources

IF HISTORY CAN BE OUR GUIDE, GLOBAL MINERAL SUPPLIES WILL be a major factor in future world relations, as well as the focus of important environmental and economic controversies. At the heart of these disputes will be the fundamental questions of how much is left and how long it will last. Throughout this book, we have quoted reserve and resource figures, which are half of this equation, and it is time now that we looked into just how such information is obtained and used. To do this, we will first review the ways that mineral reserves and resources are estimated and then review the factors that influence mineral consumption.

RESERVE AND RESOURCE ESTIMATION METHODS

Recall from our first chapter that Earth's mineral endowment is referred to as a *resource* and that the smaller fraction that we have delineated and can extract at a profit is known as a *reserve* (Figure 1-3). Estimates of reserves and resources rely on geology and statistics (Singer and Mosier, 1981; McLaren and Skinner, 1987). Geological approaches range from the simple enumeration of known deposits and their probable extensions to more complex assessments based on specific ore deposit types or favorable geological environments. Statistical methods range from those based on assumptions regarding the population

of mineral deposits to others that extrapolate exploration, production, or consumption trends.

Geological Estimates

Geological estimates are the most traditional and probably the most dependable because they are based on actual observations about rocks. In the simplest method, reserve estimates made at individual deposits are collected, evaluated, and compiled to determine the total reserve. Such estimates are useful only for measured and perhaps indicated reserves and can therefore tell us only how much material has actually been observed and quantified. Geological estimates of resources are much more complex. In most cases these estimates involve the search for *favorable geological environments* likely to host deposits of interest. If more than one type of mineral deposit is involved, then each environment must be evaluated separately.

Basin analysis and the search for oil provides a good example of geological methods. As we have discussed, oil is released by the thermal maturation of organic matter in source rocks and some of it is trapped in porous, permeable reservoir rocks from which it can be extracted. Thus, the search for oil involves the search for kerogen-rich source rocks in a basin that was buried deeply enough to enter the oil window and that contains good reservoir rocks and traps. This search can be assisted by any number of measurements on the rock. For instance, organic material can be extracted from shales and analyzed to assess its favorability for the generation of oil and whether or not it has passed through the oil window. We can study possible traps and processes that might modify them. We know, for example, that a sandstone consisting of large quartz grains makes a very porous reservoir rock. Thus, basins containing sediment eroded from coarse-grained, quartz-rich igneous rock such as granite would be likely to contain good sandstone reservoirs. Because granite makes up much of the continental crust but is absent from the ocean crust and many oceanic island arcs, we would expect these reservoirs to be in basins adjacent to or on continents. Reefs that form on the margins of some basins, even in oceanic areas, can also be good reservoirs, but only if a source rock was available to generate oil.

Metal resources can be estimated in the same way. Consider nickel, for instance. As we have learned, nickel deposits are associated with ultramafic igneous rocks, either as laterites that formed when these rocks weathered or as nickel-sulfide deposits that formed when immiscible metal-sulfide magmas separated as the ultramafic silicate magma cooled. Thus, the search for nickel deposits involves a search for ultramafic rocks that are (or were) in the right climatic and topographic setting to generate thick laterite soil, or that would have generated an immiscible sulfide melt during crystallization.

Once the processes that form a mineral deposit have been identified, geologic information can be used to rank areas in terms of the probability that they contain one or more deposits of a given type. These rankings can be made by a computer if the exact criteria are spelled out clearly and if it is given the necessary geologic information. In order to make rankings less subjective, or at least to make them less dependent on a single person, it is common to incorporate criteria developed by several experts in these assessments, an approach that has been used to evaluate the mineral potential of wilderness areas in the United States (Figure 5-10).

Although this approach provides an estimate of resources in deposits of known type, it is almost useless for estimating resources in deposits that have not yet been recognized. For instance, unconformity-type uranium deposits were not even known as a distinct class when the uranium exploration boom began in the 1970s. Thus, early estimates of uranium resources failed to include them. Unknown deposit types can be incorporated in the estimates by imaginative geological hypotheses, but are more commonly evaluated by methods that emphasize a statistical approach, as discussed next. Although these methods do not identify the type of unknown deposit, they can suggest that large resources of some type remain to be found.

Statistical Estimates

The most common statistical estimates involve the assumption that the mineral reserve can be treated as a single homogeneous population and that characteristics of the entire population can be estimated from partial information. Unlike the geologic approach, these methods do *not* provide information on the location of the reserves or resources, only on their magnitude.

One of the most interesting applications of this approach is the *crustal abundance relation* (McKelvey, 1960), which holds that reserves for some elements exhibit a constant ratio to their average crustal abundance (Figure 13-1A). If this relation holds true for all mineral commodities, reserves for less explored commodities can be estimated from reserves for well-explored ones. Furthermore, if the relation is true for reserves, it might also be true for resources, thus permitting the estimation of resources for a wide range of minerals from resource estimates for a few better known commodities. Another approach, known as the *Lasky relation* (DeYoung, 1981), is based on the observation that there is a logarithmic increase in the volume of ore for some commodities with an arithmetic decrease in grade (Figure 13-1B). Although the exact form of this relation differs for different deposit types and areas, it provides a comforting indication that the volume of mineral material available will increase greatly for even a small decrease in grade. As can be seen in Figure 13-1C, we have taken advantage of this relation in practice by gradually decreasing the grade of ore from which we extract minerals.

Reckless extrapolation of the Lasky relation might even lead us to postulate the existence of an infinite resource with an infinitely low grade, but such a conclusion runs afoul of common sense in two ways. First, our ability to extract minerals from lower and lower grade ores depends on the availability of cheap energy. Increased energy costs increase the cost of some of the technological developments that permit our use of low grade ores, rendering them uneconomic. As can be seen in Figure 13-1C, the effective grade, or yield, of copper from United States mines actually increased through the 1980s as the increased price of oil affected the cost of mining and processing. Second, such an extrapolation assumes that the mineral form of the element or commodity of interest remains the same in all types of rocks. As we have seen, most elements form their own ore minerals when they are present in high concentrations, but substitute in the crystal structure of other minerals when they are present in trace abundances. Skinner (1976) has suggested that this differ-

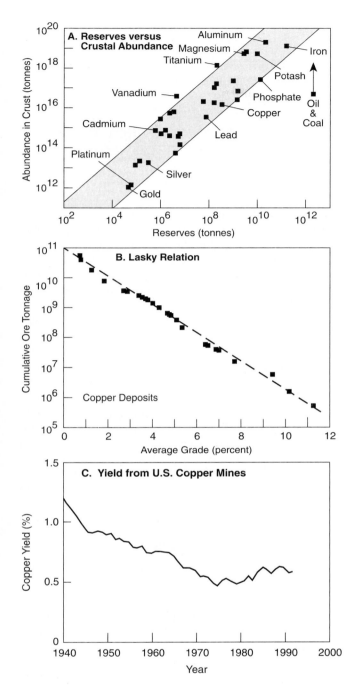

FIGURE 13-1

Geological relations used to estimate mineral reserves. **A.** Relation of world mineral reserves to abundance of various elements in the crust. Oil and coal fall below this line because they come only from sedimentary rocks in which the abundance of carbon is higher than the 200 ppm estimated for the entire crust. **B.** Lasky relation for 267 copper deposits showing linear relation between arithmetic decrease in grade and logarithmic increase in tonnage of ore (from DeYoung, 1981). **C.** Copper yield, as a percentage of the weight of the ore, from United States copper mines, showing the gradual decrease with time since 1940 and the increase since about 1980 caused by increased energy costs (drawn from data supplied by Janice Jolly, U.S. Bureau of Mines).

ence could greatly limit our ability to exploit low-grade ores and places an important upper limit on world mineral resources.

Statistical methods can make use of *exploration drilling data.* Cargill et al. (1981) suggested that there is a logarithmic relation between the amount of exploration carried out (as measured by the length of exploration drill holes) and the amount of ore or oil discovered, and that we should find larger deposits with early exploration holes than with later ones. For instance, the average size of oil discoveries in the lower 48 states decreased from 23 Mbbl in 1941–42 to 2 Mbbl in 1980. This relation permits us to estimate the ultimate amount of ore or oil that might be found by extrapolating to some intense level of exploration drilling. The tendency to extrapolate to unrealistically large amounts of drilling, thereby arriving at unrealistically large reserve estimates similar to those mentioned previously for the Lasky relation, can be constrained by studies of *target geometry.* For instance, statistical studies of the size and shape of the targeted mineral deposits can provide an estimate of the area of influence of a single drill hole (Singer and Drew, 1976; Singer and Kouda, 1988). This information can be used, in turn, to constrain the spacing and therefore the number of holes that must be drilled in an area to fully assess its mineral potential.

Combined Estimates

The most effective reserve and resource estimates use all available geologic data, exploration success rates, and insights about the nature of the population containing the deposits. Estimates of this type by the U.S. Geological Survey were used as a basis for the settlement of many of the conflicting land claims that arose during the distribution of federal land in Alaska (Singer and Ovenshine, 1979). More recent efforts attempt to estimate the gross in-place value of *undiscovered deposits* of each metal or commodity in an area, along with a level of certainty for the estimate (Brew et al., 1992).

It was an estimate of this type by M. K. Hubbert that focused public attention on our dwindling oil reserves (Hubbert, 1985). The estimate was based on the prediction that mineral production will increase smoothly, reach a peak and then decline at roughly the same rate that it increased, as we saw in Figure 1-7A. Application of this concept to estimate ultimately recoverable reserves of a mineral commodity requires that the peak of the curve be identified. For a commodity that is consumed as rapidly as it becomes available, such as oil, the consumption rate should parallel the discovery rate (as indicated by the amount of reserve available each year), with the two curves separated by the time it takes to put the discovery into production, that is, the *lead time*. Hubbert found that this was indeed the case for U.S. oil, where production followed discovery by about 10.5 years. On the basis of this relation, he estimated that oil production from the lower 48 states would peak in 1971 and decline steadily after that. A curve of this form indicated that oil reserves from the lower 48 states were about 170 Bbbl. As can be seen in Figure 13-2, this prediction was amazingly precise. Even the huge Prudhoe Bay discovery in Alaska, an area not included in the original study, had little effect on the precipitous drop in U.S. production and its rapid transition from a self-sufficient producer to an importer with a growing balance-of-trade problem.

Roper (1978) showed that many other mineral commodities *in the United States* can be fit to curves of the type used by Hubbert and that some have already passed through maxima. This very important fact has not been widely recognized, largely because *world* mineral production continued to increase through the early 1970s (Petersen and Maxwell, 1979) and the subsequent leveling off was attributed to the economic impact of rising oil prices, as discussed earlier. It remains to be seen whether this is true. For instance, it is not comforting that world oil production has been relatively constant since the early 1970s in spite of intense frontier exploration (Plate 5-1). The real problem arises because it is difficult to distinguish a peak from a simple fluctuation in a rising trend until after the fact, particularly because so many other factors affect world mineral supplies.

FACTORS THAT AFFECT THE ADEQUACY OF WORLD RESERVES

The adequacy of world mineral reserves and resources is strongly affected by the shape of the consumption curve, particularly factors that cause deviations in the theoretical smooth curve, such as public opinion, market action, and government regulation (Singer, 1977). Public opinion is a powerful but unpredictable force in world mineral demand. Its effectiveness has been seen most recently in the precipitous drop in uranium demand caused by strongly negative public opinion about the safety of nuclear power. Market action, including all forces working to change demand for a product, has been discussed throughout this book. Of most importance is *substitution*, by which one commodity takes the place of another. As pointed out earlier, energy markets have evolved through successive substitution of coal for wood, then oil for coal, then natural gas for oil. Metals have experienced similar substitution, with copper losing ground to aluminum and then to glass in fiber optics, and steel losing markets to cement and plastic. Government actions also have a strong effect on consumption, largely through their approach to stockpiling and recycling.

Stockpiles

There are many types of mineral stockpiles. Stockpiles maintained by the London Metal Exchange and other commodity exchanges are used to supply immediate industrial demand. Others, such as the large silver holdings in India, are held by private individuals for personal or speculative reasons. The most important stockpiles are those held by governments, which started as accumulations of *strategic minerals* containing commodities essential for national defense that would be in short supply if world trade patterns were interrupted by conflict (van Rensburg, 1986; Kessel, 1990). The present U.S. stockpile was authorized by the *Strategic and Critical Stockpiling Act of 1939* and augmented by a related act of 1946. Its usefulness was clearly demonstrated during World War II, when mineral supplies to the United States were greatly disrupted by German submarines, which sank 96 ships carrying Australian bauxite, as well as many others.

In spite of its origin as a badly needed source of materials during international conflict (Eckes, 1979), attitudes toward the stockpile have changed greatly over the years and its original role has been largely forgot-

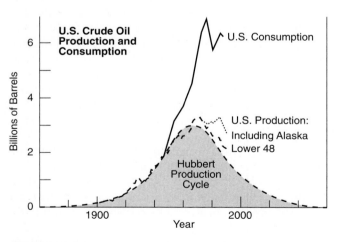

FIGURE 13-2

Predicted reserves and production cycle for oil in the lower 48 U.S. states (originally made by Hubbert in the early 1970s on the basis of data available through 1971) compared to actual production (including Alaskan production) and U.S. consumption through 1990 (from Hubbert, 1985 and U.S. Department of Energy).

ton. The stockpile was intended to last for five years in 1942, but that was cut to three years by 1958. Part of the motivation for these changes was economic because the stockpile represented a major investment that paid no dividends or interest. By 1970, amounts required for the stockpile were lowered for many commodities, thus permitting further savings. In 1973, President Nixon asked to reduce the stockpile to a one-year limit, and to dispose of 90% of its existing material. In 1985, President Reagan proposed to reduce the goals of the stockpile by over 50% (Bullis and Mielke, 1985; Mikesell, 1986).

The present U.S. stockpile contains about 35 mineral commodities, including oil, worth about $20 billion. Nonoil minerals purchased between 1950 and 1992 for about $6 billion are worth $9.7 billion, even though materials worth $6.5 billion have been sold. Oil became part of the stockpile when the *Strategic Petroleum Reserve* (SPR) was authorized as part of the *Energy Policy and Conservation Act of 1975*. Early plans envisioned a stockpile of 500 million barrels, at a cost of about $8 billion, to prevent a recurrence of the embargo of 1973. By 1991, the stockpile contained 585 Mbbl in five salt domes in the Gulf of Mexico (Figure 13-3), representing a supply adequate for 83 days, worth almost $12 billion.

With the relaxation of east-west tensions and the dissolution of the Soviet Union, calls have arisen for the stockpile to be sold. Early proposals of this type drew support from the fact that materials such as chromium and manganese ores were not in the best form for current consumption patterns, although, most of this material has since been converted to more useable ferrochromium and ferromanganese. Nevertheless, almost $500 million worth of stockpile materials were sold in 1993 versus purchases of only $22 million. Revenues from stockpile sales are sometimes earmarked for specific new government projects. They are most welcome, of course, in times of large government deficits when the stockpile can be mistaken for a rainy day savings account rather than a supply of minerals for a national emergency.

The possibility of sales from the stockpile acts as a major deterrent to exploration and, in particular, to development of new deposits. On the other hand, exploration can be stimulated by purchases for the stockpile, such as the uranium boom that was supported by purchases for nuclear weapons. In view of our failure to predict most wars and international crises, especially smaller troubles in countries such as Zaire where we get most of the world's supply of strategic cobalt, it is likely that there will be pressure to retain a stockpile, but it will continue to have short-term impacts on overall mineral consumption patterns.

Recycling

Recycling is the most perfect form of mineral use, a state in which we use the same material over and over again. The impact of this on mineral consumption would be overwhelming. We would only have to extract enough new material to provide for population increases and improvements in the standard of living. But some commodities simply cannot be recycled. They are consumed when they are used. Foremost among these are the fossil fuels, oil, natural gas, and coal, which are converted to water and carbon dioxide when they burn. Uranium is also depleted in ^{235}U during use in reactors and cannot be used further (although breeder reactors can create new nuclear fuels). The fertilizer minerals and salt are commonly dissolved and dispersed during use, preventing recycling. Thus, recycling can actually affect world mineral supplies only for the metals and some industrial minerals and products such as glass and plastics.

There are two major settings in which recycling can be undertaken, industrial and municipal. Industrial

FIGURE 13-3
U.S. Strategic Petroleum Reserve showing location **(A)** and change in oil content of reserve since 1977 **(B)** (from U.S. Department of Energy).

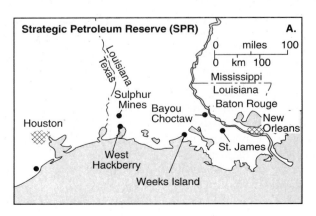

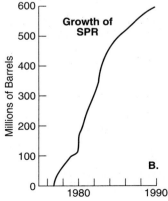

recycling is simply good sense and it is practiced widely (Noll, 1985). Many industrial facilities produce concentrated wastes that can be processed easily to recover metals and other compounds. In a few cases, these wastes have escaped into municipal systems, as when the city of Palo Alto, California, found that ash from the incineration of its sewage sludge contained gold, silver, and other metals, apparently from electronics and photographic manufacturers in the area (Gulbrandsen et al., 1978). Among the metals, steel, aluminum, copper, and lead are most widely recycled and trade in them constitutes a major industry. U.S. scrap metal exports, exclusive of precious metals, were valued at about $3.2 billion in 1989 and Canadian exports had a value of $0.8 billion.

Interestingly, the use of steel scrap is running afoul of environmental regulations because it contains PCBs, lead, and other elements that are considered hazardous (Coppa, 1991). Recently, much attention has centered on "fluff," the plastic material in cars and appliances, some of which contains potentially toxic organic compounds. Regulations in most European countries, Canada, and the United States now require that fluff and other materials be removed more carefully and are likely to have a short-term negative impact on the use of scrap. The use of scrap to make new forms of steel is also complicated by the wide range of other elements in or associated with scrap. Recycled steel cans have also been used to precipitate copper from solutions that leach low-grade copper ores. Although this type of recycling is a sensible use for steel scrap, it was recently reported to San Francisco area residents that their cans were ". . . turned into a toxic, molten brew and sprayed atop mounds of dirt. . . ." (Lapin, 1991, p. B-1).

Municipal wastes are the frontier of recycling and the arena that will have the greatest effect on consumption patterns. Almost 150 million tonnes of trash and garbage are produced in the United States each year. In spite of a tremendous increase in the volume of waste, the number of landfills in which it is put has actually declined from 18,500 in 1979 to 6,000 in 1989, with an estimated 2,000 of these slated to close within a few years (Porter, 1989). The main alternatives to landfill disposal are incineration and recycling, each of which currently accounts for only about 10% of municipal wastes. Although incinerators that burn unprocessed wastes can be fitted with scrubbers and particulate removal systems to limit volatile metal and organic emissions to acceptable levels, they are expensive and have been the subject of strong public skepticism. The future probably lies in recycling followed by incineration. The need for this approach is underscored by the growing shortage of space for landfills and the recognition that many plastic and pa-

per items do not degrade as rapidly as had been thought.

Recycling of municipal wastes is a complicated business. Recent estimates suggest that average U.S. municipal wastes consist of 41% paper and paperboard, 18% yard wastes, 9% metals, 8% each glass, durable goods, and food, and 7% plastic. In theory, it should be possible to separate these components by beneficiation processes such as we use on ores. The simplest system would involve a crusher/shredder to break up the material, a magnetic circuit to remove iron and steel, and some sort of gravity-based circuit to separate glass and other metals from paper and plastics (Figure 13-4). Further separation could be undertaken by flotation or other gravity methods. But there are some very real differences between ore beneficiation and municipal recycling facilities. The nature of the feed (i.e., the waste) can shift rapidly as a result of seasonal changes, holidays, or market forces. This results in inefficient separations because most plants are designed for a feed of roughly constant composition. The volume of waste also varies greatly, far outside the limits that can be accommodated by most plants, which always operate near capacity to be economic. But municipal wastes spoil too rapidly to be stored until the volume of feed drops enough to

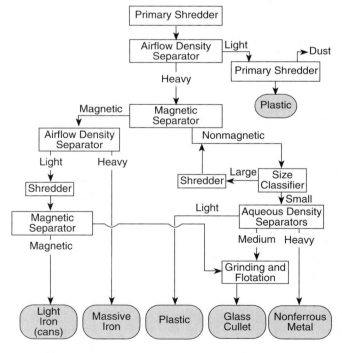

Municipal Waste Separation Process

FIGURE 13-4

Schematic processing plant for recycling of municipal wastes (modified from Phillips, 1977).

accommodate extra material. As a result of these complications, recent emphasis has fallen on the sorting of materials such as paper, cans, bottles, and plastics before collection. Attention is now shifting to the need to collect and separate mercury batteries and other hazardous wastes, as well. Although this approach is widely used and very satisfying, its economics are questionable and it is further complicated by the scarcity of reprocessing facilities for many of the wastes, a problem that is exacerbated by an unwillingness to invest in facilities when there is no way of being sure that the feed will continue to be available. These factors will probably limit recycling unless larger-scale efforts are sponsored at the state or federal level.

At present, the only real success story on the municipal waste recycling front is the recovery of natural gas. It has been known for many years that the decomposition of organic matter in municipal wastes generates large volumes of methane. By installing perforated pipe systems in landfills, this gas can be collected. In most cases, the gas is not pure enough to be used directly in natural gas systems, but it can be used in local heating or other projects.

WORLD RESERVES AND THE CHALLENGE FOR THE FUTURE

A quick overview of our mineral supply situation can be obtained by dividing reserve figures (Appendix I) by present consumption for each commodity of interest, giving us the period of years for which current reserves are adequate. As shown in Figure 13-5, this approach indicates that reserves of 18 of the 53 mineral commodities for which data are available are adequate for more than a century. Coal, iron ore, bauxite, chromium, vanadium, and platinum-group elements are among this group. Unfortunately, another ten minerals, including diamonds, silver, lead, zinc, and sulfur, have reserves that will last for only 10 to 25 years. Just above them in the 25- to 50-year category are copper, manganese, and oil. Comparisons using the reserve base provide us with a longer-term perspective and suggest that another one to five times more material is out there somewhere for us to find and use. With world population and the standard of living increasing rapidly, the time periods obtained using current consumption rates are almost certainly overly optimistic, however, possibly by a factor of two or three. In other words, we have only a decade or so before reserves for many mineral commodities will be exhausted.

The real significance of these numbers to society depends on just how long it takes to find reserves and put them into production. Reserves are of no use whatsoever if they remain in the ground. As we have seen, exploration requires long time periods and access to large areas, simply because we do not know exactly where some of these deposits are; in fact, we do not even know if they exist at all, although geologic estimates suggest that they do. Even after a deposit is discovered, it can take a decade to get it into production. By making environmental aspects of mineral exploration and production matters for public scrutiny, we have involved more people in the process and lengthened the time required to find and develop mineral deposits (Wellmer, 1992). In most cases, these delays have led to better engineering and operations that are more acceptable environmentally. In some cases, the lack of agreement among different groups has led to excessive delays or even cancellation, such as at the Crandon copper-zinc VMS deposit and the oil fields at Point Arguello. As pointed out earlier, time is money. Thus, each delay adds to the cost of a project, gradually making profitable operation more difficult. This affects us all, of course, by making minerals more expensive and less available.

The problem of converting our reserve base into reserves is more acute than it might seem. In fact, many deposits in both the reserve and reserve-base categories are there because exploration or development on them was stopped for environmental reasons. Unless opinions change, these reserves will never be produced and they are therefore of no use to us. Some ardent conservationists actually consider the reserve base to be the material that must *not* be produced. Thus, we can derive little comfort from our reserve and reserve-base figures until we know they might be of use to us in the future.

As noted earlier, any decision not to produce a mineral deposit necessarily requires that another be put into production. To be sure, some deposits should never be produced, but for most deposits we will probably have to accept that careful disruption of the land for a few decades is a necessary price for the benefits that accrue to society from the minerals that are extracted. But which deposits will we leave in the ground? To answer this, we must know as much as possible about the mineral endowment of Earth.

This puts us in a difficult position. On the one hand, we need to find minerals and produce them, and on the other, we want to keep the world as clean and undisturbed as possible. Our messy history is a warning that past ways are not adequate for the job. Mineral wastes are among our worst legacies from earlier times and we will spend many decades cleaning them up. It is important to keep in mind that we all made these messes. Individual prospectors, big companies, and even the government at all levels created pollu-

FIGURE 13-5
Adequacy of world mineral
reserves illustrated as ratio
between reserves and annual
production for 1992 (from data
of Appendix I).

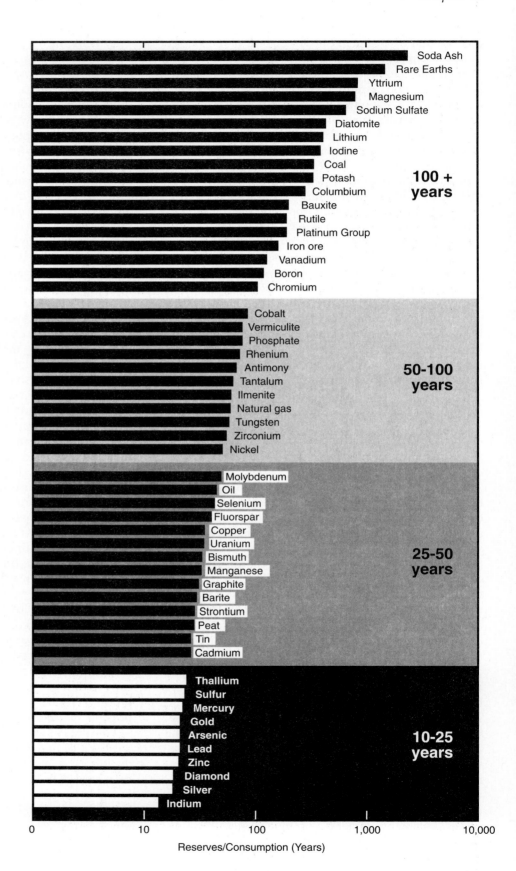

tion as we dug for Earth's riches. Now, we must all clean it up.

We should not let this clean-up effort cause us to forget the need to continue producing minerals. Modern society is based on minerals and we cannot recycle enough of them to supply a growing demand. Those of us who have learned more about mineral matters must become responsible sources of information in the impending debate about global mineral resources, keeping in mind always that we are simply stewards of the planet for forthcoming generations.

References

Abrahamson, D., 1992, "Aluminum and global warming," *Nature,* v. 356, p. 484.

Ackerman, J.M., 1992, "Main Pass—Frasch sulphur mine development," *Mining Engineering,* v. 44, pp. 222–226.

ACS, 1989, *Zeolite synthesis.* Washington, D.C.

Adams, D.F., Farwell, S.O., Robinson, E., Pack. M.R., and Bamesberget, W.L., 1981, *Biogenic sulfur strengths,* Proc. Air Pollution Control Assoc., Philadelphia, PA, 81-15.3, pp. 51–66.

Adams, S.S., and Hite, R.J., 1983, "Potash" in Lefond, S.J., *Industrial minerals and rocks, 5th ed.* Society of Mining Engineers, New York, pp. 1049–1078.

Adams, W.T., 1985, "Zirconium and hafnium," *U.S. Bureau of Mines Minerals Facts and Problems,* pp. 941–949.

Aieta, E.M., Singley, J.E., Trussel, A.R., Thorbjarnarson, K.W., and McGuire, M.J., 1987, *Radionuclides in drinking water: an overview.* J. American Water Works Association, v. 79, pp. 144–152.

Akinsanya, A.A., 1980, *The expropriation of multinational property in the Third World,* Praeger, New York.

Al-Chalabi, F.J., 1986, *OPEC at the crossroads,* Pergamon, Oxford, 248 pp.

Allcroft, R., 1956, "Copper deficiency disorders in sheep and cattle in Britain," *Journal British Grassland Society,* v. 11, pp. 182–184.

Altman, L.K., 1992, "High levels of iron are tied to an increased risk of heart disease," *The New York Times,* Sept. 8, 1992.

Ames, J.A., and Cutcliffe, W.E., 1983, "Cement and cement raw materials" in Lefond, S.J., *Industrial minerals and rocks, 5th ed.* Society of Mining Engineers, New York, pp. 133–1160.

Ampian, S.G., 1985, "Barite," *U.S. Bureau of Mines Minerals Facts and Problems,* pp. 65–74.

Ampian, S.G., 1985, "Clays," *U.S. Bureau of Mines Minerals Facts and Problems,* pp. 157–170.

Anders, G., 1993, "What price diamonds?" *Ontario Geological Survey Update,* January, pp. 34–39.

Anderson, D.L., Mogk, D.W., and Childs, J.F., 1970, "Petrogenesis and timing of talc formation in the Ruby Range, southwest Montana," *Economic Geology,* v. 85, pp. 585–600.

Anderson, G.M., and Macqueen, R.W., 1982, "Mississippi Valley-type lead-zinc deposits," *Geoscience Canada,* v. 9, no. 2.

Anderson, M.S., 1961, "History of selenium toxicity," *U.S. Dept. of Agriculture Handbook 200.*

Andrews, A.J., Owsiacki, L., Kerrich, R., and Strong, D.F., 1986, "The silver deposits at Cobalt and Gowganda, Ontario. I: Geology, petrography and whole rock geochemistry," *Canadian Journal of Earth Science,* v. 23, pp. 1480–1506.

Andrews, M.J., and Fuge, R., 1986, "Cupriferous bogs of the Coed y Brenin area, North Wales, and their significance in mineral exploration," *Applied Geochemistry,* v. 1, pp. 519–525.

Ansah, I., 1990, "Problems of trade in certain natural resource products—background study on nickel and nickel products," *General Agreement on Tariffs and Trade,* Washington, D.C., 133 pp.

Anstett, T.F., Bleiwas, D.I., and Hurdelbrink, R.J., 1985, "Tungsten availability—Market economy countries," *U.S. Bureau of Mines Information Circular 9025,* 51 pp.

Anstett, T.F., Krauss, U.H., Ober, J.A., and Schmidt, H.W., 1990, "International strategic minerals inventory summary report—lithium," *U.S. Geological Survey Circular 930-I,* 28 pp.

Appleyard, F.C., 1983, "Gypsum and anhydrite" in Lefond,

S.J., *Industrial minerals and rocks, 5th ed.* Society of Mining Engineers, New York, pp. 183–192.

Arehart, G.B., Chryssoulis, S.L., and Kesler, S.E., 1993, "Gold and arsenic in iron sulfides from sediment-hosted disseminated gold deposits: implications for depositional processes," *Economic Geology,* v. 88, pp. 171–185.

Arora, H.S., Pugh, C.E., Hossner, L.R., and Dixon, J.B., 1980, "Forms of sulfur in east Texas lignitic coal," *J. Environ. Qual.,* v. 9, pp. 383–390.

Arslan, M., and Boybay, M., 1990, "A study of the characterization of dustfall," *Atmospheric Environment,* v. 24A, no. 10, pp. 2667–2671.

Aston, R.L., 1993, "Surface vs. mineral owners claim near-surface uranium deposit," *Engineering and Mining Journal,* June, pp. 16RR–16UU.

Atwood, G., 1975, "The strip-mining of western coal," *Scientific American,* December, pp. 34–39.

Austin, G.T., 1991, "Gemstones," *Minerals Yearbook, 1990.* U.S. Bureau of Mines.

Austin, G.T., 1992, "Beyond beauty: high-tech uses for gemstones," *Minerals Today,* April, pp. 6–10.

Axtman, R.C., 1975, "Environmental impact of a geothermal power plant," *Science,* v. 187, pp. 795–799.

Axtmann, E.V., and Luoma, S.N., 1991, "Large-scale distribution of metal contamination in the fine-grained sediments of the Clark Fork River, Montana, U.S.A.," *Applied Geochemistry,* v. 6, pp. 75–88.

Baker, M., 1975, *Inactive and abandoned underground mines, water pollution prevention and control.* EPA 440/9-75-007.

Baklanoff, E.N., 1975, *Expropriation of U.S investment in Cuba, Mexico and Chile.* Praeger, New York.

Baldwin, W.L., 1983, *The world tin market.* Duke University Press, Durham, 273 pp.

Ballard, R.D., and Bischoff, J.L., 1984, "Assessment and scientific understanding of hard mineral resources in the EEZ," *U.S. Geological Survey Circular 929,* pp. 185–208.

Ballhaus, C.G., 1988, "Potholes of the Merensky reef at Brakspruit shaft, Rustenburg platinum mines," *Economic Geology,* v. 83, pp. 1140–1158.

Balzhiser, R.E., 1989, "Meeting the near-term challenge for power plants" in Ausuble, J.H., and Sladovich, H.E., eds., *Technology and Environment.* National Academy Press, Washington, D.C., pp. 95–113.

Bardossy, G., and Aleva, G.J.J., 1990, *Lateritic bauxites.* Elsevier, Amsterdam, 618 pp.

Barker, J.M., 1983, "Sulfur" in Lefond, S.J., *Industrial minerals and rocks, 5th ed.* Society of Mining Engineers, New York, pp. 1235–1274.

Barley, M.E., and Groves, D.I., 1992, "Supercontinent cycles and the distribution of metal deposits through time," *Geology,* v. 20, pp. 291–294.

Barnes, H., 1983, "Irish Wilderness: how the story started," *St. Louis Post Dispatch,* June 20, p. D-1.

Barnes, P.W., and Lien, R., 1988, "Icebergs rework shelf sediments to 500 m off Antarctica," *Geology,* v. 16, pp. 1130–1133.

Barney, G.O., 1980, *The global 2000 report to the President of the United States.* Pergamon Press, New York, 360 pp.

Barnola, J.M., 1987, "Vostok ice core provides 160,000-year record of atmospheric CO_2," *Nature,* v. 329, p. 410.

Barringer, F., 1992, "Researchers ponder the mystery of America's stroke belt in South," *The New York Times,* July, 29, p. B8.

Bartlett, A.A., 1980, "Forgotten fundamentals of the energy crisis," *J. of Geol. Education,* v. 28, pp. 4–12.

Bartlett, R.V., 1980, *The Reserve Mining controversy: science, technology and environmental quality.* Indiana University Press, Bloomington, 293 pp.

Bath, A.R., 1991, "Deep sea mining technology: recent developments and future projects," *Mining Engineering,* v. 43, no. 1, pp. 125–126.

Baturin, G.N., 1987, *The geochemistry of manganese and manganese nodules in the ocean.* Kluwer Academic Publishers, Hingham, MA, 356 pp.

Beaudoin, G., and Sangster, D.F., 1992, "A descriptive model for silver-lead-zinc veins in clastic metasedimentary terranes," *Economic Geology,* v. 87, pp. 1005–1021.

Bearden, S.D., 1988, "Celestite resources of Mexico" in Zupan, A-J.W., and Maybin, A.H., *24th Forum on the Geology of the Industrial Minerals.* South Carolina Geological Survey, Columbia, SC, pp. 13–16.

Beeby, D.J., 1988, "Aggregate resources—California's effort under SMARA to ensure their continued availability," *Mining Engineering,* v. 38, pp. 42–49.

Bekkum, H. van, Flanigen, E.M., and Jansen, J.C., 1991, *Introduction to zeolite science and practice.* Elsevier, Amsterdam, 754 pp.

Berger, B.R., and Bethke, P.M., 1986, "Geology and geochemistry of epithermal systems," *Reviews in Economic Geology,* v. 2, 298 pp.

Berkowitz, N., and Brown, R.A.S., 1977, "In-situ coal gasification: the Forestburg (Alberta) field test," *CIM Bulletin,* December, pp. 92–97.

Bernier, L., 1984, "Ocean mining activity shifting to exclusive economic zones," *Engineering Mining Journal,* July, pp. 57–62.

Berry, L.G., Mason, B., and Deitrich, R.V., 1983, *Mineralogy.* W.H. Freeman, San Francisco.

Bertram, B.M., 1993, "Salt," *Mining Engineering,* v. 45, pp. 583–584.

Bise, C.J., 1981, "Pennsylvania's subsidence-control guidelines: should they be adopted by other states?" *Mining Engineering,* v. 33, pp. 1623–1628.

Biviano, M.B., and Owens, J.F., 1992, "A tax on primary lead production?" *Minerals Today,* October, pp. 6–12.

Bjørlykke, A., and Sangster, D.F., 1981, "An overview of sandstone-hosted lead deposits and their relationship to red-bed copper and carbonate-hosted lead-zinc deposits," *Economic Geology 75th Anniversary Volume,* pp. 179–213.

Blaskett, D.R., 1990, *Lead and its alloys.* Horwood, New York, 145 pp.

Blatt, H.G., Middleton, G.V., and Murry, R.C., 1980, *Origin of sedimentary rocks.* Prentice-Hall, Englewood Cliffs, NJ.

Blechman, B.M., 1985, *National security and strategic minerals: an analysis of U.S. dependence on foreign sources of cobalt.* Westview Press, Boulder, 89 pp.

Bleimeister, W.C., and Brison, R.J., 1960, "Beneficiation of rock salt at the Detroit mine," *Mining Engineering,* August, pp. 918–922.

Bliss, N.W., 1976, "Non-bauxite sources of alumina: a survey of Canadian potential," *CIM Bulletin*, October, pp. 75–87.

Blossom, J.W., 1985, "Molybdenum," *U.S. Bureau of Mines Minerals Facts and Problems*, pp. 521–534.

Blossom, J.W., 1985, "Rhenium," *U.S. Bureau of Mines Minerals Facts and Problems*, pp. 665–672.

Blossom, J.W., 1992, "Molybdenum," *U.S. Bureau of Mines Minerals Yearbook*.

Bockman, O.C., Kaarstad, O., Lie, O.H., and Richards, I., 1990, *Agriculture and fertilizers*. Norsk Hydro, Olso, 245 pp.

Boercker, S.W., 1979, "Energy use in the production of primary aluminum," *Materials and Society*, v. 3, pp. 153–161.

Boffess, W.R., and Wixson, B.G., eds., 1978, *Lead in the environment*. National Science Foundation, RA-770241, 272 pp.

Boice, L.P., 1986, "CMA: an alternative to road salt?" *Environment*, v. 28, pp. 5ff.

Bolin, B., Doos, B.R., Jager, J., and Warrick, R.A., eds., 1986, "The greenhouse effect, climatic change and ecosystems," *Scope Report No. 29*. John Wiley, New York, 435 pp.

Bolivar, S.L., 1986, "An overview of the Prairie Creek diamond-bearing intrusion, Arkansas," *Transactions of Soc. Mining Eng.*, v. 280, pp. 1988–1994.

Borchert, H. and Muir, R.O., 1964, *Salt deposits—the origin, metamorphism and deformation of evaporites*. Van Nostrand, Princeton, New Jersey, 581 pp.

Borgese, E.M., 1985, *The mines of Neptune: metals and minerals from the sea*. Abrams, New York.

Both, R.A., and Stumpfl, E.F., 1987, "Distribution of silver in the Broken Hill orebody," *Economic Geology*, v. 82, pp. 1037–1043.

Boudreau, A.E., Mathex, E.A., and McCallum, I., 1986, "Halogen geochemistry of the Stillwater and Bushveld Complexes: evidence for transport of the platinum-group elements by Cl-rich fluids," *J. Petrology*, v. 27, pp. 346–357.

Bourne, H.L., 1989, annual contribution, "Mineral royalties," *Mining Engineering*.

Boutron, C.F. and Patterson, C.C., 1983, "The occurrence of lead in Antarctic recent snow, firn deposited over the last two centuries and prehistoric ice," *Geochimica Cosmochimica Acta*, v. 47, pp. 1355–1368.

Bowen, R., 1989, *Geothermal resources*. Elsevier, London, 485 pp.

Bowen, R., and Gunatilaka, A., 1977, "Copper: its geology and economics," *Applied Science*, London, 366 pp.

Bowker, R.P.G., 1990, *Phosphorus removal from wastewater*. Noyes, Park Ridge, NJ.

Boyd, F.R., and Meyer, H.O.A., eds., 1979, *Kimberlites, diatremes and diamonds: their geology, petrology and geochemistry*. American Geophysical Union, Washington D.C., 399 pp.

Boyle, R.W., 1979, "The geochemistry of gold and its deposits," *Geological Survey of Canada Bulletin 280*, 584 pp.

Boyle, R.W., 1968, "The geochemistry of silver and its deposits," *Geological Survey of Canada Bulletin 160*, 264 pp.

Boyle, R.W., Alexander, W.M., and Aslin, G.E.M., 1975, "Some observations on the solubility of gold," *Geological Survey of Canada Paper 75-24*.

Boyle, R.W., Brown, A.C., Jefferson, C.W., Jowett, E.C., and Kirkham, R.V., 1990, "Sediment-hosted stratiform copper deposits," *Geological Association of Canada Special Paper 36*, 710 pp.

Boynton, R.S., 1980, *Chemistry and technology of lime and limestone*. John Wiley, New York, 277 pp.

Boynton, R.S., Gutschick, K.A., Freas, R.C., and Thompson, J.L., 1983, "Lime" in Lefond, S.J., *Industrial minerals and rocks, 5th ed*. Society of Mining Engineers, New York, pp. 809–832.

Brady, N.C., 1990, *The nature and properties of soils*. Macmillan, New York, 590 pp.

Bray, C.J., and Spooner, E.T.C., 1983, "Sheeted vein Sn-W mineralization and greisenization associated with economic kaolinization, Goonbarrow china clay pit, St. Austell, Cornwall, England: geologic relationships and geochronology," *Economic Geology*, v. 78, pp. 1064–1089.

Breit, G.N., 1992, "Vanadium—resources in fossil fuels," *U.S. Geological Survey Bulletin 1877K*, pp. 1–8.

Brew, D.A., Drew, L.J., and Ludington, S.D., 1992, "The study of the undiscovered mineral resources of the Tongass National Forest and adjacent lands, southeastern Alaska," *Nonrenewable Resources*, v. 1, pp. 303–322.

Brierley, C.L., 1982, "Microbiological mining," *Scientific American*, v. 247, no. 2, pp. 44–53.

Brobst, D.A., 1983, "Barium minerals" in Lefond, S.J., *Industrial minerals and rocks, 5th ed*. Society of Mining Engineers, New York, pp. 485–499.

Broecker, W.S., 1983, "The ocean," *Scientific American*, v. 249, no. 3, pp. 146–161.

Broecker, W.S., 1985, *How to build a habitable planet*. Eldigio Press, Palisades, NY, 291 pp.

Brown, M., 1988, *World phosphates supply to the year 2000*. Organization for Economic Cooperation and Development, Washington, DC.

Bubeck, R.C., Diment, W.H., Dick, B.O., Baldwin, A.L., and Lipton, S.D., 1971, "Runoff of deicing salt: effect on Irondequoit Bay, Rochester, New York," *Science*, v. 172, pp. 1128–1131.

Bues, A.A., 1986, "Geology of tungsten," *International Geological Correlation Programme Project 26*. UNESCO, Paris, 280 pp.

Bullis, L.H., and Mielke, J.E., 1985, *Strategic and critical minerals*. Westview Press, Boulder, CO.

Buol, S.W., Hole, F.D., and McCracken, R.J., 1989, *Soil genesis and classification, 3rd ed*. Iowa State University Press, Ames, IA, 447 pp.

Burkin, A.R., 1987, *Production of aluminum and alumina*. John Wiley, New York, 241 pp.

Burnett, W.M., and Ban, S.D., 1989, "Changing prospects for natural gas in the United States," *Science*, v. 244, pp. 305–310.

Burns, R.L., 1986, "Controls on observed variations in surface geochemical exploration: a case study—the Albion-Scipio trend oilfield," M.Sc. Thesis, Wayne State University, 132 pp.

Burros, M., 1990, "Eating well," *The New York Times*, December 12.

Callahan, W.H., 1977, "The history of the discovery of the zinc deposit at Elmwood, Tennessee, concept and consequence," *Economic Geology,* v. 72, pp. 1382–1392.

Cameron, E.N., 1978, "The lower zone of the Eastern Bushveld Complex in the Olifants River trough," *J. Petrology,* v. 19, pp. 437–462.

Cameron, E.N., 1986, *At the crossroads: The mineral problems of the United States.* Wiley-Interscience, New York, 320 pp.

Cammarota, V.A., Jr., 1992, "Market transparency in the worldwide minerals trade," *Minerals Today,* June, pp. 6–12.

Campbell, J., 1975, "Oil-impregnated sandstone deposits of Utah," *Mining Engineering,* v. 27, May, pp. 47–52.

Campbell, W.J., Blake, R.L., Brown, L.L., Cather, E.E., and Sjoberg, J.J., 1977, "Selected silicate minerals and their asbestiform varieties," U.S. Bureau of Mines IC 8751.

Canney, F.C., Cannon, H.L., Cathrall, J.B., and Robinson, K., 1979, "Autumn colors, insects, plant disease and prospecting," *Economic Geology,* v. 74, pp. 1673–1676.

Cannon, W.F., and Force, E.R., 1983, "Potential for high-grade, shallow-marine manganese deposits in North America" in Shanks, W.C., ed., *Unconventional mineral deposits.* Society of Mining Engineers, Littleton, CO, pp. 175–189.

Cannon, W.F., Force, E.R., and Owens, J., 1982, "Potential for high-grade manganese deposits in North America," *Society for Mining Engineers Preprint 82-61,* 11 pp.

Capone, C.A., Jr., and Elsinga, K.G., 1987, "Technology and energy use before, during and after OPEC: the U.S. Portland cement industry," *Energy Journal,* v. 8, pp. 93–112.

Cargill, S.M., Root, D.H., and Bailey, E.H., 1981, "Estimating usable resources from historical industry data," *Economic Geology,* v. 76, pp. 1081–1095.

Carlin, J.F., 1985, "Tin," *U.S. Bureau of Mines Minerals Facts and Problems,* pp. 847–858.

Carlin, J.F., 1985A, "Bismuth," *U.S. Bureau of Mines Minerals Facts and Problems,* pp. 83–90.

Carlin, J.F., 1985B, "Indium," *U.S. Bureau of Mines Minerals Facts and Problems,* pp. 369–376.

Carlin, J.F., 1990, "Tin," *U.S. Bureau of Mines Minerals Yearbook,* pp. 651–660.

Carlson, C.L., and Swisher, J.H., 1987, *Innovative approaches to mined land reclamation.* Southern Illinois University Press, Carbondale, IL, 752 pp.

Carlson, E., 1986, "To salt or not to salt: that is the snow belt states' question," *The Wall Street Journal,* April 22, p. 31.

Carmalt, S.W., and St. John, B., 1986, "Giant oil and gas fields," *American Association of Petroleum Geologists Memoir 40,* pp. 11–53.

Carpenter, R.H., and Carpenter, S.F., 1991, "Heavy mineral deposits in the upper coastal plain of North Carolina and Virginia," *Economic Geology,* v. 86, pp. 1657–1671.

Carr, D.D., and Rooney, L.F., 1983, "Limestone and dolomite" in Lefond, S.J., *Industrial minerals and rocks, 5th ed.* Society of Mining Engineers, New York, pp. 833–868.

Carrico, L.C., 1985, "Mercury," *U.S. Bureau of Mines Minerals Facts and Problems,* pp. 499–508.

Cathcart, J.B., Sheldon, R.P., and Gulbrandsen, R.A., 1984,

"Phosphate-rock resources of the United States," *U.S. Geological Survey Circular 888,* 48 pp.

Cathles, L.M., 1981, "Fluid flow and genesis of hydrothermal ore deposits," *Economic Geology 75th Anniversary Volume,* pp. 424–457.

Cecil, C.B., and Dulong, F.T., 1986, "Sulfur content of the coal resources of the United States: Current Status," *Society of Mining Engineers Preprint 86-84,* Littleton, CO.

CEQ, 1989, *Environmental trends.* Council on Environmental Quality, Executive Office of the President, Washington, D.C., 152 pp.

Chaffee, M.A., 1982, "A geochemical study of the Kalamazoo porphyry copper deposit" in Titley, S.R., *Advances in geology of the porphyry copper deposits, southwestern North America.* University of Arizona Press, pp. 211–226.

Champigny, N. and Abbott, R.M., 1992, "Understanding the real world of environmental management in the Canadian mineral industry" in *Institution of Mining and Metallurgy, Minerals, Metals, and the Environment.* Elsevier, London, 654 pp.

Chander, S., ed., 1992, *Emerging process technologies for a cleaner environment.* Society for Mining, Metallurgy and Exploration, Littleton, CO, 276 pp.

Chapman, G.P., 1983, "Mica" in Lefond, S.J., *Industrial minerals and rocks, 5th ed.* Society of Mining Engineers, New York, pp. 915–927.

Chapman, N.A., and McKinley, I.G., 1987, *The geological disposal of nuclear waste.* Wiley, New York, 280 pp.

Chen, C.Y., 1983, "Subsidence control measures," *Mining Engineering,* v. 35, pp. 1547–1550.

Cheney, E.S., and Patton, T.C., 1967, "Origin of bedrock placer values," *Economic Geology,* v. 62, pp. 852–853.

Chester, E.W., 1983, *United States oil policy and diplomacy.* Greenwood Press, Westport, CT, 399 pp.

Chidester, R.A., 1964, "Talc resources of the United States," *U.S. Geological Survey Bulletin 1167,* 61 pp.

Chilingarian, G.V., and Yen, T.F., 1979, *Bitumens, asphalts and tar sands.* Elsevier, Amsterdam, 331 pp.

Cienski, T., and Doyle, D., 1992, "Energy conservation in the comminution of industrial minerals," *Canadian Institution of Mining and Metallurgy Bulletin,* v. 85, March, pp. 101–109.

Clark, L.H., and Burrill, G.H.R., 1981, "Unconformity-related uranium deposits, Athabasca area, Saskatchewan, and East Alligator Rivers areas, Northern Territory, Australia," *CIM Bulletin,* v. 75, pp. 91–98.

Clark, S.H.B., 1989, "Metallogenic map of zinc, lead and barium deposits and occurrences in Paleozoic sedimentary rocks, east-central United States," *U.S. Geological Survey Miscellaneous Investigation I-1773.*

Clifton, R.A., 1985, "Asbestos," *U.S. Bureau of Mines Minerals Facts and Problems,* pp. 53–64.

Clifton, R.A., 1985A, "Talc and pyrophyllite," *U.S. Bureau of Mines Minerals Facts and Problems,* pp. 799–810.

Cloud, P.E., 1983, "The biosphere," *Scientific American,* v. 249, no. 3, pp. 176–189.

CNP, 1991 Annual, *Commercial Nuclear Power.* Department of Energy, DOE/EIA-0438(9X).

Cohen, B.L., 1990, "An experimental test of linear no-threshold theory of radiation carcinogenesis" in Cothern, C.R., and Rebers, P.A., eds., *Radon, radium and uranium in drinking water*. Lewis Publishers, Chelsea, MI, pp. 69–81.

Cohen, K., 1984, "Nuclear power" in Simon, J.L., and Kahn, H., *The resourceful earth*. Basil Blackwell, Oxford, pp. 387–414.

Cohen, N., 1991, "Regulation of in-place asbestos-containing material," *Environmental Research*, v. 55, pp. 97–105.

Collings, J., 1985, "The environmental challenge of deep mining," *Optima*, v. 32, no. 1, pp. 16–23.

COM, 1992, *102nd Annual Report*. Chamber of Mines of South Africa, Johannesburg, 64 pp.

Comer, J.B., 1974, "Genesis of Jamaican bauxite," *Economic Geology*, v. 69, pp. 1251–1264.

Conger, H.M., 1992, "Environmental impact of gold ore processing" in Chander, S., ed., *Emerging process technologies for a cleaner environment*. Society for Mining, Metallurgy and Exploration, pp. 9–12.

Constantine, T.A., 1985, "Treating gold and silver mine tailings pond effluents," *Water Pollution Control*, v. 123, pp. 29ff.

Cook, D.R., 1986, "Analysis of significant mineral discoveries in the last 40 years and future trends," *Mining Engineering*, v. 38, pp. 87–94.

Cook, J., 1985, "Nuclear follies," *Forbes*, February 11, pp. 82–100.

Coolbaugh, M.J., 1967, "Special problems of mining in deep potash," *Mining Engineering*, May, pp. 68–73.

Coppa, L.V., 1991, "Recycled scrap metal trade in jeopardy?" *Minerals Today*, December, pp. 24–28.

Couturier, G., 1992, "Magnesium," *Canadian Minerals Yearbook*, pp. 21.1–27.20.

Cox, D.P., and Singer, D.A., 1992, "Distribution of gold in porphyry copper deposits," *U.S. Geological Survey Bulletin 1877-C*, C1–C25.

Crabbe, L., 1989, "The international gold standard and U.S. monetary policy from World War I to the New Deal," *Federal Reserve Bulletin*, v. 75, pp. 423–440 (June).

Craig, P.J., 1980, "Metal cycles and biological methylization" in Hutzinger, O., ed., *The handbook of environmental chemistry*, v. 1, part A. Springer-Verlag, Heidelberg, pp. 169–228.

Craig, J.R., Vaughn, D.J., and Skinner, B.J., 1988, *Resources of the earth*. Prentice Hall, Englewood Cliffs, NJ, 395 pp.

Crandall, R.W., 1981, *The U.S. steel industry in recurrent crisis*. Brookings Institution, Washington, DC, 184 pp.

Crocker, I.T., 1985, "Volcanogenic fluorite-hematite deposits and associated pyroclastic rock suite at Vergenoeg, Bushveld Complex," *Economic Geology*, v. 80, pp. 1181–1200.

Crockett, R.N., Chapman, G.R., and Forrest, M.D., 1987, "International strategic minerals inventory summary report—cobalt," *U.S. Geological Survey Circular 930-F*, 54 pp.

Cronan, D.S., 1992, *Marine minerals in the Exclusive Economic Zones*. Chapman and Hall, London, 209 pp.

Crooke, R.C., and Otteman, L.G., 1984, "Offshore oil and gas technology assessment," *U.S. Geological Survey Circular 929*, pp. 209–245.

Crookston, J.A., and Fitzpatrick, W.D., 1983, "Refractories"

in Lefond, S.J., *Industrial minerals and rocks*, 5th ed. Society of Mining Engineers, New York, pp. 373–379.

Crow, P., 1991, "The windfall profits tax," *Oil and Gas Journal*, v. 89, pp. 33.

Crozier, S.A., 1992, "Overview of environmental regulations and their impact on the United States copper industry" in Chander, S., ed., *Emerging process technologies for a cleaner environment*. Society for Mining, Metallurgy and Exploration, Littleton, CO, pp. 17–36.

Crush, J., Jeeves, A., and Yudelman, D., 1991, *South Africa's labor empire: a history of black migrancy to the gold mines*. Westview Press, Boulder, CO, 266 pp.

Culhane, P.J., 1981, *Public lands politics*. Johns Hopkins University Press, Baltimore, 225 pp.

Cumberlidge, J.T., and Chace, F.M., 1967, *Geology of the Nickel Mountain mine, Riddle, Oregon*. American Inst. Mining, Metal, and Petrol. Eng. Graton-Sales Volume, pp. 1933–1965.

Cummings, A.B., and Given, I.A., 1973, *SME Mining Engineers Handbook*. American Institute of Mining, Metallurgical and Petroleum Engineers, New York, 1034 pp.

Cunningham, L.D., 1985, "Columbium," *U.S. Bureau of Mines Minerals Facts and Problems*, pp. 185–196.

Cunningham, L.D., 1985A, "Tantalum," *U.S. Bureau of Mines Minerals Facts and Problems*, pp. 811–822.

Cunningham, L.D., 1990, "Columbium (niobium) and tantalum," *U.S. Bureau of Mines Minerals Yearbook*.

Cunningham, W.P., and Saigo, B.W., 1992, *Environmental science*. Wm. C. Brown, Dubuque, IA, 622 pp.

Dahlkamp, F.J., 1989, "Classification scheme for uranium deposits: state of the art review" in *Metallogenesis of Uranium Deposits*. International Atomic Energy Agency, Vienna, pp. 1–32.

Dancoisne, P.L., 1991, *Manganese supply and demand—new projects*. International Manganese Institute, Paris, 20 pp.

Danielson, V., 1992, "Alleged salting still a mystery," *Northern Miner*, v. 79, no. 20, p. 4.

Dasgupta, P., and Heal, G.M., 1980, *Economic theory and exhaustible resources*. Cambridge University Press, Cambridge, UK.

Davenport, P.H., Hornbrook, E.H.W., and Butler, A.J., 1975, "Regional lake sediment geochemical survey for zinc mineralization in western Newfoundland" in Elliott, I.L., and Fletcher, W.K., *Geochemical Exploration 1974*. Elsevier Scientific Publishing Co., Amsterdam, pp. 555–578.

David, M., 1977, *Geostatistical ore reserve estimation*. Elsevier, Amsterdam, 364 pp.

Davis, C.L., 1985, "Nitrogen (ammonia)," *U.S. Bureau of Mines Minerals Facts and Problems*, pp. 553–562.

Davis, L.L., 1985, "Mica," *U.S. Bureau of Mines Minerals Facts and Problems*, pp. 509–520.

Davis, L.L., 1992, "Construction—the essential role of minerals," *Minerals Today*, April, pp. 12–16.

Davis, L.L., and Tepordei, V.V., 1985, "Sand and gravel," *U.S. Bureau of Mines Minerals Facts and Problems*, pp. 689–703.

De Lucia, M., and Manfrida, G., 1990, "Breakdown of energy balance in a glass furnace," *Journal of Energy Resources Technology*, v. 112, pp. 124–129.

Devuyst, E.A., Robbins, G., Vergunst, R., Tandi, B., and

Iamarino, P.F., 1991, "Inco's cyanide removal technology working well," *Mining Engineering,* v. 43, pp. 205–208.

de Wit, M.J., 1985, *Minerals and mining in Antarctica.* Clarendon Press, Oxford, 123 pp.

DeYoung, J.H., Jr., 1981, "The Lasky tonnage-grade relationship—a reexamination," *Economic Geology,* v. 76, pp. 1067–1080.

DeYoung, J.H., Jr., Lee, M.P., and Lipin, B.R., 1984, "International strategic minerals inventory summary report—chromium," *U.S. Geological Survey Circular 930-B,* 41 pp.

DeYoung, J.H., Jr., Sutphin, D.M., and Cannon, W.F., 1984, "International strategic minerals inventory summary report—manganese," *U.S. Geological Survey Circular 930-A,* 22 pp.

DeYoung, J.H., Jr., Sutphin, D.M., Werner, A.B.T., and Foose, M.P., 1985, "International strategic minerals inventory summary report—nickel," *U.S. Geological Survey Circular 930-D,* 62 pp.

Dickens, G.R. and Owens, R.M., 1993, "Global change and manganese deposition at the Cenomanian-Turonian boundary," *Marine Georesources and Geotech.,* v. 11, pp. 27–43.

Dixit, S.S., Smol, J.P., and Kingston, J.C., 1992, "Diatoms: powerful indicators of environmental change," *Environmental Science and Technology,* v. 26, pp. 22–33.

Dobrin, M.B., 1976, *Introduction to geophysical prospecting.* McGraw-Hill, New York, 630 pp.

Domenico, P.A., and Schwartz, F.W., 1990, *Physical and chemical hydrology.* John Wiley, New York, 824 pp.

Dore, M.H.I., 1990, "On market structure and the taxation of exhaustible resources," *American Journal of Economics and Sociology,* v. 49, pp. 459–468.

Doull, J., Klassen, C.D., and Amdur, M.O., eds., 1980, *Casarett and Doull's toxicology: the basic science of poisons.* New York, Macmillan.

Dow, W.G., 1978, "Petroleum source beds on continental slopes and rises," *American Association of Petroleum Geologists Bulletin,* v. 62, pp. 1584–1606.

Drew, L.J., 1967, "Grid-drilling exploration and its application to the search for petroleum," *Economic Geology,* v. 62, pp. 698–710.

Drew, L.J., 1990, "Oil and gas forecasting: Reflections of a petroleum geologist," *International Association for Mathematical Geology Studies in Mathematical Geology No. 2.* Oxford University Press, Cary, North Carolina, 252 pp.

Drew, L.J., and Lore, G.L., 1992, "Field growth in the Gulf of Mexico—a progress report," *U.S. Geological Survey Circular 1074,* pp. 22–23.

Dreyfus, D.A., and Ashby, A.B., 1989, "Global natural gas resources," *Energy,* v. 14, pp. 773–784.

Driscoll, F.G., 1986, *Groundwater and wells.* Johnson Filtration Systems, St. Paul, MN, 1098 pp.

Drummond, I.M., 1987, *The gold standard and the international monetary system 1900–1939.* Macmillan, New York, 72 pp.

Duderstadt, J.J., and Kikuchi, C., 1979, *Nuclear power.* University of Michigan Press, Ann Arbor, 228 pp.

Dunn, J.R., 1983, "Sand and gravel" in Lefond, S.J., *Industrial minerals and rocks, 5th ed.* Society of Mining Engineers, New York, pp. 96–110.

Earney, F.C.F., 1990, *Marine mineral resources.* Routledge, London.

Easley, M.W., 1990, "The status of community water fluoridation in the United States," *Public Health Reports,* v. 105, pp. 348–353.

Eckert, G.F., 1992, "Sulfur," *Mining Engineering,* v. 44. pp. 574–575.

Eckes, A.E., 1979, *The United States and the global struggle for minerals.* University of Texas Press, Austin, 353 pp.

Economides, M.J., and Ungemach, P.O., 1987, *Applied geothermics.* John Wiley, New York, 238 pp.

Edelstein, D.L., 1985, "Arsenic," *U.S. Bureau of Mines Minerals Facts and Problems,* pp. 43–52.

EIA, 1991, *Annual energy review: Energy Information Agency,* U.S. Dept. of Energy.

Eiden, R., 1990, "The atmosphere: physical properties and climate change" in Hutzinger, O., ed., *The handbook of environmental chemistry,* v. 1, part E. Springer-Verlag, Heidelberg, pp. 147–188.

El-Hinnawi, E.E., 1981, "The environmental impacts of production and use of energy," *United Nations Environment Programme,* Tycooly Press, London, 319 pp.

Elliot, T.C., and Schwieger, R.G., eds., 1985, *The acid rain sourcebook.* McGraw-Hill, New York, 290 pp.

Elliott, J.E., 1992, "Tungsten—geology and resources of deposits in southeastern China," *U.S. Geological Survey Bulletin 1877I,* pp. 1–34.

Ellis, R., 1987, "Aredor makes the grade," *Mining Magazine,* September, pp. 206–213.

Ely, N., 1964, "Mineral titles and concessions" in Robie, E.H., ed., *Economics of the mineral industries.* American Institute of Mining, Metallurgical and Petroleum Engineers, New York, pp. 81–130.

Emigh, G.D., 1983, "Phosphate rock" in Lefond, S.J., ed., *Industrial minerals and rocks, 5th ed.* Society of Mining Engineers, New York, pp. 1017–1048.

EMJ, 1981, "Rundle: a pause for reflection and possible downsizing to 125,000 bbl/d," *Engineering and Mining Journal,* June, pp. 106–110.

EMJ, 1982, "Exxon shelves the most advanced oil shale project in U.S.," *Engineering and Mining Journal,* June, pp. 33–35.

EMR, 1973, *Report on the administration of the emergency gold mining assistance act.* Department of Energy, Mines and Resources, Ottawa, 101 pp.

Emsley, J., 1980, "The phosphorus cycle" in Hutzinger, O., ed., *The handbook of environmental chemistry,* v. 1, part A. Springer-Verlag, Heidelberg, pp. 147–168.

Environmental Pollution Agency, 1991, *National Air Pollution Emission Estimates 1940–1989.* EPA-450/4-91-004.

Ericksen, G.E., 1981, "Geology and origin of the Chilean nitrate deposits," *U.S. Geological Survey Professional Paper 1188,* 37 pp.

Ericksen, G.E., 1983, "The Chilean nitrate deposits," *American Scientist,* v. 71, pp. 366–374.

Erickson, R.L., 1973, "Crustal abundance of elements and mineral reserves and resources," *U.S. Geological Survey Professional Paper 820,* pp. 21–25.

Erling, G., and Stark, D., 1990, "Petrography applied to concrete and concrete aggregates," *American Society for Test-*

ing Materials Special Publication 1061, Philadelphia, 208 pp.

Estes, J.E., Crippen, R.E., and Star, J.L., 1985, "Natural oil seep detection in the Santa Barbara Channel, California, with shuttle imaging radar," *Geology,* v. 13, pp. 282–284.

Eugster, H.P., and Chou, I-M., 1979, "A model for the deposition of Cornwall-type magnetite deposits," *Economic Geology,* v. 74, pp. 763–774.

Eugster, H.P., and Hardie, L.A., 1978, "Saline lakes" in Lerman, A., ed., *Lakes—chemistry, geology and physics.* Springer-Verlag, Heidelberg, pp. 195–232.

Evans, A.M., 1980, "An introduction to ore geology," *Elsevier Geoscience Texts Vol. 2.* Elsevier, New York, 231 pp.

Evans, D.J.I., Shoemaker, R.S., and Veltman, H., eds., 1979, *International laterite symposium.* Society of Mining Engineers of AIME, New York, 688 pp.

Farnesworth, C.H., 1992, "Canada to divide its northern land," *The New York Times,* May 7, p. 11.

Faure, G., 1986, *Principles of isotope geology.* John Wiley, New York, 452 pp.

Feichtinger, F., Lammer, A., and Riess, M., 1988, "Platinum: the view from South Africa," *Mining Engineering,* February, pp. 91–95.

Felsenfeld, A.J., 1991, "A report on fluorosis in the United States secondary to drinking well water," *Journal of the American Medical Association,* v. 265, pp. 486–488.

Fergusson, J.E., 1982, *Inorganic chemistry of the Earth.* Pergamon Press, Oxford, 400 pp.

Fergusson, J.E., 1990, *The heavy elements: chemistry, environmental impact and health effects.* Pergamon Press, Oxford, 614 pp.

Ferrara, R., Maserti, B.E., Edner, H., Ragnarson, P., Svanberg, S., and Wallinder, E., 1992, "Mercury emissions into the atmosphere from a chlor-alkali complex measured with the lidar technique," *Atmospheric Environment,* v. 26A, pp. 1253–1258.

Ferrell, J.E., 1985, "Lithium," *U.S. Bureau of Mines Minerals Facts and Problems,* pp. 461–470.

Fettweis, G.B., 1979, *World coal resources.* Elsevier, Amsterdam, 415 pp.

Flawn, P.T., 1966, *Mineral resources.* Rand McNally, Chicago, 426 pp.

Fleisher, V.D., Garlick, W.G., and Haldane, R., 1976, "Geology of the Zambian Copperbelt" in Wolf, K.H., ed., *Handbook of stratabound and stratiform ore deposits,* v. 6. Elsevier, New York.

Foell, E.J., Thiel, H., and Schriever, G., 1992, "DISCOL: a long-term, large-scale disturbance-recolonization experiment in the abyssal eastern tropical South Pacific ocean," *Mining Engineering,* v. 44, January, pp. 90–94.

Force, E.R., 1991, "Geology of titanium-mineral deposits," *Geological Survey of America Special Paper 259,* 112 pp.

Force, E.R., and Cannon, W.F., 1988, "Depositional model for shallow-marine manganese deposits around black shale basins," *Economic Geology,* v. 83, pp. 93–117.

Fortescue, J.A.C., 1992, "Landscape geochemistry: retrospect and prospect—1990," *Applied Geochemistry,* v. 7, pp. 1–54.

Forstner, U., and Wittmann, G.T.W., 1981, *Metal pollution in the aquatic environment.* Springer-Verlag, Berlin, 489 pp.

FOS, 1990 *Annual, Federal Offshore Statistics.* Minerals Management Service.

Foss, P., 1987, *Federal lands policy.* Greenwood Press, New York, 145 pp.

Foth, H.D., 1984, *Fundamentals of soil science, 7th ed.* John Wiley, New York, 435 pp.

Fournier, R.O., 1983, "Active hydrothermal systems as analogues of fossil systems," *Geothermal Resources Council Special Report,* no. 13, pp. 263–284.

Frakes, L., and Bolton, B., 1992, "Effects of ocean chemistry, sea level and climate on the formation of primary sedimentary manganese ore deposits," *Economic Geology,* v. 87, pp. 1207–1217.

Franklin, B.A., 1988, "U.S. aid to uranium mining meets political resistance," *The New York Times,* August 22, p. 21.

Franklin, J.M., Sangster, D.M., and Lydon, J.W., 1981, "Volcanic-associated massive sulfide deposits," *Economic Geology 75th Anniversary Volume,* pp. 485–627.

Fraser, D.C., 1978, "Geophysics of the Montcalm township copper-nickel discovery," *CIM Bulletin,* January, pp. 99–104.

Freeston, D.H., 1990, "Direct uses of geothermal energy in 1990," *Geothermal Resources Council Bulletin,* pp. 188–198.

Freeze, R.A., and Cherry, J.A., 1979, *Groundwater.* Prentice-Hall, Englewood Cliffs, NJ, 603 pp.

Frey, R.W., Howard, J.D., and Dorjes, J., 1989, "Coastal sequences, eastern Buzzards Bay, Massachusetts, negligible record of an oil spill," *Geology,* v. 17, pp. 461–465.

Fuge, R., Pearce, N.J.G., and Perkins, W.T., 1992, "Mercury and gold pollution," *Nature,* v. 357, p. 369.

Fukuzaki, N., Tamura, R., Hirano, Y., and Mizushima, Y., 1986, "Mercury emission from a cement factory and its influence on the environment," *Atmospheric Environment,* v. 20, pp. 2291–2299.

Fulkerson, W., Judkins, R.R., and Sanghvi, M.K., 1990, "Energy from fossil fuel," *Scientific American,* September, pp. 129–135.

Fyfe, W.S., 1978, *Fluids in the earth's crust: their significance in metamorphic, tectonic and chemical transport processes.* Elsevier, Amsterdam.

Gambogi, J., 1990, "Silicon," *U.S. Bureau of Mines Minerals Yearbook.*

Garnar, T.W., 1983, "Zirconium and hafnium minerals" in Lefond, S.J., *Industrial minerals and rocks, 5th ed.* Society of Mining Engineers, New York, pp. 1433–1476.

Garnish, J.D., ed., 1987, "Proceeding of the 1st EEC/US Workshop on Hot Dry Rock Geothermal Energy," *Geothermics,* v. 16.

Garrett, D.E., 1992, *Natural soda ash: occurrences, processing and use.* Van Nostrand Reinhold, New York.

GEMS Monitoring and Assessment Research Centre, 1989, *United Nations Environmental Data Report.* Basil Blackwell Inc., Cambridge, MA, 547 pp.

Gentry, D.W., and O'Neil, T.J., 1984, *Mine investment analysis.* Society of Mining Engineers, New York, 502 pp.

George, A.C., and Hinchliffe, L., 1972, "Measurements of uncombined radon daughters in uranium mines," *Health Physics,* v. 23, pp. 791–803.

Gershey, E.L., Klein, R.C., Party, E., and Wilkerson, A., 1990,

Low-level radioactive waste. Van Nostrand, New York, 212 pp.

Getschow, G., and Petzinger, T., 1984, "Louisiana marshlands, laced with oil canals, are rapidly vanishing," *The Wall Street Journal,* October 24, p. 1.

Ghosh, J. A., 1983, *OPEC, the petroleum industry and United States energy policy.* Quorum, Westport, CT, 206 pp.

Gilchrest, J.D., 1980, *Extractive metallurgy.* Pergamon, Oxford, 455 pp.

Gilfillan, S.C., 1965, "Lead poisoning and the fall of Rome," *Journal of Occupational Medicine,* v. 7, pp. 53–60.

Girvan, N., 1972, *Copper in Chile.* Unwin Brothers Ltd., Gresham Press, Surrey, UK, 86 pp.

Gladstone, B., 1992, "British coal privatization," *Engineering and Mining Journal,* November, pp. 22–24.

Glas, J.P., 1988, "Protecting the ozone layer: a perspective from industry" in Ausuble, J.H., and Sladovich, H.E., eds., *Technology and the environment.* National Academy Press, Washington, DC, pp. 137–155.

Glenn, W.E., and Hohmann, G.W., 1981, "Well logging and borehole geophysics in mineral exploration," *Economic Geology 75th Anniversary Volume,* pp. 850–862.

Glover, J.E., and Harris, P.G., 1985, "Kimberlite occurrence and origin," *University of Western Australia Extension Publication No. 8,* 298 pp.

Gold, T., and Soter, S., 1980, "The deep-earth gas hypothesis," *Scientific American,* May, pp. 154–164.

Goldberg, I., Hammerbeck, E.C.I., Labuschagne, L.S., and Rossouw, C., 1992, "International strategic minerals inventory summary report—vanadium," *U.S. Geological Survey Circular 930-K,* 45 pp.

Goldfinch, C., 1991, "Panning gold for arthritis," *Health,* v. 23, p. 22 (June).

Goldman, H.B., and Reining, D., 1983, "Sand and gravel" in Lefond, S.J., ed., *Industrial minerals and rocks, 5th ed.* Society of Mining Engineers, New York, pp. 1151–1166.

Golightly, J.P., 1981, "Nickeliferous laterite deposits," *Economic Geology, 75th Anniversary Volume,* pp. 710–735.

Gordon, R.L., 1987, *World coal: economics, policies and prospects.* Cambridge University Press, Cambridge, 144 pp.

Govett, G.J.S., and Govett, M.H., 1977, "The inequality of the distribution of world mineral supplies," *CIM Bulletin,* August, pp. 59–68.

Gowland, J.S., 1992, "Effluent control at Rustenburg Base Metals Refiners" in *Institution of Mining and Metallurgy; Mineral, Metals and the Environment.* Elsevier, London, pp. 571–584.

Grabowski, R.B., Wetzel, N., and Raney, R.G., 1991, "Gold," *Minerals Today,* December, pp. 14–19.

Graffin, G.D., 1983, "Graphite" in Lefond, S.J., ed., *Industrial minerals and rocks, 5th ed.* Society of Mining Engineers, New York, pp. 757–769.

Gray, E., 1982, *The great uranium cartel.* McClelland and Stewart, Toronto, 303 pp.

Grayson, L.E., 1981, *National oil companies.* John Wiley, New York, 269 pp.

Greeley, M.N., 1990, "Mining scams—still a threat," *Minerals Today,* September, pp. 12–16.

Green, T., 1981, *World of diamonds.* Weidenfeld and Nicholson, London, 338 pp.

Greenhouse, S., 1993, "Punitive tariffs raised against foreign steel," *The New York Times,* June 23, p. 21.

Grisafe, D.A., Angino, E.E., and Smith, S.M., 1988, "Leaching characteristics of a high-calcium fly ash as a function of pH: a potential source of selenium toxicity," *Applied Geochemistry,* v. 3, pp. 601–608.

Groffman, A., Peterson, S., and Brookins, D., 1992, "Removing lead from wastewater using zeolite," *Water Environment and Technology,* May, pp. 54–59.

Grogan, R.M., and Bradbury, J.C., 1968, "Fluorite-lead-zinc deposits of the Illinois-Kentucky mining district" in Ridge, J.D., ed., *Ore deposits of the United States.* AIME, New York, pp. 370–399.

Gross, P., 1984, *Toxic and biomedical effects of fibers: asbestos, talc, inorganic fibers, man-made vitreous fibers, and organic fibers.* Noyes, Park Ridge, NJ.

Gross, W.H., 1975, "New ore discovery and the source of silver-gold veins, Guanajuato, Mexico," *Economic Geology,* v. 70, pp. 1175–1189.

Groves, D.I., Ho, S.E., and Rock, N.M.S., "1987, Archean cratons, diamond and platinum: evidence for coupled, long-lived crust-mantle systems," *Geology,* v. 15, pp. 801–805.

Groves, D.I., Phillips, N., and Ho, S.E., 1987, "Craton-scale distribution of Archean greenstone gold deposits: predictive capacity of the metamorphic model," *Economic Geology,* v. 82, pp. 2045–2058.

Guider, J.W., 1981, "Iron ore beneficiation—key to modern steelmaking," *Mining Engineering,* v. 33, pp. 410–417.

Guilbert, J.M., and Park, C.F., Jr., 1986, *The geology of ore deposits.* W.H. Freeman, New York, 985 pp.

Guiliani, G., Silva, L.J.H.C., and Couto, P., 1990, "Origin of emerald deposits of Brazil," *Mineralium Deposita,* v. 25, pp. 57–64.

Gulbrandsen, R.A., Rait, N., Krier, D.J., Baedecker, P.A., and Childress, A., 1978, "Gold, silver and other resources in the ash of incinerated sewage sludge at Palo Alto, California," *U.S. Geological Survey Circular 784C,* 7 pp.

Gunn, J.M., and Keller, W., 1990, "Biological recovery of an acid lake after reduction in industrial emissions of sulphur," *Nature,* v. 345, pp. 431–433.

Gupta, H.K., 1980, *Geothermal resources: an energy alternative.* Elsevier, Amsterdam, 227 pp.

Gutentag, E.D., Heimes, F.J., Krothe, N.C., Luckey, R.R., and Weeks, J.B., 1984, "Geohydrology of the High Plains aquifer in parts of Colorado, Kansas, Nebraska, New Mexico, Oklahoma, South Dakota, Texas and Wyoming," *U.S. Geological Survey Professional Paper 1400-B,* 64 pp.

Guthrie, G.D., Jr., 1992, "Biological effects of inhaled minerals," *American Mineralogist,* v. 77, pp. 225–243.

Guy-Bray, J., 1989, "The mineral resources program of the United Nations Department of Technical Cooperation for Development," *Ontario Geological Survey Special Volume 3,* pp. 798–801.

Hafele, W., 1990, "Energy from nuclear power," *Scientific American,* v. 263, pp. 136–145.

Hallwood, P., and Sinclair, S.W., 1981, *Oil, debt and development.* George Unwin, London, 139 pp.

Hamblin, W.K., 1989, *The earth's dynamic systems, 5th ed.* Macmillan, New York.

Hamilton, J.M., 1990, "Earth science in Canada from a user's viewpoint," *Geoscience Canada,* v. 16, pp. 213–220.

Hamilton, M., 1992, "Water fluoridation: a risk assessment perspective," *Journal of Environmental Health,* v. 54, pp. 27–31.

Hammond, O.H., and Baron, R.E., 1976, "Synthetic fuels: prices and prospects," *American Scientist,* v. 64, pp. 407–412.

Hancock, K.R., 1983, "Mineral pigments" in Lefond, S.J., ed., *Industrial minerals and rocks, 5th ed.* Society of Mining Engineers, New York, pp. 349–359.

Hanor, J.S., 1987, "Origin and migration of subsurface sedimentary brines," Society of Economic Paleontologists and Mineralogists Short Course no. 21, Tulsa, OK, 247 pp.

Harben, P.W., and Bates, R.L., 1984, "Geology of the nonmetallics," *Metal Bulletin,* New York, 392 pp.

Hargrove, E.C., 1989, *Foundations of environmental ethics.* Prentice-Hall, Englewood Cliffs, NJ, 229 pp.

Harre, E.A., and Young, R.D., 1983, "Nitrogen compounds" in Lefond, S.J., ed., *Industrial minerals and rocks, 5th* ed. Society of Mining Engineers, New York, pp. 961–988.

Harris, D.P., 1979, "World uranium resources," *Annual Review of Energy,* v. 4, pp. 403–432.

Harris, D.P., 1990, *Mineral exploration decisions.* Wiley-Interscience, New York, 436 pp.

Hartman, H.L., 1987, *Introductory mining engineering.* Wiley-Interscience, New York, 633 pp.

Hartman, K.J., 1987, "Kaolin and paper: quality of both products will improve with more sophisticated printing techniques," *Mining Engineering,* v. 37, pp. 247–252.

Haryett, C.R., 1982, "Innovations over the last decade in the Saskatchewan potash industry," *Mining Engineering,* August, pp. 1225–1227.

Hasselrot, B., and Hultberg, H., 1984, "Liming of acidified lakes and streams and its consequences for aquatic ecosystems," *Fisheries,* v. 9, pp. 4–9.

Hawtrey, R.G., 1980, *The gold standard in theory and practice.* Greenwood Press, Westport, CT, 226 pp.

Hayes, T.C., 1990, "Horizontal drilling, or how to revive oilfields," *The New York Times,* July 4.

Haynes, B.W., and Kramer, G.W., 1983, "Characterization of U.S. cement kiln dust," U.S. Bureau of Mines IC 8885.

Haynes, V., 1988, *The Chernobyl disaster.* Hogarth, London, 256 pp.

Heath, G.R., 1978, "Deep-sea manganese nodules," *Oceanus,* v. 21, pp. 60–68.

Heath, G.R., 1981, "Ferromanganese nodules of the deep sea," *Economic Geology, 75th Anniversary Volume,* pp. 736–765.

Hedrick, J.B., 1985, "Rare earth elements and ytrrium," *U.S. Bureau of Mines Minerals Facts and Problems,* pp. 647–664.

Hedrick, J.B., and Templeton, J., 1990, "Rare earth elements and yttrium," *U.S. Bureau of Mines Minerals Yearbook,* pp. 918–924.

Heinrich, E.W., 1980, "Geological spectrum of glass-sand deposits in eastern Canada and their evaluation" in Dunn, J.R., Fakundiny, R.H., and Rickard, L.V., *Proceedings of Fourteenth Annual Forum on the Geology of Industrial Minerals, New York State Museum Bulletin 436,* pp. 106–112.

Heinrich, E.W., 1981, "Geologic types of glass-sand deposits and some North American representatives," *Geological Society of America Bulletin,* v. 92, pp. 611–613.

Henderson, G.V., and Katzman, H., 1978, "Aggregate resources vs. urban development and multiple use planning, San Gabriel valley, California," *Society of Mining Engineers Preprint 78-H-32,* 10 pp.

Herkenhoff, E.C., 1972, "When are we going to mine oil?" *Engineering and Mining Journal,* June, pp. 132–138.

Hessley, R.K., Reasoner, J.W., and Riley, J.T., 1986, *Coal Science.* John Wiley and Sons, New York, 269 pp.

Hessling, M., 1992, "German hard-coal industry," *Engineering and Mining Journal,* October, pp. 16GG–16NN.

Heunis, R., 1985, "Rock bursts and the search for an early warning system," *Optima,* v. 26, pp. 30–42.

Hewett, D.F., 1929, "Cycles in metal production," *American Institute of Mining and Metallurgical Engineers Transactions,* v. 85, pp. 65–92.

Hickock, W.O., 1933, "The iron ore deposits at Cornwall, Pennsylvania." *Economic Geology,* v. 28, pp. 193–255.

Hickson, R.J., 1991, "Grasberg open-pit: Ertsberg's big brother comes onstream in Indonesia," *Mining Engineering,* v. 43, pp. 385–380.

Hight, R.P., 1983, "Abrasives" in Lefond, S.J., ed., *Industrial minerals and rocks, 5th ed.* Society of Mining Engineers, New York, pp. 11–21.

Hilliard, H.W., 1990, "Vanadium," *U.S. Bureau of Mines Minerals Yearbook,* pp. 1015–1018.

Hilts, P.J., 1990, "U.S. opens a drive on lead poisoning in nation's young," *The New York Times,* December 20.

Hilts, P.J., 1991, "U.S. fines 500 mining companies for tampering," *The New York Times,* April 5.

Hiskey, J.B. and DeVries, F.W., 1992, "Environmental considerations for alternates to cyanide processing," in Chander, S., ed., *Emerging process technologies for a cleaner environment.* Society for Mining, Metallurgy and Exploration, Littleton, CO, pp. 73–80.

Hoerr, J.P., 1988, *And the wolf finally came—the decline of the American steel industry.* University of Pittsburgh Press, Pittsburgh, 736 pp.

Hofstra, A.H., Leventhal, J.S., and Northrop, H.R., 1991, "Genesis of sediment-hosted disseminated-gold deposits by fluid mixing and sulfidation, Jerrit Canyon, Nevada," *Geology,* v. 19, pp. 36–40.

Hogan, W.T., 1991, *Global steel in the 1990s.* Lexington Books, Lexington, MA, 241 pp.

Hohmann, G.W., and Ward, S.H., 1981, "Electrical methods in mining geophysics," *Economic Geology, 75th Anniversary Volume,* pp. 806–828.

Holden, C., 1990, "Cyanide in them thar hills?" *Science,* v. 250, pp. 1514.

Holland, H.D., 1972, "Granites, solutions and base metal deposits," *Economic Geology,* v. 67, pp. 281–294.

Holland, H.D., 1984, *The chemical evolution of the atmosphere and oceans.* Princeton University Press, Princeton, N.J., 582 pp.

Holland, H.D., and Schidlowski, M., eds., 1982, *Mineral de-*

posits and the evolution of the biosphere. Springer-Verlag, New York, 323 pp.

Holton, J.R., 1990, "Global transport processes in the atmosphere" in Hutzinger, O., ed., *The handbook of environmental chemistry,* v. 1, part E. Springer-Verlag, Heidelberg, pp. 97–146.

Hook, C.O., and Russell, P.L., 1982, "World oil shale deposits," *Mining Engineering,* January, pp. 37–42.

Horne, R.A., 1978, *The chemistry of our environment.* Wiley-Interscience, New York, 869 pp.

Hossner, L.R., 1988, *Reclamation of surface mined lands.* CRC Press, Boca Raton, FL.

Hosterman, J.W., Patterson, S.H., and Good, E.E., 1990, "World nonbauxite aluminum resources excluding alunite," *U.S. Geological Survey Professional Paper 1076-C,* pp. 1–73.

Howell, J.M., and Gawthorne, J.M., 1987, *Copper in animals and man.* CRC Press, v. 1, 125 pp. and v. 2, 140 pp.

Hsieh, C.L., 1983, "Evaluating the energy performance of a lime kiln," *Tappi Journal,* v. 66, pp. 77–79.

Hubbard, G.D., 1912, "Gold and silver mining as a geographic factor in the development of the United States," Ph.D. Dissertation, Cornell University, Ithaca, NY.

Hubbert, M.K., 1985, "The world's evolving energy system" in Perrine, R.L., and Ernst, W.G., eds., *Energy: for ourselves and our posterity.* Prentice-Hall, Englewood Cliffs, NJ, pp. 44–100.

Hunt, J.M., 1979, *Petroleum geochemistry and geology.* W.H. Freeman, San Francisco, 616 pp.

Hurlbut, C.S., and Kammerling, R.C., 1991, *Gemology.* John Wiley, New York, 337 pp.

Hurt, H., 1981, *Texas rich: the Hunt dynasty from the early oil days through the silver crash.* Norton, New York.

Hutchinson, T.C., and Meema, K.M., 1987, *Lead, mercury, cadmium and arsenic in the environment.* John Wiley, New York, 360 pp.

Huttrer, G.W., 1990, "Geothermal electric power—a 1990 world status update," *Geothermal Resources Bulletin,* pp. 175–187.

Ibrahim, Y.M., 1992, "Kuwait struggles with oil damage," *The New York Times,* April 21, p. 8.

ICSU, 1987, International Council of Scientific Unions—Scientific Committee on Problems of the Environment, *Lead, mercury, cadmium and arsenic in the environment.* Wiley, New York.

IES, Annual, *International Energy Annual.* Energy Information Administration, U.S. Department of Energy.

Ingersall, A.P., 1983, "The atmosphere," *Scientific American,* v. 249, no. 3, pp. 162–174.

Ingram, G.M., 1974, *Expropriation of U.S. property in South America.* Praeger, New York.

International Petroleum Encyclopedia, 1991 Annual, PennWell Publishing, Tulsa, OK.

IPCS, 1991, *International Program on Chemical Safety—Platinum.* World Health Organization, Geneva, 87 pp.

Irving, P.M., 1991, *Acidic deposition: state of science and technology.* U.S. National Acid Precipitation Assessment Program, Washington, D.C., 275 pp.

Isherwood, R.J., Smith, R.C., and Coppa, L.V., 1989, "Regu-

lation in the domestic mineral industry," *Mineral Issues, Paper 5,* pp. 50–56.

Isherwood, R.J., Smith, R.C., Kiehn, O.A., and Daley, M.R., 1988, "The impact of existing and proposed regulations upon the domestic lead industry," U.S. Bureau of Mines OFR 55-88, 33 pp.

Ishihara, S., 1992, "The changing face of economic geology in Japan," *SEG Newsletter,* no. 10, July, pp. 24–27.

Itokawa, Y., and Durlach, J., eds. 1988, *Magnesium in health and disease.* Libbey, London, 432 pp.

Jacobs, P.A., and van Santen, R.A., 1989, "Zeolites: facts, figures, future," *Proceedings of the 8th zeolite conference.* Elsevier, Amsterdam. 1466 pp.

Jacoby, N.H., 1974, *Multinational oil.* Macmillan, New York, 323 pp.

Jahns, R.H., 1983, "Gem materials" in Lefond, S.J., ed., *Industrial Minerals and Rocks, 5th ed.,* Society of Mining Engineers, New York, pp. 279–338.

James, H.L., and Sims, P.K., eds., 1973, "Precambrian iron-formations of the world," *Economic Geology,* v. 68, pp. 913–1179.

Jan, J., and Roe, L.A., 1983, "Iodine" in Lefond, S.J., ed., *Industrial minerals and rocks, 5th ed.* Society of Mining Engineers, New York, pp. 793–798.

Jennett, J.C., Wixson, B.G., Lowsley, I.H., Purushothaman, K., Botter, E., Hemphill, D.D., Gale, N.L., and Tanter, W.H., 1977, "Transport and distribution from mining, milling and smelting operations in a forest ecosystem" in Boggess, W.R., ed., *Lead in the Environment,* National Science Foundation RA-770214 (Supt. Docs. Stock # 038-000-0038-1).

Jensen, J.H., Treckoff, W.E., and Anderson, A P., 1983, "Bromine" in Lefond, S.J., ed., *Industrial minerals and rocks, 5th ed.* Society of Mining Engineers, New York, pp. 561–566.

Jensen, N.L., 1985, "Selenium," *U.S. Bureau of Mines Minerals Facts and Problems,* pp. 705–712.

Jensen, N.L., 1985, "Tellurium," *U.S. Bureau of Mines Minerals Facts and Problems,* pp. 823–831.

Jentleson, B.W., 1986, *Pipeline politics: the complex political economy of East-West energy trade.* Cornell University Press, Ithaca, 276 pp.

Johany, A.D., 1980, *The myth of the OPEC cartel.* John Wiley, New York, 107 pp.

Johnson, C.A., and Thornton, I., 1987, "Hydrological and chemical factors controlling the concentrations of Fe, Cu, Zn and As in a river system contaminated by acid mine drainage," *Water Research,* v. 21, pp. 359–365.

Johnson, W., 1985, "Cement," *U.S. Bureau of Mines Minerals Facts and Problems, Bulletin 675,* pp. 121–131.

Johnson, W., and Paone, J., 1982, "Land utilization and reclamation in the mining industry, 1930–1980," *U.S. Bureau of Mines, Information Circular 8862,* 22 pp.

Jolly, J.H., 1985, "Zinc," *U.S. Bureau of Mines Minerals Facts and Problems,* pp. 923–940.

Jolly, J.H., 1990, "Zinc," *U.S. Bureau of Mines Minerals Yearbook.*

Jolly, J.W.L., 1985, "Copper," *U.S. Bureau of Mines Minerals Facts and Problems,* pp. 197–222.

Jolly, J.W.L., and Edelstein, D.L., 1989, 1992, "Copper," *U.S. Bureau of Mines Minerals Yearbook.*

Jones, K.C., Symon, C.J., and Johnston, A.E., 1987, "Retrospective analysis of archival soil collection," *Science of the Total Environment,* v. 67, pp. 75–89.

Jones, T.S., 1985, "Manganese," *U.S. Bureau of Mines Minerals Facts and Problems,* pp. 483–498.

Jones, T.S., 1990, "Manganese," *U.S. Bureau of Mines Minerals Yearbook.*

Joseph, R.A., 1982, "Firms that didn't mine or sell asbestos are also caught in the tide of litigation," *The Wall Street Journal,* December 14, p. 46.

Joyce, P., and Scannell, T., 1988, *Diamonds in South Africa:* Struik Publishers, Cape Town, 24 pp.

Juhlin, C., 1991, *Scientific summary report of the deep gas drilling project in the Siljan Ring impact structure.* Swedish State Power Board, Stockholm.

Juszli, M.P., 1989, "Creative distribution techniques can help market penetration of industrial minerals," *Mining Engineering,* v. 41, pp. 1009–1112.

Kadey, F.L., 1983, "Diatomite" in Lefond, S.J., ed., *Industrial minerals and rocks, 5th ed.* Society of Mining Engineers, New York, pp. 677–689.

Kaltenborn, B.P., 1990, "The Wilderness Act—catalyst for international action: a Norwegian perspective" in Lime, D.W., ed., *Managing America's Enduring Wilderness Resources.* Tourism Center, University of Minnesota, pp. 418–423.

Karr, A.R., 1983, "Coal-slurry pipeline measure is rejected by House as railroad lobby succeeds," *The Wall Street Journal,* September 28, p. 6.

KCC, 1971, *Expropriation of the El Teniente copper mine by the Chilean government.* Kennecott Copper Corporation, 8 pp.

Keating, J., 1992, "Silver," *Canadian Minerals Yearbook, 1991.* Energy, Mines and Resources Canada, pp. 40.1–40.7.

Kelly, M., 1988, *Mining and the freshwater environment.* Elsevier, Amsterdam, 231 pp.

Kennedy, J.L., 1983, *Fundamentals of drilling.* PennWell Press, Tulsa, OK, 213 pp.

Kennedy, J.L., 1984, *Oil and gas pipeline fundamentals.* PennWell Publishing, Tulsa, OK.

Kennedy, P., 1990, "Nevada flexes regulatory muscle with hefty fine," *Northern Miner,* November, p. 19.

Kennett, J.P., 1982, *Marine geology.* Prentice-Hall, Englewood Cliffs, NJ, 623 pp.

Kesler, S.E., 1976, *Our finite mineral resources.* McGraw-Hill, New York, 120 pp.

Kesler, S.E., 1978, "Economic lead deposits" in Nriagu, J.O., ed., *The biogeochemistry of lead in the environment.* Elsevier, Amsterdam, pp. 73–97.

Kesler, T.L., 1950, "Geology and mineral deposits of the Cartersville district, Georgia," *U.S. Geological Survey Professional Paper 224,* 97 pp.

Kesler, T.L., 1956, "Environment and origin of the Cretaceous kaolin deposits of Georgia and South Carolina," *Economic Geology,* v. 51, pp. 541–554.

Kesler, T.L., 1970, "Hydrothermal kaolinization in Michoacan, Mexico," *Clays and Clay Minerals,* v. 18, pp. 121–124.

Kessel, K.A., 1990, *Strategic minerals: U.S. alternatives.* National Defense University Press (Supt. of Documents), Washington, D.C., 293 pp.

Kessler, G., 1987, *Nuclear fission reactors.* Springer-Verlag, New York, 257 pp.

Khalil, M.A.K., and Rasmussen, R.A., 1987, "Atmospheric methane: trends over the last 10,000 years," *Atmospheric Environment,* v. 21, pp. 2445–2452.

Kimberley, M.M., 1978, "Paleoenvironmental classification of iron formations," *Economic Geology,* v. 73, pp. 215–229.

Kirk, W.S., 1985, "Cobalt," *U.S. Bureau of Mines Minerals Facts and Problems,* pp. 171–184.

Kirk, W.S., 1990, "Nickel," *U.S. Bureau of Mines Minerals Yearbook,* pp. 621–633.

Kirkley, M.B., Gurney, J.J., and Levinson, A.A., 1992, "Age, origin and emplacement of diamonds: a review of scientific advances in the last decade," *CIM Bulletin,* v. 85, no. 956, pp. 46–59.

Kistler, R.B., and Smith, W.C., 1983, "Boron and borates" in Lefond, S.J., ed., *Industrial minerals and rocks, 5th ed.* Society of Mining Engineers, New York, pp. 533–549.

Klein, C., and Buekes, N.J., 1992, "Time distribution, stratigraphy and sedimentologic setting, and geochemistry of Precambrian iron-formations" in Schopf, J.W., and Klein, C., eds., *The Proterozoic biosphere: A multidisciplinary study.* Cambridge University Press, New York, pp. 140–146.

Klein, D., and Goldscheider, I., 1986, "Glass in the environment," *Neues Glas,* no. 3, pp. 215–217.

Klemme, H.D., 1980, "Types of petroliferous basins" in Foster, N.H., and Beaumont, E.A., *Geologic basins: American Association of Petroleum Geologists Reprint Series,* No. 1, pp. 87–101.

Kleppie, J.D., Boyle, R.W., and Haynes, S.J., 1986, "Turbidite-hosted gold deposits," *Geological Association of Canada Special Paper 32,* 186 pp.

Klitgord, K.D. and Watkins, J.S., 1984, "Geologic studies related to oil and gas development in the EEZ," *U.S. Geological Survey Circular 929,* pp. 107–184.

Kobayashi, J., 1971, "Relation between the 'itai-itai' disease and the pollution of river water by cadmium from a mine," *Adv. Water Pollution Res., Proc. 5th Int. Conf.,* San Francisco, I-25, pp. 1–7.

Koederitz, L.F., Harvey, A.H., and Honarpour, M., 1989, *Introduction to petroleum reservoir analysis.* Gulf Publishing, Houston, TX, 251 pp.

Kohl, W.L., ed., 1991, *After the oil price collapse: OPEC, the United States, and the world oil market.* Johns Hopkins Press, Baltimore, 230 pp.

Kolata, G., 1992, "New Alzheimer's study questions link to metal," *The New York Times,* November 10.

Koljonen, T., Gustavsson, N., Noras, P., and Tanskanen, H., 1989, "Geochemical atlas of Finland: preliminary aspects," *Journal of Geochemical Exploration,* v. 32, pp. 231–242.

Konig, H.P., Hertel, R.F., and Kock. W., 1992, "Determination of platinum emissions from a three-way catalyst equipped gasoline engine," *Atmospheric Environment,* v. 26A, pp. 741–745.

Koren, E., 1992, "Mercury," *Canadian Minerals Yearbook—*

1991. Energy, Mines and Resources Canada, pp. 29.1–29.12.

Koschmann, A.H., and Bergendahl, M.H., 1968, "Principal gold-producing districts of the United States," *U.S. Geological Survey Professional Paper 610*, 283 pp.

Kostick, D.A., 1985, "Soda ash and sodium sulfate," *U.S. Bureau of Mines Minerals Facts and Problems*, pp. 741–756.

Kostick, D.A., 1992, "Sodium sulfate," *Mining Engineering*, v. 44, pp. 573–574.

Kramer, D.A., 1985, "Magnesium," *U.S. Bureau of Mines Minerals Facts and Problems*, pp. 471–482.

Kramer, D.A., and Plunkert, P.A., 1991, "Lightweight materials for new cars," *Minerals Today*, August, pp. 8–15.

Krause, C.A., 1987, *Refractories: The hidden industry.* The Refractories Institute, Pittsburgh, 268 pp.

Krauskopf, K.B., 1967, *Introduction to geochemistry.* McGraw-Hill, New York.

Krauskopf, K.B., 1988, *Radioactive waste disposal.* Chapman Hall, London, 145 pp.

Krauss, U.H., Saam H.G., and Schmidt, H.W., 1984, "International strategic minerals inventory summary report —phosphate," *U.S. Geological Survey Circular 930-C*, 41 pp.

Krauss, U.H., Schmidt, H.W., Taylor, H.A., Jr., and Sutphin, D.M., 1988, "International strategic minerals inventory —natural graphite," *U.S. Geological Survey Circular 930-H*, 29 pp.

Krishnan, E.R., and Hellwig, G.V., 1982, "Trace emissions from coal and oil combustion," *Environmental Prog.*, v. 1, pp. 290–295.

Kuck, P.H., 1988, "Iron ore," *U.S. Bureau of Mines Minerals Yearbook*, pp. 412–423.

Kuck, P.H., 1985, "Vanadium," *U.S. Bureau of Mines Minerals Facts and Problems*, pp. 895–916.

Kunasz, I.A., 1983, "Lithium raw materials" in Lefond, S.J., ed., *Industrial minerals and rocks, 5th ed.* Society of Mining Engineers, New York, pp. 869–880.

Kvenvolden, K.S., 1988, "Methane hydrate-a major reservoir of carbon in the shallow geosphere?" *Chemical Geology*, v. 71, pp. 41–51.

Kvenvolden, K.S., Ginsburg, G.D., and Soloviev, V.A., 1993, "Worldwide distribution of subaquatic gas hydrates," *Marine Letters*, v. 13, pp. 32–40.

Kwak, T.A.P., 1987, *W-Sn skarn deposits.* Elsevier, New York, 468 pp.

Labaton, S., 1991, "Judges see a crisis in heavy backlog of asbestos cases," *The New York Times*, February 6.

Lane, K.L., 1988, "The economic definition of ore: cut-off grades in theory and practice," *Mining Journal Books*, London, England, 149 pp.

Lang, R.D., 1986, "Development of Australia's first major diamond discovery," *Mining Engineering*, v. 38, pp. 13–17.

Langer, W.H., 1988, "Natural aggregates of the conterminous United States," *U.S. Geological Survey Bulletin 1594*, 33 pp.

Lankford, W.T., et al., 1985, *The making, shaping and treating of steel.* Assoc. of Iron and Steel Engineers, Pittsburgh.

Lapin, L., 1991, "Recycling reveals toxic surprise," *Press Democrat*, March 31.

Leachman, W.D., 1988, "Helium," *U.S. Bureau of Mines Minerals Yearbook*, v. 1, pp. 471–478.

Lee, C.O., Bartlett, Z.W., and Feierabend, R.H., 1960, "The Grand Isle mine," *Mining Engineering*, June, pp. 578–584.

Lee, T.H., 1989, "Advanced fossil fuel systems and beyond" in Ausabel, J.H., and Sladovich, H.E., eds., *Technology and the environment.* National Academy Press, Washington, D.C., pp. 114–136.

Lee, Y.W., Dehab, M.F., and Bogardi, I., 1992, "Nitrate risk management under uncertainty," *Journal of Water Resources Planning and Management:* v. 118, pp. 151–165.

Lefond, S.J., ed., *Industrial minerals and rocks, 5th ed.* Society of Mining Engineers, New York, 1478 pp.

Lefond, S.J., and Jacoby, C.H., 1983, "Salt" in Lefond, S.J., ed., *Industrial minerals and rocks, 5th ed.* Society of Mining Engineers, New York, pp. 1119–1150.

Lehmann, B., 1990, *Metallogeny of tin.* Springer-Verlag, Heidelberg, 211 pp.

Leney, G.W., 1964, "Geophysical exploration for iron ore," *Transactions of Society of Mining Engineering*, v. 231, pp. 355–372.

Lessmark, O., and Thornelof, E., 1986, "Liming in Sweden," *Water, Air and Soil Pollution*, v. 31, pp. 809–815.

Levinson, A.A., 1983, *Introduction to exploration geochemistry, 2nd ed.* Applied Publishing, Calgary.

Lewis, A.E., 1985, "Oil from shale: The potential, the problems, and a plan for development" in Perrine, R.L., and Ernst, W.G., eds., *Energy for ourselves and our posterity.* Prentice-Hall, Englewood Cliffs, NJ, pp. 101–122.

Lewis, J., and Benedict, C.P., 1981, "Icebergs on the Grand Banks: oil and gas considerations," *World Oil*, January, pp. 109–111.

Lime, D.W., 1989, *Managing America's enduring wilderness resource.* Minnesota Extension Service, University of Minnesota, 706 pp.

Lind, C.H., and Hem, J.D., 1993, "Manganese minerals and associated fine particulates in the streambed of Pinal Creek, Arizona, U.S.A.: a mining-related acid drainage problem," *Applied Geochemistry*, v. 8, pp. 67–80.

Link, P.K., 1982, *Basic petroleum geology.* Oil and Gas Consultants International, Tulsa, OK, 234 pp.

Link, W.K., 1952, "Significance of oil and gas seeps in world oil exploration," *American Assoc. Petrol. Geologists*, v. 36, pp. 1505–1529.

Lipin, B.R., 1982, "Low-grade chromium resources," *Society of Mining Engineers Preprint 82-62*, 17 pp.

Llewellyn, D.T., 1992, *Steels: metallurgy and applications.* Butterworth-Heinemann, Oxford, U.K., 302 pp.

LMV, 1985, *The nuclear waste primer.* Nick Lyons Books, New York, 90 pp.; 1991, *Federal offshore statistics, 1990.* Minerals Management Service, U.S. Dept. of the Interior.

Loebenstein, J.R., 1985, "Platinum-group metals," *U.S. Bureau of Mines Minerals Facts and Problems*, pp. 595–615.

Loen, J.S., 1992, "Mass balance constraints on gold placers possible solutions to source area problems," *Economic Geology*, v. 87, pp. 1624–1634.

Logue, O.T., 1991, "Safety and pneumoconioses: abrasive blasting and protective respiratory equipment," *Materials Performance, v. 30, September, pp. 32–37 and December, pp. 5–6.

Lohr, S., 1985, "Settlement is imposed in tin crisis," *The New York Times,* March 8.

Longhurst, J.W.S., 1989, *Acid deposition: sources, effects and controls.* British Library Science Reference and Information Service, London, 348 pp.

Lovelock, J., 1971, "Atmospheric fluorine compounds as indicators of air movements," *Nature,* v. 230, p. 379.

Lovelock, J., 1988, *The ages of Gaia: a biography of our living Earth.* Norton, New York, 214 pp.

Lowell, J.D., 1968, "Geology of the Kalamazoo orebody, San Manuel district, Arizona," *Economic Geology,* v. 63, pp. 645–654.

Lowell, J.D., and Guilbert, J.M., 1970, "Lateral and vertical alteration-mineralization zoning in porphyry ore deposits," *Economic Geology,* v. 65, pp. 373–408.

Lucas, J.M., 1985, "Gold," *U.S. Bureau of Mines Minerals Facts and Problems,* pp. 323–338.

Lund, R.J., and Goldstone, S.E., 1969, "Striking similarity of product price patterns," *Engineering and Mining Journal,* v. 170, no. 10, pp. 84–88.

Luxbacher, G.W., Sell, D.P., Skaggs, G.L., and McPhee, G.T., 1992, "The Clean Air Act Amendments of 1990—an eastern coal producer's view," *Society for Mining, Metallurgy and Exploration Preprint 92-229,* Littleton, CO, 7 pp.

Lyday, P.A., 1985, "Boron," *U.S. Bureau of Mines Minerals Facts and Problems,* pp. 91–102.

Lyday, P.A., 1985A, "Bromine," *U.S. Bureau of Mines Minerals Facts and Problems,* pp. 103–110.

Lyday, P.A., 1985B, "Iodine," *U.S. Bureau of Mines Minerals Facts and Problems,* pp. 377–384.

Lyday, T.Q., 1988, "Australia and Oceania," *U.S Bureau of Mines Minerals Yearbook,* v. 3, pp. 497–512.

Lynd, L.E., 1985, "Titanium," *U.S. Bureau of Mines Minerals Facts and Problems,* pp. 859–880.

Lyon, A.B., 1989, "The effect of changes in the percentage depletion allowance on oil firm stock prices," *Energy Journal,* v. 10, pp. 101–116.

Lyons, J.I., 1988, "Volcanogenic iron oxide deposits, Cerro de Mercado and vicinity, Durango, Mexico," *Economic Geology,* v. 83, pp. 1886–1906.

MacAvoy, P.W., 1988, *Explaining metals prices.* Kluwer Academic Publishers, Boston, 111 pp.

Macdonald, K.C., and Luyendyk, B.P., 1981, "The crest of the East Pacific Rise," *Scientific American,* v. 244, pp. 100–120.

Mackay, A.L., and Terrones, H., 1991, "Diamond from graphite," *Nature,* v. 352, pp. 762.

Magon, K.M., 1990, "The asbestos dilemma: make the workplace safe without endangering workers," *Safety and Health,* v. 142, pp. 50–53.

Mahoney, M.C., Naasca, P.C., and Burnett, W.S., 1991, "Bone cancer incidence rates in New York state: time trends and fluoridated drinking water," *American Journal of Public Health,* v. 81, pp. 475–479.

Malinconico, L.L., 1987, "On the variation of SO_2 emission from volcanoes," *J. Volc. Geoth. Res.,* v. 33, pp. 231–237.

Manahan, S.E., 1990, *Environmental chemistry.* Lewis, New York, 612 pp.

Mann, E.L., 1983, "Asbestos" in Lefond, S.J., ed., *Industrial minerals and rocks, 5th ed.* Society of Mining Engineers, New York, pp. 435–449.

Mannion, L.E., 1983, "Sodium carbonate deposits" in Lefond, S.J., ed., *Industrial minerals and rocks, 5th ed.* Society of Mining Engineers, New York, pp. 1187–1206.

Marchetti, C., and Nakicenovic, N., 1979, "The dynamics of energy systems and the logistic substitution model: Luxemburg, Austria," *International Institute of Applied Systems Report RR-79-13,* 79 pp.

Marr, I.L., and Cresser, M.S., 1983, *Environmental chemical analysis.* Chapman and Hall, New York, 258 pp.

Marshall, E., 1989, "Gasoline: the unclean fuel?" *Science,* v. 246, pp. 199–201.

Masters, C.D., Attanasi, E.D., Dietzman, W.D., Meyer, R.F., Mitchell, R.W., and Root, D.H., "World resources of crude oil, natural gas, natural bitumen and shale oil," *Twelfth World Petroleum Congress Proceedings.* John Wiley, Chichester, UK, v. 5, pp. 3–27.

Masters, C.D., Root, D.H., and Attanasi, E.D., 1991, "Resource constraints in petroleum production potential," *Science,* v. 253, pp. 146–152.

Masters, C.D., Root, D.H., and Attanasi, E.D., 1992, "World resources of crude oil and natural gas," *Thirteenth World Petroleum Congress Proceedings.* John Wiley and Sons, Chichester, UK, pp. 1–14.

Maynard, J.B., 1983, *Geochemistry of sedimentary ore deposits.* Springer-Verlag, Heidelberg.

Maynard, J.B., and Okita, O.M., 1991, "Bedded barite deposits in the United States, Canada, Germany, and China: two major types based on tectonic setting," *Economic Geology,* v. 86, pp. 364–376.

Mays, G.C., 1992, *Durability of concrete structures: investigation, repair, protection.* E&F Spon., London.

Mayuga, M.N., 1970, "Geology and development of California's giant—Wilmington oil field" in Halbouty, M.T., in *Geology of giant petroleum fields: American Association of Petroleum Geologists Memoir 14,* pp. 158–184.

McCarl, H.N., 1985, "Lightweight aggregate," in Lefond, S.J., ed., *Industrial minerals and rocks, 5th ed.* Society of Mining Engineers, New York, pp. 81–92.

McCawley, F.X., and Baumgardner, L.H., 1985, "Aluminum," *U.S. Bureau of Mines Minerals Facts and Problems,* pp. 9–32.

McConville, L.B., 1975, "The Athabasca tar sands," *Mining Engineering,* v. 27, January, pp. 19–25.

McFarlin, R.F., 1992, "Florida phosphate and the environment: practices, problems and emerging technologies" in Chander, S., ed., *Emerging process technologies for a cleaner environment.* Society for Mining, Metallurgy and Exploration, Littleton, CO, pp. 29–38.

McKelvey, V.E., 1960, "Relations of reserves of the elements to their crustal abundance," *American Jour. Sci.,* v. 258A, pp. 234–241.

McKelvey, V.E., 1973, "Mineral resource estimates and public policy," *U.S. Geological Survey Professional Paper 820,* pp. 9–19.

McKibben, M.A., Andes, J.P., Jr., and Williams, A.E., 1988, "Active ore formation at a brine interface in metamorphosed deltaic lacustrine sediments, the Salton Sea geothermal system, California," *Economic Geology,* v. 83, pp. 511–523.

McLaren, D.J., and Skinner, B.J., 1987, *Resources and world development.* Wiley-Interscience, New York, 940 pp.

McNamee, K., 1990, "Canada's endangered spaces: preserving the Canadian wilderness" in Lime, D.W., *Managing America's Enduring Wilderness Resource.* Minnesota Extension Service, University of Minnesota, St. Paul, pp. 425–434.

ME, 1990, "Manville trust nearly out of money as judge freezes payments," *Mining Engineering,* v. 42, p. 1054.

Meadows, D.H., Meadows, D.L., Randers, J., and Behrens, W.W., III, 1972, *The limits to growth.* Universe, New York.

Megaw, P.K.W., Ruiz, J., and Titley, S.R., 1988, "High-temperature carbonate-hosted Ag-Pb-Zn(Cu) deposits of northern Mexico," *Economic Geology,* v. 83, pp. 1856–1885.

Megill, R.E., 1981, "Problems in estimating the cost of finding oil and gas," *World Oil,* May, pp. 171–179.

Megill, R.E., and Wightman, R.B., 1984, "The ubiquitous overbid," *American Assoc. Petrol. Geol. Bull.,* v. 68, pp. 417–425.

Meisinger, A.C., 1985, "Diatomite," *U.S. Bureau of Mines Minerals Facts and Problems,* pp. 249–254.

Meriläinen, J., 1986, *Diatoms and lake acidity.* Kluwer Academic Publishers, Dordrecht, 307 pp.

Metal Statistics, annual, *American Metal Market.* Fairchild Publications, New York.

Meyer, C., 1981, "Ore-forming processes in Earth history," *Economic Geology 75th Anniversary Volume,* pp. 6–41.

Meyer, H.O.A., 1985, "Genesis of diamond: a mantle saga," *American Mineralogist,* v. 70, pp. 344–355.

Meyers, R.A., ed., 1981, *Coal handbook.* Marcel Dekker, New York, 854 pp.

Mikami, H.M., 1983, "Chromite" in Lefond, S.J., ed., *Industrial minerals and rocks, 5th ed.* Society of Mining Engineers, New York, pp. 567–584.

Mikdashi, A., 1986, *Transnational oil: Issues, policies and perspectives.* New York, St. Martin's Press.

Mikesell, R.F., 1986, *Stockpiling strategic materials: an evaluation of the national program.* American Enterprise Institute for Public Policy Research, Washington, DC.

Mikesell, R.F., 1987, *Nonfuel minerals: foreign dependency and national security.* University of Michigan Press, Ann Arbor, 257 pp.

Mikesell, R.F., 1988, *The global copper industry.* Croom Helm, New York, 160 pp.

Mikesell, R.F., and Whitney, J.W., 1987, *The world mining industry.* Allen and Unwin, London, 187 pp.

Milberg, R.P., Lagerwerff, J.V., Brower, D.L., and Biersdorf, G.T., 1980, "Soil lead accumulation alongside a newly constructed roadway," *J. Environmental Quality,* v. 9, pp. 6–8.

Miller, M., 1993, "Lime," *Mining Engineering,* v. 45, pp. 573–574.

Miller, M., 1992, "Fluorspar," *Mining Engineering,* v. 44, pp. 558–559.

Miller, R.W., Honarvar, S., and Hunsaker, B., 1980, "Effects of drilling fluids on soils and plants: I. and II.," *J. Environ. Qual.,* v. 9., pp. 547–560.

Miller, R.W., and Pesaran, P., 1980, "Effects of drilling fluids on soil and plants: II. Complete drilling fluid mixtures," *J. Environ. Qual,* v. 9, pp. 552–558.

Mills, H.N., 1983, "Glass raw materials," in Lefond, S.J., ed., *Industrial minerals and rocks, 5th ed.* Society of Mining Engineers, New York, pp. 339–351.

Mineral Commodities Survey, annual, U.S. Bureau of Mines, Washington, D.C.

Minerals Yearbook, annual, U.S. Bureau of Mines, Washington, D.C.

Minter, W.E.L., 1991, "Paleoplacers of the Witwatersrand basin," *Mining Engineering,* v. 42, pp. 195–199.

Mishra, C.P., Sheng-Fogg, C.D., Christiansen, R.G., Lemons, J.F., Jr., and De Giacomo, D.L., 1985, *Cobalt availability-market economy countries.* U.S. Bureau of Mines IC 9012, 33 pp.

Mitchell, A.W., 1984, "Barite in the western Ouachita Mountains, Arkansas," *Arkansas Geological Commission Guidebook 84-2,* pp. 124–131.

Mitchell, R.H., 1986, *Kimberlites: mineralogy, geochemistry and petrology.* Plenum Press, New York, 442 pp.

Mitchell, R.H., 1991, *Petrology of lamproites.* Plenum Press, New York, 447 pp.

MMMU, 1992, *Mines and minerals update.* Ministry of Northern Development and Mines, July, p. 5.

MMS, 1986, *Managing oil and gas operations on the outer continental shelf.* U.S. Dept. of the Interior, Minerals Management Service, 60 pp.

MMS, 1992, *Federal offshore statistics.* U.S. Department of the Interior, Minerals Management Service, 63 pp.

Mohide, T., 1992, *The international silver trade.* Silver Institute, Washington, D.C., 190 pp.

Moller, P., Cerny, P., and Saupe, F., eds., 1989, *Lanthanides, tantalum and niobium.* Springer-Verlag, Heidelberg, 380 pp.

Moore, C., and Marshall, R.I., 1991, *Steelmaking.* Institute of Metals, London, 172 pp.

Moore, J.W., 1991, *Inorganic contaminants of surface water.* Springer-Verlag, Heidelberg, 334 pp.

Morell, J.B., 1992, *The law of the sea: an historical analysis of the 1982 treaty and its rejection by the United States.* McFarland, Jefferson, NC.

Morgernstern, R., and Tirpak, D., 1990, "Environmental Protection Agency," *EPA Journal,* v. 16, no. 2.

Morse, D.E., 1985A, "Salt," *U.S. Bureau of Mines Minerals Facts and Problems,* pp. 679–688.

Morse, D.E., 1985B, "Sulfur," *U.S. Bureau of Mines Minerals Facts and Problems,* pp. 783–798.

Mossman, B.T., Bignon, J., Corn, M., Seaton, A., and Gee, J.B.L., 1989, "Asbestos: scientific developments and implications for public policy," *Science,* v. 247, pp. 294–301.

Mowatt, T., 1992, "Turning Bogota into the emerald city," *The New York Times,* June 22.

Muhn, J., and Stuart, H.R., 1988, *Opportunity and challenge: the story of BLM.* U.S. Department of the Interior, Bureau of Land Management, 303 pp.

Multhauf, R.P., 1978, *Neptune's gift: a history of common salt.* Johns Hopkins Press, Baltimore, 325 pp.

Mumpton, F.A., 1977, "Mineralogy and geology of natural

zeolites," *Mineralogical Society of America Reviews in Mineralogy,* v. 4, 225 pp.

Munro, R.G., 1982, "First Davis Strait discovery overcomes offshore hazards," *World Oil,* April, pp. 85–91.

Muntean, J.L., Kesler, S.E., and Russell, N., 1990, "Evolution of the Monte Negro acid-sulfate Au-Ag deposit, Pueblo Viejo, Dominican Republic," *Economic Geology,* v. 85, pp. 1738–1758.

Murozumi, M., Chow, T.J., and Patterson, C., 1969, "Chemical concentrations of pollutant lead aerosols, terrestrial dusts, and sea salts in Greenland and Antarctic snow strata," *Geochimica Cosmochimica Acta,* v. 33, pp. 1247–1294.

Murphy, G.F., and Brown, R.E., 1985, "Silicon," *U.S. Bureau of Mines Minerals Facts and Problems,* pp. 713–728.

Murray, R.L., 1989, *Understanding radioactive waste.* Battelle Press, Columbus.

Murray, S., Crisp, K., Klapwijk, P., Sutton–Pratt, T., and Green, T., 1991, *Gold, 1991.* Gold Fields Mineral Services, London, 64 pp.

Murray, T.H., 1988, "North American drilling activity in 1987," *American Association of Petroleum Geologists Bulletin,* v. 72, no. 10B, pp. 7–36.

Naldrett, A.J., 1981, "Nickel sulfide deposits: classification, composition and genesis," *Economic Geology 75th Anniversary Volume,* pp. 628–685.

Naldrett, A.J., 1989, "Magmatic sulfide deposits," *Oxford Monographs on Geology and Geophysics No. 14,* Oxford University Press, 186 pp.

NAPAP, 1991, *National acid precipitation assessment program, 1990 integrated assessment report.* National Acid Prevention Assessment Program, Washington, D.C., 520 pp.

Nash, J.T., Granger, H.C., and Adams, S.S., 1981, "Geology and concepts of genesis of important types of uranium deposits," *Economic Geology 75th Anniversary Volume,* pp. 63–116.

National Academy of Science–National Research Council, 1980, *Trace element geochemistry of coal resource development related to environmental quality and health:* Washington, D.C., 153 pp.

NEA, 1981, *The environmental and biological behaviour of plutonium and some other transuranium elements.* Organization for Economic Cooperation and Development, Nuclear Energy Agency, Paris, 116 pp.

Neftel, A., Moore, E., Oeschger, H., and Stuaffer, B., 1985, "Evidence from polar ice cores for the increase in atmospheric CO_2 in the past two centuries," *Nature,* v. 315, pp. 45–47.

Nehring, R., 1982, "Prospects for conventional world oil reserves," *Annual Review of Energy,* v. 7, pp. 175–200.

Nell, J.G., 1992, *Manganese: South Africa's Mineral Industry—1990.* Department of Mineral and Energy Affairs, pp. 97–102.

Nesbitt, B.E., Muehlenbachs, K., Murowchick, J.B., 1989, "Genetic implications of the stable isotope characteristics of mesothermal Au deposits and related Sb and Hg deposits in the Canadian Cordillera," *Economic Geology,* v. 84, pp. 1489–1506.

Neville, A.M., 1987, *Concrete technology.* John Wiley, New York.

Newbury, C.W., 1989, *The diamond ring: business, politics and precious stones in South Africa, 1867–1947.* Clarendon Press, Oxford, 311 pp.

Ng, A., and Patterson, C., 1981, "Natural concentrations of lead in ancient Arctic and Antarctic ice," *Geochimica et Cosmochimica Acta,* v. 45, pp. 2109–2121.

Noll, K.E., 1985, *Recovery, recycle, and reuse of industrial wastes.* Lewis, Chelsea, MI.

Norton, D., 1978, "Sourcelines, sourceregions and pathlines for fluids in hydrothermal systems related to cooling plutons," *Economic Geology,* v. 73, pp. 21–28.

Notholt, A.J.G., Sheldon, R.P., and Davidson, D.F., 1989, *Phosphate deposits of the world.* Cambridge University Press, Cambridge.

NPC, 1980, National Petroleum Council, *Unconventional gas sources,* U.S. Dept. of Energy, v. 1–5.

NRC, 1977, *Arsenic.* National Research Council–National Academy of Science, 332 pp.

NRC, 1978, *Manganese.* National Research Council, National Academy of Sciences, Washington, D.C., 191 pp.

NRC, 1985, *The competitive status of the U.S. steel industry.* National Research Council, National Academy Press, Washington, D.C., 158 pp.

NRC, 1990, *Surface coal mining effects on groundwater recharge.* National Academy Press, Washington, D.C., 159 pp.

Nriagu, J.O., 1978, *The biogeochemistry of lead in the environment.* Elsevier, Amsterdam, 422 pp.

Nriagu, J.O., ed., 1978, *Sulfur in the environment.* John Wiley, New York, 903 pp.

Nriagu, J.O., 1979, *The biogeochemistry of mercury in the environment.* Amsterdam, Oxford, New York, Elsevier, 231 pp.

Nriagu, J.O., 1979, *Copper in the environment.* John Wiley, New York, 489 pp.

Nriagu, J.O., 1983, *Lead and lead poisoning in antiquity.* Wiley Interscience, New York, 437 pp.

Nriagu, J.O., ed., 1987, *Zinc in the environment.* Wiley Interscience, New York.

Nriagu, J.O., 1990a, "The rise and fall of leaded gasoline," *The Science of the Total Environment,* v. 92, pp. 13–28.

Nriagu, J.O., 1990b, "Global metal pollution," *Environment,* v. 32, no. 7, pp. 7–33.

Nriagu, J.O., 1992, "Human lead exposure" in Needleman, H.L., *Human lead exposure.* CRC Press, Boca Raton, FL, pp. 3–21

Nriagu, J.O., and Pacyna, J.M., 1988, "Quantitative assessment of worldwide contamination of air, water and soils by trace metals," *Nature,* v. 333, pp. 134–139.

Nriagu, J.O., Pfeiffer, W.C., Malm, O., and Mierle, G., 1992, "Mercury pollution in Brazil," *Nature,* v. 356, pp. 389.

Nriagu, J.O., and Rao, S.S., 1987, "Response of lake sediments to changes in trace metal emission from the smelters at Sudbury, Ontario," *Environ. Pollution,* v. 44, pp. 211–218.

Nwoke, C., 1987, *Third world minerals and global pricing.* Zed Books, London, 229 pp.

NYT, 1991, "Unocal to halt shale project," *The New York Times,* March 27.

Ober, J.A., 1992, *Strontium.* U.S. Bureau of Mines Minerals Yearbook.

OECD, 1973, *Air pollution by fluorine compounds from primary aluminum smelting.* Organization for Economic Cooperation and Development, Paris, 108 pp.

OECD, 1975, *Cadmium and the environment.* Organization for Economic Cooperation and Development, Paris, 88 pp.

OECD, 1983, *Aluminum industry, energy aspects for structural change.* Organization for Economic Cooperation and Development, Paris, 139 pp.

OECD, 1986, *Water pollution by fertilizers and pesticides.* OECD 97 86 02 1, 144 pp.

OECD, 1991, *Coal information 1991.* International Energy Agency; Organization for Economic Cooperation and Development (OECD), Paris, 538 pp.

OGJ, 1991 annual, *Oil and Gas Journal Data Book.* PennWell Books, Tulsa, OK.

Ohle, E.L., 1991, "Lead and zinc deposits" in Gluskoter, H.J., Rice, D.D., and Taylor, R.B. eds., *The Geology of North America, Economic Geology, U.S.* Geological Society of America, Boulder CO, pp. 43–61.

Ohle, E.L., and Bates, R.L., 1981, *Geology, geologists, and mineral exploration.* 75th Anniversary Volume, Society of Economic Geologists, pp. 766–774.

Olson, D.K., 1986, "Michigan silver: native silver occurrences in the copper mines of Upper Michigan," *Mineralogical Record,* v. 17, pp. 37–48.

Olson, K.W., and Kleckley, J.W., 1989, "Severance tax stability," *National Tax Journal,* v. 42, pp. 69–78.

Olson, R.H., 1983, "Zeolites" in Lefond, S.J., ed., *Industrial minerals and rocks, 5th ed.* Society of Mining Engineers, New York, pp. 1391–1422.

O'Neil, T.J., 1974, "The minerals depletion allowance, parts I and II," *Mining Engineering,* October and November, pp. 61–64 and 39–42.

Optima, 1985, "Diamond trading over fifty years," v. 32, pp. 40–48.

Orris, G.J., 1992, "Barite—a comparison of grades and tonnages for bedded barite deposits with and without associated base-metal sulfides," *U.S. Geological Survey Bulletin 1877,* pp. B1–B12.

Osborne, D.G., 1988, *Coal preparation technology.* Kluwer Academic Publishers: Lancaster, UK, 1200 pp.

OSM, 1987, *Surface coal mining reclamation: 10 years of progress, 1977-1987.* Office of Surface Mining Reclamation and Enforcement, Washington, D.C., 48 pp.

Osterman, J.W., 1990, "Evaluating the impact of municipal water fluoridation on the aquatic environment," *American Journal of Public Health,* v. 80, pp. 1230–1235.

OTA, 1978, *Management of fuel and nonfuel minerals in federal lands: current status and issues.* Office of Technology Assessment, U.S. Congress, 435 pp.

OTA, 1984, *Acid rain and transported air pollutants.* Office of Technology Assessment, Unipub, New York, 323 pp.

OTA, 1987, *Marine minerals: exploring our new ocean frontier.* Congress of the United States, Office of Technology Assessment.

OTA, 1988, *Copper: Technology and Competitiveness.* U.S. Congress, Office of Technology Assessment, OTA-E-367, Washington, D.C., U.S. Government Printing Office, 272 pp.

Overstreet, W.C., and Marsh, S.P., 1981, *Some concepts and techniques in geochemical exploration.* 75th Anniversary Volume, Society of Economic Geologists, pp. 775–805.

Page, D.S., Boehm, P.D., Douglas, G.S., and Bence, A.E., 1993, "Identification of hydrocarbon sources in the benthic sediments of Prince William Sound and the Gulf of Alaska following the *Exxon Valdez* oil spill, *Third Symposium on Environmental Toxicology and Risk Assessment: Aquatic, Plant, and Terrestrial.* American Society for Testing and Materials, Philadelphia, PA.

Page, N.J., et al., 1985, "Exploration and mining history of the Stillwater Complex and adjacent rocks," *Montana Bureau of Mining Geology Special Paper 92,* pp. 77–92.

Palencia, C.M., 1985, *Molybdenum availability—market economy countries.* U.S. Bureau of Mines Information Circular 9044, 30 pp.

Palmer, A.R., 1984, "Decade of North American Geology geologic time scale," *Geological Society of America, Map and Chart Series MC-50.*

Papke, K.G., 1984, "Barite in Nevada," *Nevada Bureau of Mines and Geology Bulletin 98,* 125 pp.

Papp, J.F., 1985, "Chromium," *U.S. Bureau of Mines Minerals Facts and Problems,* pp. 139–156.

Papp, J.F., 1990, "Chromium," *U.S. Bureau of Mines Minerals Yearbook.*

Papp, J.F., 1993, "Chromite," *Mining Engineering,* v. 45, pp. 566–567.

Parker, R.H., 1989, *Hot dry rock geothermal energy.* Pergamon Press, Oxford, 1096 pp.

Parkinson, G., 1990, "Glass making," *Chemical Engineering,* v. 97, pp. 30–31.

Parsons, R.B., 1981, "Earned depletion and the mining industry," *CIM Bulletin,* v. 74, no. 829, pp. 132–138.

Patterson, C.C., 1987, "Lead in ancient bones and its relevance to historical development of social problems with lead," *The Sci. Total Environment,* v. 61, pp. 167–200.

Patterson, S.H., 1977, "Aluminum from bauxite: are there alternatives?" *American Scientist,* v. 65, pp. 345–351.

Patterson, S.H., Kurtz, H.F., Olson, J.C., and Neeley, C.L., 1986, "World bauxite resources," *U.S. Geological Survey Professional Paper 1076-B,* pp. 1–151.

Patterson, S.M., and Murray, H., 1983, "Clays," in Lefond, S.J., ed., *Industrial minerals and rocks, 5th ed.* Society of Mining Engineers, New York, pp. 585–599.

Paul, A., 1990, *Chemistry of glasses.* Chapman and Hall, New York.

Peabody, C.E., and Einaudi, M.T., 1992, "Origin of petroleum and mercury in the Culver-Baer cinnabar deposit, Mayacmas district, California," *Economic Geology,* v. 87, pp. 1078–1103.

Peacy, J.G., and Davenport, W.G., 1979, *The iron blast furnace, theory and practice.* Pergamon, Oxford, 251 pp.

Pearce, G., 1990, "Falconbridge budgets millions for anomaly," *Northern Miner,* v. 76, September 24.

Pearson, D.G., Davies, G.R., and Nixon, P.H., 1989, "Graphitized diamonds from a peridotite massif in Morocco and implications for anomalous diamond occurrences," *Nature,* v. 338, pp. 60–62.

Peck, J., Landberg, H.H., and Tilton, J.E., eds., 1992, *Competitiveness in metals: the impact of public policy.* Mining Journal Books, London, 310 pp.

Peck, M.C., ed., 1988, *The world aluminum industry in a changing energy era.* Resources for the Future, Washington, D.C., 231 pp.

Peeling, G.R., Kendall, G., Shinya, W., and Keyes, R., 1992, "Canadian policy perspective on environmental aspects of minerals and metals," in *Institution of Mining and Metallurgy, Minerals, Metals, and the Environment.* Elsevier, London, pp. 123–157.

Pelham, 1985, "Fluorspar," *U.S. Bureau of Mines Minerals Facts and Problems,* pp. 277–290.

Peng, S.S., 1985, "Longwall mining in the U.S.: where do we go from here?" *Mining Engineering,* v. 37, pp. 232–237.

Perry, H., 1974, "The gasification of coal," *Scientific American,* v. 230, no. 3, pp. 19–25.

Peters, A.T., 1990, "Iron and steel," *U.S. Bureau of Mines Minerals Yearbook.*

Petersen, U., 1971, "Laterite and bauxite formation," *Economic Geology,* v. 66, pp. 1070–1072.

Petersen, U., and Maxwell, R.S., 1979, "Historical mineral production and price trends," *Mining Engineering,* January, pp. 25–34.

Petkof, B., 1985, "Beryllium," *U.S. Bureau of Mines Minerals Facts and Problems,* pp. 75–82.

Phillips, G.N., Meyers, R.E., and Palmer, J.A., 1987, "Problems with the placer model for Witwatersrand gold, *Geology,* v. 15, pp. 1027–1030.

Phillips, T.A., 1977, *An economic evaluation of a process to separate raw urban refuse into its metal, mineral, and energy components.* U.S. Bureau of Mines IC 8732, 25 pp.

Pitman, F., 1987, "The west: severance tax cuts offer hope," *Coal Age,* v. 92, pp. 43–45.

PLS, 1988, *Public Land Statistics.* Bureau of Land Management, U.S. Department of Interior, 122 pp.

Plunkert, P.A., 1985A, "Antimony," *U.S. Bureau of Mines Minerals Facts and Problems,* pp. 33–42.

Plunkert, P.A., 1985B, "Cadmium," *U.S. Bureau of Mines Minerals Facts and Problems,* pp. 111–120.

Plunkert, P.A., 1985C, "Germanium," *U.S. Bureau of Mines Minerals Facts and Problems,* pp. 317–323.

Plunkert, P.A., 1985D, "Thallium," *U.S. Bureau of Mines Minerals Facts and Problems,* pp. 829–834.

Pocock, S.J., Shaper, A.G., and Powell, P., 1985, "The British regional heart study: cardiovascular disease and water quality" in Calabrese, E.J., Tuthill, R.W., and Condie, L., *Inorganics in drinking water and cardiovascular diseases.* Princeton Science Publishing Co., Princeton, NJ.

Polamski, J., Fluehler, H., and Blaser, P., 1982, "Accumulation of airborne fluoride in soils," *Journal of Environmental Quality,* v. 11, pp. 451–461.

Polmear, I.J., 1981, *Light alloys.* American Society of Metals, Metals Park, Ohio, 214 pp.

Popkin, R., 1985, "Fighting waste from gold mining," *EPA Journal,* v. 11, pp. 30ff.

Porcella, D., 1990, "Mercury in the environment," *EPRI Journal,* April/May, pp. 46–48.

Poreda, R.J., 1990, "Hydrothermal vents: ocean ridges in hot water," *Nature,* v. 346, p. 516.

Porter, J.W., 1989, "A national perspective on municipal solid waste management" in *The solid waste management prob-*

lem. Council for Solid Waste Solutions, Washington, D.C.

Porter, K.E., and Thomas, P.R., 1989, "International competitiveness of U.S. copper production, 1981–1987," *U.S. Bureau of Mines Mineral Issues, Paper 2,* pp. 8–21.

Posey, H.H., Pendleton, J.A., and Long, M.B., 1993, "The Summitville mine: a state's perspective," *SEG Newsletter,* no. 14, pp. 5-6.

Powell, J.D., 1988, "Origin and influence of coal mine drainage on streams of the United States," *Environmental Geology and Water Science,* v. 11, pp. 141–152.

Power, W.R., 1983, "Dimension and cut stone" in Lefond, S.J., ed., *Industrial minerals and rocks, 5th ed.* Society of Mining Engineers, New York, pp. 161–182.

Prain, R., 1975, *Copper: the anatomy of an industry.* Mining Journal Books, London, 298 pp.

Prather, B.E., 1991, "Petroleum geology of the Upper Jurassic and Lower Cretaceous, Baltimore Canyon Trough, Western North Atlantic Ocean," *American Assoc. Petrol. Geol. Bulletin,* v. 75, pp. 258–277.

Presser, T., and Barnes, I., 1985, "Dissolved constituents including selenium in water in the vicinity of Kesterton National Wildlife Refuge and West Grassland, Fresno and Merced Counties, California," *U.S. Geological Survey Water Res. Inv. Rpt. 85-4220.*

Pressler, J.W., 1985, "Gem stones," *U.S. Bureau of Mines Minerals Facts and Problems,* pp. 305–316.

Pressler, J.W., 1985, "Gypsum," *U.S. Bureau of Mines Minerals Facts and Problems,* pp. 349–356.

Pressler, J.W., and Pelham, L., 1985, "Lime, calcium and calcium compounds," *U.S. Bureau of Mines Minerals Facts and Problems,* pp. 453–460.

Pretorius, D.A., 1981A, "Gold, geld, gilt: future supply and demand," *Economic Geology,* v. 76, pp. 2032–2042.

Pretorius, D.A., 1981B, "Gold and uranium in quartz-pebble conglomerates," *Economic Geology 75th Anniversary Volume,* pp. 117–138.

Prieto, C., 1973, *Mining in the New World.* McGraw-Hill, New York, 239 pp.

Prindle, D.F., 1981, *Petroleum politics and the Texas Railroad Commission.* University of Texas Press, Austin, 230 pp.

Prior, D.B., Doyle, E.H., and Kaluza, M.J., 1989, "Evidence for sediment eruption on deep sea floor, Gulf of Mexico," *Science,* v. 243, pp. 517–520.

Prugger, F.F., 1979, "The flooding of the Cominco potash mine and its rehabilitation," *CIM Bulletin,* July, pp. 86–91.

Prugger, F.F. and Prugger, A.A., 1991, "Water problems in Saskatchewan potash mining-what can be learned from them?," *Canadian Institution of Mining and Metallurgy Bulletin,* v. 84, pp. 58–65.

Pura, R., 1986, "Malaysia's tin scheme stuns the industry," *The Wall Street Journal,* September 25.

Pye, E., Naldrett, A.J., and Giblin, P., 1984, "The geology and ore deposits of the Sudbury structure," *Ontario Geological Survey Special Volume 1,* 603 pp.

Rabachevsky, G.A., 1988, "The tungsten industry of the U.S.S.R.," *U.S. Bureau of Mines Mineral Issues,* 50 pp.

Radetsky, M., 1985, *State mineral enterprises in developing countries: Their impact on international mineral markets.* Johns Hopkins University Press, Baltimore.

Rampino, M.R., Self, S., and Stothers, R.B., 1988, "Nuclear winters," *Ann. Rev. Earth Planet Sci.,* v. 16, pp. 73–99.

Randall, R.W., 1972, *Real del Monte, a British Mining Venture in Mexico.* University of Texas Press, Austin, 257 pp.

Ratten, S., and Eaton, D., 1976, "Oil shale: The prospects and problems of an emerging energy industry," *Annual Review of Energy,* v. 1, pp. 183–212.

Ray, D.L., 1987, "The great acid rain debate," *American Spectator,* January, pp. 21–25.

Reading, H.G., ed., 1986, *Sedimentary environments and facies.* Blackwell, Oxford, 613 pp.

Reckling, K., Hoy, R.B., and Lefond, S.J., 1983, "Diamonds" in Lefond, S.J., ed., *Industrial minerals and rocks, 5th ed.* Society of Mining Engineers, New York, pp. 653–671.

Red Book, 1991 Annual, *Uranium resources, production and demand.* Organization for Economic Cooperation and Development, Paris, 358 pp.

Reed, M.H., and Spycher, N., 1985, "Boiling, cooling, and oxidation in epithermal systems: a numerical approach," *Reviews in Economic Geology,* v. 2, pp. 249–272.

Reed, P.C., 1990, "Preparing to manage wilderness in the 21st century," *U.S. Forest Service, Southeast Forest Experiment Station General Technical Report SE-66,* 172 pp.

Reese, R.G., 1985, "Silver," *U.S. Bureau of Mines Minerals Facts and Problems,* pp. 729–740.

Reese, R.G., 1990, "Mercury," *U.S. Bureau of Mines Minerals Yearbook,* pp. 752–771.

Rendu, J.M., and Guzman, J., 1991, "Study of the size distribution of the Carlin Trend gold deposits," *Mining Engineering,* v. 43, pp. 139–140.

Rice, D.R., 1986, "Oil and gas assessment: methods and applications," *American Assoc. Petrol. Geol. Studies in Geology 21.*

Rice, P.C., 1987, *Amber, the golden gem of the ages.* Geoscience Press, Boulder, CO, 289 pp.

Rich, R.A., Holland, H.D., and Petersen, U., 1977, *Hydrothermal uranium deposits.* Elsevier, Amsterdam.

Richards, H.G., Savage, D., and Andrews, J.N., 1992, "Granite-water reactions in an experimental Hot Dry Rock geothermal reservoir, Rosemanowes test site, Cornwall U.K.," *Applied Geochemistry,* v. 7, pp. 193–222.

Riding, A., 1991, "Accord bans oil exploration in the Antarctic for 50 years," *The New York Times,* October 5.

Riordan, P.H., 1981, *Geology of asbestos deposits.* Society of Mining Engineers, New York, 118 pp.

Robb, L.J., and Meyer, F.M., 1991, "A contribution to recent debate concerning epigenetic versus syngenetic mineralization processes in the Witwatersrand basin," *Economic Geology,* v. 86, pp. 396–401.

Robert, F., Sheahan, P.A., and Green, S.B., 1991, "Greenstone gold and crustal evolution," *Geological Association of Canada NUNA Conference Volume,* 252 pp.

Rodhe, H., and Herrera, R., eds., 1988, "Acidification in tropical countries," *SCOPE Report No. 36.* John Wiley, Chichester, 405 pp.

Romero, R.A., 1991, "Aluminum toxicity," *Env. Sci. Tech.,* v. 25, pp. 1658f.

Rona, P.A., 1988, "Hydrothermal mineralization at oceanic ridges," *Canadian Mineralogist,* v. 26, pp. 431–465.

Root, D.H., and Attanasi, E.D., 1992, "Oil field growth in the United States—how much is left in the barrel?" *U.S. Geological Survey Circular 1074,* pp. 68.

Roper, L.D., 1978, "Depletion categories for United States metals," *Materials and Society,* v. 2, pp. 217–231.

Rose, A.W., and Eggert, R.G., 1988, "Exploration in the United States" in Tilton, J.E., Eggert, R.G., and Landsberg, H.H., *World mineral exploration: Trends and economic issues.* Resources for the Future, Washington, D.C., pp. 331–362.

Rose, A.W., Hawkes, H.E., and Webb. J.S., 1979, *Geochemistry in mineral exploration.* Academic Press, New York, 657 pp.

Rosner, D., and Markowitz, G., 1991, *Deadly dust: silicosis and the politics of twentieth century America.* Princeton University Press, Princeton, NJ, 229 pp.

Ross, M., 1987, "Minerals and health: the asbestos problem" in Peirce, H.W., *21st Forum on Geology of Industrial Minerals: Arizona Bureau of Geology and Mineral Technology Special Paper 4,* pp. 83–89.

Rossman, M.D., Preuss, O.P., and Powers, M.B., 1991, *Beryllium: biomedical and environmental aspects.* Williams and Wilkins, Baltimore, 319 pp.

Round, F.E., Crawford, R.M., Mann, D.G., and Repak, A.J., 1990, "The diatoms," reviewed in *American Scientist,* v. 79, March/April, p. 174.

Rowe, G.W., 1988, "Well contamination by water softener regeneration discharge water," *Journal of Environmental Health,* v. 50, pp. 272–276.

Roy, S., 1981, *Manganese deposits.* Academic Press, New York, 458 pp.

Royal Dutch/Shell, 1983, *The petroleum handbook.* Elsevier, Amsterdam, 710 pp.

Ruiz, J., Kesler, S.E., Jones, L.M., and Sutter, J.F., 1985, "Geology and geochemistry of the Las Cuevas fluorite deposits, San Luis Potosi, Mexico," *Economic Geology,* v. 80, pp. 1200–1209.

Russell, O.L., 1981, "Oil shale," *Mining Engineering,* January, v. 33, pp. 29–54.

Rybach, L., and Muffler, L.J.P., 1981, *Geothermal systems: principals and case histories.* John Wiley, New York, 359 pp.

Rytuba, J.I., and Heropoulos, C., 1992, "Mercury—an important by-product of epithermal gold systems," *U.S. Geological Survey Bulletin 1877D,* pp. 1–23.

Sainsbury, C.L., 1969, "Tin resources of the world," *U.S. Geological Survey Bulletin 1301,* 55 pp.

Salter, R.S., ed., 1987, *International symposium on gold metallurgy.* Pergamon, Elmsford, NY, 396 pp.

Sampson, A., 1975, *The seven sisters.* Bantam, New York, 489 pp.

Sand, R.H., 1992, "Developments in asbestos litigation, Review Commission authority and criminal prosecutions," *Employee Relations Law Journal,* v. 17, pp. 503–509.

Sanger, D.E., 1993, "A ship carrying plutonium is met by protests in Japan," *The New York Times,* January 5, p. 14.

Sauer, J.R., 1988, *Brazil—paradise of gemstones.* Sauer and Co., Rio de Janeiro, 134 pp.

Saulnier, G.J., and Goddard, K.E., 1982, "Use of mathematical models to predict impacts of mining energy minerals on the hydrologic system in northwestern Colorado," *Mining Engineering,* v. 34, pp. 285–292.

Saupe, F., 1990, "Geology of the Almadén mercury deposit, Province of Ciudad Real, Spain," *Economic Geology,* v. 85, pp. 482–510.

Sawkins, F.J., 1989, "Anorogenic felsic magmatism, rift sedimentation, and giant Proterozoic Pb-Zn deposits," *Geology,* v. 17, pp. 657–660.

Sawkins, F.J., 1990, *Metal deposits in relation to plate tectonics.* Springer-Verlag, New York, 461 pp.

Schell, D., 1985, "Road salt contaminates well, causes health hazard," *Journal of Environmental Health,* v. 47, pp. 202–203.

Schenck, G.H.K., and Torries, T.F., Jr., 1985, "Aggregate—crushed stone" in Lefond, S.J., ed., *Industrial minerals and rocks, 5th ed.* Society of Mining Engineers, New York, pp. 60–80.

Schidlowski, M., 1980, "The atmosphere" in Hutzinger, O., ed., *The handbook of environmental chemistry,* v. 1, part A. Springer-Verlag, Heidelberg, pp. 1–16.

Schmalz, R.F., 1991, "The Mediterranean salinity crisis: alternative hypotheses," *Carbonates and Evaporites,* v. 6, pp. 121–126.

Schmidt, M.G., 1989, *Common heritage or common burden? The United States position on the development of a regime for deep sea mining in the Law of the Sea Convention.* Clarendon Press, Oxford, UK.

Schmoker, J.W., 1992, "Geologically oriented overview of horizontal drilling in the United States," *U.S. Geological Survey Circular 1074,* pp. 71.

Schneider, K., 1990, "Radiation danger found in oilfields across the nation," *The New York Times,* December 3.

Schneider, K., 1990, "U.S. wrestles with gap in radiation exposure rules," *The New York Times,* December 26.

Schneider, S.H. and Boston, P.J., 1991, *Scientists on Gaia.* MIT Press, Cambridge, MA, 433 pp.

Schobert, H.H., 1987, *Coal.* American Chemical Society, Washington, D.C., 298 pp.

Schottman, F.J., 1985, "Iron and steel," *U.S. Bureau of Mines Minerals Facts and Problems,* pp. 405–424.

Schreier, H., 1989, "Asbestos in the natural environment," *Studies in Environmental Science 37.* Elsevier, Amsterdam, 159 pp.

Scott, D., 1991, "Victor a major new discovery for Inco," *Northern Miner,* v. 77, no. 30, pp. 1–2.

Searles, J.P., 1993, "Potash," *Mining Engineering,* v. 45, pp. 579–582.

Searles, J.P., 1985, "Potash," *U.S. Bureau of Mines Minerals Facts and Problems,* pp. 617–634.

Sedgwick, J., 1991, "Strong but sensitive," *Atlantic,* April, p. 70ff.

Sehnke, E.D., and Plunkert, P.A., 1989, "Bauxite, alumina and aluminum," *U.S. Bureau of Mines Minerals Yearbook,* pp. 121–150.

Semkin, R.G., and Kramer, J.R., 1976, "Sediment geochemistry of Sudbury area lakes," *Canadian Mineralogist,* v. 14, pp. 73–90.

Severinghaus, N., Jr., 1983, "Fillers, filters and absorbents" in Lefond, S.J., *Industrial minerals and rocks, 5th ed.*

American Institute of Mining, Metallurgical and Petroleum Engineers, pp. 204–235.

Severson, R.C., and Shacklette, H.T., 1988, "Essential elements and soil amendments for plants: sources and use for agriculture," *U.S. Geological Survey Circular 1017,* 48 pp.

Shabecoff, P., 1990, "Bush cuts back areas off coasts open for drilling," *The New York Times,* June 27.

Shannon, S.S., Jr., 1983, "Rare earths and thorium" in Lefond, S.J., ed., *Industrial minerals and rocks, 5th ed.* Society of Mining Engineers, New York, pp. 1109–1119.

Sharpe, M.R., 1986, "Bushveld Complex—excursion guidebook," *Geocongress '86 Field Excursion 23C,* 143 pp.

Shawe, D.R., 1981, "U.S. Geological Survey workshop on nonfuel mineral-resource appraisal of wilderness and CUSMAP areas," *U.S. Geological Survey Circular 848,* 18 pp.

Shedd, K.B., 1992, "Cobalt," *U.S. Bureau of Mines Minerals Yearbook.*

Shelton, K.L., Chil-Sup, S., and Chang, J.-S., 1988, "Gold-rich mesothermal vein deposits of the Republic of Korea: Jungwon area," *Economic Geology,* v. 83, pp. 1221–1237.

Shen, G.T., and Boyle, E.A., 1987, "Lead in corals: reconstruction of historical industrial fluxes to the surface ocean," *Earth Planetary Science Letters,* v. 82, pp. 289–304.

Shephard, M., Golden, D., Komai, R., and Morasky, T., 1985, "Utility solid waste: Managing the by-products of coal combustion," *Electric Power Research Institute Journal,* v. 10, no. 8, pp. 20–35.

Shigley, J.E., Dirlam, D.M., Schmetzer, K., and Jobbins, E.A., 1990, "Gem review," *Gems and Gemology,* v. 26, pp. 4–12.

Sibley, S.F., 1985, "Nickel," *U.S. Bureau of Mines Minerals Facts and Problems,* pp. 535–552.

Siegel, A., 1960, "Mineral pigments" in Gilson, J.L., ed., *Industrial minerals and rocks, 3rd ed.* American Institute of Mining, Metallurgical and Petroleum Engineers, New York, pp. 585–593.

Sillitoe, R.H., Halls, C., and Grant, J.N., 1975, "Porphyry tin deposits in Bolivia," *Economic Geology,* v. 70, pp. 913–927.

Simmons, S.R., Gemmell, J.B., and Sawkins, J.F., 1988, "The Santo Niño silver-lead-zinc vein, Fresnillo district, Zacatecas, Mexico, Part II," *Economic Geology,* v. 83, pp. 1619–1641.

Simon, W., 1991, "Mexican silver mining," *Engineering Mining Journal,* v. 192, pp. 18–22.

Simons, M., 1987, "The smelter's price: jungle in ashes," *The New York Times,* May 28.

Simons, M., 1993, "West is warned of the high cost of fixing risky Soviet A-plants," *The New York Times,* June 22.

Sindeeva, N.D., 1964, *Mineralogy and types of deposits of selenium and tellurium.* John Wiley-Interscience, New York, 363 pp.

Singer, D.A., 1977, "Long-term adequacy of metal resources," *Resources Policy,* v. 3, pp. 127–133.

Singer, D.A., and Drew, L.J., 1976, "The area of influence of an exploratory drill hole," *Economic Geology,* v. 71, pp. 642–647.

Singer, D.A., and Kouda, R., 1988, "Integrating spatial and frequency information in the search for Kuroko deposits of the Hokuroko district, Japan," *Economic Geology,* v. 83, pp. 18–29.

Singer, D.A., and Mosier, D.L., 1981, "A review of regional mineral resource assessment methods," *Economic Geology,* v. 76, pp. 1006–1015.

Singer, D.A., and Ovenshine, A.T., 1979, "Assessing metallic resources in Alaska," *American Scientist,* v. 67, pp. 582–589.

Singhal, R.K., and Kolada, R., 1988, "Surface mining of oil sands in Canada: development in productivity improvement," *Mining Engineering,* v. 38, pp. 267–275.

Sinkankas, J., 1989, *Emerald and other beryls.* Geoscience Press, Boulder, CO, 665 pp.

Skinner, B.J., 1975, "A second iron age ahead?" *American Scientist,* v. 64, pp. 258–269.

Skinner, H.C.W., Ross, M., and Frondel, C., 1988, *Asbestos and other fibrous minerals.* Oxford University Press, New York, 204 pp.

Slansky, M., 1986, *Geology of sedimentary phosphates.* Elsevier, Amsterdam.

Slichter, L.B., 1960, "The need of a new philosophy of prospecting," *Mining Engineering,* v. 14, pp. 570–576.

Smith, C.L., Ficklin, W.H., and Thompson, J.M., 1987, "Concentrations of arsenic, antimony, and boron in steam and steam condensate at The Geysers, California," *Journal of Volcanology and Geothermal Research,* v. 32, pp. 329–341.

Smith, G.R., 1990, "Tungsten," *U.S. Bureau of Mines Minerals Yearbook.*

Smith, J.V., and Dawson, J.S., 1985, "Carbonado: diamond aggregates from early impacts of crustal rocks?" *Geology,* v. 13, pp. 342–343.

Smith, R.D., Campbell, J.A., and Felix, W.D., 1980, "Atmospheric trace element pollutants from coal combustion," *Mining Engineering,* v. 32, pp. 1603–1610.

Snow, G.G., and Mackenzie, B.W., 1981, "The environment of exploration: economic, organizational and social constraints," *Economic Geology, 75th Anniversary Volume,* pp. 861–896.

Spencer, S.M., Rupert, F.R., and Yon, J.W., 1990, "Fuller's earth deposits in Florida and southwestern Georgia" in Zupan, A.W., and Maybin, A.H., III, *Proceedings of 24th Forum on the Geology of Industrial Minerals.* South Carolina Geological Survey, Columbia, pp. 121–127.

St. Aubin, K., and Massie, S., 1987, *The abandoned mind land program.* Association of Abandoned Mined Land Programs, Springfield, IL, 28 pp.

Stafford, P.T., 1985, "Tungsten," *U.S. Bureau of Mines Minerals Facts and Problems,* pp. 881–894.

Stanglin, D., 1992, "Toxic Wasteland," *U.S. News and World Report,* April 13, pp. 40–46.

Stanley, G.G., ed., 1988, *The extractive metallurgy of gold in South Africa.* South African Institute of Mining and Metallurgy, Chamber of Mines of South Africa, Johannesburg, 614 pp.

Starr, C., 1971, "Energy and power," *Scientific American,* September, pp. 12–21.

Stauffer, H.C., 1981, "Oil shale, tar sands, and related materials," *American Chemical Society Symposium Series 163,* 395 pp.

Steinnes, E., 1987, "Impact of long-range atmospheric transport of heavy metals to the terrestrial environment in Norway" in Hutchinson, T.C., and Meema, K.M., eds., *Lead, mercury, cadmium and arsenic in the environment.* John Wiley, New York, pp. 107–129.

Stevenson, R.W., 1990, "Political conflict is renewed over drilling in Arctic Refuge," *The New York Times,* September 5.

Stinson, T.F., 1982, "State severance taxes and the federal system," *Mining Engineering,* April, pp. 382–386.

Stoiber, R.E., Williams, S.N., and Huebert, B., 1987, "Annual contribution of sulfur dioxide to the atmosphere by volcanoes," *J. Volc. Geoth. Res.,* v. 33, pp. 1–8.

Stollery, K.R., 1987, "Mineral processing in an open economy," *Land Economics,* v. 63, pp. 128–136.

Stowasser, W.F., 1985, "Phosphate rock," *U.S. Bureau of Mines Minerals Facts and Problems,* pp. 579–594.

Stowe, C.W., 1987, *Evolution of chromium ore fields.* Van Nostrand Reinhold, New York, 340 pp.

Strishkov, V.V., 1982, *The copper industry of the U.S.S.R.: Problems, issues and outlook.* U.S. Bureau of Mines, Mineral Issues, National Technical Information Service (Springfield, VA), PB85-1555661, 80 pp.

Strishkov, V.V., and Levine, R.M., 1987, *The manganese industry of the U.S.S.R.* U.S. Bureau of Mines, Mineral Issues, 39 pp.

Strishkov, V.V., and Stebler, W.G., 1985, *The chromium industry of the U.S.S.R.* U.S. Bureau of Mines, Mineral Issues, National Technical Information Service (Springfield, VA) PB91-106518, 33 pp.

Strong, D.F., Fryer, B.J., and Kerrich, R., 1984, "Genesis of the St. Lawrence fluorspar deposits as indicated by fluid inclusion, rare earth element and isotopic data," *Economic Geology,* v. 79, pp. 1142–1158.

Sturges, W.T., and Harrison, R.M., 1986A, "Bromine:lead ratios in airborne particles from urban and rural sites," *Atmospheric Environment,* v. 20, pp. 577–588.

Sturges, W.T., and Harrison, R.M., 1986B, "Bromine in marine aerosols and the origin, nature and quantity of natural atmospheric bromine," *Atmospheric Environment,* v. 20, pp. 1485–1496.

Sutphin, D.M., and Page, N.J., 1986, "International strategic minerals inventory summary report—platinum group metals," *U.S. Geological Survey Circular 930-E,* 34 pp.

Sutphin, D.M., Sabin, A.E., and Reed, B.L., 1990, "International strategic mineral inventory summary report—tin," *U.S. Geological Survey Circular 930-J,* 52 pp.

Suttill, K.R., 1988, "A fabulous silver porphyry: Cerro Rico de Potosi," *Engineering Mining Journal,* v. 189, March, pp. 50–53.

Szatmari, P., 1992, "Role of modern climate and hydrology in world oil preservation," *Geology,* v. 20, pp. 1143–1146.

Tang, Y.S., and Saling, J.H., 1990, *Radioactive waste management.* Hemisphere Publishing, New York, 460 pp.

Taugbol, G., ed., 1986, "The Norwegian monitoring programme for long-range transported air pollutants. Results 1980–1984," *Norwegian State Pollution Control Authority Report 230/86,* 95 pp.

Taylor, H.A., Jr., 1985, "Stone—dimension," *U.S. Bureau of Mines Minerals Facts and Problems,* pp. 769–776.

Taylor, H.A., Jr., 1991, "Chicago's Amoco building marble to be recycled," *Minerals Today,* March, p. 27.

Taylor, H.P., 1985, "Graphite," *U.S. Bureau of Mines Minerals Facts and Problems,* pp. 339–348.

Taylor, J., and Yokell, M., 1979, *Yellowcake: the international cartel.* Pergamon Press, New York, 203 pp.

Taylor, J.J., 1989, "Improved and safer nuclear power," *Science,* v. 244, pp. 318–325.

Taylor, R.B., 1979, *Geology of tin deposits.* Elsevier, Amsterdam, 543 pp.

Telford, W.M., Glendart, L.P., Sherif, R.E., and Keys, D.A., 1976, *Applied geophysics.* Cambridge University Press, London, 860 pp.

Tepordei, V.V., 1985, "Stone—crushed," *U.S. Bureau of Mines Minerals Facts and Problems,* pp. 757–769.

Tepordei, V.V., 1993, "Construction aggregate," *Mining Engineering,* v. 45, pp. 567–568.

Terry, R.C., 1974, *Road salt, drinking water, and safety: improving public policy and practice.* Ballinger, Cambridge, MA, 161 pp.

Thoburn, J., 1981, *Multinationals, mining and development: A study of the tin industry.* Gower, Westmead, Hampshire, UK, 183 pp.

Thomas, L., 1992, *Handbook of practical coal geology.* John Wiley, New York, 338 pp.

Thomas, P.R., 1984, "South African gold production: how much, for how long and at what cost?" *Society of Mining Engineers Preprint 84-110,* 11 pp.

Thompkins, R.W., 1982, "Radiation in uranium mines," *CIM Bulletin,* September, pp. 149–159.

Thompson, L.K., Sidhu, S.S., and Roberts, B.A., 1979, "Fluoride accumulations in soil and vegetation in the vicinity of a phosphorus plant," *Environmental Pollution,* v. 18, pp. 221–234.

Thornton, I., ed., 1983, *Applied Environmental Geochemistry.* Academic Press, New York, 376 pp.

Tilton, J.E., Eggert, R.G., and Landsberg, H.H., 1988, *World mineral exploration: trends and economic issues.* Resources for the Future, Washington, DC, 464 pp.

Tilton, J.E., ed., 1990, *World metal demand.* Resources for the Future: Washington, D.C., 341 pp.

Tissot, B.P., and Welte, D.H., 1978, *Petroleum formation and occurrence.* Springer-Verlag, Heidelberg.

Titley, S.R., and Beane, R.E., 1981, "Porphyry copper deposits," *Economic Geology 75th Anniversary Volume,* pp. 235–269.

Towner, R.R., Gray, J.M., and Porter, L.M., 1988, "International strategic minerals inventory summary report—titanium," *U.S. Geological Survey Circular 930-G,* 58 pp.

Travis, C.C., and Etnier, E.L., eds., 1984, *Groundwater pollution: environmental and legal problems.* Westview Press, Boulder, CO, 149 pp.

Tseng, W.-P., 1977, "Effects and dose-response relationships of skin cancer and Blackfoot disease with arsenic," *Environmental Health Perspectives,* v. 19, pp. 109–119.

Turneaure, F.S., 1971, "The Bolivian tin-silver province," *Economic Geology,* v. 66, pp. 215–225.

Uchitelle, 1992, "On the path to an open economy: a decrepit steel plant in the Urals," *The New York Times,* July 2.

UIA, 1991 Annual, *Uranium Industry Annual:* Department of Energy (DOE/EIA-0478(9X).

Ullman, F.D., 1991, Cement: *Mining Engineering,* v. 44, pp. 553–554.

U.S. Committee on Surface Mining and Reclamation, Board on Mineral and Energy Resources, 1980, *Sand and gravel mining and quarrying.* National Research Council.

USDE, 1990, *Annual Energy Review:* DOE/EIA-0384(90).

USGS, 1989, "Focus: Assessment of petroleum potential in the Arctic National Wildlife Refuge, Alaska," *Professional Paper 1850,* pp. 26–28.

Valiente, D.J., and Rosenmans, K.D., 1989, "Does silicosis still occur?" *Journal of the American Medical Association,* v. 262, pp. 3003–3007 and v. 263, p. 3025.

Van den Steen, A., Polloni, J., and Kalala, B., 1992, "Perspective of cobalt production in Zaire," available from *Metals Week,* 1221 Avenue of the America, New York, NY 10020..

VanLandingham, S.L., 1985, *Gemology of world gem deposits.* Van Nostrand Reinhold, New York, 406 pp.

van Rensburg, W.C.J., 1986, *Strategic minerals.* Prentice-Hall, Englewood Cliffs, NJ.

van Zyl, A.A., 1988, "De Beers' 100," *Geobulletin,* v. 31, pp. 24–28.

Vikre, P.G., 1989, "Fluid-mineral relations in the Comstock lode," *Economic Geology,* v. 84, pp. 1574–1613.

Viljayan, S., Melnyk, A.J., Singh, R.D., and Nutall, K., 1989, "Rare earths: their mining, processing and growing industrial usage," *Mining Engineering,* v. 41, pp. 13–18.

Vogely, W.A., ed., 1985, *Economics of the mineral industries.* American Institute of Mining, Metallurgical and Petroleum Engineers, New York, 660 pp.

Vourvopoulos, G., 1987, "Proceedings of the international conference on trace elements in coal," *Journal of Coal Quality,* v. 8, pp. 18–27.

Wagner, P.A., 1971, *The diamond fields of southern Africa.* C. Struik, Cape Town, 472 pp.

Wald, M.L., 1992, "Help for cleaner air from a mystery coal," *The New York Times,* August 10.

Wald, M.L., 1992, "It burns more cleanly, but ethanol still raises air-quality concerns," *The New York Times,* August 3.

Wallace, R.H., Kraemer, T.F., Taylor, R.E., and Wesselman, J.B., 1979, "Assessment of geopressurized-geothermal resources in the northern Gulf of Mexico basin," *U.S. Geological Survey Circular 790,* pp. 132–155.

Walston, R.E., 1986, "Western water law," *Natural Resources and Environment,* v. 1, no. 4, pp. 6–8, 48–52.

Walthier, T.N., Sirvas, E., and Araneda, R., 1985, "The El Indio gold, silver, copper deposit," *Engineering and Mining Journal,* v. 186, October, pp. 38–42.

Wanty, R.B., Johnson, S.L., and Briggs, P.H., 1991, "Radon-222 and its parent radionuclides in groundwater from two study areas in New Jersey and Maryland, U.S.A.," *Applied Geochemistry,* v. 6, pp. 305–318.

Ward, C.R., 1984, *Coal geology and coal technology.* Blackwell Scientific Publications, Palo Alto, 345 pp.

Wargo, J.G., 1989, "In situ leaching of disseminated gold deposits—geological factors," *Mining Engineering,* v. 40, pp. 973–975.

Warneck, P., 1988, *Chemistry of the natural atmosphere.* Academic Press, New York, 757 pp.

Warren, H.V., 1989, "Geology, trace elements health," *Social Science and Medicine,* v. 29, pp. 923–926.

Warren, J.K., 1989, *Evaporite sedimentology.* Prentice Hall, New York, 285 pp.

Watkin, E.M., Hassan-King, A.P., Boli, C.R., and Harvey, D.A.R., 1992, "Resource development after alluvial mining at Sierra Rutile Ltd., Sierra Leone," in *Institution of Mining and Metallurgy, Minerals, Metals, and the Environment.* Elsevier, London, pp. 277–341.

Weinberg, J., Harris, K.L., and White, G., 1987, *Steel in motor vehicles—a 25 year perspective.* U.S. Bureau of Mines IC 9175, 15 pp.

Weisenburger, D.C., 1987, "Lymphoma is linked to nitrate contamination of groundwater," *American Family Physician,* v. 35, p. 242.

Weisman, W.I., and McIlveen, S., 1983, "Sodium sulfate deposits" in Lefond, S.J., ed., *Industrial minerals and rocks, 5th ed.* Society of Mining Engineers, New York, pp. 1207–1224.

Weiss, H.V., Koide, M., and Goldberg, E.D., 1971, "Mercury in a Greenland ice sheet: evidence for recent input by man," *Science,* v. 174, pp. 692–694.

Wellmer, F-W., 1989, *Economic evaluations in exploration.* Springer-Verlag, Heidelberg, 163 pp.

Wellmer, F.-W., 1992, "The concept of lead time," *Minerals Industry International,* March, pp. 39–40.

Wentzler, T.H., and Aplan, F.F., 1992, "Neutralization reactions between acid mine waters and limestone" in Chander, S., ed., *Emerging process technologies for a cleaner environment.* Society for Mining, Metallurgy and Exploration, Littleton, CO, pp. 149–160.

Wessel, G.R., and Wimberly, B.H., 1992, *Native sulfur: Developments in geology and exploration.* Society for Mining, Metallurgy and Exploration, Littleton, CO, 192 pp.

West, E.G., 1982, *Copper and its alloys.* Halsted Press, New York, 221 pp.

Westall, J., and Stumm, W., 1980, "The hydrosphere" in Hutzinger, O., ed., *The handbook of environmental chemistry,* v. 1, part A. Springer-Verlag, Heidelberg, pp. 17–50.

Wheatcroft, G., 1985, *The Randlords.* Weidenfeld and Nicolson, London, 314 pp.

Wheeler, R.R., and Whited, M., 1975, *Oil, from prospect to pipeline.* Gulf Publishing Co., Houston, TX, 146 pp.

White, D.E., 1967, "Some principles of geyser activity, mainly from Steamboat Springs, Nevada," *American Journal of Science,* v. 265, pp. 641–684.

White, D.E., 1974, "Diverse origins of hydrothermal ore fluids," *Economic Geology,* v. 69, pp. 954–973

White, D.E., 1981, "Active geothermal systems and hydrothermal ore deposits," *Economic Geology 75th Anniversary Volume,* pp. 382–493.

White, J.A.L., 1968, "Native sulfur deposits associated with volcanic activity," *Mining Engineering,* June, pp. 47–54.

White, J.C., 1985, *Liming acidic waters: environmental and policy concerns.* Center for Environmental Information, Rochester, 82 pp.

White, W.H., Bookstrom, A.A., Kamilli, R.J., Ganster, M.W., Smith, R.P., Ranta, D.E., and Steininger, R.C., 1981, "Character and origin of climax-type molybdenum deposits," *Economic Geology 75th Anniversary Volume,* pp. 270–316.

WHO, 1980, *Environmental health criteria—Tin and organotin compounds.* World Health Organization, Geneva, 109 pp.

WHO, 1981A, *Environmental health criteria—Chromium.* World Health Organization, Geneva, 197 pp.

WHO, 1981B, *Environmental health criteria—Arsenic.* World Health Organization, Geneva, 174 pp.

WHO, 1981, *Environmental health criteria—Manganese.* World Health Organization, Geneva, 110 pp.

WHO, 1982, *Titanium.* World Health Organization, Geneva, 68 pp.

WHO, 1984, *Fluorine and fluorides.* World Health Organization, Geneva, 136 pp.

WHO, 1988, *Environmental health criteria—Vanadium.* World Health Organization, Geneva, 217 pp.

WHO, 1989, *Environmental health criteria—Lead.* World Health Organization, Geneva, 106 pp.

WHO, 1989A, *Environmental health criteria—Mercury.* World Health Organization, Geneva, 115 pp.

WHO, 1990, *Beryllium.* World Health Organization, Geneva, 106 pp.

WHO, 1990, *Environmental health criteria—Methyl mercury.* World Health Organization, Geneva, 144 pp.

WHO, 1991, *Environmental health criteria—Barium.* World Health Organization, Geneva, 148 pp.

WHO, 1991A, *Environmental health criteria—Inorganic mercury.* World Health Organization, Geneva, 168 pp.

WHO, 1991B, *Environmental health criteria—Nickel.* World Health Organization, Geneva, 383 pp.

Wicken, O.M., and Duncan, L.R., 1983, "Magnesite and related minerals" in Lefond, S.J., ed., *Industrial minerals and rocks, 5th ed.* Society of Mining Engineers, New York, pp. 881–896.

Williams, B., 1991, *U.S. petroleum strategies in the decade of the environment.* PennWell Books, Tulsa, OK, 336 pp.

Williams, O.R., Driver, N.E., and Ponce, S.L., 1990, "Managing water resources in wilderness areas" in Reed, P.C., ed., *Preparing to manage wilderness in the 21st century: U.S. Forest Service General Technical Report SE-66,* pp. 62–69.

Williams, S.J., 1992, "Sand and gravel—an enormous offshore resource within the U.S. exclusive economic zone," *U.S. Geological Survey Bulletin 1877H,* pp. 1–10.

Wilson, B.L., Schwarzer, R.R., and Etonyeaku, N., 1986, "The evaluation of heavy metals (chromium, nickel and cobalt) in the aqueous sediment surrounding a coal burning generating plant," *J. Env. Science and Health,* v. 21, pp. 791–808.

Wilson, F., 1985, "Mineral wealth, rural poverty," *Optima,* v. 33, no. 1, pp. 23–37.

Winchester, J.W., 1980, "Transport processes in air" in Hutzinger, O., ed., *The handbook of environmental chemistry,* v. 2, part A. Springer-Verlag, Heidelberg, pp. 19–30.

Winterhalder, K., 1988, "Trigger factors initiating natural revegetation processes on barren, acid, metal-toxic soil near Sudbury, Ontario, smelters," *U.S. Bureau of Mines IC 9184,* pp. 118–124.

Woodall, R., 1984, "Success in mineral exploration," *Geoscience Canada*, v. 11, pp. 41–46, 83–90, 127–132.

Woodall, R., 1988, "The role of mineral exploration toward the year 2000," *Geophysics*, v. 7, no. 2, pp. 35–37.

Woodall, R., 1992, "Challenge of minerals exploration in the 1990s," *Mining Engineering*, v. 44, pp. 679–683.

Woodbury, W.D., 1985, "Lead," *U.S. Bureau of Mines Minerals Facts and Problems*, pp. 433–453.

Woodbury, W.D., 1990, "Lead," *U.S. Bureau of Mines Minerals Yearbook*.

Wyllie, P.J., 1976, *The way the Earth works*. John Wiley, New York.

Xu, S., Okay, A.I., Ji, S., Sengor, A., Liu, Y., and Jiang, L., 1992, "Diamond from the Dabie Shan metamorphic rocks and its implication for tectonic setting, *Science*, v. 257, pp. 80–82.

Yada, K., 1971, "Study of microstructure of chrysotile asbestos by high-resolution electron microscopy," *Acta Cryst.*, A27, pp. 659–664.

Yeap, C.H., 1979, "Geology of tin deposits," *Bulletin of the Geological Society of Malaysia No. 11*, 392 pp.

Yergin, D., 1990, *The prize: The epic quest for oil, money, and power*. Simon Schuster, New York, 245 pp.

Yih, S.W.H., and Chun, T.W., 1978, *Tungsten: sources, metallurgy, properties and applications*. Elsevier, Amsterdam, 516 pp.

Young, G.M., 1992, "Late Proterozoic stratigraphy and the Canada-Australia connection," *Geology*, v. 20, pp. 215–218.

Young, R.S., 1979, *Cobalt in biology and biochemistry*. Academic Press, New York, 198 pp.

Zaihua, L., and Daoxian, Y., 1991, "Effect of coal mine waters of variable pH on springwater quality: a case study," *Environmental Geology and Water Science*, v. 17, pp. 219–225.

Zarzycki, J., 1991, *Glasses and the vitreous state*. Cambridge University Press, Cambridge, UK.

Zdunczyk, M., 1991, "Aggregate imports: is it achievable?" *Mining Engineering*, v. 43, pp. 145.

Zenhder, A.J.B., and Zinder, S.H., 1980, "The sulfur cycle" in Hutzinger, O., ed., *The handbook of environmental chemistry*, v. 1, part A. Springer-Verlag, Heidelberg, pp. 105–146.

Zhores, M., 1990, *The legacy of Chernobyl*. Basil Blackwell, Oxford, 312 pp.

Zopf, P.E., 1992, *Mortality patterns and trends in the U.S.* Greenwood, Westport, CT, 281 pp.

Zumberge, J.H., 1979, "Mineral resources and geopolitics in Antarctica," *American Scientist*, v. 65, pp. 345–349.

SUPPLEMENTARY REFERENCES

Akhter, M.S., Ali, S.M., and Madany, I.M., 1988, "Heavy metals analysis in Bahrain refinery sludge," *Nuclear and Chemical Waste Management*, v. 8, pp. 165–167.

Chestnut, A., 1990, "Filtration system cleans groundwater at refinery," *Pollution Engineering*, v. 22, pp. 65–67.

Fisher, W.L., Tyler, N., Ruthven, C.L., Birchfield, T.E., Poutz, J.F., 1992, An Assessment of the Oil Resource Base of the United States: Dept. of Energy/BC—93-1-SP. 54 pp.

Gurney, J.J., Levinson, A.A., and Smith, H.S., 1991, "Marine mining of diamonds off the west coast of southern Africa," *Gems and Gemology*, winter, pp. 206–219.

Kapoor, S., Smalley, G.A., and Norman, M.E., 1992, "Hazardous waste regulations affect refinery wastewater schemes," *Oil and Gas Journal*, v. 90, pp. 45–49.

Kirkley, M.B., Gurney, J.J., and Levinson, A.A., 1991, "Age, origin, and emplacement of diamonds," *Gems and Gemology*, spring, pp. 2–25.

Levinson, A.A., Gurney, J.J., and Kirkley, M.B., 1992, "Diamond sources and production, past, present and future," *Gems and Gemology*, winter, pp. 234–253.

Appendix I

World Mineral Production, Reserves, and Reserve Base

TABLE AI-1

Annual mineral production, reserves, and resources for 1992 with estimates for Russia and other states of the C.I.S. based largely on data for 1990 (from *U.S. Bureau of Mines Mineral Commodity Summaries, U.S. Department of Energy International Energy Annual, Organization of Economic Development Redbook, Canadian Minerals Yearbook, Australian Commodity Statistical Bulletin,* and *South Africa's Mineral Industry*).

	Mine Production	*Reserves*	*Reserve Base*
ALUMINUM (tonnes)			
United States	4,000,000		
Canada	1,850,000		
Australia	1,200,000		
Brazil	1,170,000		
China	850,000		
Norway	820,000		
Venezuela	560,000		
World	17,600,000		
World Resources: see bauxite			
ANTIMONY (tonnes of antimony content)			
Bolivia	7,500	308,000	317,000
South Africa	4,500	236,000	254,000
Mexico	2,600	181,000	227,000
World	60,600	4,200,000	4,700,000
World Resources: 5,100,000			
ARSENIC (tonnes)			
China	10,000		
Chile	7,000		
C.I.S.	7,000		
Mexico	5,000		
Philippines	5,000		
World	45,000	1,000,000	1,500,000
World Resources: 11,000,000			
ASBESTOS (tonnes)			
Canada	620,000	40,000,000	47,000,000
Brazil	210,000	Moderate	Moderate
Zimbabwe	160,000	Moderate	Moderate
China	150,000	Large	Large
South Africa	140,000	5,000,000	8,000,000
World	3,400,000	Large	Large
World Resources: 200,000,000			
BARITE (tonnes)			
China	1,800,000	40,000,000	150,000,000
India	525,000	30,000,000	32,000,000
United States	410,000	30,000,000	60,000,000
C.I.S.	400,000	10,000,000	75,000,000

	Mine Production	Reserves	Reserve Base
Morocco	350,000	10,000,000	11,000,000
World	5,200,000	170,000,000	500,000,000
World Resources: 2,000,000,000			

BAUXITE (tonnes)

	Mine Production	Reserves	Reserve Base
Australia	40,400,000	5,620,000,000	7,860,000,000
Guinea	17,000,000	5,600,000,000	5,900,000,000
Jamaica	11,100,000	2,000,000,000	2,000,000,000
Brazil	10,400,000	2,800,000,000	2,900,000,000
India	4,800,000	1,000,000,000	1,200,000,000
World	105,000,000	23,000,000,000	28,000,000,000
World Resources: 55,000,000,000 to 75,000,000,000			

BERYLLIUM (tonnes of contained beryllium)

	Mine Production	Reserves	Reserve Base
United States	240	not available	not available
China	55		
C.I.S.	50		
Brazil	30		
Argentina	3		
World	379		
World Resources: 66,000			

BISMUTH (tonnes of bismuth content)

	Mine Production	Reserves	Reserve Base
China	1,400	20,000	40,000
Mexico	650	10,000	20,000
Peru	500	11,000	42,000
Australia	300	18,000	27,000
Japan	130	9,000	18,000
World	3,200	110,000	250,000
World Resources: not available (NA)			

BORON (tonnes of boric oxide)

	Mine Production	Reserves	Reserve Base
Turkey	1,200,000	30,000,000	150,000,000
United States	575,000	57,000,000	210,000,000
Argentina	250,000	2,000,000	9,000,000
C.I.S.	160,000	28,000,000	140,000,000
Chile	130,000	8,000,000	41,000,000
World	2,400,000	160,000,000	630,000,000
World Resources: large			

BROMINE (tonnes of bromine content)

	Mine Production	Reserves	Reserve Base
United States	166,000	11,000,000	11,000,000
Israel	130,000		
C.I.S.	50,000	1,400,000	1,400,000
United Kingdom	28,000		
Japan	15,000		
World	400,000	NA	NA
World Resources: unlimited			

CADMIUM (tonnes of cadmium content)

	Mine Production	Reserves	Reserve Base
Japan	2,600	10,000	15,000
United States	1,700	70,000	210,000
Belgium	1,500		
Canada	1,500	80,000	170,000
Mexico	1,000	35,000	40,000
World	20,000	540,000	970,000
World Resources: 6,000,000			

CEMENT (tonnes)

	Mine Production	Reserves	Reserve Base
China	268,000,000	254,000,000	264,000,000
C.I.S.	130,000,000	140,000,000	142,000,000
Japan	92,700,000	90,000,000	94,500,000

	Mine Production	Reserves	Reserve Base
United States	70,900,000	74,500,000	75,500,000
India	53,600,000	54,500,000	63,600,000
World	1,241,000,000	1,301,000,000	1,416,000,000
World Resources: unlimited			

CESIUM (tonnes)

	Mine Production	Reserves	Reserve Base
Canada	NA	70,000	73,000
Zimbabwe	NA	23,000	23,000
Namibia	NA	7,000	9,000
World	NA	100,000	110,000
World Resources: NA			

CHROMIUM (tonnes, gross weight)

	Mine Production	Reserves	Reserve Base
South Africa	5,000,000	959,000,000	5,540,000,000
C.I.S.	3,700,000	129,000,000	129,000,000
India	850,000	59,000,000	77,000,000
Albania	800,000	6,000,000	6,000,000
Turkey	750,000	8,000,000	20,000,000
World	12,800,000	1,400,000,000	6,800,000,000
World Resources: 11,000,000,000			

COAL (Including lignite) (thousands of tonnes)

	Mine Production	Reserves	Reserve Base
United States	930,000	241,066,000	
China	838,000	114,708,000	
Germany	459,000	80,214,000	
Russia	400,000*	137,563,000*	
India	227,000	62,662,000	
Poland	215,000	41,274,000	
Australia	209,000	91,105,000	
South Africa	172,000	55,434,000	
Ukraine	120,000*	41,734,000*	
Kazakhstan	118,000*	40,000,000	
United Kingdom	89,000	3,807,000	
Canada	68,000	8,639,000	
World	4,744,000	1,041,140,000	

COBALT (tonnes of cobalt content)

	Mine Production	Reserves	Reserve Base
Zaire	9,000	2,000,000	2,500,000
Zambia	7,000	360,000	540,000
C.I.S.	2,200	140,000	230,000
Canada	2,100	45,000	260,000
Cuba	1,600	1,040,000	1,800,000
World	24,800	4,000,000	8,800,000
World Resources: 11,000,000			

COLOMBIUM (NIOBIUM) (tonnes)

	Mine Production	Reserves	Reserve Base
Brazil	11,000	3,311,000	3,629,000
Canada	2,400	136,000	408,000
Zaire	500	32,000	91,000
Nigeria	20	64,000	91,000
World	14,000	3,500,000	4,200,000
World Resources: large			

COPPER (tonnes of copper content)

	Mine Production	Reserves	Reserve Base
Chile	1,870,000	88,000,000	140,000,000
United States	1,720,000	45,000,000	90,000,000
Canada	780,000	11,000,000	23,000,000
C.I.S.	550,000	37,000,000	54,000,000
Poland	400,000	20,000,000	36,000,000
World	8,900,000	310,000,000	590,000,000
World Resources: 2,300,000,000			

	Mine Production	Reserves	Reserve Base
CRUDE OIL (thousands of barrels)			
Russia	3,578,000	49,000,000	
Saudi Arabia	2,948,000	257,800,000	
United States	2,707,000	24,700,000	
Iran	1,156,000	92,860,000	
Iraq	1,025,000**	100,000,000	
China	1,016,000	24,000,000	
Mexico	976,000	51,300,000	
United Arab Emirates	871,000	97,700,000	
Venezuela	850,000	59,100,000	
Nigeria	688,000	17,900,000	
United Kingdom	665,000	3,625,000	
Norway	662,000	7,600,000	
Kuwait	584,000**	94,000,000	
Canada	565,000	5,600,000	
Libya	541,000	22,800,000	
World	21,900,000	991,000,000	
World Resources: 1,469,000,000			
DIAMOND—INDUSTRIAL (carats)			
Australia	18,000,000	500,000,000	900,000,000
Zaire	17,000,000	150,000,000	350,000,000
Russia	8,000,000	40,000,000	65,000,000
Botswana	6,200,000	125,000,000	200,000,000
South Africa	5,200,000	70,000,000	150,000,000
World	58,100,000	980,000,000	1,900,000,000
World Resources: NA			
DIAMOND—GEM (carats)			
Australia	18,000,000	NA	NA
Botswana	12,600,000		
Russia	8,000,000		
South Africa	5,000,000		
Zaire	3,000,000		
World	52,000,000	NA	NA
World Resources: 300,000,000			
DIATOMITE (tonnes)			
United States	647,000	250,000,000	500,000,000
France	250,000	NA	2,000,000
C.I.S.	200,000	NA	NA
Spain	100,000	NA	NA
Denmark	65,000	NA	NA
World	1,600,000	800,000,000	Large
World Resources: large			
FELDSPAR (tonnes)			
Italy	1,500,000		
United States	696,000		
France	400,000		
Germany	330,000		
Thailand	330,000		
World	5,300,000		
World Resources: large			
FLUORSPAR (tonnes)			
China	1,300,000	27,000,000	46,000,000
Mongolia	400,000	20,000,000	30,000,000
C.I.S.	320,000	62,000,000	94,000,000

	Mine Production	Reserves	Reserve Base
Mexico	300,000	19,000,000	23,000,000
South Africa	250,000	30,000,000	36,000,000
World	3,600,000	210,000,000	310,000,000
World Resources: 762,000,000			

GARNET—INDUSTRIAL (tonnes)

	Mine Production	Reserves	Reserve Base
United States	54,304	5,000,000	25,000,000
Australia	28,000	1,000,000	7,000,000
China	18,000		
India	10,000	500,000	20,000,000
World	112,304	Moderate	Large
World Resources: NA			

GERMANIUM (tonnes)

	Mine Production	Reserves	Reserve Base
United States	13	450	500
World	65	NA	NA
World Resources: NA			

GOLD (tonnes; 1 tonne = 32,150.7 troy ounces)

	Mine Production	Reserves	Reserve Base
South Africa	600	20,000	22,000
United States	320	4,770	5,050
Australia	240	2,150	2,300
C.I.S. (Russia = 146 tonnes)	230	6,220	7,780
Canada	170	1,780	3,300
World	2,170	44,000	51,000
World Resources: 75,000			

GRAPHITE—NATURAL (tonnes)

	Mine Production	Reserves	Reserve Base
Korea	100,000	3,150,000	20,000,000
India	60,000	735,000	735,000
Brazil	30,000	500,000	1,000,000
Mexico	30,000	3,100,000	3,100,000
Madagascar	18,000	980,000	980,000
World	613,000	21,000,000	380,000,000
World Resources: 800,000,000			

GYPSUM (tonnes)

	Mine Production	Reserves	Reserve Base
United States	14,800,000	7,30,000,000	Large
Canada	8,200,000	450,000,000	Large
China	8,200,000		
Iran	8,000,000		
Thailand	7,300,000		
World	98,000,000	2,400,000	Large
World Resources: large			

HAFNIUM (tonnes of hafnium content)

	Mine Production	Reserves	Reserve Base
Australia	NA	420	464
South Africa	NA	250	275
United States	NA	60	66
India	NA	26	29
Brazil	NA	18	20
World	NA	780	850
World Resources: 1,000,000			

HELIUM (million cubic meters of contained helium gas)

	Mine Production	Reserves	Reserve Base
United States	90.7	7,017	13,064
C.I.S.	4.2	NA	9,153
Poland	1.4	NA	832
World	96	NA	31,000
World Resources: NA			

	Mine Production	Reserves	Reserve Base
ILMENITE (tonnes of contained TiO$_2$)			
Australia	915,000	24,000,000	66,000,000
South Africa (slag)	750,000	36,000,000	45,000,000
Canada (slag)	500,000	27,000,000	73,000,000
Norway (ilmenite + slag)	250,000	32,000,000	90,000,000
C.I.S.	235,000	5,900,000	13,000,000
World	3,200,000	200,000,000	430,000,000
World Resources: 1,000,000,000			
INDIUM (tonnes)			
Canada	40	600	1,500
France	25		
Japan	25	100	150
Belgium	20		
Italy	15		
World	140	2,300	4,600
World Resources: NA			
IODINE (tonnes)			
Japan	7,500	4,000,000	7,000,000
Chile	4,300	900,000	1,200,000
United States	2,000	550,000	550,000
C.I.S.	2,000	400,000	
China	500	400,000	400,000
World	16,400	6,400,000	9,700,000
World Resources: NA			
IRON ORE (thousands of tonnes of crude ore)			
C.I.S.	200,000	63,700,000	78,000,000
Brazil	156,000	11,100,000	17,300,000
China	120,000	9,000,000	9,000,000
Australia	118,000	16,000,000	28,100,000
India	58,000	5,400,000	12,100,000
World	844,500	150,000,000	230,000,000
World Resources: 800,000,000			
KYANITE AND RELATED MINERALS (tonnes)			
South Africa	220,000		
India	55,000		
France	50,000		
World	340,000		
World Resources: NA			
LEAD (tonnes of lead content)			
Australia	525,000	10,000,000	35,000,000
United States	410,000	10,000,000	22,000,000
Canada	330,000	6,000,000	13,000,000
China	320,000	7,000,000	11,000,000
Peru	185,000	2,000,000	3,000,000
World	3,200,000	63,000,000	130,000,000
World Resources: 1,400,000,000			
LIME (tonnes)			
C.I.S.	25,400,000		
China	19,100,000		
United States	16,400,000		
Germany	9,600,000		
Japan (quicklime only)	6,500,000		
Mexico	6,500,000		
World	133,600,000		
World Resources: large			

	Mine Production	Reserves	Reserve Base
LITHIUM (tonnes of contained lithium)			
Chile	1,800	1,270,000	1,360,000
Australia	1,500	372,000	435,000
Canada	400	181,000	363,000
China	310	NA	NA
Brazil	30	907	NA
World	5,600	2,200,000	8,400,000
World Resources: 12,660,000			
MAGNESIUM COMPOUNDS (tonnes)			
China	750,000	745,000,000	1,045,000,000
Korea, North	460,000	445,000,000	745,000,000
C.I.S.	350,000	650,000,000	725,000,000
Austria	320,000	15,000,000	20,000,000
Greece	250,000	30,000,000	30,000,000
World	3,090,000	2,500,000,000	3,400,000,000
World Resources: large			
MAGNESIUM METAL (tonnes)			
United States	135,000		
C.I.S.	75,000		
Norway	30,000		
Canada	20,000		
France	15,000		
World	303,000		
World Resources: large			
MANGANESE (tonnes, gross weight)			
C.I.S.	6,800,000	300,000,000	450,000,000
China	3,400,000	14,000,000	29,000,000
South Africa	2,300,000	370,000,000	4,000,000,000
Brazil	1,800,000	21,000,000	58,000,000
Gabon	1,500,000	52,000,000	160,000,000
World	18,800,000	800,000,000	4,800,000,000
World Resources: large			
MERCURY (tonnes; 1 tonne = 29.0082 flasks)			
C.I.S.	2,000	10,000	17,000
Mexico	700	5,000	9,000
Spain	500	76,000	90,000
Algeria	400	2,000	3,000
Turkey	100	3,000	7,000
World	4,800	130,000	240,000
World Resources: 600,000			
MICA (natural), SCRAP AND FLAKE (tonnes)			
United States	93,000		
C.I.S.	38,000		
Canada	10,000		
Korea, Republic of	5,000		
India	3,000		
World	182,000		
World Resources: large			
MICA (natural), SHEET (tonnes)			
India	3,600		
C.I.S.	2,000		
United States	NA		
World	6,600		
World Resources: large			

	Mine Production	Reserves	Reserve Base
MOLYBDENUM (tonnes of molybdenum content)			
United States	45,400	2,720,000	5,350,000
Peru	18,000	140,000	230,000
C.I.S.	18,000	450,000	680,000
China	15,000	500,000	1,200,000
Chile	14,500	1,130,000	2,450,000
World	108,000	5,500,000	12,000,000
World Resources: 5,000,000			
NATURAL GAS (trillion cubic feet of marketable production; not including flared, vented or reinjected gas; Nigeria flares 76% and Iraq flares 50% of total gas produced)			
Russia	22.60	1,592***	
United States	18.59	169	
Canada	4.14	97	
Netherlands	2.92	69	
United Kingdom	1.95	19	
Algeria	1.62	116	
Saudi Arabia	1.59	184	
Indonesia	1.36	65	
Mexico	1.35	71	
United Arab Emirates	1.16	198	
Venezuela	1.03	110	
Nigeria	.98	104	
Germany	.97	9	
Norway	.87	61	
Iran	.78	600	
Australia	.76	15	
World	89.33	4378	
World Resources: 9258.00			
NICKEL (tonnes of nickel content)			
Russia	225,000	6,600,000	7,300,000
Canada	214,000	6,200,000	14,000,000
New Caledonia	95,000	4,500,000	15,000,000
Australia	70,000	2,200,000	6,800,000
Indonesia	70,000	3,200,000	13,000,000
World	916,000	47,000,000	110,000,000
World Resources: 130,000,000			
NITROGEN (fixed)—AMMONIA (tonnes produced)			
United States	14,500,000		
India	7,600,000		
Canada	3,400,000		
Netherlands	3,120,000		
Mexico	2,550,000		
World	102,100,000		
World Resources: unlimited			
OIL SHALE (tonnes)			
Estonia	23,300,000		
Russia	5,000,000		
World	NA		
World Resources: 1,888,000,000,000 (in situ)			
PEAT (thousands of tonnes)			
C.I.S.	166,000	5,320,000	160,000,000
Ireland	8,500	171,000	900,000
Finland	6,000	70,000	7,000,000
Germany	1,600	46,000	50,000

	Mine Production	Reserves	Reserve Base
Canada	900	24,000	336,000,000
World	186,730	5,700,000	510,000,000
World Resources: 2,100,000,000			

PERLITE (tonnes)
United States	571,000	50,000,000	200,000,000
Greece	230,000	50,000,000	300,000,000
World	2,000,000	700,000,000	2,000,000,000
World Resources: large			

PHOSPHATE ROCK (tonnes)
United States	47,000,000	1,230,000,000	4,440,000,000
C.I.S.	24,000,000	1,330,000,000	1,330,000,000
China	20,000,000	210,000,000	210,000,000
Morocco/W. Sahara	19,000,000	5,900,000,000	21,440,000,000
Tunisia	6,500,000		270,000,000
World	141,000,000	12,000,000,000	34,000,000,000
World Resources: large			

PLATINUM-GROUP METALS (tonnes; 1 tonne = 32,150.7 troy ounces)
South Africa	150	50,000	59,000
C.I.S.	121	5,900	6,000
Canada	11	250	280
United States	7.8	250	780
World	294	56,000	66,000
World Resources: 100,000			

POTASH (thousand tonnes of K_2O equivalent)
C.I.S.	8,200,000	3,600,000,000	3,800,000,000
Canada	7,100,000	4,400,000,000	9,700,000,000
Germany	3,400,000	750,000,000	900,000,000
Israel	1,300,000	53,000,000	600,000,000
France	1,050,000	15,000,000	35,000,000
World	25,035,000	9,400,000,000	17,000,000,000
World Resources: large			

PUMICE AND PUMICITE (tonnes)
Italy	5,000,000	NA	NA
Greece	1,500,000	NA	NA
Spain	900,000	NA	NA
Turkey	450,000	NA	NA
United States	439,000	Large	Large
World	10,400,000	NA	NA
World Resources: large			

QUARTZ CRYSTAL (industrial) (tonnes)
United States	430	Moderate	Moderate

RARE EARTHS (tonnes of rare earth oxide content)
United States	16,000	13,000,000	14,000,000
China	16,000	43,000,000	48,000,000
C.I.S.	8,500	19,000,000	21,000,000
Australia	4,000	5,200,000	5,800,000
India	2,500	1,100,000	1,300,000
World	52,000	100,000,000	110,000,000
World Resources: NA			

RHENIUM (tonnes of rhenium content)
United States	13	386	4,540
Chile	5	1,310	2,540
Peru	5	45	545
Canada	3.5	32	1,540

	Mine Production	Reserves	Reserve Base
C.I.S.	2	594	771
World	29	2,500	10,000
World Resources: NA			

RUTILE (tonnes of contained TiO$_2$)

	Mine Production	Reserves	Reserve Base
Australia	180,000	5,300,000	42,000,000
Sierra Leone	130,000	2,000,000	2,000,000
South Africa	75,000	3,600,000	4,500,000
India	5,000	4,400,000	4,400,000
Sri Lanka	5,000	800,000	800,000
World	410,000	85,000,000	170,000,000
World Resources: 200,000,000			

SALT (tonnes)

	Mine Production	Reserves	Reserve Base
United States	35,500,000		
China	27,200,000		
Germany	14,500,000		
C.I.S.	14,000,000		
Canada	11,000,000		
World	186,000,000		
World Resources: unlimited			

SAND AND GRAVEL (construction) (tonnes)

	Mine Production	Reserves	Reserve Base
United States	732,400,000		
World	NA		
World Resources: large			

SAND AND GRAVEL (industrial)

	Mine Production	Reserves	Reserve Base
Netherlands	24,500,000		
United States	23,700,000		
Argentina	9,000,000		
Germany	6,800,000		
Italy	4,200,000		
World	110,400,000		
World Resources: large			

SELENIUM (tonnes of selenium content)

	Mine Production	Reserves	Reserve Base
Japan	550		
Canada	350	7,000	15,000
United States	270	12,000	19,000
Belgium	260		
Germany	100		
World	1,800	75,000	130,000
World Resources: NA			

SILICON (tonnes)

	Mine Production	Reserves	Reserve Base
C.I.S.	900,000		
China	700,000		
United States	350,000		
Norway	350,000		
Brazil	250,000		
World	3,600,000		
World Resources: large			

SILVER (tonnes; 1 tonne = 32,150.7 troy ounces)

	Mine Production	Reserves	Reserve Base
Mexico	2,000	37,000	40,000
United States	1,800	31,000	72,000
Peru	1,500	25,000	37,000
Canada	1,200	37,000	47,000
C.I.S.	1,200	44,000	50,000
World	13,700	280,000	420,000
World Resources: depends on base metal production			

	Mine Production	*Reserves*	*Reserve Base*
SODA ASH (tonnes)			
United States	9,500,000	23,130,000,000	38,736,000,000
Kenya	245,000	51,000,000	NA
Mexico	190,000	180,000,000	454,000,000
Botswana	150,000	363,000,000	NA
World	32,500,000		
World Resources: large			
SODIUM SULFATE (tonnes)			
Mexico	670,000	165,000,000	227,000,000
Spain	450,000	180,000,000	272,000,000
United States	350,000	857,000,000	1,361,000,000
Canada	300,000	84,000,000	272,000,000
C.I.S.	300,000	1,814,000,000	2,268,000,000
World	4,800,000		
World Resources: large			
STEEL (tonnes of raw steel)			
Japan	99,000,000		
Russia	93,000,000		
United States	84,000,000		
China	78,000,000		
Ukraine	55,000,000		
Germany	44,000,000		
South Korea	27,000,000		
Italy	25,500,000		
Brazil	24,000,000		
France	19,000,000		
United Kingdom	18,000,000		
Canada	12,300,000		
South Africa	8,600,000		
Australia	6,200,000		
World	725,000,000		
World Resources: see iron ore			
STONE (crushed) (tonnes)			
United States	1,050,000,000		
World	NA		
World Resources: unlimited			
STONE (dimension) (tonnes)			
United States	987,000		
World	NA		
World Resources: unlimited			
STRONTIUM (tonnes of contained strontium)			
Mexico	70,000		1,360,000
Turkey	60,000		
Iran	40,000		
Spain	40,000		
Algeria	5,400		
World	238,400	6,800,000	12,000,000
World Resources: 1,000,000,000			
SULFUR (tonnes)			
United States	10,600,000	140,000,000	230,000,000
Canada	7,200,000	158,000,000	330,000,000
C.I.S.	7,000,000	250,000,000	750,000,000
China	5,500,000	100,000,000	250,000,000
Poland	3,000,000	130,000,000	300,000,000
World	52,700,000	1,400,000,000	3,500,000,000
World Resources: 5,000,000,000			

	Mine Production	Reserves	Reserve Base
TALC AND PYROPHYLLITE (tonnes)			
China	2,300,000	Large	Large
Japan	1,300,000	132,000,000	200,000,000
United States	1,071,000	136,000,000	544,000,000
Korea, Republic of	840,000	14,000,000	18,000,000
Brazil	570,000	14,000,000	54,000,000
World	8,961,000	Large	Large
World Resources: NA			
TANTALUM (tonnes of tantalum conent)			
Australia	250	4,500	9,100
Brazil	90	900	1,400
Canada	45	1,800	2,300
Zaire	10	1,800	4,500
Nigeria	2	3,200	4,500
World	410	22,000	35,000
World Resources: NA			
TAR SAND (thousands of barrels of contained oil)			
Canada		200,000,000	1,000,000,000
World	NA		
World Resources: NA			
TELLURIUM (tonnes of tellurium content)			
United States	NA	3,600	6,000
Japan	60		
Canada	12	700	1,500
Peru	8	500	1,600
World	NA	22,000	38,000
World Resources: NA			
THALLIUM (tonnes of thallium content)			
United States	0.5	32	118
World	15.5	380	640
World Resources: 652,200			
THORIUM (tonnes of thorium oxide (ThO$_2$) equivalent)			
United States	NA	158,000	298,000
Australia	NA	300,000	340,000
Canada	NA	100,000	100,000
India	NA	292,000	300,000
Norway	NA	166,000	183,000
World	NA	1,200,000	1,400,000
World Resources: NA			
TIN (tonnes of tin content)			
China	45,000	1,560,000	1,630,000
Brazil	32,000	1,210,000	2,500,000
Indonesia	27,000	750,000	820,000
Malaysia	23,000	1,210,000	1,230,000
Bolivia	18,000	450,000	900,000
World	200,000	8,000,000	10,000,000
World Resources: NA			
TITANIUM AND TITANIUM DIOXIDE (tonnes)			
C.I.S.	20,000	52,000	100,000
Japan	15,500	30,000	320,000
China	2,000	3,000	30,000
World	39,500	110,000	3,750,000
World Resources: 1,200,000,000 (as for ilmenite and rutile)			
TUNGSTEN (tonnes of tungsten content)			
China	25,000	1,050,000	1,400,000
C.I.S.	6,000	280,000	400,000

	Mine Production	Reserves	Reserve Base
Portugal	1,400	26,000	26,000
Austria	1,100	10,000	15,000
Bolivia	1,000	58,000	110,000
World	39,800	2,300,000	3,400,000
World Resources: NA			

URANIUM (tonnes of U$_3$O$_8$; reserves are reasonably assured U$_3$O$_8$ in ore available at less than $130/kg)

	Mine Production	Reserves	Reserve Base
U.S.S.R.***	13,500	no data	
Canada	9,250	360,000	
United States	3,000	375,000	
Niger	2,960	172,720	
Namibia	2,500	100,750	
Australia	2,300	529,000	
France	2,095	37,150	
China	2,000	no data	
South Africa	1,750	344,400	
World	34,582	2,255,000	

VANADIUM (tonnes of vanadium content)

	Mine Production	Reserves	Reserve Base
South Africa	17,100	3,000,000	12,000,000
C.I.S.	9,000	5,000,000	7,000,000
China	6,000	2,000,000	3,000,000
World	32,100	10,000,000	27,000,000
World Resources: 63,000,000			

VERMICULITE (tonnes)

	Mine Production	Reserves	Reserve Base
South Africa	215,000	20,000,000	80,000,000
United States	170,000	25,000,000	100,000,000
C.I.S.	85,000	NA	NA
World	520,000	50,000,000	200,000,000
World Resources: 600,000,000			

YTTRIUM (tonnes of yttrium oxide (Y$_2$O$_3$) content)

	Mine Production	Reserves	Reserve Base
China	430	220,000	240,000
Australia	80	102,000	110,000
India	50	36,000	38,000
Brazil	22	400	1,500
Maylasia	20	13,000	21,000
World	700	500,000	560,000
World Resources: NA			

ZINC (tonnes of zinc content)

	Mine Production	Reserves	Reserve Base
Canada	1,290,000	21,000,000	56,000,000
Australia	1,000,000	17,000,000	65,000,000
China	720,000	5,000,000	9,000,000
Peru	595,000	7,000,000	12,000,000
United States	550,000	16,000,000	50,000,000
World	7,365,000	140,000,000	330,000,000
World Resources: 1,800,000,000			

ZIRCONIUM (tonnes)

	Mine Production	Reserves	Reserve Base
Australia	300,000	23,300,000	25,900,000
South Africa	200,000	2,100,000	2,400,000
United States	100,000	3,200,000	3,600,000
C.I.S.	80,000	4,000,000	6,000,000
Brazil	20,000	1,000,000	1,100,000
India	20,000	2,100,000	2,400,000
World	765,000	49,000,000	58,000,000
World Resources: >60,000,000			

*C.I.S. reserves apportioned from U.S.S.R. reserve figure on the basis of present production data
**1989 production (before disruption caused by Gulf War)
***Russian reserve estimate based on proportion of total U.S.S.R. production obtained from Russia.

Appendix II

Minerals, Rocks, Geologic Time and Composition of the Crust

TABLE AII-1

Classification of selected minerals based on anions (oxidation state of sulfur in sulfide minerals can be more complex than shown here). Abundant minerals shown in bold type. Some mineral elements or subscripts are indicated by W, X and Y, as defined in the table, and more complete mineral compositions for some ore minerals are given in Appendix III.

Name	Anion	Common Example
Native Elements	None	Gold-Au
Sulfides	S^{-2}, S^{-1}	Galena-PbS
Oxides/Hydroxides	O^{-2}, OH^-	**Magnetite-Fe_3O_4**
Halides	$F^-, Cl^-,$	**Halite-NaCl**
	Br^-, I^-	Sylvite-KCl
Carbonates, Nitrates	$CO_3^{-2}, NO_3^-,$	**Calcite-$CaCO_3$**
Borates	BO_2^{-4}, BO_4^{-5}	**Dolomite-$CaMg(CO_3)_2$**
Sulfates, Chromates,	XO_4^{-2}	**Gypsum-$CaSO_4 \cdot 2H_2O$**
Molybdates, Tungstates		Barite-$BaSO_4$
Phosphates, Arsenates	$PO_4^{-3}, AsO_4^{-3}, AsO_4^{-5}$	Apatite-$Ca_5(PO_4)_3(F,OH)$
Silicates	$Si_aO_b^{-c}$	**Quartz-SiO_2**
(W=Ca,Na; X=Mg,Fe; Y=Al,Fe;		**Kaolinite-$Al_4Si_4O_{10}(OH)_8$**
Z=Si,Al)		**Montmorillonite-**
		$Al_2Si_4O_{10}(OH)_2 \cdot xH_2O$
		Felsic Silicates
		Plagioclase Feldspar-$WZSi_2O_8$
		Potassium Feldspar-$KAlSi_3O_8$
		Muscovite-$KAl_2(AlSi_3O_{10})(OH)_2$
		Mafic Silicates
		Biotite-$KX_3(AlSi_3O_{10})(OH)_2$
		Amphibole-$(W,X,Y)_{7-8}(Z_4O_{11})_2(OH)_2$
		Pyroxene-$(W,X,Y)_2Z_2O_6$
		Olivine-X_2SiO_4

TABLE AII-2
Classification and compositions of major rock types (parentheses indicate whether rock is felsic or mafic, as defined in text).

Rock Type	Main Minerals, Grains or Characteristics
Igneous (Volcanic/Plutonic)[a]	
Rhyolite/Granite[a] (Felsic)	Quartz, Potassium Feldspar, Plagioclase Feldspar, Biotite, Muscovite
Latite/Syenite (Felsic)	Potassium Feldspar, Plagioclase Feldspar, Biotite, Hornblende
Andesite/Tonalite (Intermediate)	Plagioclase Feldspar, Biotite, Hornblende
Basalt/Gabbro (Mafic)	Olivine, Pyroxene, Plagioclase Feldspar
Komatiite/Peridotite/Kimberlite (Ultramafic)	Olivine, Pyroxene
Sedimentary	
Clastic	
Conglomerate	Gravel-sized grains (>2 mm)
Sandstone	Sand-sized grains (0.0625 to 2 mm)
Quartz Sandstone	Largely quartz grains
Graywacke	Largely rock grains
Arkose	Largely feldspar grains
Shale	Silt-clay-sized grains (<0.0625 mm)
Chemical/Organic	
Limestone	Largely calcite, often with fossils
Dolomite	Largely dolomite
Chert	Quartz and amorphous silica
Rock Salt	Halite
Gypsum	Gypsum
Coal	Plant fragments, humic material
Metamorphic	
Greenstone, Amphibolite	Metamorphosed basalt or gabbro
Marble	Metamorphosed limestone or dolomite
Slate	Metamorphosed shale
Quartzite	Metamorphosed quartz sandstone

[a]*Pumice* and *obsidian* are igneous rocks of granitic composition that consist entirely of glass; they did not cool slowly enough to form crystals. *Pegmatite* is an intrusive igneous rock with unusually large crystals. *Carbonatite* is an igneous rock (usually intrusive) made up largely of calcite. *Serpentinite* is altered ultramafic rock, commonly seen in obduction zones and along faults. *Schist* and *gneiss* are metamorphic rocks named for their textures rather than their compositions. Schist consists of many planar minerals that have the same orientation known as foliation; gneiss consists of layers of schist separated by layers of more equant minerals such as quartz and feldspar. Schist commonly forms from sedimentary rocks and gneiss forms from igneous rocks.

TABLE AII-3

Divisions of geologic time (from Palmer, 1983).

Eon	Era	Period	Epoch	Age Range (Ma)
Phanerozoic	Cenozoic	Quaternary	Holocene	0–0.01
			Pleistocene	0.01–1.6
		Tertiary	Pliocene	1.6–5.3
			Miocene	5.3–23.7
			Oligocene	23.7–36.6
			Eocene	36.6–57.8
			Paleocene	57.8–66.4
	Mesozoic	Cretaceous		66.4–144
		Jurassic		144–208
		Triassic		208–245
	Paleozoic	Permian		245–286
		Pennsylvanian		286–320
		Mississippian		320–360
		Devonian		360–408
		Silurian		408–438
		Ordovician		438–505
		Cambrian		505–570
Proterozoic				570–2,500
Archean				2,500–3,800
Hadean				3,800–4,650

TABLE AII-4

Average composition of continental crust rocks (after Brookins, 1990) showing symbol for element, rank with respect to silicon (the most abundant element in the crust), and content in wt% or ppm.

Aluminum	Al 3	8.23%	Helium	He 78	0.003 ppm	Rhodium	Rh 75	0.005 ppm
Antimony	Sb 63	0.2 ppm	Holmium	Ho 56	1.5 ppm	Rubidium	Rb 22	90 ppm
Argon	Ar 70	0.04 ppm	Hydrogen	H 10	0.14%	Ruthenium	Ru 74	0.01 ppm
Arsenic	As 52	1.8 ppm	Indium	In 66	0.1 ppm	Samarium	Sm 39	7.3 ppm
Barium	Ba 14	425 ppm	Iodine	I 60	0.5 ppm	Scandium	Sc 31	25 ppm
Beryllium	Be 47	2.8 ppm	Iridium	Ir 80	0.001 ppm	Selenium	Se 69	0.05 ppm
Bismuth	Be 65	0.17 ppm	Iron	Fe 4	5.63%	Silicon	Si 2	28.15 ppm
Boron	B 37	10 ppm	Lanthanum	La 29	33 ppm	Silver	Ag 68	0.07 ppm
Bromine	Br 49	2.5 ppm	Lead	Pb 36	15 ppm	Sodium	Na 6	2.36 ppm
Cadmium	Cd 64	0.2 ppm	Lithium	Li 33	20 ppm	Strontium	Sr 15	375 ppm
Calcium	Ca 5	4.15%	Lutetium	Lu 59	0.8 ppm	Sulfur	S 16	260 ppm
Carbon	C 17	200 ppm	Magnesium	Mg 7	2.33%	Tantalum	Ta 50	2 ppm
Cerium	Ce 25	67 ppm	Manganese	Mn 12	950 ppm	Tellurium	Te 73	0.01 ppm
Cesium	Cs 45	3 ppm	Mercury	Hg 67	0.08 ppm	Terbium	Tb 58	1.1 ppm
Chlorine	Cl 20	130 ppm	Molybdenum	Mo 55	1.5 ppm	Thallium	Tl 61	0.45 ppm
Chromium	Cr 21	100 ppm	Neodymium	Nd 28	28 ppm	Thorium	Th 38	9.6 ppm
Cobalt	Co 30	28 ppm	Nickel	Ni 23	75 ppm	Thulium	Tm 62	0.25 ppm
Copper	Cu 26	55 ppm	Niobium	Nb 34	20 ppm	Tin	Sn 51	2 ppm
Dysprosium	Dy 42	5.2 ppm	Nitrogen	N 32	25 ppm	Titanium	Ti 9	0.57 ppm
Erbium	Er 46	2.8 ppm	Osmium	Os 76	0.005 ppm	Tungsten	W 54	1.5 ppm
Europium	Eu 57	1.2 ppm	Oxygen	O 1	46.4%	Uranium	U 48	2.7 ppm
Fluorine	F 13	625 ppm	Palladium	Pd 71	0.01 ppm	Vanadium	V 19	135 ppm
Gadolinium	Gd 40	7.3 ppm	Phosphorus	P 11	0.105%	Ytterbium	Yb 43	3 ppm
Gallium	Ga 35	20 ppm	Platinum	Pt 72	0.01 ppm	Yttrium	Y 27	33 ppm
Germanium	Ge 53	1.5 ppm	Potassium	K 8	2.09%	Zinc	Zn 24	70 ppm
Gold	Au 77	0.00 ppm	Praseodmium	Pr 41	6.5 ppm	Zirconium	Zr 18	165 ppm
Hafnium	Hf 44	3 ppm	Rhenium	Re 79	0.001 ppm			

Appendix III

Ore Minerals and Materials

TABLE AIII-1

Elements commonly recovered from natural ores showing the main ore minerals and their content of the desired element (in weight percent). Some of these minerals are also used in mineral form without processing, as noted in Table AIII-2.

Element Obtained	Symbol	Ore Minerals	Composition	Maximum Content of Desired Element
Aluminum	Al	Bauxite		
		Boehmite	$Al_2O_3 \cdot H_2O$	43
		Diaspore	$Al_2O_3 \cdot H_2O$	43
		Gibbsite	$Al_2O_3 \cdot H_2O$	34
		Alunite	$KAl_3(SO_4)_2(OH)_6$	26
		Nepheline	$NaAlSiO_4$	19
Antimony	Sb	Stibnite	Sb_2S_3	72
		Tetrahedrite	$Cu_8Sb_2S_7$	25
Arsenic	As	Arsenopyrite	FeAsS	46
		Realgar	AsS	70
		Orpiment	As_2S_3	61
		Löllingite	$FeAs_2$	73
		Smaltite	$CoAs_2$	72
		Enargite	Cu_3AsS_4	19
Barium	Ba	Barite	$BaSO_4$	59
Beryllium	Be	Beryl	$Be_3Al_2Si_6O_8$	5
		Bertrandite	$Be_4Si_2O_7(OH)_2$	15
Bismuth	Bi	Native bismuth	Bi	100
		Bismuthinite	Bi_2S_3	81
Boron	B	Borax	$Na_2B_4O_7 \cdot 10H_2O$	12
		Kernite	$Na_2B_4O_7 \cdot 4H_2O$	16
		Colemanite	$Ca_2B_6O_{11} \cdot 5H_2O$	16
Cadmium	Cd	Greenockite	CdS	78
		Sphalerite	(Zn,Cd)S	<1
Cesium	Cs	Pollucite	$(Cs,Na)AlSi_2O_6$	43
Chromium	Cr	Chromite	$(Mg,Fe)(Cr,Al,Fe)_2O_4$	44
Cobalt	Co	Linneaite	Co_3S_4	58
		Cobaltite	(Co,Fe)AsS	35
		Pyrrhotite	$(Fe,Ni,Co)_{1-x}S_x$	<1
Copper	Cu	Native copper	Cu	100
		Chalcopyrite	$CuFeS_2$	35
		Chalcocite	Cu_2S	80
		Bornite	Cu_5FeS_4	63
		Enargite	Cu_3AsS_4	48
Fluorine	F	Fluorite	CaF_2	49
		Cryolite	Na_3AlF_6	54
Gold	Au	Native gold	Au	100
		Electrum	(Au,Ag)	variable
		Calaverite	$AuTe_2$	44
Hafnium	Hf	Zircon	$(Zr,Hf)SiO_4$	<2
Iron	Fe	Magnetite	Fe_3O_4	72
		Hematite	Fe_2O_3	70
		Geothite	$Fe_2O_3 \cdot H_2O$	60
		Siderite	$FeCO_3$	48
		Pyrite	FeS_2	47

Element Obtained	Symbol	Ore Minerals	Composition	Maximum Content of Desired Element
Lead	Pb	Galena	PbS	87
Lithium	Li	Spodumene	$LiAlSi_2O_6$	8
		Lepidolite	$KLi_2AlSi_4O_{10}F_2$	8
		Petalite	$LiAlSi_4O_{10}$	5
Magnesium	Mg	Olivine	$(Fe,Mg)_2SiO_4$	34
		Magnesite	$MgCO_3$	28
		Dolomite	$CaMg(CO_3)_2$	13
Manganese	Mn	Braunite	$(Mn,Si)_2O_3$	70
		Pyrolusite	MnO_2	63
		Psilomelane	$BaMn_9O_{18} \cdot 2H_2O$	52
Mercury	Hg	Native mercury	Hg	100
		Cinnabar	HgS	86
Molybdenum	Mo	Molybdenite	MoS_2	60
Nickel	Ni	Garnierite	$(Mg,Ni)_6Si_4O_{10}(OH)_8$	47
		Pentlandite	$(Fe,Ni)_9S_8$	36
		Pyrrhotite	$(Fe,Ni,Co)_{1-x}S_x$	<2
Niobium(Columbium)	Nb-Ta	Columbite	$(Fe,Mn)Nb_2O_6$	66
Phosphorus	P	Apatite	$Ca_5(PO_4)_3(F,Cl,OH)$	17
Platinum group				
Platinum	Pt	Sperrylite	$PtAs_2$	57
Palladium	Pd	Froodite	$PdBi_2$	20
Iridium	Ir	Laurite	$(Ru,Ir,Os)S_2$	61(75,75)
Osmium	Os			
Rhodium	Rh			52
Ruthenium	Ru			
Potassium	K	Sylvite	KCl	14
		Carnallite	$KCl \cdot MgCl_2 \cdot 6H_2O$	
Rare-earth metals				
Yttrium	Y	Monazite	$(Ce,La,Th,Y)PO_4$	
Important lanthanide		Bastnäsite	$CeFCO_3$	64
elements		Xenotime	YPO_4	48
Cerium	Ce			
Europium	Eu			
Lanthanum	La			
Rhenium	Re	Molybdenite	$(Mo,Re)S_2$	<1
Selenium	Se	Copper ores		
Silicon	Si	Quartz	SiO_2	46
Silver	Ag	Native silver	Ag	100
		Argentite	Ag_2S	87
		Argentiferous galena	$(Pb,Ag,Bi,Sb)S$	
		Argentiferous tennantite-tetrahedrite	$(Cu,Fe,Ag)_{12}(As,Sb)_4S_{13}$	
Sodium	Na	Halite	$NaCl$	39
Strontium	Sr	Celestite	$SrSO_4$	48
Sulfur	S	Native sulfur	S	100
		Gypsum	$CaSO_4$	24
		Pyrite	FeS_2	47
Tantalum	Ta	Tantalite	$(Fe,Mn)Ta_2O_6$	79
		Microlite	$(Na,Ca)_2Ta_2O_6(OH)$	69
Tellurium	Te	Calaverite	$AuTe_2$	56
		Copper ores	—	trace levels
Thorium	Th	Monazite	$(Ce,La,Th,Y)PO_4$	71
Tin	Sn	Cassiterite	SnO_2	79
Titanium	Ti	Rutile	TiO_2	61
		Ilmenite	$FeTiO_3$	32
Tungsten	W	Wolframite	$(Fe,Mn)WO_4$	76
		Scheelite	$CaWO_4$	80

Element Obtained	Symbol	Ore Minerals	Composition	Maximum Content of Desired Element
Uranium	U	Uraninite	UO_2	88
		Coffinite	$U(SiO_4)_{1-x}(OH)_{4x}$	60
		Carnotite	$K_2(UO_2)_2(VO_4)_2 \cdot 3H_2O$	57
Vanadium	V	Carnotite	$K_2(UO_2)_2(VO_4)_2 \cdot 3H_2O$	12
Zinc	Zn	Sphalerite	ZnS	67
		Willemite	Zn_2SiO_4	56
Zirconium	Zr	Zircon	$(Zr,Hf)SiO_4$	49 (<2)

TABLE AIII-2

Ore minerals and other solid, natural materials commonly used in mineral or natural form, with major uses for each. Some of these minerals are also processed for specific elements as noted previously and others, such as trona, sylvite, and halite, undergo some chemical processing before actual use.

Ore Mineral	Composition	Major Uses
Anhydrite (see gypsum)	—	—
Asbestos		
Chrysotile	$Mg_6Si_4O_{10}(OH)_8$	Filler in cement and plaster
Barite	$BaSO_4$	Oil drilling "mud", glass, filler
Boron minerals		Glass and chemical industries
Calcite (limestone, marble)	$CaCO_3$	Fillers; production of lime (CaO) and other calcium chemicals
Chromite	$(Mg,Fe)(Cr,Al,Fe)_2O_4$	Refractory compounds
Clays		Construction materials, fillers, carriers, and extenders
Kaolinite	$Al_4Si_4O_{10}(OH)_8$	
Montmorillonite (fuller's earth)	$Al_2Si_4O_{10}(OH)_2 \cdot xH_2O$	
Diamond	C	Abrasive
Diatomite (microscopic shell of aquatic organism)	$SiO_2 \cdot nH_2O$	Filler; also used extensively to filter liquids
Feldspar		Ceramics, abrasive, filler
Microcline	$KAlSi_3O_8$	
Plagioclase	$(Na,Ca)Al_2Si_2O_8$	
Fluorite	CaF_2	Blast furnace flux
Graphite	C	Refractories, lubricant, electrical equipment
Gypsum (anhydrite)	$CaSO_4 \cdot 2H_2O$; $(CaSO_4)$	Plaster, cement
Kyanite	Al_2SiO_5	Refractory compounds
Limestone (see calcite)	—	—
Magnesite	$MgCO_3$	Refractory compounds
Mica		Filler, electrical insulation
Muscovite	$KAl_2(AlSi_3O_{10})(OH)_2$	
Phlogopite	$KMg_3(AlSi_3O_{10})(OH)_2$	
Vermiculite	$Mg_3Si_4O_{10}(OH)_2 \cdot xH_2O$	Thermal insulation, packaging
Ocher, umber, sienna	Iron oxide with variable amounts of silica, alumina, manganese oxide, and water	Pigment
Phosphate minerals		Fertilizer, phosphorus chemicals
Apatite	$Ca_5(PO_4)_3(F,Cl,OH)$	
Potassium minerals		Fertilizer, potassium chemicals
Sylvite	KCl	
Carnallite	$KCl \cdot MgCl_2 \cdot 6H_2O$	
Pyrophyllite	$Al_2Si_4O_{10}(OH)_2$	Filler
Silica sand		
Quartz	SiO_2	Glass

Ore Mineral	Composition	Major Uses
Sodium minerals		Sodium chemicals
Halite (salt)	$NaCl$	
Trona	$Na_2CO_3 \cdot NaHCO_3 \cdot 2H_2O$	
Nahcolite	$NaHCO_3$	
Talc	$Mg_3Si_4O_{10}(OH)_2$	Filler
Titanium dioxide minerals		Pigment
Rutile	TiO_2	
Ilmenite	$FeTiO_3$	
Wollastonite	$CaSiO_3$	Filler
Zeolites	A large family of complex hydrous sodium, calcium, aluminium silicates	Ion-exchange medium for water purification, etc.

Appendix IV

Units and Conversion Factors

TABLE AIV-1
Units and Conversion Factors

Weight

1 tonne (t)	= 1 metric ton (mt)
	= 1000 kilograms (kg)
	= 1,000,000 grams (g)
	= 32150.7 troy ounces (t oz)
	= 1.102 short tons (st)
	= 0.84 long tons (lt)
	= 2024.6 pounds (lb)
kilogram	= 1000 grams (g)
	= 1,000,000 milligrams (mg)
	= 2.0246 pounds (lb)
1 troy ounce	= 31.1 grams (g)
	= 1.09714 avoirdupois ounces (oz)
1 short ton	= 2000 pounds (lb)
1 long ton	= 2240 pounds (lb)
1 pound	= 453.6 grams (g)
1 flask (mercury)	= 76 pounds (lb)
1 carat (diamond)	= 200 milligrams (mg)
1 karat (gold)	= 1 twenty-fourth part

Distance

1 kilometer	= 1000 meters (m)
	= 100,000 centimeters (cm)
	= 1,000,000 millimeters (mm)
	= 0.62 miles (mi)
1 meter	= 100 centimeters (cm)
	= 3.281 feet (ft)
1 mile	= 5280 feet (ft)
	= 1.609 kilometers (km)
1 foot	= 12 inches (in)
	= 30.48 centimeters (cm)
1 nautical mile	= 1.152 miles (statute miles) (mi)

Area

1 hectare	= 10,000 square meters (m^2)
	= 0.1 square kilometers (km^2)
	= 2.471 acres (ac)
1 square mile	= 640 acres (ac)
	= 259 hectares (hc)
1 acre-foot	= 43,560 cubic feet (ft^3)
	= 325,851 gallons (gal)

Volume

1 liter	= 1000 cubic centimeters (cc)
	= 0.264 U.S. gallons (U.S. gal)
	= 1.0566 quarts (qt)
1 cubic meter	= 35.32 cubic feet (ft^3)
	= 264.2 U.S. gallons (U.S. gal)
1 quart	= 0.9464 liters (l)

1 U.S. gallon	= 4 quarts (qt)
	= 3.785 liters (l)
1 barrel (bbl)	= 31.5 U.S. gallons (gal)
1 Imperial gallon	= 4 Imperial quarts (Imp qt)
	= 4.546 liters (l)
1 Imperial barrel	= 36 Imperial gallons (Imp gal)
1 barrel (oil)	= 42 U.S. gallons (U.S. gal)
	= 0.15899 cubic meters (m^3)
1 barrel (cement)	= 170.5 kilograms
	= 345 pounds

Energy

1 kilojoule	= 1000 joules (J)
	= 239 gram-calories (g-cal)
1 calorie	= 3.9685×10^{-3} British thermal units (Btu)
1 kilowatt-hour	= 3.6×10^6 joules
	= 3412.9 Btu
1 Btu	= 1.055056 kilojoules (kJ)
1 kilowatt	= 3412.9 Btu per hour (Btu/hr)
	= 1000 joule/second (J/sec)
	= 14.34 calories/minute (cal/min)

Energy Equivalents

1 tonne of coal	= 27,200,000 kilojoules
1 bbl oil	= 6,000,000 kilojoules
1000 cubic feet (natural gas)	= 1,278,000 kilojoules

Glossary

abiogenic gas natural gas formed by deep-earth processes not involving organic matter

accuracy degree to which an analysis approaches the correct number

acid mine drainage acidic water, usually from the oxidation of pyrite, which drains from areas disturbed by mining

acid rain acidic rain, usually caused by the dissolution of CO_2, SO_2, and other gases

activated charcoal charcoal or similar forms of carbon with a surface that has been treated to enhance its capacity to adsorb ions or molecules

ad valorem tax tax imposed on the value of property or goods

adit a roughly horizontal tunnel with one open end, used as an entrance to a mine

adsorption the process by which liquids, gases, or dissolved substances attach to the surface of solids; compare to *ion exchange*

aerobic referring to an environment in which oxygen is present

aggregate natural and synthetic rock and mineral material used separately and as a filler in cement, asphalt, plaster, and other materials

alluvium unconsolidated sediment deposited by streams (contrast with *colluvium*)

alluvial fan a fan-shaped deposit of alluvium

alpha particle the nucleus of a helium atom consisting of two protons and two neutrons; one of the products of radioactive decay

alumina Al_2O_3, the intermediate product in aluminum production

amalgamation (1) the process by which adjacent mining operations are joined to make a larger one; (2) the process by which gold is dissolved from its ore by mercury, which is then separated from the dissolved gold by heating to drive off mercury vapor

amino acid organic acids with carboxyl ($COHOH^-$) and amino (NH_2^-) groups

amorphous without crystal structure, as in a glass

amortization payment or deductions related to repayment of debt

andesite an intermediate igneous (volcanic) rock

anaerobic referring to a lack of oxygen

anion a negatively charged ion

anomalous samples that differ significantly from all others in a group or population

anorthosite an intrusive igneous rock consisting largely of plagioclase feldspar

anthracite high-grade coal with a very high C:H ratio

anthropogenic formed or contributed by human activity rather than natural Earth processes

aqueous consisting of water or pertaining to water

aquifer rock or regolith through which groundwater moves easily (i.e., with high porosity and permeability)

Archean the eon of geologic time between about 3.8 and 2.5 billion years (Ga)

artesian well a well that flows naturally, without pumping

asbestos fibrous, acicular, or thread-like minerals with minimum length:diameter ratio of 3, usually serpentine asbestos (chrysotile) and amphibole asbestos (crocidolite, amosite)

asbestosis chronic lung ailment resulting from the inhalation of asbestos

assay analysis of the grade of ore or concentrate

asthenosphere a zone of the mantle immediately below the lithosphere, beginning at a depth of 100 km and extending downward for at least 200 km

atmosphere the shell of gas and water vapor that surrounds Earth

atmospheric aerosol particles of liquids and solids suspended in the atmosphere

atom the smallest particle of an element that can enter into a chemical combination

atomic number number of protons in the nucleus of an atom

atomic weight weight of an atom (usually in relation to oxygen as 16), determined largely by number of protons and neutrons

background a term used to refer to values characteristic of the average or most common sample in a population (see *anomalous*)

backwardation future price less expensive than cash (spot, present) price

bacterial degradation changes in organic matter caused by bacterial activity, including formation of kerogen

balance statement statement of income and expenses provided by a company or other financial entity

banded iron formation (BIF) chemical sediment with alternating layers of iron and silica or silicate minerals; also called taconite, itabirite, jaspilite, or ironstone

basalt mafic, extrusive igneous rock, commonly formed at mid-ocean ridge spreading centers; formed by partial melting of mantle

base metals copper, zinc, lead, tin; so-named by ancient alchemists who could not convert them to gold

basement rock older rocks that underlie sedimentary basins and other younger rocks

basic oxygen furnace steelmaking furnace with a jet of pure oxygen

basin analysis evaluation of the likelihood that a sedimentary basin contains crude oil, natural gas, or other mineral deposits

batholith irregularly-shaped, large body of intrusive igneous rock, commonly measuring tens of kilometers in length

bauxite ore from which aluminum is obtained; consists of hydrous aluminum oxide, clay, and silica minerals

bedrock solid rock that underlies softer rocks or unconsolidated material near the surface

bench step-like zones around the walls of an open pit that prevent rockfalls and collapse

beneficiation physical separation of an ore mineral from its ore by crushing, grinding, froth flotation, and other methods

berylliosis chronic lung disease caused by the inhalation of beryllium compounds

beta particle negatively charged particle identical to electron or its positively charged equivalent; radioactive decay product

bioaccumulation accumulation of an element or compound in living organisms; extreme bioaccumulation is biomagnification

biosphere all living and dead organic matter on Earth

bittern salts salts that crystallize from highly evaporated seawater, rich in potassium and magnesium

bitumen semisolid organic material similar in composition to heavy crude oil

bituminous coal intermediate-grade coal; the most commonly used type

black lung disease (see *coal worker pneumoconiosis*)

black smoker submarine hot spring emitting water that contains suspended black particles of sulfide minerals precipitated by rapid cooling of the water

blast furnace furnace used to convert iron ore to pig iron

blister copper copper produced by smelting; commonly contains gold, silver, and other metals that must be removed in a refinery

block caving underground mining method involving the collapse of large blocks of ore that are withdrawn through tunnels below the blocks

blow-out preventer large valve on oil, gas, and other deep-drilling rigs to prevent the escape of pressurized brines and other fluids

body fluid water, blood, and other natural fluids

boil to change state from liquid to gas at the boiling point

boiling water reactor a type of nuclear reactor that uses boiling water as the working fluid

bonanza very high-grade ore, especially of gold and silver

bonus bid nonrefundable (money) bid for the right to explore a tract of land

bottom ash ash from coal combustion, which falls to the bottom of the combustion chamber

brass an alloy of copper and zinc

breccia a rock consisting of angular fragments in a matrix of finer-grained rock or chemically precipitated minerals

breeder reactor a nuclear reactor that produces more fissionable fuel than it consumes (see *converter reactor*)

brine an aqueous solution with a salinity higher than that of seawater (about 3.5% by weight)

bronze an alloy of copper and tin

bullion strictly speaking, metal with a high content of gold or silver; sometimes used to refer to lead metal

by-product in mineral resources, any secondary product that results from the production of the primary one; examples are molybdenum from some copper production and natural gas from some oil production

calcine to heat and drive off a gas that is part of a compound; often used for process by which CO_2 is driven off from limestone ($CaCO_3$) to form lime (CaO)

CANDU reactor a type of nuclear reactor that uses unenriched fuel

caprock (1) impermeable rock that seals the top of some oil and gas traps; (2) gypsum-rich insoluble residue at the top of a rising salt dome

carbon tax a tax on CO_2 generated during the combustion of fossil fuels

carbonate pertaining to minerals or rocks containing CO_3^{-2}, such as calcite, dolomite, and limestone

carbonatite an igneous rock composed largely of calcite and dolomite

cartel organization designed to control the availability and price of a commodity

casing pipe used to line a drill hole to prevent collapse of the walls and escape or entry of unwanted fluids

catagenesis geologic process intermediate between diagenesis and metamorphism, by which oil and gas are formed

catalyst a substance that enhances the rate of a chemical reaction without participating in it

cation a positively charged ion

cellulose a carbohydrate that is the main constituent of the cell walls of woody plants

cement (1) any new mineral that is precipitated from aqueous solution in pores in a rock; (2) a mixture of powdered lime, clay, and other minerals that crystallizes to form a hard solid when water is added (hydraulic cement); (3) the binding material in concrete

Cenozoic the era of geologic time from 66.4 million years (Ma) to the present

chelate a compound with a ring-like structure, in which a central cation is connected to each surrounding anion by at least two bonds; commonly seen in organometallic molecules

chemical bond attraction of one atom, ion, or molecule to another, usually by sharing electrons

chert sedimentary rock consisting of fine-grained silica

chimney a vertical tube-shaped ore body, usually lead-zinc skarn

chromophores elements or compounds that create color in minerals

claim a parcel of land obtained under a law that permits public land to be obtained for mineral exploration and extraction; in the United States claims can be for bedrock deposits (lode) or alluvial deposits (placer)

clastic sediment sediment consisting of fragments of minerals, rocks, and other material; when lithified, forms clastic sedimentary rocks such as conglomerate, sandstone, and shale

clay (1) any of several hydrous aluminum silicate minerals, including kaolinite and smectite; (2) clastic material smaller than $\frac{1}{256}$ mm

clinker agglomeration of calcium silicate and other compounds formed by calcining limestone, clay, and other minerals in a cement kiln

coal combustible rock consisting largely of the partly decomposed remains of land plants

coal worker pneumoconiosis lung disease caused by the inhalation of coal dust (also known as black lung disease)

coalification the process by which dead land plants are converted to coal

coating clay kaolinite and other clays used to coat paper

coking coal coal used to make coke for steel plants and other industrial uses

colloid a suspension of particles with a diameter between 10^{-9} and 10^{-6} meters in size

colluvium regolith that has not been moved (contrasted with *alluvium*)

commodity future a contract for sale or purchase of a commodity at a set date for a set price

complex ion an ion that contains more than one atom

compound a substance containing more than one element and having properties different from its constituent elements

concentrate a product in which the proportion of ore mineral has been increased significantly above its concentration in ore, usually by crushing, grinding, and mineral separation by flotation or some other method

concession the right to explore for minerals or produce them from a parcel of land; commonly granted by governments

concrete   an artificial rock consisting of a combination of cement and aggregate

condensation   the process by which a gas changes to a liquid

conglomerate   a coarse-grained, clastic sedimentary rock containing rounded to semirounded pebbles and cobbles

continental crust   (see *crust*)

continuous miner   machine that advances as it digs rock in an open-pit or underground mine and transports it away from the mining face

continuous smelting   smelting process that includes roasting, matte production, and conversion to metal in a single continuous process

contour mining   a form of open-pit mining in which a flat-lying layer in hilly terrane is mined from a bench cut around the side of a hill

convergent margin   a boundary where two lithospheric plates of Earth move together; to accommodate this movement, one plate will subduct or obduct

converter   the last step in the conventional smelting of copper and other metals, involving the production of metal by blowing air through molten metal sulfide (matte)

converter reactor   a nuclear reactor that produces some fissionable material, but not enough to fuel another similar reactor

core   (1) central part of Earth, consisting largely of iron; (2) a cylinder of rock obtained by drilling

cost depletion   (see *depletion allowance*)

cotango   future price more expensive than present (spot, cash) price

cracking   the second step in the conventional refining of crude oil, involving breaking large organic molecules into smaller ones

craton   unusually thick, stable continental crust

crude oil   liquid consisting largely of organic molecules containing 5 to 30 carbon atoms

crust   outermost shell of Earth; includes continental crust (granitic rocks and sediments) with thickness of 40 to 50 km, and ocean crust (basalt and other mafic rocks) with thickness of 5 to 10 km

crustal abundance   average concentration of an element in the continental crust

crystal   solid with a regular arrangement of atoms or ions

crystal fractionation   process by which crystals sink or float away from their parent magmas, commonly changing composition of residual magma

crystal lattice   the orderly arrangement of atoms in a crystal

cullet   waste glass, commonly recycled

cumulate   a textural term referring to minerals that appear to have accumulated in layers within layered igneous complexes

cupola   protrusion on the top of an igneous intrusion; magmatic fluids and mineral deposits are thought to concentrate around cupolas

cut-off grade   minimum grade of ore that can be extracted profitably

cuttings   rock fragments cut by bit when rock is drilled

cyclone   a device that separates particles from fluids by centrifugal force

delta   deposit of clastic sediment where a stream enters a lake, river, or ocean

dental amalgam   an alloy of mercury, silver, and tin used to fill cavities in teeth

depleted uranium   uranium from which most daughter products of radioactive decay have been removed

depletion allowance   deduction from taxable income derived from mineral production, reflecting future exhaustion of deposit; variants include cost depletion (fraction of cost of property), percentage depletion (percentage of gross income), and earned depletion (fraction of funds expended in exploration)

detection limit   the level below which an analytical method cannot detect the element or compound of interest

deuterium   isotope of hydrogen with one neutron and one proton

development   the process of preparing a mineral deposit for production

diagenesis   the chemical and physical processes that occur in sediments shortly after deposition

diatom   single-celled, siliceous aquatic plant related to algae

diorite   mafic to intermediate intrusive igneous rock

direct reduction   production of iron without complete melting; produces sponge iron

discount rate   interest or inflation rate used to calculate present value of future income

dispersion   outward migration of ions, elements, compounds, minerals, rocks, or other substances from a relatively concentrated source

distill   to boil and then condense a liquid, usually to separate it from other liquids or solids

dividend   that part of after-tax income paid to owners of a business as a return on their investment

divergent margin   a boundary where two lithospheric plates move apart, commonly a mid-ocean ridge

doré   gold-silver bullion that must be purified in a refinery

dowser   a person who attempts to locate water by extrasensory measurements at the surface

dredging   open-pit mining done from a barge that floats in a flooded pit digging its way along through regolith

drill   machine used to bore holes into rock for exploration or production; it uses a drill bit to cut through the rock

drill stem test   test performed on fluid reservoirs intersected during exploration drilling to determine whether they can be produced profitably

drilling mud   high-density slurry of pulverized barite, bentonite clay, and other minerals and compounds used to wash rock drill cuttings to the surface

drive   natural pressure of rocks, water, and gas forcing oil from a reservoir toward the surface

dry deposition   deposition of minerals or compounds from the atmosphere

dumping   selling a commodity at a price below its real cost

dunite   ultramafic igneous rock consisting largely of olivine

earned depletion   (see *depletion allowance*)

electric arc furnace   steelmaking furnace that is used largely with scrap iron or sponge iron

electrolytic   (see *electrowinning*)

electron   very small, negatively-charged particles that surround the nucleus of an atom, balancing the charge of the protons

electrowinning   concentration of one or more dissolved ions onto electrodes placed in a solution; basis for hydrometallurgy; commonly referred to as solvent extraction-electrowinning (SX-EW)

element   any substance that cannot be broken into smaller parts by chemical or physical reactions

eminent domain   the right of government to appropriate privately owned land for public use

energy resources   fuel minerals, including fossil fuels and nuclear fuels

energy tax   (see *carbon tax*)

enhanced oil recovery (EOR)   methods used to increase the yield of oil from a reservoir, including CO_2 flooding, gas injection, thermal effects, and water flooding

enrichment   process by which fraction of ^{235}U in uranium is increased sufficiently to permit fission to take place

epigenetic   refers to mineral deposits that formed after their enclosing rocks (compare to *syngenetic*)

epilimnion   the upper layer of water in a stratified fresh-water lake

equilibrium   in chemistry, the conditions under which a chemical reaction proceeds both forward and backward at the same rate, resulting in constant relative abundances of reactants and products

ethanol   ethyl alcohol; grain alcohol (C_2H_5OH)

eutrophication   process by which increased organic productivity in surface waters consumes dissolved oxygen

evaporation   process by which a liquid becomes a gas

evaporite   rock consisting of minerals that precipitated from evaporated water

exchange   (see *ion exchange*)

Exclusive Economic Zone (EEZ)   200-mile zone of ocean water and floor around the United States and its territories claimed by the U.S. federal government

exhalative   refers to fluids that flow out onto the ocean floor from springs, not necessarily hot fluids

exinite   (see *liptinite*)

expropriate   most commonly used when governments take assets from private owners with limited compensation

extender   anything added to dilute a more costly material; a filler

extension   the process by which two tectonic segments of Earth move apart; creates a rift

extralateral rights   right to mine minerals outside vertical boundaries of a claim because they are a continuation of ore within the claim

extrusive rocks   igneous rocks formed where magma flows onto surface of the Earth

feasibility analysis   test made to determine whether a deposit can be produced economically

felsic   refers to rocks rich in feldspar and its common elements, sodium, potassium, calcium, and aluminum

ferroalloy   any alloy made with iron

ferrous metals   iron and metals such as manganese, nickel, chromium, cobalt, vanadium, molybdenum, and tungsten, which are used to form ferroalloys

filler   (see *extender*)

fineness   weight proportion of pure gold in an alloy

fissile   (1) describes an isotope that can undergo nuclear fission, (2) describes rock that can be split into sheets or slabs

fission   nuclear reaction in which a radioactive isotope splits apart to form two daughter isotopes of approximately equal size, subatomic particles, and energy

flash furnace   smelter in which injections of concentrate, preheated air, and flux are heated rapidly to high temperatures

flint   (see *chert*)

flocculation   electrostatic attraction of colloidal particles to one another to form larger grains

flue-gas desulfurization (FGD)   removal of sulfur from gases emitted by power plants, smelters, refineries, and other industrial installations

fluid inclusion   imperfection in a crystal that trapped part of the fluid from which the crystal grew

fluorosis   disease involving dental defects, bone lesions, and other effects of excess fluorine

flux   substance added to lower the melting temperature of a material

fly ash   small ash particles that go up the stack during the combustion of coal and other materials

fool's gold   pyrite (FeS_2)

fossil fuel   coal, crude oil, natural gas, tar sand, oil shale, and related materials consisting of organic matter that has been preserved in rocks

Frasch process   method used to extract sulfur from deeply buried deposits by melting it with superheated steam

fresh water   water with little or no dissolved solids

froth flotation   method used to make a concentrate from a slurry of pulverized ore by coating one or more minerals with an organic liquid that causes them to attach to bubbles and float to the surface

fuller's earth   any clay material that has sufficient absorbent capacity to clean and decolorize textiles

fulvic acid   humic material that remains in solution when alkaline extract (humic acid) is acidified

fumarole   vent through which steam and other gases reach Earth's surface

future   (see *commodity future*)

gabbro   mafic, intrusive igneous rock consisting of feldspar and pyroxene

galvanized steel   steel coated with a thin film of zinc to prevent corrosion

gamma rays   electromagnetic energy, similar to X-rays but with shorter wavelength, emitted during radioactive decay

gambusino   Mexican term for peasant mine worker; garimpero in Portuguese

gangue mineral   waste minerals in an ore deposit, commonly quartz and calcite

garimpero   (see *gambusino*)

gas   a state of matter in which the atoms or molecules are essentially unrestricted by cohesive forces

gas hydrate   a solid clathrate compound containing eight CH_4 molecules in a cage of 46 water molecules ($CH_4 \cdot 5.75 H_2O$)

gas injection   pumping natural gas into an oil field to maintain drive on the oil or store oil

gasification   the process by which coal is converted to natural gas

geiger counter   a device used to measure alpha particles emitted by radioactive decay

geochemical exploration   mineral exploration based on analysis of the chemical composition of rocks, soils, water, gas, and living organisms

geochronology   determination of the age of a rock or mineral based on its content of radioactive and daughter isotopes

geomedicine   study of the effects of rocks and minerals on health

geophysical exploration   mineral exploration by measurement of magnetic, gravity, electrical, radioactive, or other physical features of rocks and minerals

geopressured gas   gas held under pressure in water in or below impermeable sedimentary rocks

geostatistics   statistical treatment of spatially arranged data such as assay values from a mineral deposit

geothermal gradient   natural increase in Earth's temperature downward in the crust and mantle (usually 20° to 40°C/km in upper crust)

geyser   hot spring that boils intermittently

ghost town   town that has been deserted, often after exhaustion of a nearby mineral deposit

glass   amorphous solid

global change   geologically rapid changes in climate caused by natural and anthropogenic changes in the composition of the atmosphere and hydrosphere

goiter   disease of the thyroid gland, manifested by swelling of the neck

gold standard   use of gold to determine the value of a currency

graben   valley formed when the crust is pulled apart (by extension)

granite   a common felsic, intrusive, igneous rock consisting of quartz, feldspar, and mica minerals

greenhouse gas   a gas (especially H_2O, CO_2, CH_4) that prevents incoming solar energy from being radiated back into space, thus warming Earth

groundwater   water stored in the ground; commonly refers to water that can be extracted by wells for domestic, industrial, and agricultural use

guano   accumulations of bat and bird excrement

gusher   oil well that flows to the surface uncontrollably

Hadean   eon of geologic time prior to about 3.8 billion years (Ga)

half-life   period of time necessary for half of the amount of an isotope to undergo radioactive decay

hard water water that does not lather easily with soap; rich in dissolved calcium and magnesium

heavy media slurry of heavy minerals in water with a high bulk density; used to separate minerals on the basis of density

heavy water water that is enriched in deuterium

hemoglobin iron-containing protein in red blood cells that reversibly binds oxygen

heterogeneous mixture mixture containing more than one phase, mineral, or substance

horst a linear mountain or hill that separates two grabens

humic acid humic material that is soluble in dilute alkali solutions but precipitates when the solution is acidified

hydraulic cement (see *cement*)

hydraulic conductivity the rate at which water can move through porous rocks (equivalent to permeability for water)

hydraulic mining use of a high-pressure stream of water to disaggregate gravels for mining and processing

hydrocarbon trap hydrocarbon-bearing reservoir rock that is isolated by caprock from overlying porous, permeable rocks and the surface

hydrometallurgy the separation of a desired metal from an ore or concentrate by dissolution and later precipitation or electrowinning

hydrosphere that part of Earth consisting of water at and near the surface, including the ocean

hydrothermal alteration mineral changes caused by the reaction between a rock and a hydrothermal solution

hydrothermal solution natural hot water that circulates through Earth

hypolimnion deeper water level (below epilimnion) in a stratified fresh-water lake

igneous rock rock that formed by the crystallization or cooling of magma; includes volcanic (extrusive) rocks that form when magma reaches the surface and plutonic (intrusive) rocks that form below the surface

immiscibility process by which a homogeneous fluid separates into two fluids

impermeable term used to refer to rocks through which fluids pass very slowly or not at all

import quota limitation placed on the amount of a product that can be imported

inertinite complex mixture of fungal remains, unoxidized wood or bark, and other altered plant material in coal

intermediate a term used for igneous rocks that have compositions between mafic and felsic

intrusive rock igneous rock that formed below Earth's surface

ion atom or group of atoms that has lost (cation) or gained (anion) one or more electrons

ion exchange exchange of ions between a solution and a solid

ionizing radiation radiation that ionizes atoms by removing electrons

island arc arc of volcanic islands above a convergent margin (volcanic arc)

isotope atoms with the same number of protons but different numbers of neutrons

itabirite (see *banded iron formation*)

kerogen complex organic molecules resulting from the modification of organic matter preserved in sediments

kiln long, horizontal cylindrical furnace that is tilted slightly to allow material to progress through it; widely used in the production of cement, iron ore pellets, and other mineral products

kimberlite porphyritic peridotite with high sodium or potassium content, containing olivine, mica, and garnet; source of diamonds

kinetics refers to rate at which chemical reactions take place

komatiite ultramafic volcanic rock consisting largely of olivine and pyroxene; commonly associated with nickel deposits

lacustrine term used to refer to water, sediments, and other features of lakes

Lake Copper copper produced from mines in the Keweenaw district of northern Michigan, valued for its small amounts of silver

lamproite porphyritic mafic to ultramafic rock commonly containing biotite and amphibole, similar to kimberlite

land withdrawal restriction of the uses to which public land can be put

latency period period between exposure (to the cause of a health problem) and appearance of first symptoms

laterite type of soil formed by extremely intense weathering that removes all but the most immobile of elements

layered igneous complex (LIC) mafic, intrusive igneous rocks containing (cumulate) mineral layers

lava magma that extrudes onto Earth's surface

leach pad pile of ore onto which a reactive solution is sprayed to leach the element of interest; commonly used to recover gold from low-grade ores using cyanide solutions

leaded gasoline gasoline to which lead has been

added, usually in the form of tetraethyl lead [$Pb(C_2H_5)_4$]

light metals aluminum, magnesium, titanium, beryllium, and other metals that have relatively low densities

lightweight aggregate aggregate with a relatively low density, including natural materials such as perlite and pumice, as well as shale and other rocks that have been expanded by heating

lignin hard material embedded in the vascular matrix of plant cells

lignite a form of low rank coal

lime CaO; produced by calcining limestone or calcite

limestone a sedimentary rock consisting largely of the minerals calcite and aragonite, which have the same composition ($CaCO_3$)

liptinite remains of spores, algae, resins, needles, and leaf cuticles in coal

liquid state of matter in which molecules can change position freely but maintain a relatively fixed volume by mutual attraction

liquified natural gas (LNG) natural gas that has been liquified by cooling to temperatures of $-160°C$

liquified petroleum gas (LPG) ethane, ethylene, propane, propylene, normal butane, butylene, and isobutane produced at refineries or natural gas processing plants

lithosphere (1) entire solid Earth (environmental geochemistry); (2) the rigid, outer 100 km of Earth including crust and upper mantle (geology)

lode a vein or lenticular ore body

lode claim (see *claim*)

longwall mining underground mining in which flat-lying, tabular ore bodies (usually coal) are removed, usually by continuous miners, along long fronts and transported to a shaft by conveyor

maceral grains or subdivisions of coal, including vitrinite, liptinite, and inertinite

mafic rocks rich in magnesium and iron and minerals containing these elements, such as olivine and pyroxene

magma molten igneous rock with some crystals and dissolved gases

magmatic water water that was dissolved in a magma and later expelled (exsolved) when the magma crystallized to form water-poor silicate minerals

manganese nodule ball of manganese oxide, usually consisting of concentric shells, found on the floor of the ocean and some lakes

mantle middle shell of Earth, below crust and above core; consists of ultramafic rocks

margin term referring to the use of credit in commodity and stock purchases

marginal demand extra demand for a commodity; usually related to short-term market changes

marine related to seawater or the ocean

marl limestone containing a significant proportion of clay or silt

matte molten metal sulfide, formed by melting sulfide minerals in the early stages of smelting, usually involving loss of some S as SO_2

maturation progressive alteration of natural organic material by burial and reaction with microbes and water

mean sum of all values in a group divided by number of values; average

median central value in a group, with same number of values above and below

mesothelioma tumor in the mesothelium, the tissues that encase the lung; can be benign or malignant

Mesozoic era of geologic time between 245 and 66 million years (Ma)

metamorphic rock rock formed by the recrystallization of preexisting igneous, sedimentary, or metamorphic rock, without significant melting

metamorphic water water driven off from micas and other hydrous minerals during metamorphism, when they are recrystallized to form minerals that contain less water

metastable refers to minerals or other substances that persist at temperature, pressure, and other conditions outside their range of stability because their breakdown reactions are too sluggish to reach a new equilibrium state

meteorite fragment of extraterrestrial material that reaches Earth's surface

meteoric water atmospheric precipitation; in ore deposits, precipitation that seeps into the ground and descends to levels at which it is heated and incorporated into the hydrothermal systems

methaemoglobinaemia (blue-baby syndrome) condition caused when nitrite (NO_2^-) oxidizes Fe^{+2} in hemoglobin, forming methemoglobin that cannot carry oxygen to tissue; causes oxygen deficiency

methanol CH_3OH; wood alcohol

methylmercury CH_3Hg^+, an organometallic compound containing mercury

micrite a very fine-grained limestone

microbiological mining mining aided by microbes that decompose ore minerals releasing elements of interest to solution

mineral a naturally occurring, inorganic, crystalline solid with a regular chemical composition

mineral deposit any unusual mineral concentration, regardless of whether it can be extracted at a profit (see *ore deposit*)

mineral right ownership or control over real or potential minerals on a parcel of land

microalloy alloy, usually with iron, in which only very small concentrations of the additive are used

mobile (mobility) in geochemistry, the term refers to the degree to which an element or compound can be dispersed from its rock or mineral source, often by dissolution in water

mode the most common number in a population

moderator substance used to slow the flux of neutrons in a nuclear reactor, thus controlling the rate of fission

molecule the smallest unit of matter that can exist by itself and exhibit all the properties of the original substance

monosaccharide (see *sugar*)

moose pasture worthless land promoted for discovery or production of minerals

moraine ridge of gravel and sand deposited at the front of a stationary glacier by conveyor-like action of the ice, or along the sides of the glacier as it moves

native adjective used to refer to elements that occur alone as solids or liquids in nature (without being combined with other elements)

natural gas methane (CH_4) with small, but locally important amounts of ethane (C_2H_6) and heavier hydrocarbons, H_2S, He, CO_2, and N_2

natural gas liquid hydrocarbons in natural gas that are separated as liquids (includes ethane, propane, butanes, and pentane) at surface conditions

neutron a subatomic particle with a mass of one and no electrical charge; part of the nucleus of an atom

nitrogen fixation processes by which microbes convert nitrogen in air or fertilizers to forms that can be taken up by plants

nonferrous metals metals that are not commonly used in or with steel; commonly refers to the base and light metals

nonrenewable resources minerals or other resources that are exhausted more rapidly than they can be regenerated naturally; usually refers to mineral deposits

normal population a group of samples or population in which the mean, median, and mode are the same

nuclear reactor apparatus in which a nuclear chain reaction can be initiated, maintained, and controlled, releasing energy at a specific rate

nuclide a species of atom distinguished by its number of protons and neutrons; a specific isotope

nugget unusually large piece of native metal, usually found in a placer deposit

obduction movement of one lithospheric plate onto another at a collisional tectonic margin

ocean crust (see *crust*)

ocher iron oxide pigment with a yellowish-orange color

oil (see *crude oil*)

oil shale kerogen-bearing shale that yields oil or gas when distilled

oil window the range of depths (and related temperatures and pressures) to which sediments must be buried to generate oil from organic material

oolite a rock consisting of small (<1 cm) spheres, usually formed by chemical precipitation during sedimentation or diagenesis

open-cast mine open-pit mine

operating profit profit before tax but after deduction of operating and financial expenses

ore deposit mineral deposit that can be extracted at a profit

organic material material consisting largely of carbon and hydrogen

organometallic compound compound in which metals are connected to complex molecules consisting largely of carbon and hydrogen

orogeny deformation and metamorphism during large-scale lithospheric plate collisions; mountain building

overburden rock or regolith covering a mineral or ore deposit

overthrust belt belt of rocks in which one block of rocks has been pushed or thrust over another, concealing those beneath

oxidation loss of electrons from elements; commonly characterized by the availability of free oxygen

ozone O_3; a blue gas

paleobauxite old bauxite preserved by burial beneath younger sediments

particulates governmentese for "particles"

patent to transfer ownership in a mineral claim from government to claim holder

patio process gold recovery process in which ore is spread onto a patio, crushed, and covered with mercury (see *amalgamation*)

payback period the amount of time needed to recover the initial investment in a project plus relevant interest

peat a sedimentary layer of unconsolidated plant material, commonly deposited in a swamp, from which coal is derived

pegmatite a very coarse-grained, intrusive igneous rock thought to form from water-rich magma

pellets spheres of iron oxide and clay made during beneficiation to facilitate steelmaking; natural spheres of phosphate in sediments, probably formed during diagenesis

percentage depletion (see *depletion allowance*)

peridotite ultramafic igneous rock consisting largely of olivine and pyroxene

perlite rhyolitic volcanic glass with a concentric crack pattern

permeability the capacity of a rock to transmit fluid (see *hydraulic conductivity*)

Phanerozoic eon of geologic time between 570 million years (Ma) and the present

phenocryst large crystal in a matrix of smaller ones in an igneous rock

phosphogypsum calcium phosphate waste product produced during processing of phosphate ore

pig iron iron with several percent carbon as well as other impurities such as phosphorus, silicon, and sulfur

placer claim (see *claim*)

plaster usually refers to plaster of Paris ($CaSO_4 \cdot \frac{1}{2} H_2O$), which is prepared by partial calcining of gypsum, and which crystallizes as gypsum when water is added

plate tectonics theory proposing that Earth's outer shell consists of plates of lithosphere that move about causing earthquakes, volcanoes, and other geologic processes

playa flat central area of an enclosed desert basin with no river outlet

plutonic rock (see *igneous rock*)

pneumoconiosis any of several diseases of the lungs characterized by fibrous hardening caused by the inhalation of irritating particles

podiform lenticular, commonly refers to layers of chromite that have been sheared during deformation

point bar a sand or gravel bar formed on the inside of a river bend; common site of placer accumulations

polysaccharides large molecules or polymers consisting of hundreds or thousands of monosaccharides (sugar, CH_2O)

polygon a closed plane figure with several sides and angles, usually more than four

population in statistics, a group of samples

porosity volume percent pore space in a rock

porphyritic textural term referring to rock with large crystals (phenocrysts) in a matrix or ground mass of smaller crystals; rocks with this texture are porphyries

portland cement (see *cement*)

pothole (1) rounded depression in bedrock over which a stream flows, formed by abrasion by rocks in the stream; (2) depression in the platinum-bearing Merensky Reef of the Bushveld layered igneous complex, probably formed by chemical corrosion

polymer a large molecule consisting of connected simple molecules of the same type

Precambrian period of Earth history before about 570 million years (Ma) ago

precious metals gold, silver, and platinum; sometimes includes other platinum-group elements, palladium, rhodium, ruthenium, iridium, and osmium.

precipitate substance formed when dissolved ions or compounds combine to form a solid substance

precision agreement between replicate analyses of the same sample by the same method

property tax (see *ad valorem tax*)

prospector an individual who searches for mineral deposits largely by examination and analysis of rocks at the surface, using little scientific theory to select areas for further work

protein a biological polymer based on amino acids

Proterozoic eon of geologic time between 2.5 and 0.57 billion years (Ga)

proton elementary particle with a positive charge equal to that of an electron, but with a mass about 1,837 times greater

pseudomorph mineral that has the external form of another mineral

pumice felsic volcanic rock with a large proportion of vesicles; often buoyant enough to float

pyroclastic refers to fragments of magma, crystals, or other rocks ejected during a volcanic eruption

pyrolysis chemical decomposition by heat

pyrometallurgy smelting processes based on thermal decomposition of ore minerals

quicksilver mercury

radioactive capable of spontaneously giving off alpha and beta particles, gamma and other radiation

radon a radioactive chemical element formed by the decay of radium (Rn, atomic number 86); forms gas in Earth's crust

random drilling mineral exploration conducted by drilling holes at locations that have been selected randomly in a specified area

rank degree of maturation of coal

rate of return (see *return on investment*)

recharge flow of precipitation into an aquifer

reclamation remedial action taken to restore areas of mineral production to conditions that prevailed before production

reduction chemical process in which valence electrons are added to elements; commonly characterized by the scarcity of free oxygen

reef (1) accumulation of marine skeletal material; (2) layer of rock consisting of this material; (3) planar ore layer in some clastic sedimentary rocks and layered igneous rocks

refining conversion of crude oil to gasoline, heating oil, asphalt, and other products; removal of trace amounts of impurity elements from a smelted metal

reforming the third step in conventional oil refining, involving reconstitution of organic molecules

refractory refers to substances that are difficult to melt

regalian pertaining to royalty

regolith unconsolidated rock and soil material overlying bedrock

reserve that part of the reserve base that can be extracted at a profit at the time of determination, whether or not facilities are available on the property

reserve base reserves plus some resources, including marginally economic reserves and subeconomic reserves

reserve tax ad valorem tax on mineral reserves

reservoir (1) any part of Earth or its surroundings with a specified composition or characteristics; (2) porous, permeable rock that can store and produce oil, gas, brine, water, or other fluids

residence time average period of time during which a substance remains in a reservoir

residuum insoluble material remaining after intense weathering

resource concentration of natural material that can be extracted now or in the future; includes reserves and is divided into demonstrated resources that have been measured to some degree and inferred resources that are only thought to be present

retained earnings profits that are retained by a business rather than being distributed as dividends to shareholders or paid in tax

return on investment profit from an investment expressed as a fraction of the original investment

revenues income from business operations

reverberatory furnace a type of furnace used in copper and other types of smelting

rhizosphere the area of soil surrounding the root of a plant

rift linear valley formed by extension of the crust; graben

roast to heat a rock, mineral, or ore to drive away gases

rock consolidated mixture of grains of one or more minerals

rock burst spontaneous, explosive disintegration of rock exposed in a mine; commonly caused by stresses in deep mines

rock cycle idealized relation between igneous, metamorphic, and sedimentary rocks

room and pillar a type of underground mining involving removal of rooms of ore leaving intervening walls or pillars to support the roof or back

rotary kiln a type of kiln that rotates causing mineral material to flow from one end to the other

salar a playa or salt flat

salt dome salt intrusion that rises from an underlying layer of salt

salt flat a playa in which evaporites have been deposited

saltwater encroachment movement of salt water into an aquifer

sand clastic material between $\frac{1}{16}$ and 2 mm in diameter

sand spit a bar or pile of sand formed by shallow wave and current action, often at the mouth of a river or bend in the coast

sandstone a clastic sedimentary rock consisting of sand-sized grains

saturated (1) in chemistry, refers to a solution that has dissolved the maximum amount possible of a specific substance; (2) in groundwater, refers to porous rock that is filled by water (below water table)

scintillometer a device for measuring particles and energy emitted by radioactive decay

scoria hardened lava with abundant vesicles, usually basalt

scrubber device to recover SO_2 and other acid-forming gases produced by combustion of coal and other fossil fuels and by smelters

seam layer of ore, usually coal or chromite

sedimentary rock rock formed by the deposition of grains of preexisting rock material (clastic sedimentary rock), precipitation of dissolved material (chemical sedimentary rock), or accumulation of organic matter (organic sedimentary rock)

seismic (1) refers to movement of natural or artificial shock waves through rock; (2) refers to a form of geophysical exploration using artificial shock waves to delineate buried rock features

severance tax a tax imposed on the value of minerals removed from the ground

shaft a vertical or inclined hole used to reach mineral deposits below the surface

shale a clastic sedimentary rock consisting of fine-grained particles, many of which are clay minerals

share refers to part interest in a business, usually

sold to raise capital for the enterprise (also known as stock in the United States and Canada)

shatter cone a conical fracture pattern formed in rocks by the passage of shock waves; commonly caused by meteorite impact

shield the exposed part of a craton

silicate tetrahedron the basic unit of many silicate minerals; a tetrahedron with a central silicon cation and oxygen anions at each corner

silicon carbide SiC_4, a synthetic hard substance used as an abrasive

silicosis a lung disease caused by inhalation of angular fragments rich in silicate rocks or minerals

silt clastic material between $\frac{1}{16}$ and $\frac{1}{256}$ mm in diameter

sink in geochemistry, a reservoir into which an element, compound, or substance is concentrated

sinter silica-rich deposit from hot-spring waters that reach the surface

skarn metamorphic rock rich in calcium-bearing silicate minerals, commonly formed at or near intrusive rock contacts by addition of silica to limestone and dolomite; often hosts metal ore minerals

skeletal material framework for some living organisms; usually consists of calcite, apatite, or silica in animals and cellulose in plants

skewed refers to a population that is not normal, (mean, mode and median are not the same)

slag calcium-rich silicate waste material from smelting of metal ore concentrates

slate shale that has been metamorphosed to produce a durable rock that splits easily into slabs

slimes very fine-grained rock and mineral material produced by excessive grinding during beneficiation

slurry pipeline pipeline that transports pulverized solid material as a mixture (slurry) with water or other fluid

smelt to separate a metal from its ore mineral by pyrometallurgy

smog mixture of smoke and fog, usually containing SO_2

smoke particles from combustion suspended in heated air

soapstone soft, magnesium-rich rock containing large amounts of the mineral talc $[Mg_3Si_4O_{10}(OH)_2]$

soda ash Na_2CO_3; a chemical compound prepared from hydrous sodium carbonate minerals such as trona ($NaCO_3 \cdot NaHCO_3 \cdot 2H_2O$)

soft water water in which soap lathers easily; contains relatively small amounts of dissolved calcium and magnesium

soil upper part of the regolith (geological definition);

commonly shows some compositional zoning reflecting effects of weathering

solid substance that resists pressure or efforts to change its shape, as contrasted with a liquid or gas

solid solution capacity of a solid substance (usually crystalline) to exchange one element or ion for another of similar properties; has variable chemical composition within certain limits

solubility degree to which a substance will dissolve, usually in water or an aqueous solution

solute substance that dissolves in a solvent

solvent solution in which a solute dissolves

solvent extraction (SX) concentration of a desired element (such as gold) from a primary solution by dissolving it in a smaller volume of a second solution that is mixed with the primary solution but separates from it because it is immiscible; first step in solvent extraction-electrowinning (SX-EW) process

sour gas natural gas in which the partial pressure of H_2S is more than 0.01 bar

source in geochemistry, a reservoir that releases or emits a substance of interest

source rock rock containing sufficient organic material to produce crude oil and/or natural gas when buried beneath other sediments

spent fuel nuclear fuel that can no longer support a fission reaction

sponge iron (see *direct reduction*)

stable (1) refers to minerals or other substances that are present at temperature, pressure, and other conditions under which they do not react to form other minerals or compounds; (2) refers to isotopes that do not undergo spontaneous radioactive decay

stainless steel corrosion-resistant steel containing 12% to 50% Cr

standard in chemistry, a sample against which other samples are compared

standard deviation for any population, the square root of the mean of the sum of squares of deviations of all samples from the population mean

statement of income summary of revenues received by a corporation or other taxable entity, with expenses that are deductible from taxable income

steel iron-based alloy containing up to 2% C

stock small, irregularly-shaped body of intrusive igneous rock, commonly measuring a few kilometers in diameter; also see *share*

stock exchange public (i.e., open) market in which shares (stock) can be bought and sold

stockpile materials stored for later use

stockwork system of anastamosing (intersecting) small veins

stope underground opening or room from which ore has been removed

strata layers (plural of stratum, a single layer)

stratabound confined to a single layer or stratum

strategic mineral mineral material considered necessary to national defense; usually not available domestically

stratiform having the form of a flat layer or stratum

strip mining form of open-pit mining in which flat-lying ore bodies are mined in linear zones (strips) and overburden is replaced after mining

stripping ratio ratio of volume of overburden to volume of mineable ore in an open-pit mine

subbituminous coal low-rank coal

subduction plate tectonic process in which lithospheric plate returns to the mantle at a convergent margin

sublevel caving type of underground mining in which panels of ore are mined from a series of parallel tunnels known as sublevels

sublimation process by which a solid vaporizes without passing through a liquid state

subsidence lowering of the land surface caused by collapse of underlying rock; sometimes related to withdrawal of fluids from reservoirs or ore from mines

sugar monosaccharide with the general formula CH_2O (most commonly, glucose [$C_6H_{12}O_6$])

sulfate-reducing bacteria bacteria that derive energy from the conversion of SO_4^{-2} to H_2S or some other form of reduced sulfur

superphosphate see triple superphosphate

superalloys steel alloys that retain strength at temperatures above 800°C

surface chemistry chemical processes that occur at surfaces of substances, largely involving adsorption and ion exchange related to broken chemical bonds at surfaces

surface rights ownership or control over surface of a parcel of land

suspended material particles suspended in a fluid, usually water or air

sweet gas natural gas containing H_2S with a partial pressure of less that 0.01 bar

swelling clay clay mineral (usually smectite group) in which the space between individual clay layers increases when water is adsorbed, thereby increasing volume

syncrude "synthetic" oil recovered by processing of tar sand

syngenetic formed at the same time as the enclosing rocks

synroc substance synthesized by cooling molten material with the composition of average rock; used as a host for nuclear wastes

taconite (see *banded iron formation*)

tailings waste material that remains after processing of pulverized ore to make a concentrate of the desired ore mineral(s)

talus accumulation of rock debris at the base of a cliff

tar semisolid bitumen

tar sand sand or sandstone in which much of the interstitial pore space is filled with bitumen or tar

tax holiday period of time during which a tax is not imposed

tertiary production methods (see *enhanced oil recovery*)

tetraethyl lead organometallic compound [$Pb(C_2H_5)_4$] added to some gasoline to prevent premature combustion

thermal coal coal used for production of heat, usually in electric power plants

thermogenic decomposition decomposition of organic material in response to increasing temperature

thiol organic compound containing sulfhydryl (-SH) functional group

thixotropy pertaining to substances that behave like solids when not stirred but convert to liquids when stirred; desirable property of drilling mud

tidewater land land between high- and low-tide levels

tiger's eye a gem stone consisting of silicified amphibole asbestos

timber rights ownership or control over timber on a parcel of land

tin plate narrow-gauge steel plate with a coating of tin on both sides

torbanite Australian term for oil shale

toxic poisonous; detrimental to health

trace metal metal present in very small (trace) amounts

turnover time time required for the entire content of a specific substance in a reservoir to be replaced with new material of the same substance; see *residence time*

trap (see *hydrocarbon trap*)

triple superphosphate (TSP) fertilizer material containing 44% to 46% P_2O_5 prepared by treating phosphate minerals with phosphoric acid

tuff pyroclastic material of approximately sand size

ultramafic refers to igneous rock consisting largely of magnesium and iron silicates and oxides

unconformity surface that represents a break in the sedimentary rock record, caused by erosion or nondeposition of sediment

underground mine mine that removes ore from beneath the surface through a shaft or adit

unit train train that carries a single product from one

location; largely trains hauling coal from western U.S. mines to eastern U.S. consumers

unitization process by which production from an oil or gas field on several land parcels is apportioned to each land owner

unleaded gasoline (see *leaded gasoline*)

unsaturated (see *saturated*)

unstable (see *stable*)

vaporization process by which a liquid converts to a vapor

vascular plant plant with cells joined into tubes that carry water and nutrients throughout the plant; includes modern species except mosses and related plants

vein planar fracture in rock that has been sealed with minerals precipitated from hydrothermal solutions or cooler water

vertical crater retreat system of underground mining in which successively higher, horizontal blocks of ore are blasted from the upper part of a stope and recovered at the bottom

vesicle hole in volcanic rock preserving the form of a bubble of gas that was originally dissolved in the magma, but separated from it

vitrinite remains of wood and bark in coal and other carbon-rich deposits

volatile organic compounds (VOC) organic compounds that are easily vaporized; includes gasoline

volcanic rocks rocks formed from lava or pyroclastic material

water flooding a method of enhanced oil recovery involving injection of water into peripheral parts of a field to increase drive on remaining oil

water rights ownership or control of water on or below the surface of a parcel of land

water table top of the zone that is saturated with groundwater

water washing dissolution and decomposition of organic material by groundwater

weathering disintegration and decomposition of rocks at or near Earth's surface

well completion action taken to bring an oil or gas discovery well into production

well logging process of lowering measuring devices into exploration drill holes to measure rock characteristics, such as magnetism, conductivity, and radioactivity

wet deposition precipitation of new compounds formed by reactions among suspended particles, aerosol, and gases in air

working fluid fluid(s), usually water, that transport heat from a source of thermal energy to turbines in electric power generating plants

Index*

*(see Appendices and Glossary for further information)

383

386

389